Social Manufacturing on Industrial Internet

Social Manufacturing on Industrial Internet

Editors

Pingyu Jiang
Gang Xiong
Timo R. Nyberg
Zhen Shen
Maolin Yang
Guangyu Xiong

MDPI • Basel • Beijing • Wuhan • Barcelona • Belgrade • Manchester • Tokyo • Cluj • Tianjin

Editors

Pingyu Jiang
Xi'an Jiaotong University
China

Gang Xiong
Chinese Academy of Sciences
China

Timo R. Nyberg
Aalto Advisors
Finland

Zhen Shen
Chinese Academy of Sciences
China

Maolin Yang
Xi'an Jiaotong University
China

Guangyu Xiong
Ningbo Intelligent
Manufacturing Institute
China

Editorial Office
MDPI
St. Alban-Anlage 66
4052 Basel, Switzerland

This is a reprint of articles from the Special Issue published online in the open access journal *Machines* (ISSN 2075-1702) (available at: https://www.mdpi.com/journal/machines/special_issues/Social_Manufacturing).

For citation purposes, cite each article independently as indicated on the article page online and as indicated below:

LastName, A.A.; LastName, B.B.; LastName, C.C. Article Title. *Journal Name* **Year**, *Volume Number*, Page Range.

ISBN 978-3-0365-7756-2 (Hbk)
ISBN 978-3-0365-7757-9 (PDF)

Contents

About the Editors

Pingyu Jiang

Pingyu Jiang is a professor of the state key laboratory for manufacturing systems engineering at Xi'an Jiatong University (XJTU), China. He received his B.E. degree in Mechanical Engineering from Nanjing Institute of Technology (now Southeast University), China, in 1983, and M.E. and Ph.D. degrees in Mechanical Engineering from XJTU in 1988 and 1991. Dr. Jiang held Humboldt and JSPS international research fellowships from 1995 to 1999, respectively, in WZL of RWTH Aachen, IPK of TU Berlin, Germany, and the Department of Production Engineering of Tokyo Metropolitan University, Japan. He was also an ASTAR visiting staff of Nanyang Technological University in 2022, Singapore, JSPS, and a Humboldt short-term visiting professor, respectively, in the Department of Production Engineering of Tokyo Metropolitan University in 2003, Japan, and LPS of TU Dortmund and IWB of TU Munich in 2011, Germany. Dr. Jiang has been a faculty member at XJTU since 1991 and was promoted to Full Professor in 1999. He also served as the Vice Dean of the School of Mechanical Engineering, XJTU, from 2003 to 2006. He is both an active trustee and a deputy head in several divisions of nation-wide academic societies such as China Mechanical Engineering Society, China Artificial Intelligence Society, China Automation Society, etc. He also serves as an Associate Editor of Journal of Xi'an Jiaotong University and is the Editorial Board Member of several journals, such as the Journal of Engineering Design, Machines, Systems, etc. Dr. Jiang is author or co-author of more than 300 papers and 6 monographs. His current research interest is systems engineering for manufacturing and services, including social manufacturing, cyber-physical systems, MES/smart factory, product design with artificial intelligence, product service systems, open collaborative product development, etc.

Gang Xiong

Gang Xiong is a professor at the Institute of Automation, Chinese Academy of Sciences (CAS), China. He received B.Eng. and M.Eng. degrees from Xi'an University of Science and Technology, China, in 1991 and 1994, respectively, and a Ph.D. degree from Shanghai Jiao Tong University, China. From 1996 to 1998, he was a Postdoctoral Scientist and an Associate Professor for Zhejiang University, China. From 1998 to 2001, he was a Senior Research Fellow with the Tampere University of Technology, Finland. From 2001 to 2007, he was a Specialist and the Project Manager with Nokia Corporation, Finland. In 2007, he was a Senior Consultant and the Team Leader with Accenture and Chevron, USA. In 2008, he was the Deputy Director of the Informatization Office, CAS, China. In 2011, he became the Deputy Director of Cloud Computing Center, CAS, China. Between January 2015 and December 2017, he worked as FinDiPro Professor for Aalto University, Finland. He has successfully finished more than 50 projects funded by Chinese NSFC/MOST/MIIT, Finnish TEKES, the Academy of Finland etc. He is the author or co-author of about 450 refereed journal and conference papers, including 70 SCI papers, and 320 EI papers. He is editor or co-editor of 2 academic books, the author of 10 book chapters. He is author or co-author of about 60 Chinese patents, 6 PCT, and 60 Chinese Software Copyrights. He received 5 awards from the IEEE and the government. His research interests include artificial intelligence, intelligent control and management, 3D printing and social manufacturing.

Timo R. Nyberg

Timo R. Nyberg received M.Sc. (Tech) degrees in Mechanical Engineering from the Helsinki University of Technology, Finland, in 1988, and the Dr. (Tech) degrees in Automation from Helsinki University of Technology, Finland, in 1993. From 1996 to 2002, he was Professor of Automation and Director of Graduate School, Tampere University of Technology, Finland. From 2001 to 2002, he was Senior Vice President, Novo Astra Oy, Finland. From 2002 to 2005, he was Laboratory Director and Management Board member, Helsinki University of Applied Sciences, Finland. Since 2010, he has worked as a Senior Research Fellow and Research Director of the School of Science at Aalto University, Finland. Since 2012, he has also been a Visiting Professor of the Cloud Computing Center, CAS, China. His research covers enterprise automation systems, business strategies in software, and social manufacturing. He has finished more than 30 research projects funded by Finnish TEKES, the Academy of Finland, the European Union, and international companies. He is the author or co-author of about 100 refereed journal and conference proceeding papers, and 35 EU/US/Finnish patents. He has experience from both the academic world and industries (Valmet, Nokia). He has over 15 years of experience from R&D cooperation with Chinese and American universities (Tsinghua University, Tongji University, MIT, and Stanford University). He is a board member in Finland–China Trade association and Finnish Foundation for Inventions.

Zhen Shen

Zhen Shen received his B. E. and Ph.D. in 2004 and 2009, respectively, both from the Department of Automation, Tsinghua University. Currently, he is an Associate Professor at the Institute of Automation, Chinese Academy of Sciences. His research interests are intelligent manufacturing and complex systems. He has authored about 100 referred journal and conference papers, and holds more than 30 national invention patents and 5 US patents. He is a recipient of the 2005 "Outstanding Achievement Award" from the United Technology Research Center, a recipient of the Second Class Award of China Industry-University-Research Cooperation Innovation Achievement in 2018 and the Wu Wenjun Second Class Award of Artificial Intelligence Science and Technology in 2019. He is a member of the IEEE, and was selected into the CAS Key Technical Talents program in 2019.

Maolin Yang

Maolin Yang obtained the Bachelor's degree in Mechanical Engineering from Tongji University in 2012, Master's degree in Industrial Engineering from Shanghai University of Engineering Science in 2015, and Ph.D. degree in Mechanical Engineering From Xi'an Jiaotong University in 2021. Currently, he is an Assistant Professor in the School of Mechanical Engineering at Xi'an Jiaotong University (XJTU), China. His research interests include social manufacturing, innovation product development, deep learning-based data-driven intelligent product design, etc.

Guangyu Xiong

Guangyu Xiong is a SCOR Professional (SCOR-P) and SOCR-DS specialist certified by ASCM, and an ASCM (Association for Supply Chain Management) instructor. She has worked for one of the top 500 enterprises, the ABB group, for many years as Scientist, Senior specialist, Chief Specialist and Quality and OpX manager. She has led improvement projects to meet customers' expectation and business targets in previous years. Meanwhile, she has had more than 40 publications, including academic papers and books. Dr. Xiong is skilled and knowledgeable at SCM and OpX (Operations Excellence) with Lean/TPS, 6-sigma and TOC (theory of constraints), both in implementation and training/coaching. Her work is focused on process development or improvement through business supply chain/value chain. She is extending her subject on certain innovation areas, such as lean cleaner production and lean energy efficiency. Her new interests are industrial 4.0, intelligent technology in SCM and industrial management. She is experienced in supply chain/logistics and operations. Her subjects cover SCM/VCM (Supply Chain Management/Value Chain Management), Supplier Development, Quality Improvement, Change Management, and OpX (Operational Excellence) with related methodologies and tools.

machines

MDPI

Editorial

Editorial: Social Manufacturing on Industrial Internet

Pingyu Jiang [1,*], Gang Xiong [2], Timo R. Nyberg [3], Zhen Shen [2], Maolin Yang [1] and Guangyu Xiong [4]

1 State Key Laboratory for Manufacturing Systems Engineering, Xi'an Jiaotong University, Xi'an 710049, China; maolin@mail.xjtu.edu.cn
2 The State Key Laboratory of Management and Control for Complex Systems, Institute of Automation, Chinese Academy of Sciences, 95 Zhongguancun East Road, Beijing 100190, China; gang.xiong@ia.ac.cn (G.X.); zhen.shen@ia.ac.cn (Z.S.)
3 Aalto Advisors, FI-02150 Espoo, Finland; timo.nyberg@aaltoadvisors.com
4 Laboratory-Operations Development, Ningbo Intelligent Manufacturing Institute, Ningbo 315151, China; xiongguangyu@hotmail.com
* Correspondence: pjiang@mail.xjtu.edu.cn

Citation: Jiang, P.; Xiong, G.; Nyberg, T.R.; Shen, Z.; Yang, M.; Xiong, G. Editorial: Social Manufacturing on Industrial Internet. *Machines* **2023**, *11*, 383. https://doi.org/10.3390/machines11030383

Received: 9 March 2023
Accepted: 11 March 2023
Published: 14 March 2023

The fast development of the industrial internet is boosting the evolution of the manufacturing industry to a new stage of socialization, servitization, universal interaction and connection, and platformization. Under such a background, social manufacturing has emerged as a new kind of manufacturing paradigm characterized by self-driven, self-organization, self-adaptive, and cyber–physical–social interaction among huge numbers of socialized manufacturing resource providers. The most prominent advantage of social manufacturing is its capability of completing production/service orders with the limited internal manufacturing resource of an enterprise by utilizing socialized manufacturing resources from the outside, and this can be applied in both large and small enterprises and trigger value co-creation for both resource providers and demanders. In this regard, this Special Issue is established to explore how exactly the newly emerged social manufacturing paradigm influences the trends of mass customization and the configuration/operation patterns during order delivering, and how advanced information technologies such as industrial internet, cloud computing, blockchain, etc., can boost the development and application of social manufacturing. In total, 14 papers, including 2 review articles, have been collected in this Special Issue, and the key topics explored in these papers are briefly introduced below.

The paper in [1] introduced the issue of product customization under the context of social manufacturing. It introduced a framework that utilizes advanced deep learning models and cloud computing technologies to transfer the text/image data generated during manufacturing process into customized 3D contents that can be directly used for 3D printing. Supported by the framework, more effective product customization and more efficient social manufacturing operation and optimization can be realized.

The papers in [2–5] provided insights from the perspective of production configuration under the context of social manufacturing. Specifically, a graph convolutional network-based method for socialized designer team configuration was proposed in [2]. By utilizing the graph convolution network embedded with the graph matching algorithm, it can identify the socialized designers with the suitable design resources for a certain socialized design project, and thus provide decision making support for designer team configuration. In [3], a fast manufacturing system configuration model was established under the context of industrial 4.0 and social manufacturing, and the model could be particularly useful for small enterprises to improve their existing manufacturing system to meet the new requirements of wider customized product varieties, a shorter response time to new orders, a faster manufacturing system configuration process, etc., from the new market environment. Aiming at the situation that one large 3D printing order could be collaboratively carried out by multiple factories under the context of social manufacturing, a multi-part production planning system particularly for 3D printing orders was established in [4]. The system utilizes multiple types of intelligent algorithms to support printing order separation, printing part orientation, and nesting during order collaboration. A novel method for configurating the data acquisition network for socialized intelligent factory was proposed in [5]. The data acquisition network can realize real-time data collection and pre-processing for energy consumption analysis, and thus support further intelligent optimization on energy consumption.

The papers in [6–9] provided ideas from the perspective of supporting production operations with advanced techniques such as industrial internet, cloud-edge collaboration technologies, etc., which are fundamental for enabling the social manufacturing paradigm. For example, production data management and application during cloud manufacturing is considered in [6]. In this paper, a kind of cloud manufacturing system enhanced with an industrial internet of things and cloud-edge collaboration was established. The system can describe the characteristics of heterogeneous manufacturing resources, the operational data of the resources, and their relations with the service-oriented manufacturing system. In addition, a middleware and AI edge gate way model was established, and it utilizes real-time sensor data to realize the remote monitoring and controlling of cloud manufacturing resources. Supported by the system, companies can better utilize the distributed cloud manufacturing resources and improve their response speed to personalized orders. In [7], a kind of equipment asset management model was established based on industrial internet platform techniques and a fuzzy DEMATEL-TOPSIS algorithm under the general framework of system engineering. By collecting the required data for asset management from the industrial internet platform established for the target industrial, the fuzzy DEMATEL-TOPSIS algorithm is then applied to identify the relations among the requirements from customers and the characteristics of the assets registered in the platform. In this way, the model helps to establish the solutions for asset management for the entire product lifecycle. The one review article in [8] reviewed the related research on utilizing advanced information technologies (e.g., edge/cloud/fog computing, big data collecting and processing, artificial intelligence, digital twin, etc.) for equipment or production line maintenance. The other review paper in [9] established a blockchain-based method to support a trustworthy operation environment for manufacturing activities under the context of industrial 5.0 and social manufacturing. By applying an industrial Internet of Things network enhanced with blockchain networks, the method can protect the confidential and private data of stakeholders, and thus support the realization of resilient and trustworthy manufacturing operation.

The papers in [10–13] provided technical road maps from the perspective of a few fundamental information and data techniques that can support social and intelligent manufacturing realization. For example, a kind of high precision synchronous control method for a fieldbus control system is established in [10], and it can improve the control accuracy of multi-axis collaborative machining tasks. A novel method for liquid crystal display module alignment and particle detection in anisotropic conductive film bonding was established in [11]. By applying only one camera instead of multiple to obtain images of multiple locations, the method can realize the transformation from image space to world coordinates. Compared with the traditional methods which apply multiple cameras, the method can accurately identify the rotation center, the position, and angle deviation of the target object with a relatively lower economic and time cost. A kind of deep learning-driven defect image generation method was established in [12]. The method was for solving data enhancement problems in industrial defect detection. By applying a masked defect image generation adversarial network, the method can solve the problems of a loss of background information, an insufficient consideration of complex defects, and a lack of accurate annotation image data which usually occurred during data-driven defect detection. A kind of online dimensional error prediction method to predict the errors occurred during grinding process was proposed in [13]. The method was driven by principal component analysis, extreme learning machine, genetic algorithm, and ensemble strategy (bagging algorithm). By applying the method, grinding errors can be detected in a real-time manner with our extra devices and space occupation.

Finally, a case study of social manufacturing from the entry point of the value chain was proposed in [14]. First, a kind of value chain model under the context of the social manufacturing paradigm was established; it then utilized the cases from the crane industry to demonstrate the advantages of the established model.

All the papers above provide a reference for academic research and the industrial application of social manufacturing. However, as a newly emerged research topic, there are still unexplored issues for more optimal interaction/configuration/operation under the context of social manufacturing such as the implementation of collective and social 3D printing factories, realizing collective intelligence during social manufacturing operation, applying advanced cross-modal data processing techniques for extracting and utilizing production data for social manufacturing optimization, etc.

Author Contributions: Conceptualization, P.J., G.X. (Gang Xiong), T.R.N., Z.S., M.Y. and G.X. (Guangyu Xiong); writing—original draft preparation, M.Y.; writing—review and editing, P.J. and M.Y. All authors have read and agreed to the published version of the manuscript.

Funding: This research received no external funding.

Acknowledgments: We express thanks to the editors of MDPI for their excellent support for this Special Issue; it would have been impossible without their persistence and great effort. Many thanks to all our colleagues who have made contributions and thank you especially to all the reviewers for their professional and valuable comments, helping us finalize these high-quality manuscripts.

Conflicts of Interest: The authors declare no conflict of interest.

References

1. Li, Y.; Wu, H.; Tamir, T.S.; Shen, Z.; Liu, S.; Hu, B.; Xiong, G. An Efficient Product-Customization Framework Based on Multimodal Data under the Social Manufacturing Paradigm. *Machines* **2023**, *11*, 170. [CrossRef]
2. Liu, D.; Wu, D.; Wu, S. A Graph Matching Model for Designer Team Selection for Collaborative Design Crowdsourcing Tasks in Social Manufacturing. *Machines* **2022**, *10*, 776. [CrossRef]
3. Xie, S.; Zhang, W.; Xue, F.; Li, D.; Liu, Y.; Fleischer, J.; Ehrmann, C. Industry 4.0-Oriented Turnkey Project: Rapid Configuration and Intelligent Operation of Manufacturing Systems. *Machines* **2022**, *10*, 983. [CrossRef]
4. Makanda, I.L.D.; Yang, M.; Shi, H.; Guo, W.; Jiang, P. A Multi-Part Production Planning System for a Distributed Network of 3D Printers under the Context of Social Manufacturing. *Machines* **2022**, *10*, 605. [CrossRef]
5. Zhang, C.; Zhang, J.; Ji, W.; Peng, W. Data Acquisition Network Configuration and Real-Time Energy Consumption Characteristic Analysis in Intelligent Workshops for Social Manufacturing. *Machines* **2022**, *10*, 923. [CrossRef]
6. Zhang, Y.; Tang, D.; Zhu, H.; Zhou, S.; Zhao, Z. An Efficient IIoT Gateway for Cloud–Edge Collaboration in Cloud Manufacturing. *Machines* **2022**, *10*, 850. [CrossRef]
7. Bao, Y.; Zhang, X.; Zhou, T.; Chen, Z.; Ming, X. Application of Industrial Internet for Equipment Asset Management in Social Digitalization Platform Based on System Engineering Using Fuzzy DEMATEL-TOPSIS. *Machines* **2022**, *10*, 1137. [CrossRef]
8. Li, Q.; Yang, Y.; Jiang, P. Remote Monitoring and Maintenance for Equipment and Production Lines on Industrial Internet: A Literature Review. *Machines* **2023**, *11*, 12. [CrossRef]
9. Leng, J.; Chen, Z.; Huang, Z.; Zhu, X.; Su, H.; Lin, Z.; Zhang, D. Secure Blockchain Middleware for Decentralized IIoT towards Industry 5.0: A Review of Architecture, Enablers, Challenges, and Directions. *Machines* **2022**, *10*, 858. [CrossRef]
10. Chen, L.; Zheng, J.; Fan, D.; Chen, N. Research on the High Precision Synchronous Control Method of the Fieldbus Control System. *Machines* **2023**, *11*, 98. [CrossRef]
11. Li, T.; Zhang, F.; Yang, H.; Luo, H.; Zhang, Z. A Novel Method for LCD Module Alignment and Particle Detection in Anisotropic Conductive Film Bonding. *Machines* **2023**, *11*, 49. [CrossRef]
12. Wei, J.; Zhang, Z.; Shen, F.; Lv, C. Mask-Guided Generation Method for Industrial Defect Images with Non-uniform Structures. *Machines* **2022**, *10*, 1239. [CrossRef]
13. Yang, L.; Jiang, Y.; Liu, H.; Yang, X. Dimensional Error Prediction of Grinding Process Based on Bagging–GA–ELM with Robust Analysis. *Machines* **2023**, *11*, 32. [CrossRef]
14. Xiong, G.-Y.; Helo, P.; Ekstrom, S.; Tamir, T.S. A Case Study in Social Manufacturing: From Social Manufacturing to Social Value Chain. *Machines* **2022**, *10*, 978. [CrossRef]

machines

Article

An Efficient Product-Customization Framework Based on Multimodal Data under the Social Manufacturing Paradigm

Yanpeng Li [1,2], Huaiyu Wu [1], Tariku Sinshaw Tamir [1,2], Zhen Shen [1,3,*], Sheng Liu [1], Bin Hu [1,3] and Gang Xiong [1,4]

1 State Key Laboratory for Management and Control of Complex Systems, Beijing Engineering Research Center of Intelligent Systems and Technology, Institute of Automation, Chinese Academy of Sciences, Beijing 100190, China
2 School of Artificial Intelligence, University of Chinese Academy of Sciences, Beijing 100049, China
3 Intelligent Manufacturing Center, Qingdao Academy of Intelligent Industries, Qingdao 266109, China
4 Guangdong Engineering Research Center of 3D Printing and Intelligent Manufacturing, The Cloud Computing Center, Chinese Academy of Sciences, Dongguan 523808, China
* Correspondence: zhen.shen@ia.ac.cn

Abstract: With improvements in social productivity and technology, along with the popularity of the Internet, consumer demands are becoming increasingly personalized and diversified, promoting the transformation from mass customization to social manufacturing (SM). How to achieve efficient product customization remains a challenge. Massive multi-modal data, such as text and images, are generated during the manufacturing process. Based on the data, we can use large-scale pre-trained deep learning models and neural radiation field (NeRF) techniques to generate user-friendly 3D contents for 3D Printing. Furthermore, by the cloud computing technology, we can achieve more efficient SM operations. In this paper, we propose an efficient product-customization framework that can provide new ideas for the design, implementation, and optimization of collaborative production, and can provide insights for the upgrading of manufacturing industries.

Keywords: social manufacturing; multimodal data; product customization; 3D content generation; blockchain; cloud manufacturing

Citation: Li, Y.; Wu, H.; Tamir, T.S.; Shen, Z.; Liu, S.; Hu, B.; Xiong, G. An Efficient Product-Customization Framework Based on Multimodal Data under the Social Manufacturing Paradigm. *Machines* **2023**, *11*, 170. https://doi.org/10.3390/machines11020170

Academic Editor: Kai Cheng

Received: 30 November 2022
Revised: 31 December 2022
Accepted: 2 January 2023
Published: 26 January 2023

1. Introduction

With the rapid development of social computing and social manufacturing (SM), traditional manufacturing mode is transforming to customization, small batch, digital production [1]. The cost of rapid prototyping equipment, represented by 3D printers, is decreasing, and the technology is becoming increasingly popular. Desktop and industrial-grade 3D printers make it possible for individuals to produce products independently. Users can personalize and manufacture physical objects in small batches without relying on complex factories as long as they have valid 3D model data (or have other types of manufacturing instructions). Each 3D printer is an independent manufacturing node with a naturally distributed, decentralized character. Anyone with an idea for a product design, even if they do not have the knowledge of 3D modeling, can design a digital model by relying on AI technology, and then realize it with manufacturing facilities. With a well-developed design platform, innovative ideas can be realized on the Internet. Using distributed manufacturing tools such as 3D printers, personalized and customized products can be provided to users worldwide.

Industry 4.0 utilizes the cyber-physical system (CPS) in combination with additive manufacturing, the Internet of things, machine learning, and artificial intelligence technologies to a realize and smarten manufacturing [2]. It aims to improve the intelligence level of the manufacturing industry and establish a smart manufacturing model with adaptability and resource efficiency. Cyber-physical-social systems (CPSS) further enhance the integration of Internet information and social manufacturing systems and fully integrate industry with the human society [3]. In the CPSS industrial environment, social manufacturing seamlessly connects the Internet of Things with 3D printers. The public can fully participate in the entire lifecycle of the product manufacturing process, to achieve real-time personalized, large-scale innovation, and "agile mobile intelligence".

Social manufacturing is a paradigm that uses 3D printing to involve the community in the process of producing products. During the 3D printing process, the first step is to obtain a 3D model. Currently, most 3D models are generated by computer-aided design (CAD), but the time and human resources costs are high. It is essential to design a user-friendly end-to-end 3D content generation model so that users can personalize the 3D model using text representation. Futhermore, a cloud platform can be built and put it into practical use.

The rest of the paper is organized as follows. In Section 2, we present the literature review of SM. Section 3 introduces the theoretical concepts. Section 4 reviews the current difficulties in product customization. Finally, in Section 5, the multimodal data-based product customization manufacturing framework is proposed and illustrated with practical application examples, followed by concluding remarks in Section 6.

2. Literature Review

Utilizing SM is a practical way to achieve intelligent, social, and personalized manufacturing. SM allows for the active participation of all facets of social customization and the intense personalization of every component of every product. The term SM was proposed by Professor Fei-Yue Wang at the "Workshop on Social Computing and Computational Social Studies" in 2010 [3], and its formal definition is given in his article in 2012 [4]: "The social manufacturing refers to the personalized, real-time and economic production and consumption mode for which a consumer can fully participate in the whole lifetime of manufacturing in a crowdsourcing fashion by utilizing the technologies, such as 3D printing, network, and social media." Thus, it is noted that SM is inspired by the social computing concept that has been hailed as a cutting-edge manufacturing solution for the coming era of personalized customization. In the same way that the social computing [5] enables everyone in the society to obtain computing capacity, the SM enables everyone to obtain innovation and manufacturing capacity.

In the context of the SM, the critical feature is that everyone participates in product design, manufacturing and consumption in social production. Everyone can turn their ideas into reality. SM is a further the development and continuation of the existing crowdsourcing model. Xiong et al. [6] summarize this feature as "from mind to product". Based on this concept, Shang et al. [7–9] provided a vision of connected manufacturing services that are smart and engaging online. Mohajeri et al. [10–12] argued for using social computing and other Internet technologies to realize personalized, real-time, and socialized production. Jiang et al. [13–16] focused on the mechanisms for crowdsourcing and outsourcing services throughout the entire lifespan of individualized production. As for the SM service mode, Xiong et al. [17] presented a social value chain system that applies the SM mode to the entire value chain and makes a contribution to bringing more potential to value-adding for all involved participants while reducing waste. Cao et al. [18] provided a manufacturing capability estimation model for the SM mode's service level that is based on the ontology, the semantic web, the rough set theory, and a neural network. It includes models for production service capability and machining service capability. In the evaluation system proposed by Xiong et al. [19], the best supplier is selected by using the hierarchical analysis method (AHP) and fuzzy comprehensive method. The suggested techniques can assist prosumers in assessing their suitability for a certain manufacturing requirement. Additionally, it can assist prosumers in their search and ability matching. Ming et al. [20] proposed a data-driven view in the context of SM. Xiong et al. [21] summarized the five layers that make up an SM system architecture, and a detailed discussion of each of the layers is presented next.

(1) Resource Layer: The production and service resources found in this layer can be combined to form a social resource network. A few of the resources include 3D printers,

sophisticated logistics networks, information networks using 3D intelligent terminals, and heterogeneous operating systems. In this layer, 3D intelligent terminals can offer interaction and perception capabilities, and a modern logistics network can offer the logistical capability. Along with a range of physical links, the logistics network offers transparent information transmission options. For terminal perception, information communication, and production and product logistics, this layer is essential.

(2) Support Layer: Service module encapsulation, service discovery, service registration, service management, service data management, and middleware management are all part of this layer. Enterprise information can be transferred reliably, efficiently, and safely in accordance with user needs using the service registration and service discovery modules.

(3) Environment Layer: This layer includes the computer environment, information analysis environment and monitoring management environment that works in coordinated operation.

(4) Application Layer: This layer includes 3D modeling, operational monitoring tools, a platform for collaborative manufacturing management, a data collection and analysis platform, an optimization tool for allocating socialized manufacturing resources (SMRs), an assessment mechanism for system management, real-time monitoring, and resource scheduling.

(5) User Layer: This layer is composed of manufacturers, prosumers, and SMRs. Crowdsourcing is used to connect manufacturers and prosumers, and SMRs can be tailored for production.

3. Concepts

3.1. 3D Modeling

Additive manufacturing (AM) is a fabrication method that uses a digital prototype file as the foundation for constructing objects by printing layer by layer using material, such as powdered metal or plastic. Without the need for traditional tools, fixtures, or multiple machining processes, this technology enables the rapid and precise manufacturing of parts with a complex shape in a single machine, thereby allowing "free-form" manufacturing. Moreover, the more complex the structure of the product, the more significant the accelerating effect of its manufacturing. Currently, AM technology has many applications in the aerospace, medicine, transportation and energy, civil engineering, and other fields. With the development of AM technology and the popularity of 3D printers, there is an increasing demand for 3D modeling. The traditional modeling approach uses forward modeling software—3DMax, AutoCAD, etc. [22]—which is highly technical. It takes a long time to design a model, making it difficult to build complex, arbitrarily shaped objects. Roberts et al. [23] proposed a method to obtain 3D information from 2D images, known as reverse modeling. Since then, vision-based 3D reconstruction developed rapidly, and many new methods have emerged. Image-based rendering (IBR) methods could be divided into two categories depending on whether they rely on a geometric prior or not. Those methods that rely on geometric priors generally require a multi-view stereo (MVS) algorithm first to calculate the geometric information of the scene, and then to guide the input image for view composition. The methods that do not rely on geometric priors can generate new views directly from the input images. In 2018, Yao proposed an unstructured multi-view 3D reconstruction network (MVSNet) [24], an end-to-end framework using deep neural networks (DNNs). IBR methods that rely on geometric a priori information perform better when geometric information is abundant and accurate. Still, when the geometry is missing or incorrect, it produces artifacts and degrades the quality of the view. The light field is the main area of research that does not rely on geometric a priori methods. Traditional methods for drawing light fields require dense and regular image captures, making them difficult to apply in practice. In recent years, with the development of DNN, researchers have discovered that it is possible to synthesize new views by fitting a neural network to a light sample of the scene, thereby implicitly encoding the light field of the input image. Compared to traditional light fields, the neural reflectance field (NeRF) [25] method can be used for handheld captures with a small number of input images, greatly extending the applicability. The main advantages and disadvantages of the mainstream 3D modeling approach are shown in Table 1. As shown in Figure 1, NeRF can be briefly summarized as using a multi-layer preceptron (MLP) neural network that learns a static 3D scene implicitly. While the input is the spatial coordinates, the view direction output is the bulk density of the spatial location under that view direction and the view-related camera light radiation field. In its basic form, a NeRF model represents scenes as a radiance field approximated by a neural network. The radiance field describes color and volume density for every point and for every viewing direction in the scene. This is written as,

$$F(\mathbf{x}, \theta, \phi) \rightarrow (\mathbf{c}, \sigma) \tag{1}$$

where $\mathbf{x} = (x, y, z)$ are the in-scene coordinates, (θ, ϕ) represent the azimuthal and polar viewing angles, $\mathbf{c} = (r, g, b)$ represents color, and σ represents the volume density. This 5D function is approximated by one or more MLP, sometimes denoted as F. The two viewing angles (θ, ϕ) are often represented by $\boldsymbol{d} = (dx, dy, dz)$, a 3D Cartesian unit vector. This neural network representation is constrained to be multi-view consistent by restricting the prediction of σ, the volume density (i.e., the content of the scene), to be independent of viewing direction, whereas the color $\mathbf{c}$ is allowed to depend on both viewing direction and in-scene coordinates. NeRF utilizes a neural network as an implicit representation of a 3D scene instead of the traditional explicit modeling of point clouds, meshes, voxels, etc. Through such a network, it is possible to directly render a projection image from any angle at any location. For this purpose, NeRF introduces the concept of radiation field, which is a very important concept in graphics, and here we give the definition of the rendering equation,

$$L_o(\boldsymbol{x}, \boldsymbol{d}) = L_e(\boldsymbol{x}, \boldsymbol{d}) + \int_{\Omega} f_r(\boldsymbol{x}, \boldsymbol{d}, \omega_i) L_i(\boldsymbol{x}, \omega_i) \cos\theta d\omega_i \tag{2}$$

The neural radiation field represents the scene as volumetric density and directional radiation brightness at any point in space. Using the principles of classical stereoscopic rendering, we can render the color of any ray passing through the scene. The bulk density $\sigma(x)$ can be interpreted as the derivable probability that the ray stays at position X for an infinitesimal particle. Under the conditions of the nearest boundaries t_n and farthest boundaries t_f, the color $C(\mathbf{r})$ of the camera light is:

$$C(\mathbf{r}) = \int_{t_n}^{t_f} T(t)\sigma(\mathbf{r}(t))\mathbf{c}(\mathbf{r}(t), \boldsymbol{d})dt, \text{ where } T(t) = \exp\left(-\int_{t_n}^{t} \sigma(\mathbf{r}(s))ds\right) \tag{3}$$

In practical applications, a fully automated, full-flow 3D reconstruction tool chain is urgently needed in many areas across industries, which also opens up many ideas and opportunities for our 3D reconstruction research. There are two key issues in this process that deserve our consideration. The first is that in the construction process of a 3D reconstruction system, the traditional geometric vision has clear interpretability, so how can we integrate an end-to-end deep learning method? Whether it is an end-to-end replacement or one embedded in our current process seems to be inconclusive and needs further exploration. The second is, from this practicality, how can we combine the reverse reconstruction in computer vision with the forward reconstruction in graphics so as to truly realize the highly structured and highly semantic 3D model required by industry from massive images. Generation is also an essential trend in the future. Only by solving these problems can we indeed use our image 3D reconstruction system to effectively support various business requirements in practical application settings, such as digital city planning, VR content production, and high-precision mapping.

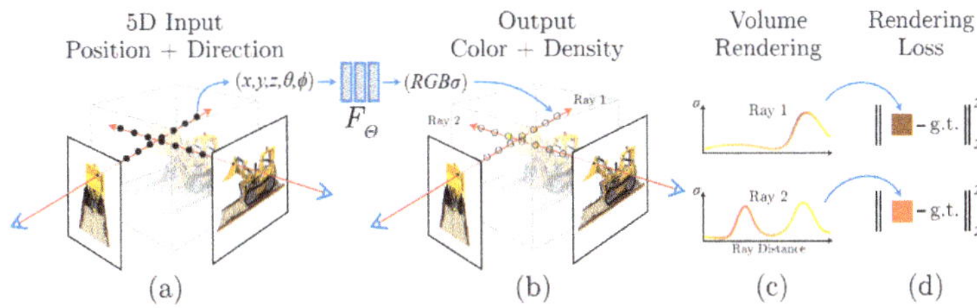

Figure 1. An overview of NeRF model presented by Ben Mildenhall et al. [25]. We create images using the sampling of 5D coordinates (location and viewing direction) along camera rays (**a**), feeding those locations into an MLP to create a color and volume density (**b**), and then utilizing volume rendering algorithms to merge these values into an image (**c**). Because of the differentiability of this rendering function, we can improve the scene representation by decreasing the difference between the synthetic and actual ground truth observed images (**d**).

Table 1. Advantages and disadvantages of the mainstream 3D modeling approach.

Methods	Advantages	Disadvantages
Forward Modeling (CAD)	Ideal for showing details of objects	Complicated and challenging to parameterize accurately
MVS 3D Reconstruction	Better performance when geometric information is abundant and accurate	Missing or incorrect geometry reduces the quality of the view.
NeRF-based 3D Reconstruction	Higher quality models	Long training time and poor generalization.

3.2. Cloud Manufacturing

Cloud manufacturing is an intelligent, efficient, and service-oriented new manufacturing model proposed in recent years. We aim to organically combine the 3D content generation model and cloud manufacturing mode so that this technology can be successfully implemented and applied to actual production. Cloud manufacturing is a cutting-edge manufacturing model that combines cloud computing, big data, and the Internet of Things to enable service-oriented, networked, and intelligent manufacturing. It provides a customized approach to production based on the use of widely distributed, on-demand manufacturing services to meet dynamic and diverse individualized needs and to support socialized production [26–29]. Cloud manufacturing emphasizes the embedding of computing resources, capabilities, and knowledge into networks and environments, allowing the center of attention of manufacturing companies to shift or return to users' needs themselves. Cloud manufacturing is committed to building a communal manufacturing environment where manufacturing companies, customers, intermediaries, and others can fully communicate. In the cloud manufacturing model, user involvement is not limited to the traditional user requirement formulation and user evaluation but permeates every aspect of the entire manufacturing lifecycle. In the cloud manufacturing model, the identity of the customer or user is not unique, i.e., a user is a consumer of cloud services but also a provider or developer of cloud services, reflecting a kind of user participation in manufacturing [30]. The advanced 3D-printing cloud model integrates 3D printing technology with the cloud manufacturing paradigm. Furthermore, the advanced 3D-printing cloud-model-based 3D-printing cloud platform has a service-oriented architecture, personalized customization technology, and a scalable service platform [31]. The 3D-printing cloud platform is not a standalone information-sharing platform, but rather an open, shared, and scalable platform that offers both 3D printing and other types of high-value-added knowledge services. The 3D-printing cloud platform can power a 3D printing community based on collaborative innovation and promote the growth of 3D-printing-related industries. Manufacturing resources (such as 3D printers and robots) from various regions and enterprises are encapsulated into various types of manufacturing services in the cloud platform, with the goal of providing service demanders with on-demand service compositions. Yang et al. [32] provided a cloud-edge cooperation mechanism for cloud manufacturing to offer customers on-demand manufacturing services, greatly enhancing the usage of distributed manufacturing resources and the responsiveness to the needs of customized products. Tamir et al. [33] proposed a new robot-assisted AM and control system framework which effectively combines 3D printing and robotic arm control to better support cloud printing tasks.

4. The Difficulties in Product Customization

Offering products with their colors, components, and features changed to suit consumers' preferences is known as product customization. The practical implementations of NeRF-based product customization currently face several difficulties due to the limitations of technology and resources. This section evaluates the research and divides the challenges into three groups,

(1) The accuracy of 3D modeling: We need to understand and render 3D scenes, because we human beings already live in such a 3D world, and we need to interact with others. When making a virtual scene, we hope to reconstruct the object from different perspectives and then analyze and observe the object. In the medical field, we hope to reconstruct the parts of each person and then guide the doctor's decisions. At the same time, we hope to have a seamless interaction with this virtual world, hoping to get realistic enjoyment in the virtual world. For the next generation of artificial intelligence, we hope that it can understand 3D scenes so that it can better serve humans. Hopefully, we can make artificial intelligence have the ability to interact. This will form a pipeline. After reconstructing the 3D representation from the 2D world, a realistic rendering can be made. On top of this, it is hoped that a 3D scene could be generated so that the generated model could be used to learn the entire process. However, there are specific difficulties in realizing this. In fact, the generation of 3D scenes is a very complicated process. Take NeRF as an example, it requires multi-view images. However, after all, the amount of data in multi-view images is far from enough. This kind of data is actually a lot of this single-view images on the Internet, which lacks perspective information. We hope to use the existing pre-trained model to provide prior knowledge.

(2) Security and supervision: AM files can be easily transferred from the AM design stage to the shop floor during final production. Digital supply networks and chains can be created thanks to the ease with which parts and products can be shared and communicated thanks to AM's digital nature. In addition, these digital advantages come with a few drawbacks. A digital design-and-manufacturing process increases the likelihood of data theft or tampering without a robust data protection framework. Data leakage and identity theft will pose significant security risks as the scale of the SM production system grows, and the participation of dishonest and malicious nodes will put the interests of honest nodes in jeopardy. At the same time, when physical products are transported via product data, it is essential to secure, store, and share data containing all important information. The digital thread for AM, also known as DTAM, creates a single, seamless data link between initial design concepts and finished products in order to mitigate risks. The integration of multiple printers and printing technologies and a number of distinct and disjointed physical manufacturing facilities is the primary obstacle in this challenge. Additionally, because parts must be inspected throughout the process rather than just at the end, businesses can need help to keep track of events that take place during the additive process. This may be necessary for part certification and qualification.

(3) Production efficiency: The uneven distribution of manufacturing resources in the manufacturing industry and the low utilization rate of resources have seriously affected the development of the manufacturing industry. To effectively integrate scattered manufacturing resources and improve the utilization rate of manufacturing resources, cloud manufacturing, a manufacturing model that uses information technology for services,

appears in everyone's vision. As one of the key research issues of the cloud manufacturing platform, the scheduling of manufacturing resources in the cloud manufacturing environment will affect the overall operational efficiency of the cloud manufacturing platform. In both academia and industry, cloud computing resource scheduling problems are considered to be as difficult as non-deterministic polynomial optimization problems, that is, NP problems. Therefore, algorithms that solve relatively conventional scheduling problems may suffer from dimensional damage when the scale of the problem increases. With the development of cloud computing and the increase in complexity, this problem has become more challenging.

5. The Efficient Product Customization Framework

Considering that the characteristics of NeRF and other artificial intelligence technology have great potential for solving the difficulties of SM, a customization production framework of 3D printing using NeRF and ultra-large-scale pre-trained multimodal models as the baseline is proposed in this paper. As shown in Figure 2, the framework is constructed in the order of the multimodal data-based customization production process and value flow, specifically, from the bottom to the top, it is divided into three parts: 3D modeling service, blockChain encryption service, and cloud management service.

5.1. 3D Modeling Service

3D content customization is a very challenging task as it is a 3D representation. The existing stylized methods are 2D, so we can use the 2D stylized form to provide a kind of supervision. Still, this kind of supervision does not have 3D information, so we need to use the mutual learning between 2D and 3D. Two-dimensional neural networks can provide a stylized reference. More importantly, we offer this 3D-based spatial consistency information to 3D NeRF and finally stylize NeRF. Therefore, we can convert the scenes taken by mobile phones into this 3D stereo-stylized effect.

To solve the difficulties in 3D modeling, as shown in Figure 3, the service includes two modules: image-based 3D digital-asset reconstruction and text-based 3D digital-asset generation. In the image-based module, the end-to-end model can automatically generate a 3D geometric model without human intervention by taking multiple 2D pictures of the object from different angles with a device such as a mobile phone. The entire model is implemented in the following steps. First, a 360° surrounding image needs to be captured at a fixed focal length. After that, we use this tool for pose estimation. After recovering the poses of the photographed objects, the poses obtained from the sparse reconstruction are converted into local light field fusion (LLFF)-format data and then rendered by NeRF. The NeRF combines light-field sampling theory with neural networks, using an MLP to implicitly learn the scene in the sampling of all light, and it achieves good view synthesis results. In addition, NeRF uses the images themselves for self-supervised learning, which is applicable to a wide range of datasets, from synthetic to real-world. Finally, the corresponding mesh model is exported. To make NeRF practical for 3D modeling in AM, we build it based on instant neural graphics primitives (Instant-NGP) [34], a fast variant of NeRF. In addition, in order to achieve the realization of the reconstruction process, only the target object is modeled. We use image segmentation algorithms in the pre-processing process, which focuses on the accurate extraction of the reconstructed target and can reduce redundant information in the image during the reconstruction process.

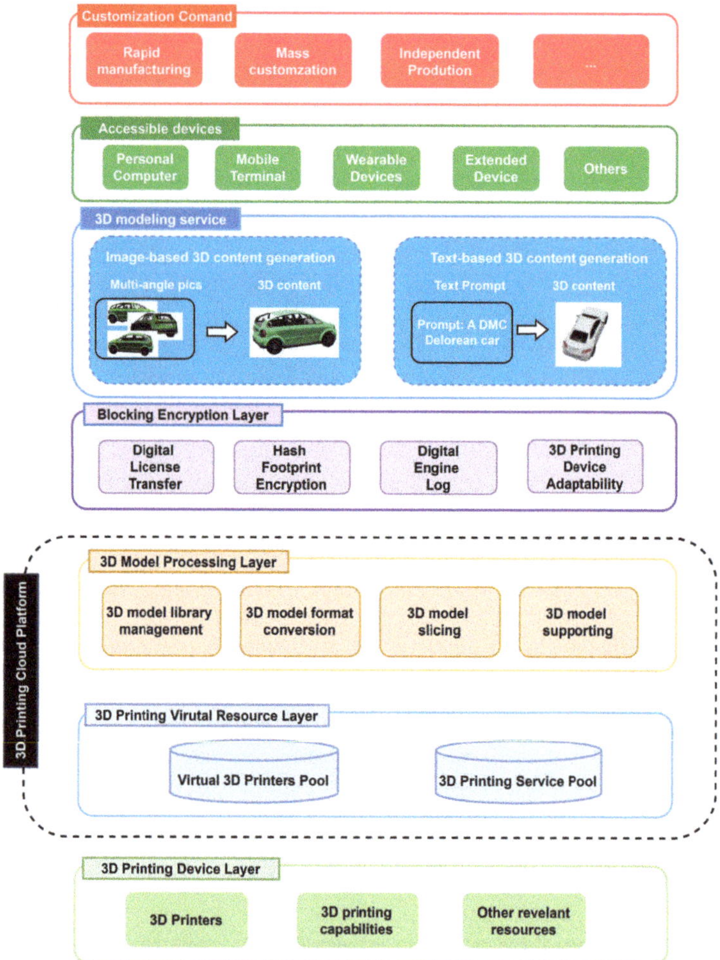

Figure 2. The Efficient Product Customization Framework.

The process of 3D reconstruction based on sequenced images is rather complex. In order to further enhance the convenience and versatility of product customization, we propose a text-guided 3D digital-asset generation module. Recent breakthroughs in text-to-image synthesis have been driven by multimodal models trained on billions of image-to-text pairs. The CLIP multimodal pre-training model is called contrastive language-image pre-training [35], i.e., a pre-training method based on contrasting text-image pairs. CLIP uses text as a supervised signal to train the visual model, which results in a very good zero-shot effect and good generalization of the final model. The training process is as follows: The input of CLIP is a pair of picture–text pairs. The text and images are output with corresponding features by the text encoder and image encoder, respectively. The text features and image features are then compared and learned on these outputs. If the input to the model is **n** pairs of image–text pairs, then this pair of mutually paired image–text pairs are positive samples (the parts marked in blue on the diagonal of the output feature matrix in the Figure 4), and the other pairs are negative samples. The training process of the model is thus to maximize the similarity of the positive samples and minimize the similarity of the negative samples. However, applying this approach to 3D synthesis requires large-scale datasets of labeled 3D assets and efficient methods for denoising 3D data.

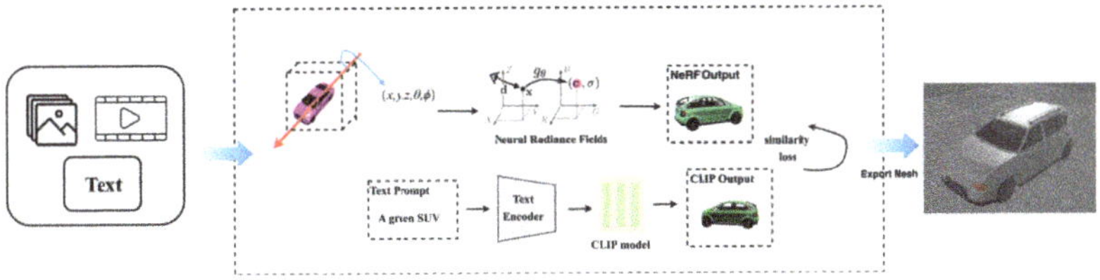

Figure 3. The pipeline of 3D content generation service.

The traditional NeRF scheme for generating 3D scenes requires multiple 3D photographs to achieve 360° visual reconstruction. In contrast, the Dreamfields [36] algorithm used in this paper does not require photos to generate 3D models and can generate entirely new 3D content. In fact, the algorithm is guided by a deep neural network (DNN) that can display geometric and color information based on the user's textual description of the 3D object and some simple adjustments. When training the algorithm, a multi-angle 2D photo is required as a supervised signal, and once trained, a 3D model is generated and a new view is synthesized. The role of the CLIP multimodal pre-training model is to evaluate the accuracy of the text-generated images. After the text is fed into the network, the untrained NeRF model generates random views from a single perspective, and then the CLIP model is used to evaluate the accuracy of the generated images. In other words, as shown in Figure 3, NeRF renders the image from a random position, and finally uses CLIP as a measure the similarity between the text description and the composite image of given parameters θ and position s orientation p. This process is repeated 20,000 times from different views until a 3D model is generated that matches the text description. The corresponding loss function for this training process is:

$$\mathcal{L}_{\text{CLIP}}\,(\theta,\ \text{pose}\ \mathbf{p},\ \text{caption}\ y) = -\mathbf{g}(\mathbf{I}(\theta,\mathbf{p}))^{\mathrm{T}}\mathbf{h}(\mathbf{y}) \tag{4}$$

$$\mathcal{L}_T = -\min(\tau, \text{mean}(T(\theta,\mathbf{p}))) \tag{5}$$

$$\mathcal{L}_{\text{total}} = \mathcal{L}_{\text{CLIP}} + \lambda\mathcal{L}_T \tag{6}$$

where $g(\cdot)$ is the image encoder and $h(\cdot)$ is the text encoder. The L_{clip} objective is to maximize the cosine similarity of the rendered image embedding to the text embedding, and the L_t objective is to maximize the average transmittance. Inspired by Dreamfields algorithm, this paper leverages a priori knowledge from large pre-trained models to generate 3D digital assets better. Using Dreamfields as the baseline, the images are encoded by a CLIP image encoder and compared with the text input encoded by a CLIP text encoder. To implement support for Chinese prompts, we replaced CLIP with Taiyi-CLIP [37], a visual-language model using Chinese-Roberta-wwm [38] as the language encoder, and applied the vision transformer (ViT) [39] in CLIP as the vision encoder. They froze the vision encoder and tuned the language encoder to speed up and stabilize the pre-training process, and applied Noah-Wukong [40] and Zero-Corpus [41] as the pre-training datasets. There are two main limitations of existing dataset collection methods. Specifically, a dataset can contain 100 million image pairs collected from the web. The Wukong dataset was compiled from a query list of 200,000 words in order to cover a sufficiently diverse set of visual concepts. Furthermore, the CLIP model transforms this encoded text input into an image embedding, the output of which is also used for a loss function. In addition, this encoded text input is transformed into an image embedding by the CLIP model. This output is also used for a loss function. The specific operation of the model for each input uses a text prompt. It is necessary to retrain a NeRF model, which will require multi-angle 2D photos, and after completing the training can generate a 3D model, and thus, synthesis of the new perspective. The role of CLIP is still to evaluate the accuracy of text-generated images.

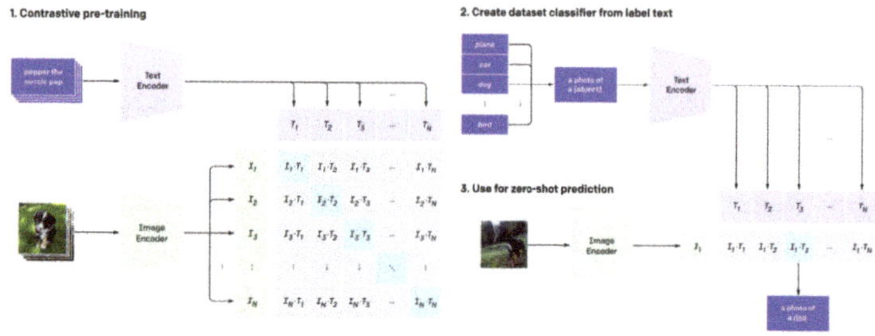

Figure 4. The schematic diagram of the principle of CLIP model.

5.2. Blockchain Encryption Service

3D Printing has gone through these stages: from concept to CAD file to generating design (if available) to actual 3D Printing. All these steps represent a loophole where 3D Printing could be compromised or even stolen, putting the company's intellectual property at risk. All projects, from start to finish, can be done in a blockchain, from the communication of the project to the production and transfer of data, and 3D printing and delivery. Everything could be accessible in the chain, and each party would have all the data. Off-chain, cross-chain, off-chain management, and off-chain certification facilities are primarily referred to as "blockchain infrastructure" in this context. They serve as the foundation for communication between blockchains [42]. As shown in Figure 5, firstly, the 3D-printing files are encrypted using a hashing algorithm. Based on the content of the file, the encryption algorithm generates a fingerprint of a unique hash value. The blockchain stores a hash value that verifies the authenticity of the file rather than the file itself. At the second step, we upload the hash value generated after encryption, which is the so-called digital fingerprint, to the blockchain. Finally, if someone wants to print the 3D content, the service will upload the key printing information, such as operator, and location, to the blockchain. Then the blockchain transaction ID is generated, which can further trace the source of the 3D content operation. Trusted printers can communicate with the blockchain by installing so-called Secure Elements [43] onto AM machines.

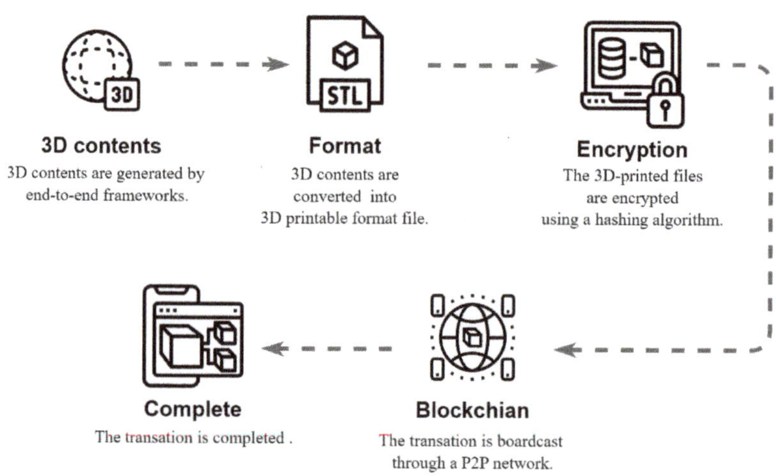

Figure 5. Process of blockchain encryption of 3D contents.

The characteristics and their relevance to AM are shown in Table 2. The most significant advantage of applying blockchain technology to AM is the development of trust. The designer would no longer need to be concerned about his designs being used illegally and could instead focus on getting better at making models. The individual user could also choose to print the files in a variety of ways, either by the number of times they are printed or by gaining access to the files and downloading them directly, both of which are now available at a relatively lower price and with full disclosure regarding the models' origin. On-demand production is also environmentally friendly and carbon neutral, and manufacturers could produce locally, close to their customers, thereby reducing storage and transportation costs. In addition to the actual goods, brand manufacturers could offer their customers printable files for customization, accessories, and replacement parts. In short, the blockchain for 3D printing will make it possible to trade 3D printing in the form of "tokens" and promote 3D printing as a technology to a greater extent.

Table 2. Blockchain characteristics and their relevance to AM.

Blockchain Characteristic	Relevance to AM
Distributed data maintained within and among stakeholders	Helps in the management of activity in distributed supply chain expected to be found with AM
Near real time-settlement and exchange are nearly instant	Changes to a design are made instantly, which facilitates efficient AM processes
Trustless environment—cryptographic validation of transactions	Designed to protect against risks of unauthorized data access
Irreversibility the transaction history is append-only	Helps with cyber risk and IP protection as it is intended to provide an indelible and traceable record of changes

5.3. Cloud Management Service

Theoretically, 3D printing is the ideal method of production for cloud manufacturing. Digital files can be printed anywhere with just a 3D printer and suitable materials, thanks to their ease of transfer and no geographic restrictions. The cloud-based 3D printing product personalization service platform uses a browser/server model that allows users to use a browser to access the platform. The Internet serves as the medium between the browser and the server, enabling the cloud service platform to provide a variety of cloud 3D-Printing services. This cloud platform's architecture consists primary of three layers [30]: the virtual resources layer, the 3D-printing-resource layer, and the technology support layer. The foundation for the cloud services platform is the primary technical support layer. The cloud administration stage must include the framework as a management of the executives model, which intends to provide consistent and smooth specialized assistance for the cloud-producing administration stage's activity. The primary function of the virtual resources layer is to abstract and simplify cloud service platform-connected 3D-printing resources: The cloud service platform's cloud computing technology is used to describe the various physical 3D-printing resources as virtual resources [31], resulting in virtual data resources. A virtual cloud pool is created when virtual data resources are encapsulated and published to the cloud platform's resource service center module. Users can select the printing resources they require from the cloud. For the personalized service platform, the 3D-printing resource layer provides software, material, and equipment resources. The purpose of the user interface layer is to provide the cloud service platform with user-friendly application interfaces, allowing users to invoke various cloud services freely.

6. Case Study

Recreational equipment manufacturing is selected as the practice object to verify the effectiveness of the product customization framework proposed in this paper. The toy manufacturing industry is an integral part of the traditional manufacturing industry. It has a high demand for labor and a large export volume of products. Surveys show that the 2020 toy and game market is projected to be US $135 billion [44]. 3D Printing allows the creation of physical objects from geometric representations by continuously adding material. It can effectively respond to customers' needs and help achieve service-oriented manufacturing. In China, the traditional toy market has interactive electronic toys with high-technology content, high-tech intelligent toys, and educational toys. They can foster children's imagination, creativity, and hands-on skills. These toys are mainstream. The traditional peak season for the toy industry is June to October each year. Thanks to government measures to promote consumption, and urban consumption upgrades, China's toy and retail scale have maintained steady growth. The main export markets are the United States, the United Kingdom, Japan, and other countries, of which exports to the United States amounted to US $8.57 billion in 2021, an increase in 6.8% over the previous year, accounting for 25.6% of China's total exports. According to the research study of Made-in-China, China's Toys Export in November Amounted to US $ 2.44 Billion, up by 21.18%.

In this paper, the pictures used in the production of datasets were obtained by taking video with mobile phones and then drawing frames. The dataset consists of 86 different scenes, which are mainly composed of scenes with four rotations and scans at 90-degree intervals, thereby realizing a 360-degree model. The production of the dataset can be summarized in three steps. First, do feature matching of pictures to obtain the camera pose. Second, convert the matching pose into LLFF format. Finally, upload the required files to the corresponding folder of NeRF and set up the configuration file. LLFF-format data can be the corresponding picture parameters. Camera position and camera parameters can be stored in a simple and effective file to facilitate Python reading. And the NeRF model's source code has the necessary configuration and modules for direct training on LLFF-format datasets, making it easy for researchers to use. COLMAP is a universal motion structure (SfM) and multi-view stereo (MVS) processing tool with graphical and command-line interfaces. It provides a wide range of functions for the reconstruction of ordered and disordered image collections. We used this tool to estimate the photo position. Through the sparse reconstruction of COLMAP, the position and posture of the photo are restored. The next step is to generate the data format used for NeRF training and select the data in LLFF format. The whole process is end-to-end trainable using the dataset as input. The trained NeRF-based model is used to build an end-to-end module. Users can provide multiple photos at different angles for the same object. For example, take one image at each angle, and provide 3–4 angled pictures. If you need to generate a position that cannot be photographed, such as the bottom, you need to provide further pictures of the model lying on its side. When moving around an object, adjacent photos must overlap by at least 70%. It is recommended to take at least 30 photos to provide more pictures, which is helpful in generating higher-quality 3D models.

As for the text-to-3D module, we use a Pytorch implementation of an original Dreamfields algorithm as baseline. We mainly changes the back end of the original Dreamfields from the original NeRF to our model and replaces the origin CLIP encoder with the Taiyi-CLIP [37] encoder. In addition, to improve the performance of the generated content, we apply R-Drop [45] for regularization in the forward propagation of the CLIP pre-trained model; these methods enhance the expressiveness and generalization ability of our model. We used search engines to crawl about 10,000 image–text pairs of data related to the toy manufacturing industry to form fine-tuned datasets. Each image has up to five descriptions associated with it. The input for the model is a batch of subtitles and a batch of images passed through the CLIP text encoder and image encoder, respectively. The training process uses contrast learning to learn joint embedded representations of images and subtitles. In this embedded space, the images are very close to their respective descriptions, as are similar images and similar descriptions. Conversely, images and descriptions of different images may be pushed further. In order to standardize our dataset and prevent overfitting due to the size of the dataset, we used both image and text enhancement. Image enhancement was done online using the built-in conversion in the Pytorch Torchvision package. The transformations used were random cropping, random resizing and cropping, color jitter, and random horizontal and vertical flipping.

As shown in Tables 3 and 4, we have performed some experiments to test the print time, quality, and the corresponding actual printing time of the generated models with different inputs, and we can see that the models generated within this framework are reliable in terms of efficiency and quality. Some results are shown in Figure 6.

Table 3. Some examples of 3D content generation based on Surrounding photos.

Product Name (Numbers of Photos)	Generate Time	Print Time
"Gundam (74)"	17 min 58 s	2 h 38 min 30 s
"Plane (102)"	13 min 12 s	1 h 5 min 22 s
"Princess (194)"	40 min 23 s	3 h 43 min 28 s
"Doll Cow (140)"	21 min 25 s	4 h 48 min 52 s

Table 4. Some examples of 3D content generation based on text prompt.

Text Prompt	Generate Time	Print Time
"An unlighted candle."	45 min 37 s	1 h 55 min 26 s
"A science fiction style gear"	41 min 35 s	3 h 48 min 4 s
"A chair"	28 min 47 s	3 h 58 min 11 s
"A plush toy of a corgi"	33 min 12 s	3 h 50 min 37 s

Figure 6. 3D content generation examples.

The resource scheduling interface of the cloud platform adopts the development mode of separating the front and back ends. Springboot, Mybatis, and Springcloud frameworks were applied to develop the back-end of the cloud platforms. The persistent operation of data storage uses MongoDB and MySQL databases. The cloud platform is equipped with a cloud server with 16 G memory and 8 NVIDIA V100 GPUs graphics cards. We deployed the pre-trained large model described above on TF-Serving's server. TF-Serving will automatically deploy it according to the incoming port and model path. The server where the model is located does not need a Python environment (take Python training as an example). And then the application service directly launches a service call to the server where the model is located. The call can be made through grpc.

The 3D models of the 3D model file library come from the uploads of various designers on the cloud platform. The designers of the 3D printing cloud platform designed 3D models with a certain market value according to their own capabilities and market needs and then uploaded them to the cloud platform. The cloud platform is responsible for optimizing and managing these model files. Users can quickly search for suitable 3D model files. The management of the 3D model file library mainly includes the removing repeat, classification, evaluation, and security management of model files.

The cloud platform's overall framework is shown in Figure 7. It supports data processing in the 3D printing process, including data format conversion, support design, slice calculation, print path planning, structural analysis, model optimization, etc. These steps depend on each other and affect each other, and together determine the the efficiency of printing and the quality of the finished product. In addition, in order to improve the computational efficiency of the support-generation process in the 3D-printing process, the parallelized slicing method, the model placement pointing optimization and its parallelization method [46], and GPU-based parallelized support generation were successfully applied to this platform.

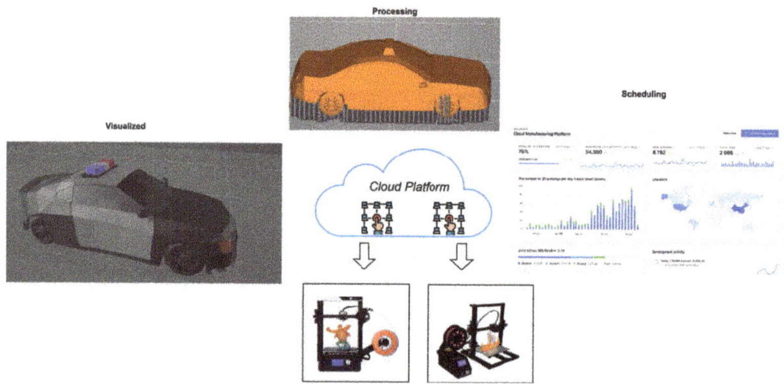

Figure 7. The product customization framework built on a cloud platform.

7. Conclusions

In SM, providing a user-friendly approach to product customization has been the focus of research. As an early attempt to apply NeRF and large pre-trained vision-language models such as CLIP, we propose an efficient product customization framework in the SM paradigm. It provides a new method of collaborative product design, which can efficiently solve the problems existing in current manufacturing systems, such as the accuracy of 3D modeling, security, supervision, anti-counterfeiting, and efficient allocation of resources. However, the key technologies used in this paper, such as neural-volume rendering and multimodal data processing technology, still need to be further studied and improved. The main limitations can be summarized as follows. Firstly, choosing a suitable resolution is very important for printing a high-quality model. A too low resolution will inevitably affect the quality of the finished print. A low resolution results in a non-smooth surface for the finished 3D printing. It is currently difficult to obtain high-resolution geometry or textures for 3D models with the model presented in this paper. To address this issue, a coarse-to-fine optimization approach could be proposed, in which multiple diffusion priors at different resolutions would optimize the 3D representation, which would result in view-consistent geometry and high-resolution details. Secondly, in the current implementation of the 3D model generation algorithm, a NeRF network needs to be retrained for each text prompt, which results in low efficiency of model generation and requires a large amount of GPU resources. In the future, we can consider loading the existing generic mesh model and then iteratively modifying the 3D model by text prompts to generate an ideal 3D model. A priori forward-looking production framework for future integration would be instructive. Accordingly, this work is proved to be a revolutionary break through for solving the core problems of 3D modeling in SM.

Author Contributions: Conceptualization, Y.L. and Z.S.; methodology, Y.L.; software, Y.L. and S.L.; validation, Y.L., Z.S. and G.X.; formal analysis, H.W.; investigation, Y.L. and T.S.T.; writing—original draft preparation, Y.L. and T.S.T.; writing—review and editing, Y.L., Z.S. and B.H.; visualization, Y.L. and Z.S.; funding acquisition, Z.S. and G.X. All authors have read and agreed to the published version of the manuscript.

Funding: This work was supported in part by the National Key Research and Development Program of China under Grant 2018YFB1700602; in part by the National Natural Science Foundation of China under Grants 92267103, U1909204 and 61872365; in part by the Scientific Instrument Developing Project of the Chinese Academy of Sciences (CAS) under Grant YZQT014; in part by the Guangdong Basic and Applied Basic Research Foundation under Grant 2021B1515140034; in part by the Foshan

Science and Technology Innovation Team Project under Grant 2018IT100142; and in part by the Collaborative Innovation Center of Intelligent Green Manufacturing Technology and Equipment, Shandong, under Grant IGSD-2020-015. The work of Zhen Shen was supported by the CAS Key Technology Talent Program.

Institutional Review Board Statement: Not applicable.

Informed Consent Statement: Not applicable.

Data Availability Statement: Not applicable.

Conflicts of Interest: The authors declare no conflict of interest.

References

1. Ding, K.; Jiang, P. Incorporating social sensors, cyber-physical system nodes, and smart products for personalized production in a social manufacturing environment. *Proc. Inst. Mech. Eng. Part B J. Eng. Manuf.* **2018**, *232*, 2323–2338. [CrossRef]
2. Zheng, T.; Ardolino, M.; Bacchetti, A.; Perona, M. The applications of Industry 4.0 technologies in manufacturing context: A systematic literature review. *Int. J. Prod. Res.* **2021**, *59*, 1922–1954. [CrossRef]
3. Wang, F.Y. The emergence of intelligent enterprises: From CPS to CPSS. *IEEE Intell. Syst.* **2010**, *25*, 85–88. [CrossRef]
4. Wang, F.Y. From social computing to social manufacturing: The coming industrial revolution and new frontier in cyber-physical-social space. *Bull. Chin. Acad. Sci.* **2012**, *6*, 658–669.
5. Wang, F.Y. Social computing: Concepts, contents, and methods. *Int. J. Intell. Control Syst.* **2004**, *9*, 91–96.
6. Xiong, G.; Wang, F.Y.; Nyberg, T.R.; Shang, X.; Zhou, M.; Shen, Z.; Li, S.; Guo, C. From mind to products: Towards social manufacturing and service. *IEEE/CAA J. Autom. Sin.* **2017**, *5*, 47–57. [CrossRef]
7. Shang, X.; Wang, F.Y.; Xiong, G.; Nyberg, T.R.; Yuan, Y.; Liu, S.; Guo, C.; Bao, S. Social manufacturing for high-end apparel customization. *IEEE/CAA J. Autom. Sin.* **2018**, *5*, 489–500. [CrossRef]
8. Shang, X.; Shen, Z.; Xiong, G.; Wang, F.Y.; Liu, S.; Nyberg, T.R.; Wu, H.; Guo, C. Moving from mass customization to social manufacturing: A footwear industry case study. *Int. J. Comput. Integr. Manuf.* **2019**, *32*, 194–205. [CrossRef]
9. Shang, X.; Chen, X.; Niu, L.; Xiong, G.; Shen, Z.; Dong, X.; Shen, Z.; Liu, C.; Xi, B. Blockchain-based social manufacturing for customization production. *IFAC-PapersOnLine* **2020**, *53*, 53–58. [CrossRef]
10. Mohajeri, B.; Nyberg, T.; Karjalainen, J.; Tukiainen, T.; Nelson, M.; Shang, X.; Xiong, G. The impact of social manufacturing on the value chain model in the apparel industry. In Proceedings of the 2014 IEEE International Conference on Service Operations and Logistics, and Informatics, Qingdao, China, 8–10 October 2014; pp. 378–381.
11. Mohajeri, B.; Nyberg, T.; Karjalainen, J.; Nelson, M.; Xiong, G. Contributions of social manufacturing to sustainable apparel industry. In Proceedings of the 2016 IEEE International Conference on Service Operations and Logistics, and Informatics (SOLI), Beijing, China, 10–12 July 2016; pp. 24–28.
12. Mohajeri, B.; Kauranen, I.; Nyberg, T.; Ilen, E.; Nelson, M.; Xiong, G. Improving sustainability in the value chain of the apparel industry empowered with social manufacturing. In Proceedings of the 2020 15th IEEE Conference on Industrial Electronics and Applications (ICIEA), Kristiansand, Norway, 9–13 November 2020; pp. 235–240.
13. Jiang, P.; Leng, J.; Ding, K. Social manufacturing: A survey of the state-of-the-art and future challenges. In Proceedings of the 2016 IEEE International Conference on Service Operations and Logistics, and Informatics (SOLI), Beijing, China, 10–12 July 2016; pp. 12–17.
14. Ding, K.; Jiang, P.Y.; Zhang, X. A framework for implementing social manufacturing system based on customized community space configuration and organization. In *Advanced Materials Research*; Trans Tech Publications Ltd.: Bäch, Switzerland, 2013; Volume 712, pp. 3191–3194.
15. Jiang, P.; Ding, K.; Leng, J. Towards a cyber-physical-social-connected and service-oriented manufacturing paradigm: Social Manufacturing. *Manuf. Lett.* **2016**, *7*, 15–21. [CrossRef]
16. Leng, J.; Jiang, P.; Xu, K.; Liu, Q.; Zhao, J.L.; Bian, Y.; Shi, R. Makerchain: A blockchain with chemical signature for self-organizing process in social manufacturing. *J. Clean. Prod.* **2019**, *234*, 767–778. [CrossRef]
17. Xiong, G.; Helo, P.; Ekstrom, S.; Tamir, T.S. A Case Study in Social Manufacturing: From social manufacturing to social value chain. *Machines* **2022**, *10*, 978. [CrossRef]
18. Cao, W.; Jiang, P.; Jiang, K. Demand-based manufacturing service capability estimation of a manufacturing system in a social manufacturing environment. *Proc. Inst. Mech. Eng. Part B J. Eng. Manuf.* **2017**, *231*, 1275–1297. [CrossRef]
19. Xiong, G.; Chen, Y.; Shang, X.; Liu, X.; Nyberg, T.R. AHP fuzzy comprehensive method of supplier evaluation in social manufacturing mode. In Proceedings of the the 11th World Congress on Intelligent Control and Automation, Shenyang, China, 29 June–4 July 2014; pp. 3594–3599.
20. Yin, D.; Ming, X.; Zhang, X. Understanding data-driven cyber-physical-social system (D-CPSS) using a 7C framework in social manufacturing context. *Sensors* **2020**, *20*, 5319. [CrossRef]
21. Xiong, G.; Tamir, T.S.; Shen, Z.; Shang, X.; Wu, H.; Wang, F.Y. A Survey on Social Manufacturing: A Paradigm Shift for Smart Prosumers. *IEEE Trans. Comput. Soc. Syst.* **2022**, 1–19. [CrossRef]

22. Brière-Côté, A.; Rivest, L.; Maranzana, R. Comparing 3D CAD models: Uses, methods, tools and perspectives. *Comput.-Aided Des. Appl.* **2012**, *9*, 771–794. [CrossRef]
23. Roberts, L.G. Machine Perception of Three-Dimensional Solids. Ph.D. Thesis, Massachusetts Institute of Technology, Cambridge MA, USA, 1963.
24. Yao, Y.; Luo, Z.; Li, S.; Fang, T.; Quan, L. Mvsnet: Depth inference for unstructured multi-view stereo. In Proceedings of the the European Conference on Computer Vision (ECCV), Munich, Germany, 8–14 September 2018; pp. 767–783.
25. Mildenhall, B.; Srinivasan, P.P.; Tancik, M.; Barron, J.T.; Ramamoorthi, R.; Ng, R. Nerf: Representing scenes as neural radiance fields for view synthesis. *Commun. ACM* **2021**, *65*, 99–106. [CrossRef]
26. Mai, J.; Zhang, L.; Tao, F.; Ren, L. Customized production based on distributed 3D printing services in cloud manufacturing. *Int J. Adv. Manuf. Technol.* **2016**, *84*, 71–83. [CrossRef]
27. Adamson, G.; Wang, L.; Holm, M. The state of the art of cloud manufacturing and future trends. In Proceedings of the International Manufacturing Science and Engineering Conference, Madison, WI, USA, 10–14 June 2013; American Society of Mechanical Engineers: New York, NY, USA, 2013; Volume 55461, p. V002T02A004.
28. Xu, X. From cloud computing to cloud manufacturing. *Robot. Comput.-Integr. Manuf.* **2012**, *28*, 75–86. [CrossRef]
29. Tao, F.; Zhang, L.; Venkatesh, V.; Luo, Y.; Cheng, Y. Cloud manufacturing: A computing and service-oriented manufacturing model. *Proc. Inst. Mech. Eng. Part B J. Eng. Manuf.* **2011**, *225*, 1969–1976. [CrossRef]
30. Guo, L.; Qiu, J. Combination of cloud manufacturing and 3D printing: Research progress and prospect. *Int. J. Adv. Manuf. Technol.* **2018**, *96*, 1929–1942. [CrossRef]
31. Cui, J.; Ren, L.; Mai, J.; Zheng, P.; Zhang, L. 3D printing in the context of cloud manufacturing. *Robot. Comput.-Integr. Manuf.* **2022**, *74*, 102256. [CrossRef]
32. Yang, C.; Wang, Y.; Tang, R.; Lan, S.; Wang, L.; Shen, W.; Huang, G.Q. Cloud-edge-device Collaboration Mechanisms of Cloud Manufacturing for Customized and Personalized Products. In Proceedings of the 2022 IEEE 25th International Conference on Computer Supported Cooperative Work in Design (CSCWD), Hangzhou, China, 4–6 May 2022; pp. 1517–1522.
33. Tamir, T.S.; Xiong, G.; Dong, X.; Fang, Q.; Liu, S.; Lodhi, E.; Shen, Z.; Wang, F.Y. Design and optimization of a control framework for robot assisted additive manufacturing Based on the Stewart Platform. *Int. J. Control Autom. Syst.* **2022**, *20*, 968–982. [CrossRef]
34. Müller, T.; Evans, A.; Schied, C.; Keller, A. Instant neural graphics primitives with a multiresolution hash encoding. *arXiv* **2022**, arXiv:2201.05989.
35. Radford, A.; Kim, J.W.; Hallacy, C.; Ramesh, A.; Goh, G.; Agarwal, S.; Sastry, G.; Askell, A.; Mishkin, P.; Clark, J.; et al. Learning transferable visual models from natural language supervision. In Proceedings of the International Conference on Machine Learning(ICML), Online, 18–24 July 2021; pp. 8748–8763.
36. Jain, A.; Mildenhall, B.; Barron, J.T.; Abbeel, P.; Poole, B. Zero-shot text-guided object generation with dream fields. In Proceedings of the the IEEE/CVF Conference on Computer Vision and Pattern Recognition (CVPR), New Orleans, LA, USA, 21–24 June 2022; pp. 867–876.
37. Wang, J.; Zhang, Y.; Zhang, L.; Yang, P.; Gao, X.; Wu, Z.; Dong, X.; He, J.; Zhuo, J.; Yang, Q.; et al. Fengshenbang 1.0: Being the foundation of chinese cognitive intelligence. *arXiv* **2022**, arXiv:2209.02970.
38. Cui, Y.; Che, W.; Liu, T.; Qin, B.; Yang, Z. Pre-training with whole word masking for chinese bert. *IEEE/ACM Trans. Audio Speech Lang. Process.* **2021**, *29*, 3504–3514. [CrossRef]
39. Dosovitskiy, A.; Beyer, L.; Kolesnikov, A.; Weissenborn, D.; Zhai, X.; Unterthiner, T.; Dehghani, M.; Minderer, M.; Heigold, G.; Gelly, S.; et al. An image is worth 16x16 words: Transformers for image recognition at scale. *arXiv* **2020**, arXiv:2010.11929.
40. Gu, J.; Meng, X.; Lu, G.; Hou, L.; Niu, M.; Xu, H.; Liang, X.; Zhang, W.; Jiang, X.; Xu, C. Wukong: 100 Million large-scale Chinese cross-modal pre-training dataset and a foundation framework. *arXiv* **2022**, arXiv:2202.06767.
41. Xie, C.; Cai, H.; Song, J.; Li, J.; Kong, F.; Wu, X.; Morimitsu, H.; Yao, L.; Wang, D.; Leng, D.; et al. Zero and R2D2: A large-scale Chinese cross-modal benchmark and a vision-Language framework. *arXiv* **2022**, arXiv:2205.03860.
42. Ouyang, L.; Yuan, Y.; Wang, F.Y. A blockchain-based framework for collaborative production in distributed and social manufacturing. In Proceedings of the 2019 IEEE International Conference on Service Operations and Logistics, and Informatics (SOLI), Zhengzhou, China, 6–8 November 2019; pp. 76–81.
43. Holland, M.; Nigischer, C.; Stjepandić, J. Copyright protection in additive manufacturing with blockchain approach. In *Transdisciplinary Engineering: A Paradigm Shift*; IOS Press: Amsterdam, The Netherlands, 2017; pp. 914–921.
44. Petersen, E.E.; Kidd, R.W.; Pearce, J.M. Impact of DIY home manufacturing with 3D printing on the toy and game market. *Technologies* **2017**, *5*, 45. [CrossRef]
45. Liang, X.; Wu, L.; Li, J.; Wang, Y.; Meng, Q.; Qin, T.; Chen, W.; Zhang, M.; Liu, T.-Y. R-drop: Regularized dropout for neural networks. *Adv. Neural Inf. Process. Syst.* **2021**, *34*, 10890–10905.
46. Li, Z.; Xiong, G.; Zhang, X.; Shen, Z.; Luo, C.; Shang, X.; Dong, X.; Bian, G.B.; Wang, X.; Wang, F.Y. A GPU based parallel genetic algorithm for the orientation optimization problem in 3D printing. In Proceedings of the 2019 International Conference on Robotics and Automation (ICRA), Montreal, QC, Canada, 20–24 May 2019; pp. 2786–2792.

machines

Article

A Graph Matching Model for Designer Team Selection for Collaborative Design Crowdsourcing Tasks in Social Manufacturing

Dianting Liu [1,2], Danling Wu [1,*] and Shan Wu [1]

1 College of Mechanical and Control Engineering, Guilin University of Technology, Guilin 541004, China
2 College of Information Science and Engineering, Guilin University of Technology, Guilin 541004, China
* Correspondence: 1020190643@glut.edu.cn; Tel.: +86-18226928138

Abstract: In order to find a suitable designer team for the collaborative design crowdsourcing task of a product, we consider the matching problem between collaborative design crowdsourcing task network graph and the designer network graph. Due to the difference in the nodes and edges of the two types of graphs, we propose a graph matching model based on a similar structure. The model first uses the Graph Convolutional Network to extract features of the graph structure to obtain the node-level embeddings. Secondly, an attention mechanism considering the differences in the importance of different nodes in the graph assigns different weights to different nodes to aggregate node-level embeddings into graph-level embeddings. Finally, the graph-level embeddings of the two graphs to be matched are input into a multi-layer fully connected neural network to obtain the similarity score of the graph pair after they are obtained from the concat operation. We compare our model with the basic model based on four evaluation metrics in two datasets. The experimental results show that our model can more accurately find graph pairs based on a similar structure. The crankshaft linkage mechanism produced by the enterprise is taken as an example to verify the practicality and applicability of our model and method.

Keywords: graph matching; Graph Convolutional Network (GCN); attention mechanism; fully connected neural network

Citation: Liu, D.; Wu, D.; Wu, S. A Graph Matching Model for Designer Team Selection for Collaborative Design Crowdsourcing Tasks in Social Manufacturing. *Machines* **2022**, *10*, 776. https://doi.org/10.3390/machines10090776

Academic Editors: Kai Cheng, Pingyu Jiang, Gang Xiong, Timo R. Nyberg, Zhen Shen, Maolin Yang and Guangyu Xiong

Received: 19 July 2022
Accepted: 5 September 2022
Published: 6 September 2022

1. Introduction

The traditional method of product innovation is to conduct market research first, then design new products according to market feedback, and finally produce new products according to market demand. However, this mode is high-cost and low-reward. With the continuous innovation of science and technology, the diversified development of market demand, and the increasing complexity of products, more and more companies are aware of the need to use social resources for the innovative development of new products in social manufacturing. Crowdsourcing mode [1] emerges as the times require. This kind of open product design mode can make full use of social manpower, software and hardware resources, shorten the product development cycle, and improve product design quality. Crowdsourcing tasks are generally selected and completed by individuals based on crowdsourcing platforms, but as the complexity of product design increases, multi-person collaborative design is needed, so as to achieve design results that are both partially and overall satisfactory. Therefore, how to find a suitable designer team has become a key problem that needs to be resolved.

To solve this problem, we describe the product design task under the crowdsourcing mode as a graph, which is called the collaborative design crowdsourcing task network graph. The nodes represent the tasks in the crowdsourcing design, and the edges represent the relationships between the tasks. The designer team is also described as a graph, called the designer network graph. The nodes represent the designers, and the edges represent the connections between the designers. Therefore, we can consider the matching problem between the collaborative design crowdsourcing task network graph and the designer network graph to find a suitable designer team. However, due to the difference in the nodes and edges of the two types of graphs, we can only consider graph matching based on graph structure similarity. The problem to be studied is transformed into a graph similarity matching problem [2,3]. Graph similarity matching takes similarity of the query graph as the query criterion. In the target graph dataset, all graphs or subgraphs that are similar to the query graph are searched.

On the basis of viewpoints mentioned above, we propose a graph matching model based on graph structure similarity to calculate the similarity score between the collaborative design crowdsourcing task network graph and the designer network graph in this paper. The model first uses the Graph Convolutional Network (GCN) to extract features of the graph structure to obtain the node-level embeddings. Secondly, an attention mechanism considering the differences in the importance of different nodes in the graph assigns different weights to different nodes to aggregate node-level embeddings into graph-level embeddings. Finally, the two graph vectors are input into a multilayer fully connected neural network for feature extraction after they are obtained from the concat operation, and the binary cross-entropy loss function is used to predict the similarity score of the graph pair. The main contributions of this paper are summarized as follows:

- We regard the graph similarity calculation problem as a learning problem, i.e., to learn a function based on GNN. When the structure of the two graphs is input, the function outputs the similarity score (between 0~1) of the two graphs;
- We construct sample labels to train the model. The construction method of the sample label is innovative;
- The improved model and the basic model in this paper are compared and tested based on the accuracy ratio, precision ratio, recall ratio, and AUC index on two real graph datasets to verify the effectiveness of the improved model;
- We conduct a case study to prove the practical application of the improved model.

The sections of this paper are organized as follows. Section 2 is the related work that introduces the literature review. Section 3 is related materials and methods to describe in detail the graph matching model based on the graph structure similarity proposed in this paper. In Section 4, we compare the improved model and the basic model based on four evaluation metrics in two graph datasets to perform experiments. Section 5 is the case study. Section 6 gives the conclusions.

2. Related Work

Graph similarity calculation is a calculation method to predict the similarity score between one pair of graphs, including classification and clustering of graphs [4], social group network similarity recognition [5], object recognition in computer vision [6], biomolecular similarity search [7,8], etc. In graph deep learning, the models for graph similarity calculation can generally be divided into two categories. These two types of models use the GCN to learn embeddings and use embeddings to calculate the similarity score of the graph pairs [9]. One is the graph embedding model. This type of model mainly encodes each graph directly into a vector representation and then calculates the similarity score between one pair of graphs by calculating the similarity of the vectors, avoiding the difficulty of direct calculation on the graph. For example, Ktena et al. used a GCN to learn graph embedding, input the obtained graph embedding into a fully connected layer, and trained the fully connected layer to calculate the similarity score of the graph pairs [10]. Although this method has high time efficiency for graph similarity calculation, it is easy to ignore a lot of node information in the graph, which makes the results less accurate. The other one is the graph matching model. This type of model encodes each node in the graph as a vector representation, which contains not only the state information of the node itself, but also the state information of neighboring nodes, and uses different interaction strategies to calculate the similarity score of the graph pairs. For example, Riba et al. used a GCN to obtain node-level embeddings for the graph and calculated the similarity score of the graph pairs by calculating the similarity of the node-level embeddings [11]. This method is more suitable for graph similarity calculation on small graphs. Since this method considers each node in the graph, it still has the problem of low time efficiency for the similarity calculation on large graphs. It only uses node-level embeddings for graph similarity calculation, making the results less comprehensive.

In addition, there are methods that combine graph embedding models and graph matching models to calculate the similarity scores of the graph pairs. Xu et al. first divided two graphs to be matched into a set of smaller subgraphs, used a network model that combined an attention mechanism with a Graph Neural Network (GNN) [12,13] to learn the node-level embeddings and graph-level embeddings of these subgraphs, and calculated the similarity score of the graph pairs by comprehensively considering node-level embedding similarity and graph-level embedding similarity between subgraphs [14]. Although this method comprehensively considers node-level embeddings and graph-level embeddings for graph similarity calculation, it performs well only on small graphs, and it is too complicated with a low time efficiency for large graphs. Li et al. first interacted the nodes in the two graphs with each other by introducing a cross-graph attention layer and then calculated the similarity of the graph pairs by calculating the similarity of the graph embeddings [15]. Compared with the methods mentioned above, although this method improves the performance of graph similarity calculation on large graphs, it still has the problem of a high computation cost. That is, since each cross-graph matching step requires the computation of the full attention matrices, this is expensive for large graphs. As the matching model operates on graph pairs, it cannot be directly used for indexing and searching through large graph databases. Bai et al. first used a GCN to obtain node-level embeddings and proposed a new attention mechanism to aggregate node-level embeddings into graph-level embeddings. Secondly, a Neural Tensor Network (NTN) [16] was used to calculate the similarity of two graph-level embeddings. When bypassing the NTN module, the node-level similarity could be directly calculated. That is, the similarity score of the graph pairs was calculated by comprehensively considering graph-level embedding similarity and node-level embedding similarity [17,18]. This method not only considers node-level embeddings and graph-level embeddings for graph similarity computation, but an attention mechanism is also added for node aggregation, making it perform better on large graphs compared to the methods mentioned above. In addition, it can handle graphs with node types but cannot process edge features. There are also methods that combine a GCN with traditional neural networks to calculate the similarity scores of the graph pair. For example, Xiu et al. used the Earth Mover Distance (EMD) to obtain the one-to-one mapping between nodes of two graphs at each stage and obtained the correlation matrix according to the node-level embeddings in the two graphs, and the correlation matrices of all stages were input into a convolutional neural network, and the similarity score of the graph pairs was predicted by minimizing the mean squared error [19]. Although this method does not have high time efficiency for graph similarity calculation on large graphs, it improves the accuracy of the calculation by using a small increase in complexity, so as to achieve a reasonable trade-off between accuracy and efficiency. Bai et al. used a GCN with different layers to construct multiple correlation matrices between nodes of two graphs and used the correlation matrices based on a convolutional neural network to predict the similarity scores of the graph pairs [20]. This method achieved the state-of-the-art performance on four real-world graph datasets compared to existing popular methods for graph similarity calculation, but it still has high computational complexity problems on large graphs.

3. Materials and Methods

Referring to the literature review and the proposed idea of problem-solving in this paper, we define an undirected and unweighted graph $G = (V, E)$, where $V = \left\{v_1, \cdots, v_{|V|}\right\}$ is a set of nodes and $E = \left\{e_1, \cdots, e_{|E|}\right\}$ is a set of edges. $\boldsymbol{H} \in \mathbb{R}^{N \times d}$ represents the features of nodes, where N is the number of nodes in graph G (or $N = |V|$) and d is the dimension of the node feature vectors. Our goal is to learn a function based on a GNN that takes the structure of two graphs as input and outputs the similarity score of the two graphs.

3.1. Materials

In this section, we introduce the related materials of our graph matching model based on graph structure similarity.

3.1.1. Graph Convolutional Network (GCN)

We use a GCN to obtain the node representations of the graphs. Traditional Convolutional Neural Networks (CNNs) [21] can only process Euclidean spatial data, such as images, text, and speech. As non-Euclidean spatial data, graph data do not satisfy translation invariance and cannot be studied using traditional CNNs. The GCN is a variant proposed on the basis of a CNN and graph embedding [22]. The main idea of the GCN is to generate the node representation in the graph by aggregating the characteristics of the node itself and the characteristics of its neighboring nodes and to use a fixed number of layers (each layer has a different weight value) to deal with cyclic interdependence in the architecture.

The GCN mainly includes two categories, one is frequency domain convolution [23], which mainly uses Fourier transform to achieve convolution. The Fourier transform can not only be used to separate noise points from normal points but also to speed up convolution. The other one is spatial convolution [23]; the core is to aggregate neighbor node information. The general framework is the Message Passing Neural Network (MPNN) [11]. The MPNN mainly includes two phases: a message passing phase and a readout phase. In the message passing phase, the state information of the node is aggregated and updated. In the readout phase, the feature vector of the whole graph is read out.

3.1.2. Graph Embedding

The GCN is usually used to map the graph data, i.e., convert a high-dimensional dense matrix into a low-dimensional dense vector. This process is usually called graph embedding. It is also called network embedding or graph representation learning. It can solve the problem that graph data are difficult to efficiently input into machine learning algorithms. Graph embedding mainly includes node-level embeddings and graph-level embeddings. Node-level embedding means that each node in the graph is encoded as a vector. In the graph similarity matching problem, node-level embedding is suitable for calculating node similarity. Graph-level embedding means that the whole graph is encoded as a vector and is often used to predict, compare, or visualize the whole graph at the graph level. In the graph similarity matching problem, graph-level embeddings are suitable for the similarity calculation of graph pairs. Figure 1 is the process of graph data being represented as graph embedding. Graph data are represented as $|V|$ vectors, and the dimension of each vector is d. Therefore, the graph embedding is a $|V| \times d$ vector, and each row of the graph embedding is a node-level embedding, which represents the state information of the node in the graph.

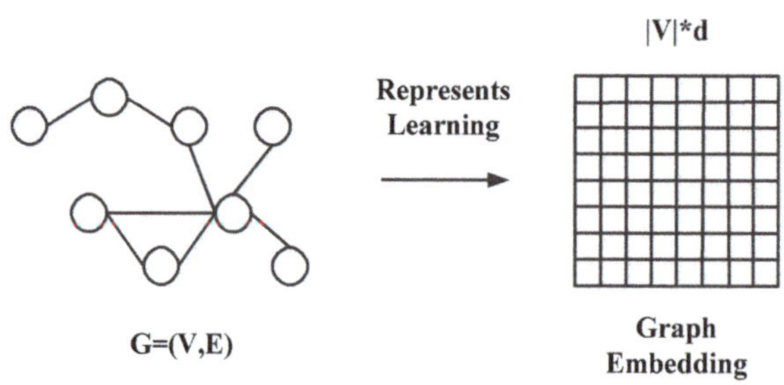

Figure 1. Graph embedding process.

For node-level embedding learning, there are currently several types of methods, including matrix factorization-based methods (NetMF [24]), skip-gram-based methods (DeepWalk [25], Node2Vec [26], LINE [27]), autoencoder-based methods (SDNE [28]), and neighbor aggregation-based methods (GCN, GraphSAGE [29]). Among them, the most popular method is based on neighbor aggregation. In this method, the GCN is the most widely used.

For graph-level embedding learning, the classic methods such as Graph2vec [30], which also use the skip-gram idea [31], encode the whole graph into a vector representation. In addition, methods such as Patchy-san [32], sub2vec (embed subgraphs) [33], and Deep WL kernels are also commonly used for learning graph-level embeddings.

Node-level embeddings can be transformed into graph-level embeddings through aggregation. The aggregation method is usually simple average aggregation, weighted average aggregation, and maximum aggregation, where the simple average and weighted average aggregation methods are also called "sum-based" methods [10].

The current method of aggregating nodes is either "sum-based" methods or maximum aggregation. Although these aggregation methods consider the importance of nodes, they do not consider the differences in the importance of different nodes in a graph, for example, the GMN model [15]. Aiming at this problem, we propose an attention mechanism that assigns different weights to different nodes in aggregate node-level embeddings.

3.1.3. Graph Edit Distance (GED)

The Maximum Common Subgraph (MCS) [34] and the GED [35] are usually the criteria for measuring the similarity of graph structure, and the MCS is equivalent to the GED under the same cost function. Therefore, in this paper, we use the GED to measure the structural similarity between two graphs.

The GED that measures the similarity between two graphs is defined as the smallest editing operation for mutual transformation between two graphs. The GED is the operation required to transform graph G_1 into graph G_2 including an edge insertion or deletion in the graph, an isolated node insertion or deletion in the graph, and the label modification of the node or edge. However, the calculation of the GED is usually regarded as an NP-Hard problem, so we cannot calculate the exact value of the GED every time. Sometimes we can only use some heuristic algorithms to calculate the approximate value of the GED.

3.2. Methods

We establish a graph matching model based on graph structure similarity. Aiming at the problem that the basic model (gcn-pool) does not consider the difference of the importance of different nodes, we propose an attention mechanism to aggregate nodes to improve the basic model, called gcn-attention (gcn-attn). Gcn-attn is mainly composed of 3 parts: (1) The GCN obtains the feature vector of the graph structure, i.e., GCN extracts node-level embeddings based on the graph structure; (2) An attention mechanism aggregates node-level embeddings into graph-level embeddings; (3) The two graph vectors are input into the multilayer fully connected neural network for feature extraction to predict the similarity score. The framework of the graph matching model is shown in Figure 2.

The difference between the basic model and the improved model is the process of aggregating node-level embeddings into graph-level embeddings. The aggregation method is average aggregation or maximum aggregation in the basic model. The aggregation method is to use the attention mechanism for aggregation in the improved model. The detailed formula description of each part is as follows.

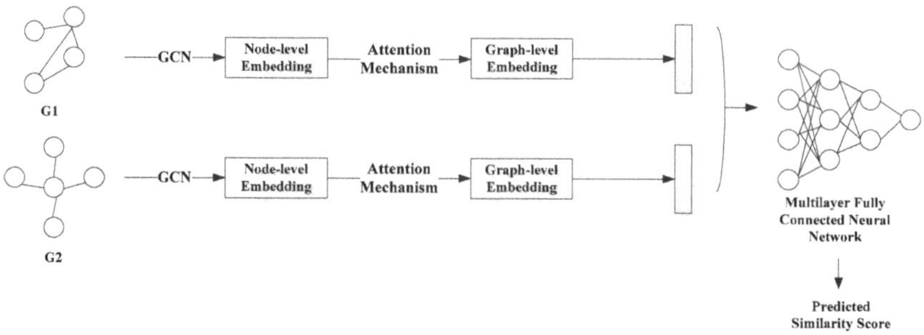

Figure 2. The framework of the graph matching model.

3.2.1. Node Embedding

We use 2 identical GCNs to obtain the node representations of the two graphs independently and in parallel, and the model parameters of each layer are not shared. The GCN updates the node state of the whole graph referring to Equation (1):

$$H^{(l+1)} = \delta\left(D^{-\frac{1}{2}}\tilde{A}D^{-\frac{1}{2}}H^{(l)}W_1^{(l)}\right) \tag{1}$$

where $\delta(\cdot)$ is the activation function such as $\mathrm{ReLU}(x) = \max(0, x)$, $A \in \mathbb{R}^{N\times N}$ is the adjacency matrix of graph G. $\tilde{A} = A + I_N$ denotes the adjacency matrix with an inserted self-loop. The degree matrix D is a diagonal matrix, $D(i,i) = \sum_{j=1}^{N} A(i,j)$. $D^{-\frac{1}{2}}\tilde{A}D^{-\frac{1}{2}}$ is the normalized adjacency matrix. $H^{(l)}W_1^{(l)}$ is equivalent to a linear transformation for all node embeddings in the l layer. Each node is left multiplied by the adjacency matrix to show that the features of the nodes are the result of adding the feature of the neighbor node, where $W_1^{(l)} \in \mathbb{R}^{d^l\times d^{l+1}}$ is a learnable mapping matrix, and d^{l+1} is the dimension of the embedding for $H^{(l+1)}$.

The node-level embeddings in the graph can be described as Equation (2):

$$h_i^{(l+1)} = \delta\left(\frac{1}{n(i)}\sum_{k\in n(i)} W_1^{(l)}h_k^{(l)} + b^{(l)}\right) \tag{2}$$

where the node-level embedding $h_i \in \mathbb{R}^d$ of node i is equivalent to the i row in the whole graph embedding $H \in \mathbb{R}^{N\times d}$. $k \in n(i)$ represents the neighbor node k of node i. $n(i)$ represents the number of neighbor nodes of node i. $b^{(l)} \in \mathbb{R}^{d^{l+1}}$ represents bias. Intuitively, the graph convolution operation aggregates the first-order neighbor features of the node.

3.2.2. Attention Mechanism

In the step of aggregating nodes, we consider the differences in the importance of different nodes and propose an attention mechanism. The principle of the attention mechanism is briefly described as follows: the average value of the states of all nodes in the graph is calculated, and the inner product of the average value and the state of each node in the graph is used as the attention weight. The attention weight is used to aggregate node-level embeddings into graph-level embeddings.

Firstly, the averaging method is used to obtain the representation characteristics of the information of the whole graph referring to Equation (3):

$$c = \tanh\left(\frac{1}{N}W_2\sum_{i=1}^{N} h_i\right) \tag{3}$$

where $W_2 \in \mathbb{R}^{d\times d}$ is a learnable weight matrix.

Secondly, the result of the inner product of the whole graph information and the node information is used as the attention coefficient of each node referring to Equation (4):

$$\alpha_i = \sigma\left(h_i^T c\right) \tag{4}$$

where $\sigma(\cdot)$ is a sigmoid activation function, $\sigma(x) = \frac{1}{1+\exp(-x)}$ ensures that the attention coefficient is between 0 and 1.

Finally, after obtaining the attention weight of each node, the graph-level embedding is described as Equation (5):

$$h_G = \sum_{i=1}^{N} \alpha_i h_i \tag{5}$$

where $h_G \in \mathbb{R}^d$ represents the graph-level vector of the whole graph.

3.2.3. Loss Function

After obtaining the graph-level embeddings, the two graph-level embeddings are input into a multilayer fully connected neural network to obtain the similarity score of the graph pairs after performing the concat operation. In order to enable the model to accurately calculate the similarity score of the graph pairs, we need to train the model. We construct sample labels for training, and the specific construction method is as follows: we randomly split 80% and 20% of all graphs in the dataset as training set and testing set, respectively. The data are generated by sampling and are controlled by the *sample* parameter. For example, *sample* = 50 means that 50 graphs are randomly selected to match each graph in the training set to form a training sample.

In order to train the model, firstly, the GED function in the Pytorch framework is used to calculate the GED of the graph pairs in the training samples of the dataset. Secondly, the *median* function in the framework is directly called to obtain the median of the GED of the graph pairs, and the median is used as the threshold for classification. If the GED is greater than the threshold, it is classified as 0, which means that the two graphs are not similar; if the GED is less than the threshold, it is classified as 1, which means that the two graphs are similar. Therefore, it can be regarded as a binary classification problem.

We use the binary cross-entropy loss function *BCELoss* to predict the similarity score of the graph pairs. This function is used to calculate the binary cross entropy loss between the output value (predicted value) of the graph matching model and the true value. It can be used directly by calling the *F.binary_cross_entropy* in the Pytorch framework, and the function form is simplified referring to Equation (6):

$$Loss = -2(y \cdot \log(\hat{y}) + (1-y) \cdot \log(1-\hat{y})) \tag{6}$$

where $\hat{y}$ is the probability that the model predicts that the two graphs are similar and the value of $\hat{y}$ is between 0 and 1. y is the constructed sample label. If the two graphs are similar in the training sample, the value of y is 1, otherwise the value of y is 0.

3.2.4. The Complexity Analysis

The model proposed in this paper is used to predict whether graph pairs are similar. The analysis of time complexity mainly includes two parts: (1) node-level embedding and graph-level embedding calculation steps; (2) graph-pair similarity score calculation steps.

In the simplest case, we only visit each edge once and perform two calculation operations on the two nodes it connects for the embedding model, so that the time complexity of the process of forming local topological features is usually $O(E')$, where E' is the number of edges in the larger graph in the two graphs [14]. For the node-level embedding and graph-level embedding calculation steps, the time complexity is $O(E')$. For calculating the graph pair similarity score, since the two graph-level vectors are input into a multilayer fully connected neural network after the concat operation, no time complexity analysis is performed in this step. Only the vector dimension changes from d to $2d$. To sum up, the time complexity of the whole model is $O(E')$.

4. Experiments

In this section, we mainly compare the basic model (gcn-pool) and the improved model (gcn-attn) based on four evaluation metrics on two graph datasets to verify the advancement and effectiveness of the improved model.

4.1. Datasets

We perform experiments on two real graph datasets. These two datasets have been applied to the graph similarity matching problem many times to verify the effectiveness of the method. The specific information is shown in Table 1.

Table 1. Introduction to datasets.

Dataset	Graph Meaning	#Graph	#Pairs	Avg#Nodes	Avg#Edges
AIDS	Chemical Compounds	700	490 K	8.90	8.80
LINUX	Program Dependency Graphs	1000	1 M	7.58	6.94

AIDS [14]: This dataset is an anti-HIV dataset containing a large number of sparse undirected graphs. The number of nodes ranges from 4 to 245. Each graph represents the topological structure of the compound and contains node labels and edge labels. We select 700 graphs, each of which has at least 10 nodes. LINUX [14]: The Program Dependency Graph (PDG) dataset generated by the Linux kernel contains 48,747 graphs. Each graph represents a function, where a node represents a statement and an edge represents the dependency between the two statements. The nodes are unlabeled. We select 1000 graphs, each of which has at least 10 nodes.

4.2. Parameter Settings

For the model construction, we set the number of GCN layers to 3, and use the ReLU as the activation function. For the initial node representations, we adopt a one-hot encoding strategy for the AIDS dataset to reflect the node type and a constant encoding strategy for the LINUX dataset because the nodes in the LINUX dataset have no labels. The output dimensions of the first layer, second layer, and third layer of the GCN are set to 64, 32, and 16, respectively. We use five fully connected layers to build a fully connected neural network, which are a linear combination layer with 96 inputs and 256 outputs; a nonlinear activation layer with ReLU as the activation function; a linear combination layer with 256 inputs and 256 outputs; a linear combination layer with 256 inputs and 16 outputs; and a linear combination layer with 16 inputs and 1 output to predict the graph pair similarity score. The details are shown in Table 2.

Table 2. Construction of the graph matching model.

```
Graph Matching(
    (convolution_1): GCNConv (29, 64)
    (convolution_2): GCNConv (64, 32)
    (convolution_3): GCNConv (32, 16)
    (tensor_network): Sequential (
        (0): Linear (in_features = 96, out_features = 256, bias = True)
        (1): ReLU (inplace = True)
        (2): Linear (in_features = 256, out_features = 256, bias = True)
    )
    (fully_connected_first): Linear (in_features = 256, out_features = 16, bias = True)
    (scoring_layer): Linear (in_features = 16, out_features = 1, bias = True)
)
```

According to Table 2, firstly, the graph node feature extraction is performed based on the graph structure through the three-layer GCN. Secondly, each node in the graph is aggregated into a composite vector of the graph. Finally, the torch.nn.Sequential class in Pytorch is used to build a fully connected neural network. The graph vectors of the two graphs are input into the fully connected neural network after the concat operation to calculate the similarity score of the graph pairs.

All experiments are run on a PC, using Nvidia GeForce GTX 1660 Super GPU, 16 G, RAM. The excellent Pytorch Geometric framework to implement our model in Python is used. As for training, we set the batch size to 64, dropout to 0.2, use the Adam algorithm for optimization, and set the initial learning rate to 0.01. We set the number of iterations to 100 and select the best model based on the lowest validation loss.

4.3. Evaluation Metrics

For the binary classification problem, the commonly used evaluation metrics include the accuracy ratio, precision ratio, recall ratio, and AUC index. Before introducing these metrics, some definitions are described. The prediction results of the binary classification problem include four categories according to the situation: True Positive (TP): the predicted value is 1, the true value is 1; False Positive (FP): the predicted value is 1, the true value is 0; True Negative (TN): the predicted value is 0, the true value is 0; False Negative (FN): the predicted value is 0, the true value is 1. Thus the total number of positive samples T = TP + FN, and the total number of negative samples F = FP + TN, as shown in Table 3.

Table 3. Prediction results of the binary classification problem.

Classification	The True Value: 1 (Positive Samples)	The True Value: 0 (Negative Samples)
The predicted value: 1	TP (True Positive)	FP (False Positive)
The predicted value: 0	FN (False Negative)	TN (True Negative)

The accuracy ratio is the ratio of the number of samples that the predicted value of the model is the same as the true value to the total number of samples, referring to Equation (7):

$$acc = \frac{\mathrm{TP} + \mathrm{TN}}{\mathrm{T} + \mathrm{F}} \tag{7}$$

There is a problem with using the accuracy ratio to evaluate algorithms, i.e., when there are extremely biased data in the dataset, the accuracy ratio cannot objectively evaluate the pros and cons of the algorithm.

The precision ratio is the ratio of the number of true positive samples of the model to the number of true samples, referring to Equation (8):

$$pre = \frac{\mathrm{TP}}{\mathrm{TP} + \mathrm{FP}} \tag{8}$$

The recall ratio is the ratio of the number of true positive samples to the total number of all positive samples, referring to Equation (9):

$$rec = \frac{\mathrm{TP}}{\mathrm{TP} + \mathrm{FN}} = \frac{\mathrm{TP}}{\mathrm{T}} \tag{9}$$

The precision ratio and the recall ratio are trade-offs, i.e., the higher the precision ratio, the lower the recall ratio. In some cases, these two metrics need to be taken into account at the same time, so the AUC index is proposed.

The AUC index represents the area enclosed by the ROC (Receiver Operating Characteristic) curve, which is used to comprehensively evaluate the accuracy ratio and the recall ratio. The reason for using the AUC value as the evaluation standard is that in many cases the ROC curve cannot clearly indicate which classifier performs better, and the AUC value range is generally between 0.5~1, and the classification effect corresponding to a larger AUC value is better.

4.4. Results

We run the basic model (gcn-pool) and the improved model (gcn-attn) 100 times based on the accuracy ratio, precision ratio, recall ratio as well as the AUC index on the AIDS and LINUX datasets. The results of 100 calculations for each metric are averaged to obtain a comparison of the average results of each metric of the two models on the AIDS and LINUX datasets, as shown in Tables 4 and 5.

Table 4. Average results of four evaluation metrics based on the AIDS dataset.

Model	Accuracy Ratio	Precision Ratio	Recall Ratio	AUC Index
gcn-pool	0.722	0.637	0.640	0.686
gcn-attn	0.799	0.749	0.694	0.765

Table 5. Average results of four evaluation metrics based on the LINUX dataset.

Model	Accuracy Ratio	Precision Ratio	Recall Ratio	AUC Index
gcn-pool	0.833	0.807	0.839	0.797
gcn-attn	0.907	0.913	0.901	0.889

It can be seen from the experimental data that in the AIDS dataset, compared to the basic model (gcn-pool), the accuracy ratio of the improved model (gcn-attn) has increased by 10.7%, the precision ratio has increased by 17.6%, the recall ratio has increased by 8.4%, and the AUC index has increased by 11.5%.

In the LINUX dataset, compared to the basic model (gcn-pool), the accuracy ratio of the improved model (gcn-attn) has increased by 8.9%, the precision ratio has increased by 13.1%, the recall ratio has increased by 7.4%, and the AUC index has increased by 11.5%.

In order to more intuitively see the comparison effect of the two models based on the four evaluation metrics, a curve comparison chart is shown in Figure 3.

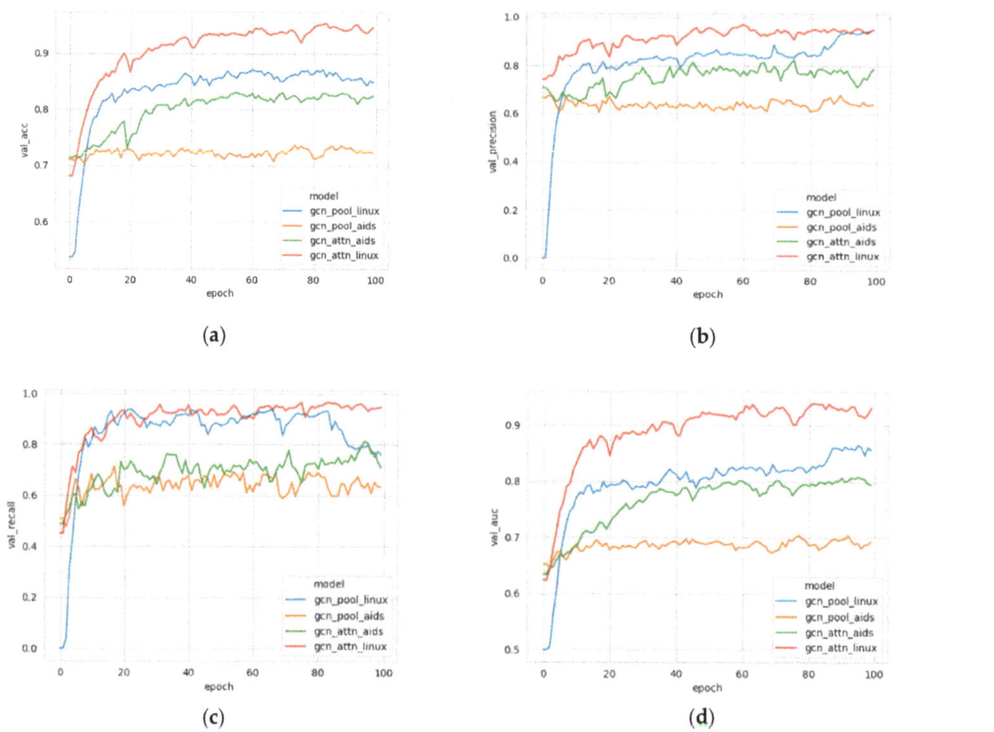

Figure 3. The curve comparison chart of the two models based on four evaluation metrics for the AIDS and LINUX datasets. (**a**) Accuracy ratio, (**b**) Precision ratio, (**c**) Recall ratio, (**d**) AUC index.

It can be seen from the curve comparison chart that the accuracy ratio, the precision ratio, the recall ratio, and the AUC index of the improved model (gcn-attn) are higher than those of the basic model (gcn-pool) for both the AIDS dataset and the LINUX dataset. The larger these evaluation metrics, the better the effect of the model.

Therefore, it can be concluded that the gcn-attn model is more advanced and effective than the gcn-pool model regarding the graph matching problem based on graph structure similarity, i.e., the improved model is able to more accurately find graph pairs based on a similar structure.

5. Case Study

In order to verify that the model proposed in this paper can be accurately applied to the matching problem between the collaborative design crowdsourcing task network graph and the designer network graph, we conduct a case study. At the same time, the Pytorch-Geometric framework is used for visualization.

5.1. Case Description

We take the crankshaft linkage mechanism produced by one enterprise as an example, where the mechanical analysis of the crankshaft linkage mechanism requires a design team. We first use a graph to describe the tasks involved in the mechanical analysis of the crankshaft linkage mechanism, then establish a designer community candidate set with several designer teams, and finally use the method proposed in this paper to find the most suitable designer team.

5.2. Use of Method Proposed

The mechanical analysis of the crankshaft linkage mechanism includes force analysis (including stress, bearing force, and crankshaft torque analyses), material analysis (including toughness and strength analyses), and motion analysis [36]. After the nine tasks are numbered, the collaborative design crowdsourcing task network graph is constructed according to their dependencies, as shown in Figure 4.

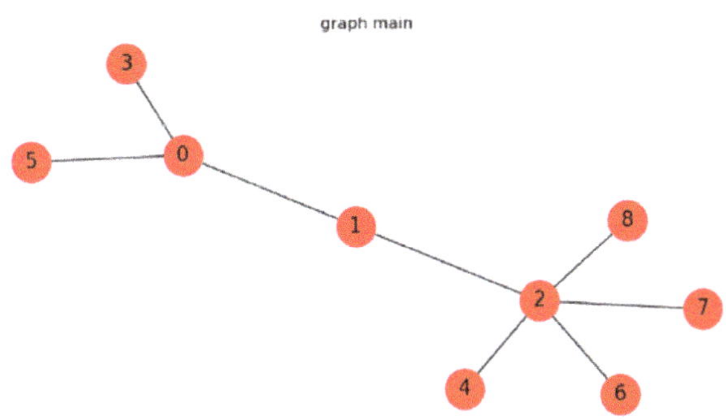

Figure 4. The collaborative design crowdsourcing task network graph.

In order to construct the designer community candidate set, five WeChat groups are studied. The members of these five WeChat groups are all designers who joined the crowdsourcing design. By collecting the conversational communication data within the WeChat group as the analysis text and numbering the designers, the data are preprocessed into a network structure, so as to construct the designer community candidate set that includes five designer network graphs, as shown in Figure 5.

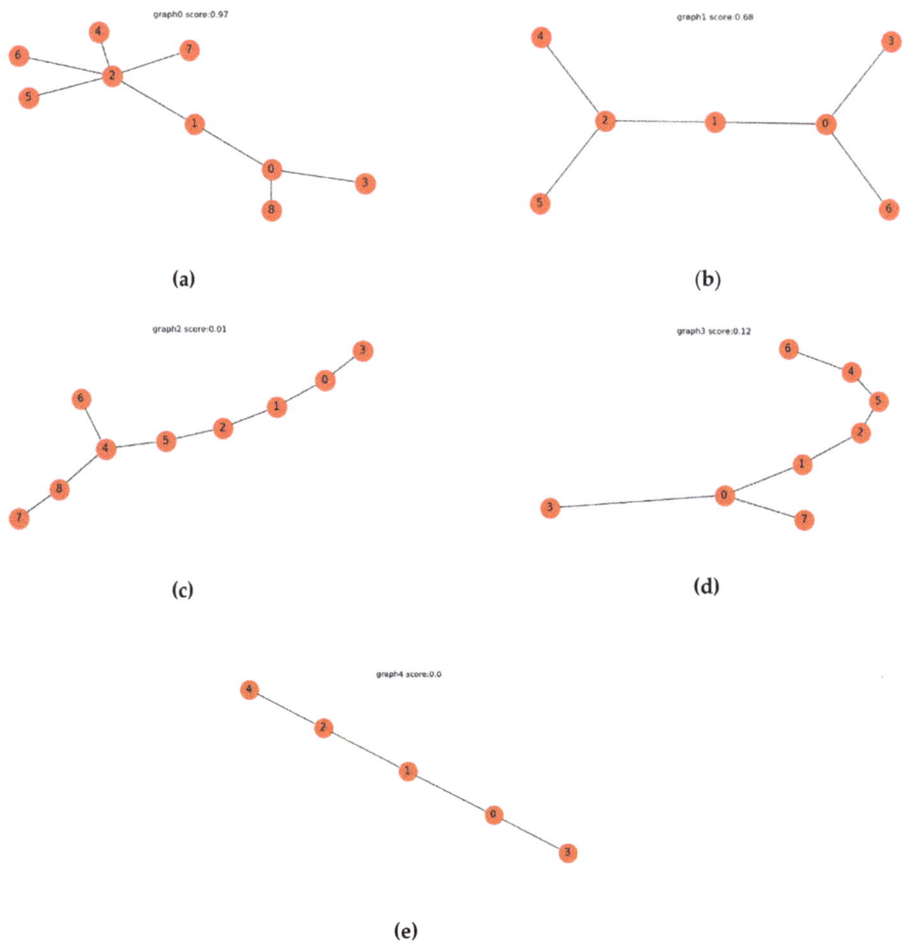

Figure 5. The designer community candidate set. (**a**) Designer network graph 1, (**b**) Designer network graph 2, (**c**) Designer network graph 3, (**d**) Designer network graph 4, (**e**) Designer network graph 5.

As shown in the framework of the graph matching model in Figure 2, we take the collaborative design crowdsourcing task network graph in Figure 4 is as G_1. Five designer network graphs in Figure 5 are successively taken as G_2. G_1 and G_2 are input into the graph matching model independently and in parallel. Since the graph matching model has been trained, it already has the best model parameters for calculating the similarity score of the graph pairs. Therefore, after G_1 and G_2 are input into the model, two identical GCNs first extract the node-level embeddings of G_1 and G_2 based on graph structure in parallel. Secondly, the node-level embedding are aggregated into graph-level embedding by the attention mechanism proposed in this paper. Finally, the similarity score between G_1 and G_2 is output through the calculation of the multi-layer fully connected neural network after the two graph-level embeddings are obtained using concat operation.

In summary, we take the collaborative design crowdsourcing task network graph in Figure 4 as the query graph, the designer community candidate set in Figure 5 as the target graph set, and use the improved graph matching model in this paper to find the target graph most similar in structure to the query graph from the target graph set.

5.3. Case Results and Discussion

It can be seen that the value above each graph in the designer community candidate set represents the matching similarity score between the collaborative design crowdsourcing task network graph in Figure 4 and the designer network graph in Figure 5. The larger the value, the more similar the two graphs are based on the graph structure. The matching similarity scores between the designer network graph and the collaborative design crowdsourcing task network graph are 0.97, 0.68, 0.01, 0.12, and 0.00. Therefore, the designer network graph 1 has the highest similarity to the collaborative design crowdsourcing network graph, so we have found the most suitable designer team for the collaborative design crowdsourcing task.

The case study describes the mechanical analysis of the crankshaft linkage mechanism as a collaborative design crowdsourcing task network graph and uses this graph as a query graph to find the designer network graph based on similar graph structure, which is actually a graph similarity matching problem. The traditional acceleration strategies of graph similarity algorithms are all at the cost of increased memory consumption, i.e., space is exchanged for time, and the efficiency is low. Therefore, the graph matching model combined with GNN proposed in this paper is more innovative. The experimental data and case results also reflect the rationality and practicality of the method and model in this paper. On the basis of finding a suitable designer team, the relevant indicators of tasks and designers can be considered for task assignment in the future, thus making up for the problem of node similarity not considered in graph matching in this paper.

6. Conclusions

Graph structures are widely used in various fields to describe complex relationships between things, such as the World Wide Web, social networks, protein interaction networks, chemical molecular structures, power grids, and road networks. With the development of these fields and the increase in data, the size and number of graphs are also increasing. How to perform efficient graph matching operations on a large number of accumulated graphs has become a new research goal of academia and industry. As a research method of artificial intelligence, deep learning is changing the world at an unprecedented speed, and it plays an important role in the field of education, image recognition, speech technology, and unmanned driving.

The increase in the scale and complexity of product design often involves tasks that require multiple people to design collaboratively, so it is necessary to find a suitable designer team. Aiming at this problem, we propose a graph similarity matching method to find a designer team suitable for the collaborative design crowdsourcing task of the product. Compared with the traditional method, this paper introduces the matching problem between the collaborative design crowdsourcing task and the designer team into the field of graph theory research and combines deep learning knowledge to make the research method more innovative and researchable. The model and method in this paper are suitable for matching between undirected and unweighted network graphs based on a similar graph structure, such as social relationship query, social security analysis, recommendation system, network attack detection, and biological data analysis.

Although the model and method proposed in this paper are feasible and practical, there are still some shortcomings. The experiments in Section 3 are performed on the public AIDS dataset and the LINUX dataset, and the graph data in these two datasets are small in scale. The graph matching model proposed in this paper performs well on small-scale graph data, but still does not perform well on large-scale graph data. The improved graph matching model in this paper is simple. In the future, we will design a more complex graph matching model for large-scale graph matching.

Author Contributions: Conceptualization, supervision, project administration, funding acquisition and methodology, D.L.; software, formal analysis, data curation, writing—original draft preparation, writing—review and editing, and visualization, D.W.; investigation and resources, S.W. All authors have read and agreed to the published version of the manuscript.

Funding: This research was funded by the National Natural Science Foundation of China, grant number 71961005, and Natural Science Foundation of Guangxi Zhuang Autonomous Region, grant number 2020GXNSFAA297024.

Data Availability Statement: The datasets used and/or analyzed during the current study are available from the corresponding author Danling Wu (1020190643@glut.edu.cn) upon reasonable request.

Conflicts of Interest: The authors declare no conflict of interest. The funders had no role in the design of the study; in the collection, analyses, or interpretation of data; in the writing of the manuscript; or in the decision to publish the results.

References

1. Qiu, D.Y. Product Design Task Recommendation Based on Crowdsourcing Mode. *J. Mach. Des.* **2017**, *34*, 48–52.
2. Xiang, Y.Z.; Tan, J.X.; Han, J.S.; Shi, H. Survey of Graph Matching Algorithms. *Comput. Sci.* **2018**, *45*, 27–45.
3. Yu, J.; Liu, Y.B.; Zhang, Y.; Liu, M.Y. Survey on Large-scale Graph Pattern Matching. *J. Comput. Res. Dev.* **2015**, *52*, 391–409.
4. Neuhaus, M.; Bunke, H. Edit Distance-based Kernel Functions for Structural Pattern Classification. *Pattern Recognit.* **2006**, *39*, 1852–1863. [CrossRef]
5. Ogaard, K.; Roy, H.; Kase, S.; Nagi, R.; Sambhoos, K.; Sudit, M. Discovering Patterns in Social Networks with Graph Matching Algorithms. In Proceedings of the 6th International Conference on Social Computing, Behavioral-Cultural Modeling, and Prediction, Washington, DC, USA, 2–5 April 2013.
6. Aghasi, A.; Romberg, J. Convex Cardinal Shape Composition and Object Recognition in Computer Vision. In Proceedings of the 49th Asilomar Conference on Signals, Systems and Computers, Pacific Grove, CA, USA, 8–11 November 2015.
7. Suzuki, H.; Kawabata, T.; Nakamura, H. Omokage Search: Shape Similarity Search Service for Biomolecular Structures in Both the PDB and EMDB. *Bioinformatics* **2016**, *4*, 619–620. [CrossRef] [PubMed]
8. Tian, Y.; Mceachin, R.C.; Santos, C.; States, D.J.; Patel, J.M. Saga: A Subgraph Matching Tool for Biological Graphs. *Bioinformatics* **2007**, *23*, 232–239. [CrossRef]
9. Kipf, T.N.; Welling, M. Semi-Supervised Classification with Graph Convolutional Networks. In Proceedings of the 4th International Conference on Learning Representations, San Juan, Puerto Rico, 2–4 May 2016.
10. Ktena, S.I.; Parisot, S.; Ferrante, E. Distance Metric Learning using Graph Convolutional Networks: Application to Functional Brain Networks. In Proceedings of the 20th International Conference on Medical Image Computing and Computer-Assisted Intervention, Quebec City, QC, Canada, 11–13 September 2017.
11. Riba, P.; Fischer, A.; Llado's, J. Learning Graph Distances with Message Passing Neural Networks. In Proceedings of the 24th International Conference on Pattern Recognition, Beijing, China, 20–24 August 2018.
12. William, L.H. Graph Representation Learning. *Synth. Lect. Artif. Intell. Mach. Learn.* **2020**, *14*, 1–159.
13. Zhou, J.; Cui, G.Q.; Hu, S.D. Graph Neural Networks: A Review of Methods and Applications. *AI Open* **2020**, *1*, 57–81. [CrossRef]
14. Xu, H.; Duan, Z.; Feng, J. Graph Partitioning and Graph Neural Network based Hierarchical Graph Matching for Graph Similarity Computation. *Neurocomputing* **2020**, *439*, 348–362. [CrossRef]
15. Li, Y.; Gu, C.; Dullien, T. Graph Matching Networks for Learning the Similarity of Graph Structured Objects. In Proceedings of the 36th International Conference on Machine Learning, Long Beach, CA, USA, 10–15 June 2019.
16. Socher, R.; Chen, D.; Manning, C.D. Reasoning with Neural Tensor Networks for Knowledge Base Completion. In Proceedings of the 26th International Conference on Neural Information Processing Systems, Lake Tahoe Nevada, CA, USA, 5–10 December 2013.
17. Bai, Y.; Ding, H.; Bian, S. SimGNN: A Neural Network Approach to Fast Graph Similarity Computation. In Proceedings of the 12th ACM International Conference on Web Search and Data Mining, Melbourne, Australia, 11–15 February 2019.
18. Bai, Y.; Ding, H.; Sun, Y. Convolutional Set Matching for Graph Similarity. In Proceedings of the 32th Conference and Workshop on Neural Information Processing, Montreal, QC, Canada, 3–8 December 2018.
19. Xiu., H.; Yan, X.; Wang, X. Hierarchical Graph Matching Network for Graph Similarity Computation. *arXiv* **2020**. [CrossRef]
20. Bai, Y.; Ding, H.; Sun, Y. Learning-based Efficient Graph Similarity Computation via Multi-Scale Convolutional Set Matching. In Proceedings of the 32th AAAI Conference on Artificial Intelligence, New Orleans, LA, USA, 2–7 February 2018.
21. Kim, P. Convolutional Neural Network. In *MATLAB Deep Learning*; George, R., Ed.; Apress: Berkeley, CA, USA, 2017; pp. 121–147.
22. Goyal, P.; Ferrara, E. Graph Embedding Techniques, Applications, and Performance: A Survey. *Knowl-Based Syst.* **2018**, *151*, 78–94. [CrossRef]
23. Xu, B.B.; Cen, K.T.; Huang, J.J. A Survey on Graph Convolutional Neural Network. *Chin. J. Comput.* **2020**, *43*, 755–780.
24. Qiu, J.Z.; Dong, Y.X.; Ma, H.; Li, J.; Wang, K.S.; Tang, J. Network Embedding as Matrix Factorization: Unifying Deepwalk, Line, Pte, and Node2vec. In Proceedings of the 11th ACM International Conference on Web Search and Data Mining, New York, NY, USA, 5–9 February 2018.

25. Perozzi, B.; Al-Rfou, R.; Skiena, S. Deepwalk: Online Learning of Social Representations. In Proceedings of the 20th ACM SIGKDD International Conference on Knowledge Discovery and Data Mining, New York, NY, USA, 24–27 August 2014.
26. Grover, A.; Leskovec, J. Node2vec: Scalable Feature Learning for Networks. In Proceedings of the 22nd ACM SIGKDD International Conference on Knowledge Discovery and Data Mining, San Francisco, CA, USA, 13–17 August 2016.
27. Jian, T.; Meng, Q.; Wang, M. Line: Large-scale Information Network Embedding. In Proceedings of the 24th International Conference on World Wide Web, Florence, Italy, 18–22 May 2015.
28. Wang, D.; Peng, C.; Zhu, W. Structural Deep Network Embedding. In Proceedings of the 22nd ACM SIGKDD International Conference on Knowledge Discovery and Data Mining, San Francisco, CA, USA, 13–17 August 2016.
29. Hamilton, W.L.; Ying, R.; Leskovec, J. Inductive Representation Learning on Large Graphs. In Proceedings of the 31st International Conference on Neural Information Processing Systems, Long Beach, CA, USA, 4–9 December 2017.
30. Narayanan, A.; Chandramohan, M.; Venkatesan, R. Graph2vec: Learning Distributed Representations of Graphs. In Proceedings of the 13th International Workshop on Mining and Learning with Graphs, Halifax, NS, Canada, 14 August 2017.
31. Lazaridou, A.; Pham, N.T.; Baroni, M. Combining Language and Vision with a Multimodal Skip-gram Model. In Proceedings of the 2015 Conference of the North American Chapter of the Association for Computational Linguistics, Denver, CO, USA, 31 May–5 June 2015.
32. Niepert, M.; Ahmed, M.; Kutzkov, K. Learning Convolutional Neural Networks for Graphs. In Proceedings of the 33rd International Conference on International Conference on Machine Learning, New York, NY, USA, 19–24 June 2016.
33. Adhikari, B.; Zhang, Y.; Ramakrishnan, N. Sub2Vec: Feature Learning for Subgraphs. In Proceedings of the 22nd Pacific-Asia Conference on Knowledge Discovery and Data Mining, Melbourne, Australia, 15–18 May 2018.
34. Horst, B.; Kim, S. A Graph Distance Metric Based on the Maximal Common Subgraph. *Pattern Recognit. Lett.* **1998**, *19*, 255–259.
35. Bunke, H. What is the distance between graphs. *Bull. EATCS* **1983**, *20*, 35–39.
36. Chen, Y.; Zuo, L.; Niu, Y. Task Assignment Method of Product Development Based on Knowledge Similarity. *J. Comput. Appl.* **2019**, *39*, 323–329.

machines

Article

Industry 4.0-Oriented Turnkey Project: Rapid Configuration and Intelligent Operation of Manufacturing Systems

Shulian Xie [1,*], Weimin Zhang [1], Feng Xue [1], Dongdong Li [2], Yangbokun Liu [1], Jürgen Fleischer [3] and Christopher Ehrmann [1,3]

1 School of Mechanical Engineering, Tongji University, Shanghai 201804, China
2 Saic Volkswagen Automotive Co., Ltd., Shanghai 201805, China
3 wbk Institute of Production Science, Karlsruhe Institute of Technology (KIT), 76131 Karlsruhe, Germany
* Correspondence: shulian00@tongji.edu.cn; Tel.: +86-182-0198-9760

Abstract: More extensive personalized product requirements and shorter product life cycles have put forward higher requirements for the rapid establishment, commissioning, and operation of corresponding manufacturing systems. However, the traditional manufacturing system development process is complicated, resulting in a longer delivery time. Many manufacturing enterprises, especially small and micro enterprises, may not have the necessary manufacturing knowledge or capabilities to meet these requirements. Therefore, it is essential to promote the construction of turnkey projects under the paradigm of Industry 4.0, parallelizing and integrating the existing manufacturing system development process based on mass manufacturing equipment to quickly provide turnkey solutions for manufacturing systems' configuration and implementation for these enterprises. This paper aims to extract and refine the configuration and operation key views of the Industry 4.0-oriented Turnkey Project (I4TP) from Reference Architecture Model Industrie 4.0 (RAMI4.0) and use it to guide the development of key functional processes of turnkey projects to achieve rapid configuration and efficient operation management of manufacturing systems. The turnkey project platform in the Advanced Manufacturing Technology Center (AMTC) is taken as a demonstration case to provide a reference idea for the rapid configuration and intelligent operation of the turnkey manufacturing system.

Keywords: industry 4.0; turnkey project; manufacturing system configuration; manufacturing system operation; key enabling tools

Citation: Xie, S.; Zhang, W.; Xue, F.; Li, D.; Liu, Y.; Fleischer, J.; Ehrmann, C. Industry 4.0-Oriented Turnkey Project: Rapid Configuration and Intelligent Operation of Manufacturing Systems. *Machines* **2022**, *10*, 983. https://doi.org/10.3390/machines10110983

Academic Editors: Pingyu Jiang, Gang Xiong, Timo R. Nyberg, Zhen Shen, Maolin Yang and Guangyu Xiong

Received: 29 September 2022
Accepted: 24 October 2022
Published: 27 October 2022

1. Introduction

In the context of Industry 4.0, as people's demands for personalized products increase and the acceleration of consumer demands changes, the speed of product upgrades is accelerating, and product life cycles are becoming shorter and shorter. What's more, the batch size of a single product is also corresponding to decreases [1,2]. Manufacturing companies are required to reduce the time of product development and manufacturing systems from set-up to operation as much as possible, realize the rapid reconfiguration of the manufacturing systems according to orders to adapt to new production tasks, and save costs. However, new manufacturing systems' traditional sequential development processes make the delivery time longer and they cannot catch up with current market changes. With the continuous development of globalization, manufacturing equipment has become abundant, and the development of manufacturing systems has gradually transformed into selecting manufacturing equipment and combining them to form a manufacturing system.

The continuously evolving computer-aided technologies, as well as the emerging ICTs (Information and Communications Technologies), such as the Internet of Things (IoT), cloud computing, mobile internet, Artificial Intelligence (AI), and so on, provide technical support for intelligent manufacturing to a large extent, including efficient product engineering and manufacturing [3]. Although engineers can use these technologies to address the challenges of the rapid development of products and the quick response of manufacturing systems, how to effectively integrate these technologies to realize the application in actual production is a problem faced by many manufacturing companies. At the same time, many manufacturing companies, especially small and micro enterprises, have certain product innovation and development capabilities but may lack the necessary knowledge or ability to plan and integrate appropriate manufacturing systems to respond to market demands [4].

Therefore, it is essential to build a turnkey project to provide turnkey services in response to market changes and manufacturing companies' turnkey needs for manufacturing systems. ICTs are comprehensively used to parallel and integrate manufacturing system development processes. Under various constraints such as cost, quality, delivery time, and batch size, a manufacturing system that quickly meets the product needs of the market is developed based on the existing manufacturing equipment, and the intelligent operation management of the manufacturing system is realized. It provides manufacturers with a turnkey solution and enables them to quickly establish or adjust manufacturing systems and successfully put them into operation according to product orders.

The construction of the Industry 4.0-oriented Turnkey Project (I4TP) involves numerous factors, such as business, assets, information, and functions under the Industry 4.0 paradigm. In the theoretical research aspect, the Reference Architecture Model Industrie 4.0 (RAMI4.0) [5] aims to create digital description rules for technical objects throughout their life cycle and related value changes and defines detailed concepts, standards, and interactions. However, there is no detailed description of the implementation and application process for specific fields, including turnkey projects [6], and there is no subdivision architecture based on RAMI4.0 to guide the formation, commissioning, and operation of manufacturing systems. Regarding the application aspect, the rapidly emerging smart factories around the world, such as Fast Radius' Cloud Manufacturing Platform [7], Phoenix Contact's Digital Factory [8], Rold's SMARTFAB [9], and some other cases [10,11], have the capability of autonomous overall management, able to carry out systematic coordination, reorganization, and expansion to provide specific solutions for product production. Although the smart factory can provide various support for the turnkey project and even expand the business service scope, it still cannot fully satisfy the core functional requirements in the whole process of the turnkey project from product to manufacturing system configuration and operation.

To cope with the above problems, the views of two core functional processes are proposed in this paper focused on the set-up, commissioning, and operation process to guide the construction of I4TP. The I4TP's characteristics, transforming the development of manufacturing systems into configuration processes and carrying out the efficient intelligent operation management to the manufacturing system, are considered in these views. The specific implementation of the turnkey project is explained with a demonstration case. The main contributions of this paper are as follows:

- In response to the growing demand for turnkey services of manufacturing systems under the trend of personalized manufacturing, key research on the configuration and operation of manufacturing systems is carried out around the construction and implementation of I4TP;
- Through the introduction of the I4TP, the core functional process of it is summarized into the configuration and the operation management of manufacturing systems, and the configuration view and the operation view are extracted and established, respectively, from RAMI4.0 to conceptually describe these two aspects;
- Guided by the configuration and operation views, a turnkey project platform is established in the Advanced Manufacturing Technology Center (AMTC), and the design, development, and application of key enabling tools for the configuration and operation of I4TP are introduced in detail through this platform case;

- The key function process of configuration and operation of the turnkey project platform is verified through product cases, which shows the feasibility of the platform and the good application prospects of the I4TP.

The rest of this paper is arranged as follows. Section 2 reviews the current research status of the two key issues of configuration and operation in I4TP. Section 3 briefly introduces the concept of I4TP, and proposes the configuration view and the operation view. Section 4 details the design and development of key enabling tools for the configuration and operation of the turnkey project platform in AMTC. Section 5 verifies the turnkey project platform through the configuration and operation cases of three product-oriented manufacturing systems. Section 6 summarizes the full text.

2. Related Work

The two key issues involved in I4TP are the rapid configuration and digital operation management of manufacturing systems. This section will briefly introduce the related works in these two aspects.

- Manufacturing system configuration

With the rise of concepts such as flexible manufacturing systems (FMSs) and reconfigurable manufacturing systems (RMSs), the research on the configuration of manufacturing systems has gradually become a hotspot. Especially RMSs, with their core features such as modularity, integrability, customization, convertibility, and scalability [12,13], make the configuration of manufacturing systems one of the key issues in the application of RMSs. In the I4TP, the RMS is also the best paradigm for its turnkey manufacturing system solution at present, which can support the manufacturing system to quickly adapt to production needs [14]. The factors involved in the configuration of the manufacturing system mainly include product modules, manufacturing processes, processing equipment, equipment layout, buffer allocation, etc. [15–18] Sabioni et al. [19] proposed a 0–1 nonlinear integer programming model for mass customization to, in parallel, optimize the configuration of modular products, processing equipment, and equipment layouts, and, based on the Genetic Algorithm (GA), to solve it. Under the social manufacturing environment, Zhang et al. [20] proposed a flexible configuration method for distributed Manufacturing Resources (MRs) based on the Non-dominated Sorting Genetic Algorithm-III (NSGA-III) algorithm and Louvain algorithm. This method can complete high-quality service composition and manufacturing communities' allocation of complex manufacturing tasks under multiple objectives, such as time, cost, and quality, and enable dynamic reconfiguration of MRs under abnormal disturbances. Khettabi et al. [21] established an environment-oriented multi-objective RMS design method, using Nonlinear Multi-objective Integer Programming (NL-MOIP) and four improved evolutionary algorithms to complete the selection of reconfigurable machines and tools. Liu et al. [22] further provided a quad-play CMCO (i.e., Configuration design-Motion planning-Control development-Optimization decoupling) design architecture based on the Digital Twin (DT). A prototype system of the DT manufacturing system design platform is developed and used for the design of flow-type smart manufacturing systems to promote customization of intelligent manufacturing in dimensions such as manufacturing equipment, operation, control, and execution.

- Manufacturing system operation

For the operation management of the manufacturing system, it is first necessary to realize the digitalization and informatization integration of the manufacturing system through technologies such as the Industrial Internet of Things (IIoT) and DT. Further, the real-time monitoring, Predictive Maintenance (PdM), and other health management functions of the manufacturing system are realized by developing data-driven models, and even the deep collaboration and integration of the virtual model and the real system are realized in order to carry out the virtual debugging of newly established manufacturing system [23–25]. Wang et al. [26] proposed a DT-based Big Data Virtual-real Fusion (DT-BDVRL) reference framework for intelligent manufacturing supported by the Industrial

Internet. The method and process for building a Big Data Learning and Analysis (BDLA) model are introduced, and the digital thread in DT-BDVRL's virtual and real fusion analysis, iteration, and closed-loop feedback in the product life cycle process is described. To ensure the reliable operation of machine tools, Luo et al. [27] studied a hybrid method based on the DT model and DT data-driven for PdM and conducted a case study on tool life prediction. Liu et al. [28] proposed a new DT-enabled collaborative data management framework for metal Additive Manufacturing (AM) systems to monitor and analyze the entire production process and presented a typical application scenario for defect analysis of metal AM layers with cloud computing and deep learning. With the continuous deepening and expansion of research on IIoT and DT technologies, the operation management of manufacturing systems can be increasingly supported by digitalization and intelligence.

From the above introduction, it can be identified that the following gaps between the existing related works and the I4TP:

The above research solves the theoretical and technical problems such as the manufacturing system configuration algorithm in a limited range, the data-driven production system design method, and the virtual-real collaboration and integration of the manufacturing system and its components. However, there is still a lack of discussion on I4TP. Compared with the existing research on the multi-factor configuration of manufacturing systems and data-driven manufacturing system operation management, the I4TP emphasizes the rapid acquisition of manufacturing system configuration and operation solutions and provides manufacturers with turnkey services for new product-oriented manufacturing systems. Although there have been some achievements for reference on the basic supporting technologies involved, there is still a lack of theoretical and technical guidance for the development of key functional processes of turnkey projects and reference cases for key enabling tools.

This paper aims to fill these gaps, provides key views of the configuration and operation of I4TP, and provides a design and development case of a turnkey project platform to provide theoretical and technical references for the construction and implementation of turnkey projects.

3. Industry 4.0-Oriented Turnkey Project and Its Configuration View and Operation View

3.1. Introduction of the Industry 4.0-Oriented Turnkey Project

While the concepts of Industry 4.0 and turnkey projects have long been around, they are usually discussed separately. For the I4TP, it is necessary to be located under the Industry 4.0 paradigm. At present, turnkey projects usually refer to projects that meet the needs of a single customer, including the delivery of a complete system [29,30], and there is no clear and unified definition. The I4TP considered in this paper sees providing manufacturing enterprises with turnkey services of manufacturing systems as the main goal, quickly building a product-oriented manufacturing system and running it stably, and efficiently completing the production of new products, including personalized products. Therefore, the definition of I4TP is proposed here: in the context of Industry 4.0, a project that a turnkey service provider uses ICTs and other technologies to quickly form manufacturing systems that satisfy the production needs of customers (manufacturing companies mainly) and implement it into manufacturing verification in order to provide customers with turnkey solutions of manufacturing systems.

As shown in Figure 1, the basic process of generating a turnkey solution is set-up, commissioning, and operation [4,31]. The customer (manufacturer) proposes the product to be processed. After being handled by the feature extractor, turnkey configurator, and turnkey builder, a new manufacturing system comes out and is put into operation under turnkey management. The prerequisite for achieving the rapid formation of the manufacturing system implementation scheme is to have a standardized business process, good synergy among stakeholders and functional modules, and bountiful functional modules that enable efficient automatic and intelligent operation. As for realizing the rapid establishment or reconstruction of the manufacturing system, it is obligatory to integrate and package all kinds of software and hardware in standard modularization and connect each other with a unified interface and data format to fulfill plug-and-play. In addition, with the degradation of production equipment performance, changes in market demands, and innovation of technologies, manufacturing companies need to develop or bring in new equipment and applications, so turnkey projects are also required to have good compatibility and scalability. Therefore, the I4TP should meet the following items:

- Standardized turnkey service business process;
- Good coordination ability;
- Rapid information exchange capability;
- A mass of functional modules for automatic and intelligent operation;
- Standard modular functional components;
- Strong compatibility and scalability.

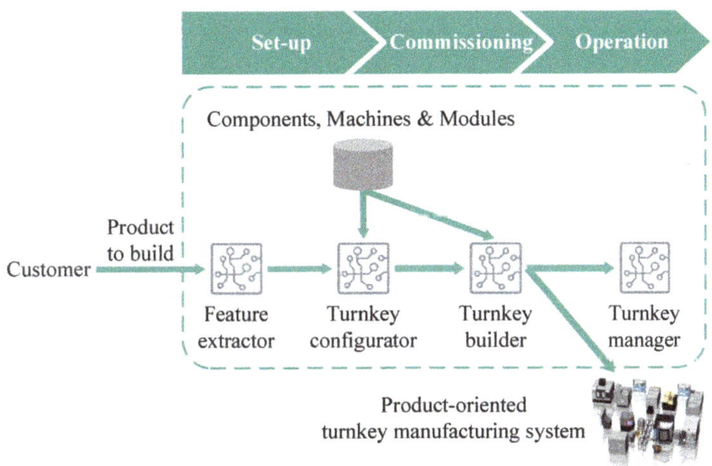

Figure 1. The basic process for generating a turnkey solution [4,31].

Thus, it can be determined that the construction goal of the I4TP is based on the existing manufacturing equipment from different suppliers and through standardized and normalized verification, to make full use of ICTs and the potential for the collaborative process from the web-based approach to carry out the needful parallelization, integration, and automated and intelligent upgrade of the existing manufacturing system development processes, and realize the information interaction with the assets, then under the various constraints, such as cost, quality, delivery time, and operation objectives, quickly establish and form manufacturing systems that adapt to the product needs of the market, and implement it into production verification, providing manufacturing companies with turnkey solutions.

Compared with the existing manufacturing system development process, the biggest feature of the turnkey project is to make full use of ICTs to transform the development process into the configuration process, and to realize the intelligent management of the manufacturing system. Therefore, under the paradigm of Industry 4.0, rapid configuration and intelligent operation of manufacturing systems are two key elements in turnkey projects. To facilitate the participation of all stakeholders involved in the turnkey service, the cloud-based turnkey engineering platform will be the core of the turnkey project. It enables the turnkey project to configure an optimized manufacturing system for a given product and to realize automatic debugging through the plug-and-play scheme. At the same time, the unified standard of the interoperability scheme is met to realize the cooperation between different platform architectures.

3.2. Configuration View and Operation View of the Turnkey Project

A turnkey project is a systematic project involving rich content, such as business, assets, information, and functions. Therefore, a systematic theoretical framework should be used as a guide, especially concerning the construction method of the two key contents of the manufacturing system's rapid configuration and intelligent operation in the I4TP.

The basic purpose of Industry 4.0 is to promote cooperation and collaboration between technical objects, while RAMI4.0 is a structured description of the basic ideas of Industry 4.0, creating digital description rules for technical objects throughout their life cycle and related value changes [5,32–34]. RAMI4.0 is a hierarchical three-dimensional model, including three dimensions: Life cycle & value stream, Hierarchy levels, and Layers. At the same time, an Industry 4.0 Component (I4.0C) reference model, namely the Asset Administration Shell (AAS), is also proposed to provide digital applications corresponding to RAMI4.0 [35–38]. In RAMI4.0 and its derived series of research reports or technical standards and other materials, there are also plug-and-play (or named plug-and-produce) use cases for field devices [39], industry 4.0 asset identification standards [40], Security access control of Industry 4.0 components [41] and other aspects are defined or introduced. In general, RAMI4.0 is committed to the top-down specification and standardization of the information interaction and service application of Industry 4.0 assets, which can provide technical asset frameworks support for various application fields under the Industry 4.0 paradigm. However, RAMI 4.0 has not systematically explored how to construct and apply engineering functional programs, and cannot provide clear guidance for the construction of I4TP, especially the development of functional processes. Therefore, this section proposes a configuration view and an operation view for the two key functional processes of manufacturing system configuration and operation management in turnkey projects, and the concepts and characteristics of these two key functional processes are described.

- Configuration view

The configuration view of the turnkey project is shown in Figure 2. The configuration view covers basic processes such as feature extractor and turnkey configurator. The turnkey project is a project that provides manufacturers with turnkey services for manufacturing systems. The levels it faces are from the manufacturing system to the cooperation between enterprises. In RAMI4.0, it corresponds to the station to the connected world. Of course, the configuration of the system still needs to be based on products, equipment, etc. Through the configuration process, what is obtained is the configuration scheme of the manufacturing system, that is, the prototype of the manufacturing system, so it corresponds to the "Type" stage of the Life Cycle & Value Stream in RAMI4.0. The stakeholders involved in the configuration of the manufacturing system include manufacturers, platform operators, equipment/component suppliers, etc. As the service object of the turnkey project, the manufacturer can designate equipment/component suppliers, etc. It is up to the manufacturer to request a turnkey manufacturing system and to provide data about the product which is to be produced. Production equipment and turnkey platform-compliant equipment data are provided by equipment/component suppliers and stored in or accessible through the equipment database. As the operator of the turnkey project, the platform operator provides core configuration services. Through the mutual matching relationship between products, processes, and equipment, under the constraints or configuration goals of feasibility, security, and economy, and comprehensive considering the layout, scheduling, simulation, and other aspects, a manufacturing system configuration scheme that meets the manufacturer's needs is given. It is worth mentioning that the same enterprise can play different roles in the configuration process at the same time. For example, the manufacturer itself can also act as an equipment/component supplier to provide its existing equipment for the configuration of the manufacturing system.

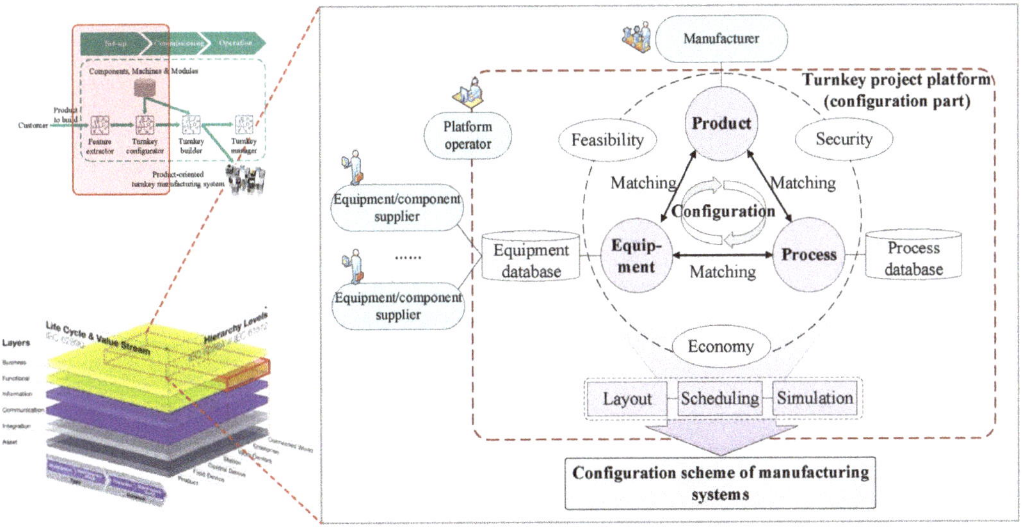

Figure 2. Configuration view.

The entire configuration process is described in detail in the form of a mathematical model. First, it is necessary to extract the processing features based on the information p of the product to be processed by using a feature recognition algorithm or tool (denoted as q). The processing features, optional processes, and optional equipment of the product are represented by x, y, and z, respectively, where

$$x = q(p) \tag{1}$$

In the configuration function of the turnkey project platform, the applicable processing feature range, process database, and equipment database are represented by X, Y, and Z, respectively. Then, the set of possible device configuration schemes can be expressed as

$$F = \{ f(x, y, z) | \alpha_x \subseteq \alpha_y, \beta_y \subseteq \beta_z, \gamma_x \subseteq \gamma_z, x \subseteq X, y \subseteq Y, z \subseteq Z \} \tag{2}$$

where $f(x, y, z)$ is the matching algorithm model. α_i, β_i, and γ_i all represent a specific attribute set in the set i. After further constraints and optimization of feasibility, security, and economy (represented by A, B, C, respectively), the near-optimal equipment configuration scheme F^* can be screened out.

$$F^* = k^* \in F | A(k^*) = 1, B(k^*) = 1, C(k^*) = \min(C(k) | k \in F) \tag{3}$$

According to the equipment configuration scheme F^* and workshop environment I, through the layout design method or generation algorithm (denoted as g), the set of optional layout schemes L is obtained. Then, through the process of simulation or mathematical model calculation (denoted as h), the corresponding scheduling scheme set S is obtained. The final manufacturing system configuration scheme R can be obtained by selecting the best combination (L^*, S^*) from L and S according to specific objectives. The configuration scheme R is the unique near-optimal solution given by the whole configuration process model for customer (manufacturer) needs, including equipment configuration, workshop layout, system scheduling, and other aspects.

$$L = \{ g(F^*, I) \} \tag{4}$$

$$S = \{ h(F^*, l) | l \in L \} \tag{5}$$

$$\boldsymbol{R} = (\boldsymbol{F}^*, \boldsymbol{L}^*, \boldsymbol{S}^*) \tag{6}$$

Of course, due to the close interaction among equipment configuration, workshop layout, and production scheduling, the above algorithm models can be further integrated to form multiple collaborative optimization models to improve overall configuration performance.

$$\boldsymbol{R} = J(q, f, A, B, C, g, h) \tag{7}$$

Through the above analysis, the key development process of manufacturing system configuration, a turnkey service-enabling tool, is shown in Figure 3. After developing standardized modular databases that provide basic support, developing algorithm models for key functions, integrating functions, and packaging software and hardware, a manufacturing system configuration enabling tool is ultimately formed.

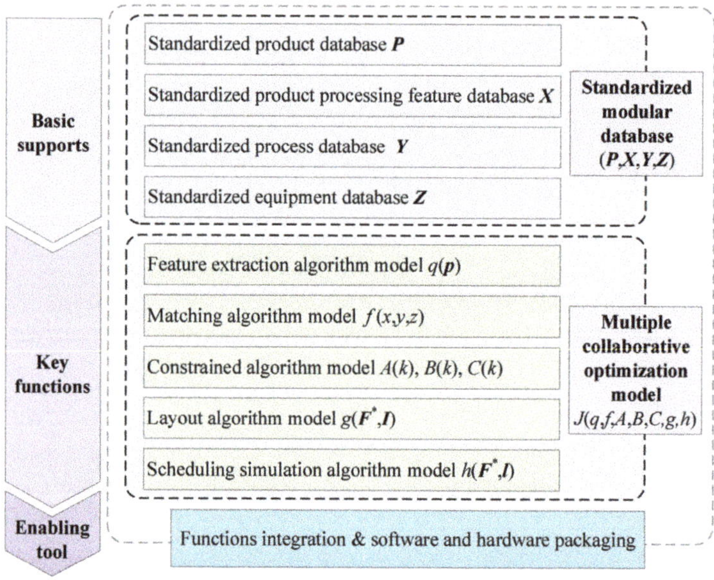

Figure 3. The key development process of the manufacturing system configuration service-enabling tool.

- Operation view

The operation view of the turnkey project is shown in Figure 4. The operation view of the turnkey project involves the construction and management of the manufacturing system, which, similarly to the configuration view, corresponds to the hierarchy level range from station to connected word in RAMI 4.0. In the turnkey project, the operation process of the manufacturing system is the instantiation of the configuration result and the intelligent management after instantiation, so it belongs to the "Instance" stage of the Life Cycle & Value Stream. The stakeholders involved in the operation of the manufacturing system mainly include manufacturers, platform operators, equipment/component suppliers, material/parts suppliers, and IoT suppliers. Similarly, manufacturers can select suppliers. The production equipment is provided by equipment/component suppliers, and the production equipment is integrated into the workshop in a standard modular form as much as possible and has advanced functions such as plug-and-play and automatic debugging, to ensure the rapid construction of the manufacturing system. Raw materials for producing products are provided by material/parts suppliers. IoT suppliers provide equipment interconnection services to achieve reliability data acquisition in the workshop. The platform operator obtains real-time data on physical assets from the IIoT platform connected to the equipment and integrates it. According to the hierarchical relationship of the "component-equipment-manufacturing system", real-time monitoring, fault diagnosis, Remaining Useful Life (RUL) prediction, PdM, and other various operation and maintenance functions are realized layer by layer. These functions are integrated upward, and the intelligent operation and health management of the entire manufacturing system are finally realized. Once the whole system has been confirmed to be operating normally and smoothly, the complete turnkey solution can be delivered to the manufacturer. When manufacturers have new production requirements, the manufacturing system can be reconfigured, and the functional structures in the IIoT platform and turnkey project platform are also updated.

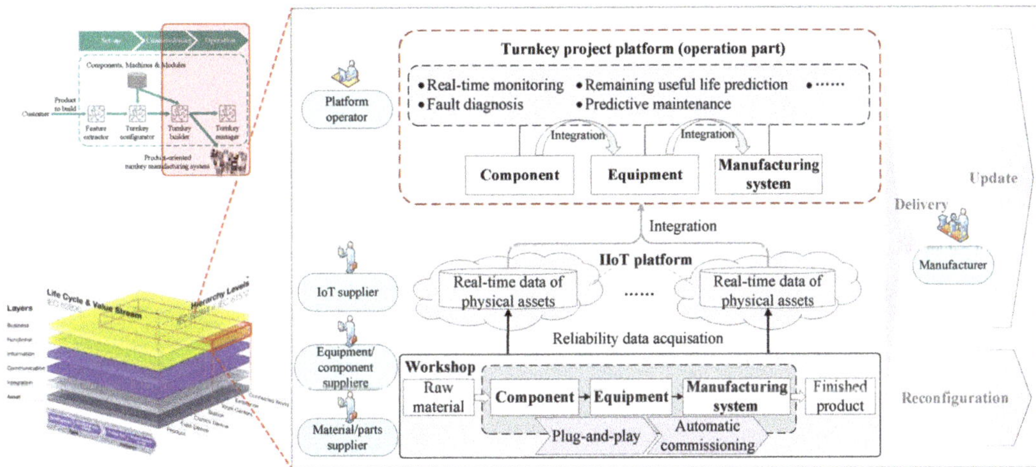

Figure 4. Operation view.

To facilitate analysis and discussion, a mathematical description of the operation process is further established. First, we integrate the real-time data obtained by using the IIoT platform to facilitate the realization of subsequent operation and health management functions.

$$\boldsymbol{D}_O = \boldsymbol{D}_{P1} \oplus \boldsymbol{D}_{P2} \oplus \cdots \oplus \boldsymbol{D}_{Pm} \tag{8}$$

where $\boldsymbol{D}_{Pi}$ represents the collection of real-time data from the supplier i. Using these real-time data, the functional set of each level of the "component-equipment-manufacturing system" can be implemented and integrated layer by layer according to the configuration scheme $\boldsymbol{R}$ of the manufacturing system. The operation and health management function sets of parts, equipment, and manufacturing systems are, respectively, denoted as $\boldsymbol{M}_F$, $\boldsymbol{M}_D$, $\boldsymbol{M}_S$, and the functional models of which are, respectively, denoted as u_i, v_i, w_i, and the function sets at each level of "component-equipment-manufacturing system" can be expressed as

$$\boldsymbol{M}_F = \{u_1(\boldsymbol{D}_{F1}|\boldsymbol{D}_{F1} \subseteq \boldsymbol{D}_O), \cdots, u_n(\boldsymbol{D}_{Fn}|\boldsymbol{D}_{Fn} \subseteq \boldsymbol{D}_O)|\boldsymbol{R}\} \tag{9}$$

$$\boldsymbol{M}_D = \{v_1(\boldsymbol{M}_{F1}, \boldsymbol{D}_{D1}|\boldsymbol{M}_{F1} \subseteq \boldsymbol{M}_F, \boldsymbol{D}_{D1} \subseteq \boldsymbol{D}_O), \cdots, v_r(\boldsymbol{M}_{Fr}, \boldsymbol{D}_{Dr}|\boldsymbol{M}_{Fr} \subseteq \boldsymbol{M}_F, \boldsymbol{D}_{Dr} \subseteq \boldsymbol{D}_O)|\boldsymbol{R}\} \tag{10}$$

$$\boldsymbol{M}_S = \{w_1(\boldsymbol{M}_{D1}, \boldsymbol{D}_{S1}|\boldsymbol{M}_{D1} \subseteq \boldsymbol{M}_D, \boldsymbol{D}_{S1} \subseteq \boldsymbol{D}_O), \cdots, w_t(\boldsymbol{M}_{Dr}, \boldsymbol{D}_{Sr}|\boldsymbol{M}_{Dr} \subseteq \boldsymbol{M}_D, \boldsymbol{D}_{Sr} \subseteq \boldsymbol{D}_O)|\boldsymbol{R}\} \tag{11}$$

Among them, $\boldsymbol{D}_{Fi}$, $\boldsymbol{D}_{Di}$, and $\boldsymbol{D}_{Si}$ represents the data subset of $\boldsymbol{D}_O$ related to the functional model u_i, v_i, and w_i. $\boldsymbol{M}_{Fi}$ represents a set of the operation and health management function set of a device's components that are related to the functional model v_i of the device level, and is integrated through the model v_i. $\boldsymbol{M}_{Di}$ represents a set of the operation and health management function of a manufacturing system's devices that are related to the functional model w_i of the manufacturing system level, and is integrated through the model w_i.

Once the manufacturing system is operational, the turnkey solution $\boldsymbol{TKS} = (\boldsymbol{R}, \boldsymbol{M}_F, \boldsymbol{M}_D, \boldsymbol{M}_S)$, i.e., the near-optimal set of configuration and operation solutions can be delivered to the manufacturer. When a manufacturer needs to produce a new product, they can realize the conversion of the turnkey solution according to a certain reconstruction and update method (denoted as T) based on the existing turnkey solution.

$$\boldsymbol{TKS}' = (\boldsymbol{R}', \boldsymbol{M}'_F, \boldsymbol{M}'_D, \boldsymbol{M}'_S) = T(\boldsymbol{R}, \boldsymbol{M}_F, \boldsymbol{M}_D, \boldsymbol{M}_S) \quad (12)$$

The key development process of the service-enabling tool for manufacturing system operation is shown in Figure 5. Based on the reliability data acquisition of components, equipment, and manufacturing systems and the standardized real-time database formed by integration, develop various levels of operation and health management models and reconfiguration and update algorithm models. Finally, we integrate these basic supports and various key functions and packaged them on software and hardware to form a manufacturing system operation service-enabling tool.

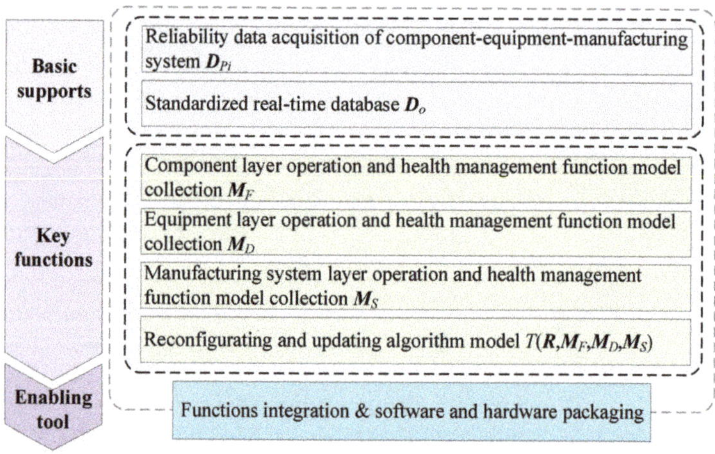

Figure 5. The key development process of manufacturing system operation service-enabling tool.

4. Design and Development of Key Enabling Tools for Manufacturing System Configuration and Operation: Taking AMTC as an Example

Under the configuration view and the operation view, different specific construction plans will be formed according to different manufacturing industries and different actual needs. Here, the mapping between key views of the turnkey project and specific applications is shown by taking the development of AMTC's turnkey project platform as an example.

4.1. Development Background of the Turnkey Project Platform in AMTC

To create an environment for the research, demonstration, and teaching of intelligent manufacturing-related technologies, and to provide technical services for related enterprises, AMTC needs to build turnkey projects to meet the processing needs of diversified products. As a research base and demonstration center of advanced manufacturing technology, AMTC has abundant equipment and technical resources to support the construction of turnkey projects.

- Equipment supports

As the basic support for the turnkey project, the main equipment in the AMTC workshop currently is shown in Figure 6. The left part is mainly large and medium-sized equipment assets. These assets can meet the needs of turnkey projects through network transformation and integration and packaging. These devices are not easy to move, so the layout usually remains unchanged during actual processing, but configuration and reconfiguration can still be achieved in terms of scheduling. The right part is a modular production unit based on the i5Blocks intelligent manufacturing demonstration line, it can be automatically combined to form a new production line according to the configuration results, and has the advantage of quick replacement, which will give full play to the advantages of turnkey projects. In addition, when serving manufacturers, i.e., configuring and building manufacturing systems, the optional assets from the corresponding manufacturers and suppliers will also be included in the asset category of the turnkey project.

Figure 6. Devices in the AMTC workshop.

- Machine communication supports

In terms of machine communication, the communication from the physical layer to the information layer is composed of three solutions, as shown in Figure 7, provided by iSESOL, Microcyber, and HUAWEI, respectively. Among them, the solution provided by iSESOL uses the iPORT protocol to communicate between devices and the cloud. It has a strong compatibility with various common industrial communication protocols. The main machine tools in the AMTC workshop are connected to collect internal data through this solution, including the DMU65 machining center, the M1.4 Machining center, etc., and a vibration sensor is installed on the spindle of the M1.4 machining center to collect its vibration data. Microcyber is more focused on the security of the workshop's local area network and adopts the data security encryption technology based on the multi-key distribution mechanism and Modbus security communication technology based on the national cryptographic algorithm and security protocol, etc. Microcyber provides vibration and temperature sensors, machine tool access points, and other equipment, which can be used for data acquisition inside and outside the machine tool and workshop environment data. Currently, this solution is mainly used for equipment vibration and temperature data acquisition in the AMTC workshop. The solution provided by HUWEI focuses on the industrial application of 5G technology, supports the most commonly used top 7 industrial communication protocols, and establishes application scenarios on the TS2 conveyor in the AMTC workshop for material transfer locations monitoring. In addition, the iSESOL cloud also provides the OpenAPI open service interface, which can be interconnected with other industrial cloud platforms or functional applications to jointly complete real-time monitoring and digital management of the entire workshop.

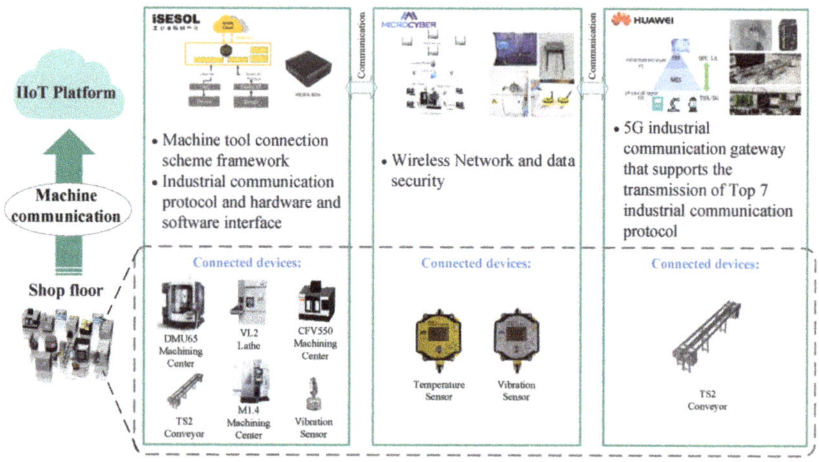

Figure 7. Machine communication solutions.

4.2. The Overall Functional Design of the Turnkey Project Platform

Based on the aforementioned analysis and the basic environment of AMTC, this section combines DT technology, artificial intelligence technology, and actual needs to design the overall function of the turnkey project platform. The functional structure of the obtained turnkey project platform is shown in Figure 8. It is mainly composed of two sub-platforms: the system configuration platform and the operation management platform. The two sub-platforms correspond to the two key functional processes of the manufacturing system configuration and operation management of the turnkey project, that is, the enabling tools for these two key functional processes. The turnkey project platform is mainly oriented to the production of designed products and takes information, such as product models, materials, drawings, and production indicators, as input. Then, the steps of processing feature extraction, manufacturing process matching, system configuration, workshop layout generation, scheduling simulation and optimization are completed in the system configuration platform. The configuration results are handed over to the workshop staff to complete the system construction or reconstruction. The operation management platform realizes real-time production monitoring, analyzes the operation status of equipment/components, and formulates PdM strategies by collecting real-time operation data of workshop equipment/components.

4.3. Design and Development of the System Configuration Platform

The detailed functional structure design of the system configuration platform is shown in Figure 9. MongoDB is used as the database for storing various data involved in the manufacturing system configuration process. First, the platform user imports the information of the product to be processed, and after the import is completed, a feature extraction instruction is issued to the backend server. The server identifies the product features and the relationship between the features based on the STEP file of the product and outputs a list of processing features. Then, the process matching instruction is issued, and the server performs the identification and modeling of the interdependence between the parameters of the product, processing features, etc. At the same time, it analyzes its impact and risk and performs process matching based on the interdependence between the processing features and the process, and the corresponding process list is outputted. Next, the user sets various requirements for product production within the allowable range of the platform, including configuration optimization goals such as the lowest production cost and highest production efficiency. After that, the user issues system configuration instructions, and the server according to equipment, tools, and other configuration resources and product requirements, configures the manufacturing system based on GA and outputs the configuration result. Finally, the user issues a workshop layout generation instruction, and the server generates the workshop layout scheme based on GA, according to the configuration result and equipment information. Plant Simulation is used to carry out scheduling simulation and optimization of the layout scheme to output the final workshop layout and scheduling scheme.

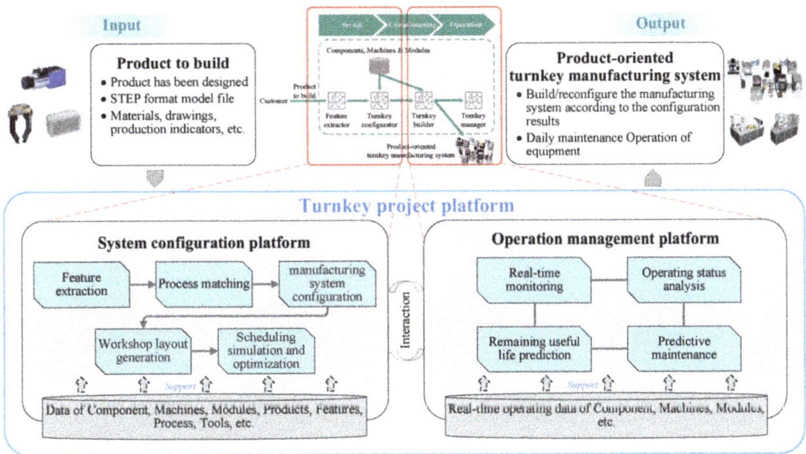

Figure 8. The functional structure of the turnkey project platform.

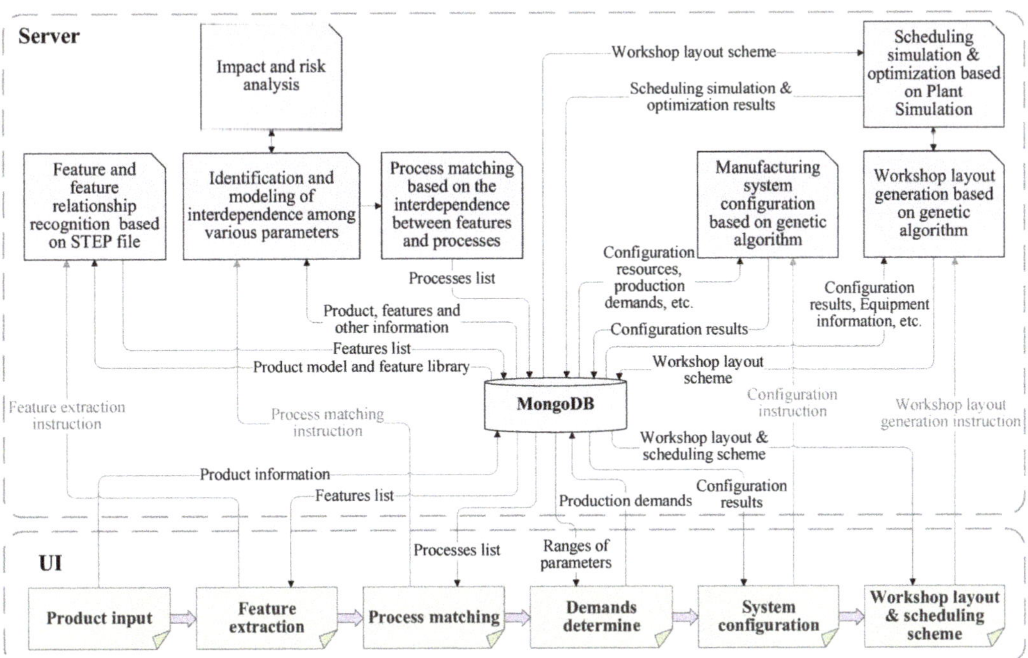

Figure 9. The detailed functional structure design of the system configuration platform.

The User Interface (UI) of the system configuration platform developed according to the design scheme is shown in Figure 10. Considering the convenience of the application, the development method adopted is a web-based method, which belongs to a website application project and is a Browser/Server (B/S) application structure. Apache is used to host web services and the XAMPP package is applied for the configuration management of Apache. The main structure of the backend server is developed with Python3, and the frontend browser pages are written with Hyper Text Mark-up Language 5 (HTML5), Cascading Style Sheets (CSS), and JavaScript.

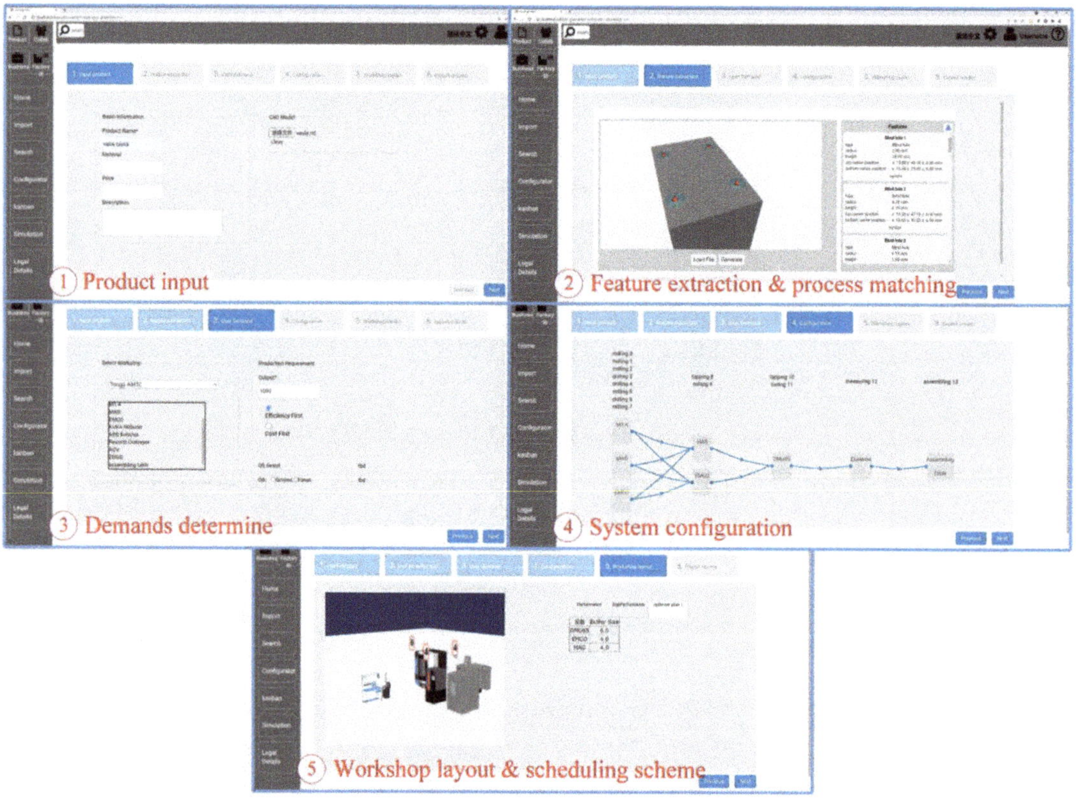

Figure 10. The UI of the system configuration platform.

During the entire system configuration process, the system configuration platform integrates feature extraction, process matching, equipment configuration, layout generation, scheduling simulation and optimization into the backend server. Platform users can quickly obtain manufacturing system configuration schemes with simple operations instead of tedious and time-consuming manufacturing system design processes. What's more, manufacturers who need to build new manufacturing systems can use the services provided by this platform to obtain solutions that meet their needs without extensive expertise in manufacturing system design.

4.4. Design and Development of the Operation Management Platform

The detailed functional structure design of the operation management platform is shown in Figure 11, and the concept of AAS [5,33,35,37] is introduced for description. In the operation management platform, according to the structural level of the manufacturing system, several machine communication solutions introduced in Figure 7 are used to realize the information interaction between the key components-equipment-system and the server in the workshop. Corresponding to the different levels of components-equipment-system in the server, a management module (i.e., AAS) is constructed to manage each key asset at each level. Compared with the current common way of independent monitoring of key equipment/components, the hierarchical operation management scheme is more conducive to the assessment and prediction of production status from the perspective of the overall manufacturing system in the production process. Thereby, the operation and maintenance decisions can be made or adjusted promptly. Combined with the actual needs of AMTC workshops, different management strategies are adopted for different assets. Taking the DMU65 machining center as an example, the RUL prediction of the bearings in its spindle and feeding system is based on the Support Vector Machine (SVM). The spindle and feeding system use different methods to realize fault diagnosis and PdM functions based on the RUL prediction results of the bearings. On this basis, data from other sources in the DMU65 machining center is also integrated to realize real-time monitoring, fault diagnosis, and PdM of the entire machine at the equipment level. Management measures such as real-time monitoring are also implemented for other equipment, such as robots and conveyors. By integrating the data of the underlying equipment with the results of fault diagnosis, PdM, and other functions, the real-time monitoring, fault diagnosis, and PdM of the manufacturing system can be further realized. The real-time operation data of assets at all levels and the analysis results of various health management functions are stored in the database and displayed to users in real time. Users can intuitively view the operating status of the manufacturing system and its equipment and components through the UI, and conduct inspection, maintenance, repair, replacement, etc. based on the analysis results of various health management functions. When the manufacturing system needs to be reconstructed, the communication relationship between the AASs in the server will also be updated synchronously.

Combining the characteristics and advantages of the three machine communication solutions in Figure 7, as well as their currently connected device assets, the developed operation management platform is shown in Figure 12, consisting of four parts. The first part adopts the machine communication solution provided by iSESOL, which connects the main equipment in the workshop and uploads the data to the cloud for real-time monitoring. The monitored equipment operation data includes spindle speed, spindle load, feed speed, number of processed workpieces, running time, etc. The second part also uses the solution provided by iSESOL to connect equipment, and then develop local functions in the iSESOL BOX, including RUL prediction of the bearing, fault diagnosis of the spindle, feeding system, and machine tool, and PdM of the spindle, feeding system, and machine tool. The third part adopts the solution provided by Microcyber to realize the temperature and vibration monitoring of the machine tool and workshop environment. The fourth part adopts the solution provided by HUAWEI, uses the 5G network to communicate, and monitors the position of the workpiece on the conveyor.

In the operation management platform shown in Figure 12, the web-based UI of the first part is provided by iSESOL directly, and the UIs of the rest parts are also developed in a web-based way. In the first part, the device operation data is collected and uploaded to the cloud, which can be accessed by remote users, providing cloud services. The rest are connected through the local area wireless networks of the workshop, and the local equipment in the workshop is used to provide computing support for various functions, and all provided are edge services. The entire operation management platform adopts a distributed management strategy combining cloud and edge, which can fully guarantee the flexibility of the manufacturing system and facilitate subsequent system reconstruction. Of course, current edge services can also be integrated into the cloud as needed. In addition, in the future, users can obtain these functional applications by downloading Applications (APPs) on mobile clients connected to the cloud.

After completing the construction or reconstruction of the manufacturing system according to the configuration scheme and starting the trial operation, the whole process of digital management of the workshop production is carried out through the operation management platform. It can quickly verify the feasibility and reliability of the new manufacturing system to ensure that it can be put into operation smoothly. Meanwhile, it also provides manufacturers with a complete set of turnkey manufacturing system solutions, including follow-up operation management, so that manufacturers can directly operate the manufacturing system without the need for separate system operation testing and network design transformation.

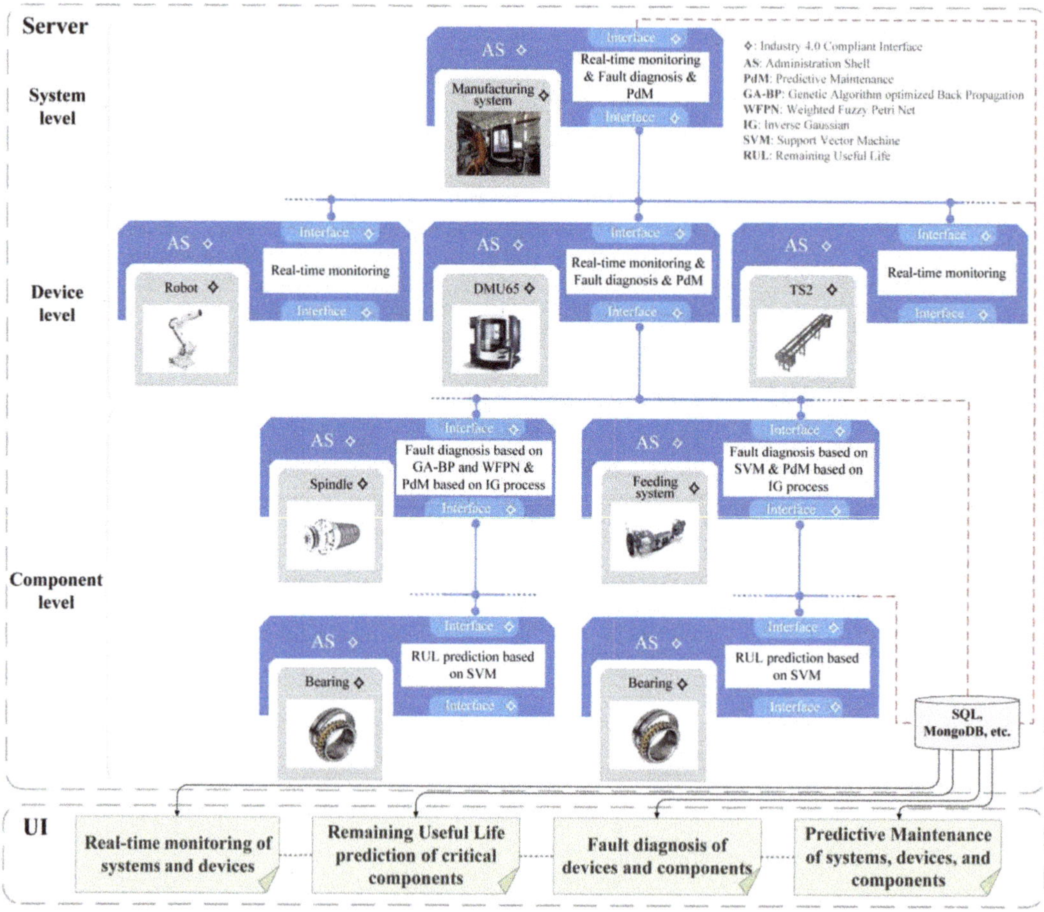

Figure 11. The detailed functional structure design of the operation management platform.

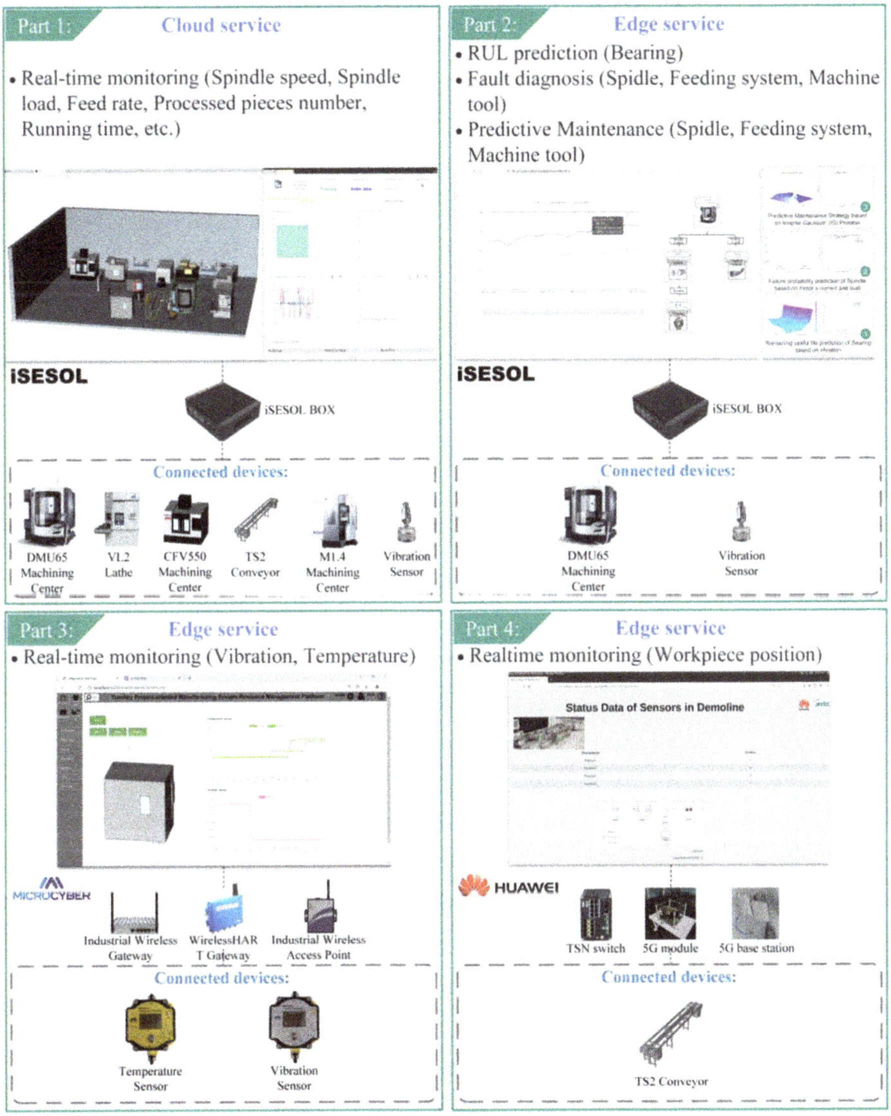

Figure 12. The devices' connection and UIs of the operation management platform.

5. Product-Oriented Manufacturing System Configuration and Operation Cases

To verify the developed turnkey project platform, three products, the hydraulic valve block, the gripper base, and the automotive fuel cell (model), are used as verification cases.

This platform has properly simplified the evaluation of production efficiency and production cost in the configuration process. The efficiency and cost indexes are obtained by weighting the sum of several key parameters, and the efficiency-cost comprehensive index is further constructed by weighting the efficiency and cost indexes. This comprehensive index is used for scheme comparison. The factors considered in the calculation of the production efficiency index include the spindle speed, the feeding speed, the power, the average tool change duration, the tool change times of the equipment, etc. While the factors considered in the calculation of the production cost index include the tool price, the equipment price, the average processing cost of the equipment to complete a single process step, etc. The weight of each parameter is obtained by the analytic hierarchy process. The smaller the efficiency-cost comprehensive index, the better the solution.

- Hydraulic valve block

The manufacturing system configuration and operation verification results of the hydraulic valve block case are shown in Figure 13. The material of the hydraulic valve block is cast iron, and its blank is a casting, which needs to complete the processing of multiple planes, smooth holes, threaded holes, and stepped surfaces. Through the functional process of the system configuration platform, the manufacturing system configuration scheme under two different optimization objectives of production efficiency priority and production cost priority is obtained. Compared with the original manufacturing system scheme, when considering efficiency first, the comprehensive index of the system configuration scheme given by the platform is 0.24, which is much better than the 0.81 of the original system. When considering cost first, the system configuration scheme can also perform better. According to the equipment configuration results and layout generation results, the Plant Simulation software is further used for simulation and optimization. The number next to the equipment model represents the optimized buffer capacity matched with the equipment. The configuration scheme that prioritizes production efficiency is selected for implementation. The operation management part in Figure 13 shows the operation of the new manufacturing system and the operation management platform at a certain time. It can be seen that the turnkey project platform can configure the equipment required for the production of new products and generate a reasonable workshop layout according to different user needs. Optimization results such as buffer capacity are given after simulation and digital management is executed when new manufacturing systems are put into service.

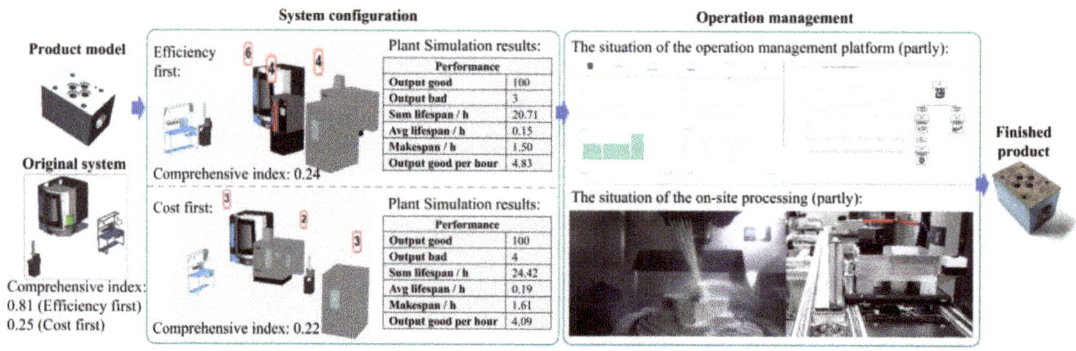

Figure 13. Configuration and operation of the hydraulic valve block manufacturing system.

- Gripper base

The configuration and operation of the manufacturing system for the gripper base case are shown in Figure 14. The material of the gripper base is aluminum alloy, and its blank is an aluminum alloy block. Multiple planes, slots, and holes are some of the features to be machined. The manufacturing system configuration scheme can be quickly obtained by using the system configuration platform. According to the efficiency-cost comprehensive index of different solutions, compared with the original manufacturing system solution, the system configuration platform can provide a better solution in the two optimization directions of efficiency first and cost first. The efficiency-first scheme is adopted for product production, and the operation management platform effectively guarantees the safety and stability of the production process.

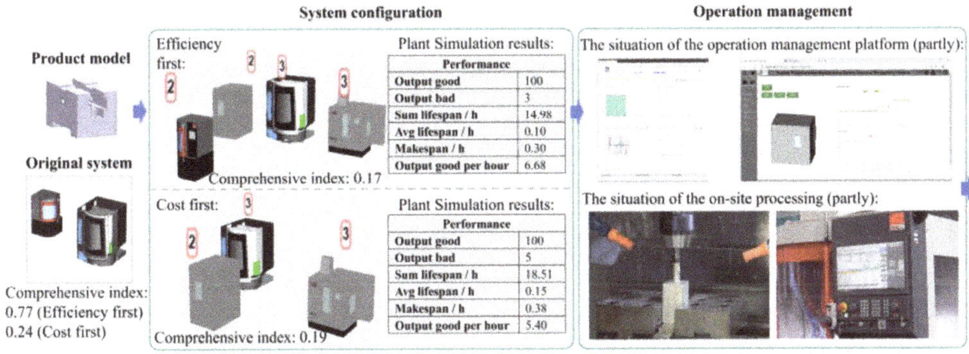

Figure 14. Configuration and operation of the gripper base manufacturing system.

- Automotive fuel cell (model)

The production of automobile fuel cells is divided into two main processes, 3D printing of air intake grilles and assembly of air intake grilles and ready-made battery stacks. To simulate this process in order to verify its feasibility and save costs, the Polylactic Acid (PLA) material is used to replace the original material to establish an imitated production scene. The manufacturing system in this case is built for the first time. The configuration scheme obtained by the system configuration platform and the on-site processing situation are shown in Figure 15, and the equipment used is modular units. The currently used modular units have not been connected to the operation management platform, but the potential of the turnkey project platform can still be reflected through the generation of configuration schemes by the system configuration platform and the on-site processing.

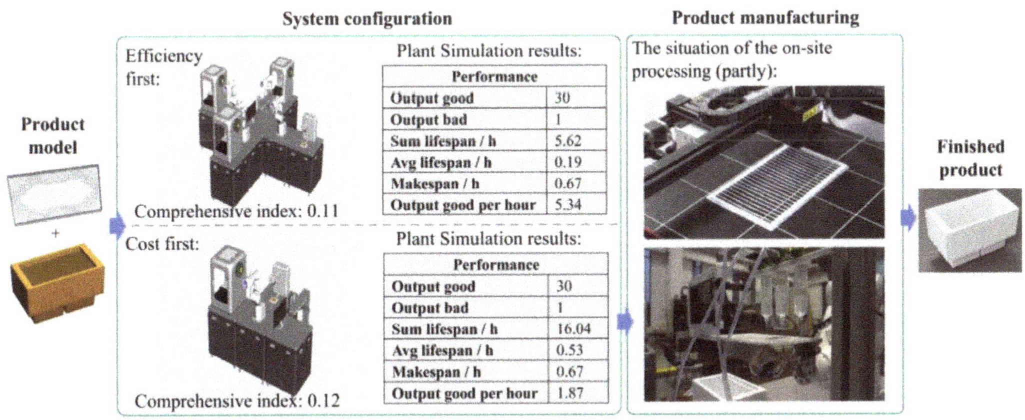

Figure 15. Configuration and field situation of the fuel cell (model) manufacturing system.

To save verification costs, only a small number of finished products have been produced for each product in practice, and all of them meet the product design requirements. This shows that the turnkey solutions for the three products are feasible.

At present, the turnkey project platform has completed the development, integration, and encapsulation of many key functional models under the guidance of the configuration and operation views, forming two key enabling tools, the system configuration platform and the operation management platform. The above three cases further prove the feasibility and effectiveness of the turnkey project platform, as well as the huge application potential of the I4TP. The I4TP makes it possible to rapidly configure and operate the new manufacturing system. Although the above cases are only the application of the turnkey project platform in the same workshop, with the continuous enrichment of built-in functions of the platform, it will be fully able to adapt to emerging manufacturing modes such as social manufacturing. It can provide a platform for various manufacturers, suppliers, and other stakeholders to jointly participate in the task allocation, docking, and communication of the product production.

6. Conclusions and Summary

In response to wider personalization and transforming market demands, manufacturing enterprises need to respond more quickly to realize the rapid establishment and operation of related manufacturing systems. However, numerous enterprises, especially small and micro enterprises, cannot cope with the rapid updating of product-oriented manufacturing system configurations alone. Therefore, it is very necessary to make full use of the existing ICTs to carry out the construction of turnkey projects and provide these enterprises with turnkey services of manufacturing systems. Focusing on the generation process of the turnkey project, this paper proposes a configuration view and an operation view for the two key contents of the system configuration and operation in the turnkey project, to provide a theoretical basis for the development of the key function process of the turnkey project. To a certain extent, it makes up for the defect that the existing reference architecture cannot provide specific guidance for the construction of turnkey projects. At the same time, the turnkey project platform in AMTC is used as a demonstration case to provide a reference for the development and application of key enabling tools for configuration and operation. The research results are summarized as follows:

- According to the basic process of generating the turnkey scheme and the characteristics of the turnkey project, the configuration view and the operation view of the turnkey project are proposed based on RAMI4.0. The configuration and operation processes are described in the form of mathematical models, and the key function development processes of configuration and operation enabling tools are given;
- Guided by the configuration and operation views, relying on AMTC's equipment and machine communication network support, the overall functional design of the turnkey project platform is carried out according to actual needs. The functional structure design and application development of the system configuration platform and the operation management platform are focused on. Various functions of configuration and operation are integrated into the servers, and simple and easy-to-use UIs are provided. This provides users with turnkey services for the rapid configuration of manufacturing systems and intelligent operation management;
- The developed turnkey project platform is verified by taking three products as case studies. The results show that the system configuration and operation management functions of the turnkey project platform can operate normally, and they have achieved good application results, which can meet the rapid configuration and efficient operation management requirements of the manufacturing system.

In general, the developed turnkey project platform has the following advantages. (1) This platform makes it possible to quickly obtain turnkey solutions for manufacturing systems that are difficult to achieve with traditional methods. (2) It can optimize the production efficiency, production cost, and other objectives of the manufacturing system in the rapid configuration process. (3) It can provide a hierarchical health and operation management scheme for the manufacturing system, and compared with the independent monitoring mechanism between equipment/components, it is more conducive to ensuring the secure and stable operation of the manufacturing system.

At present, the turnkey project platform in AMTC provides only research examples. The construction of turnkey projects from functional development to practical application still requires a lot of research work, including rapid product development, processing feature extraction, process matching, manufacturing system configuration, workshop layout generation and optimization, modular production equipment design and implementation, real-time monitoring and intelligent control of production, coordination mechanism between manufacturers and suppliers, etc., all need to be developed, improved and promoted. These factors require further organic integration in the future.

Author Contributions: Conceptualization, S.X., W.Z., J.F. and C.E.; methodology, S.X. and W.Z.; software, S.X. and Y.L.; validation, S.X. and F.X.; formal analysis, S.X.; investigation, S.X. & D.L.; resources, S.X. and W.Z.; data curation, S.X.; writing—original draft preparation, S.X.; writing—review & editing, F.X. and D.L.; visualization, S.X.; supervision, F.X. and D.L.; project administration, W.Z.; funding acquisition, W.Z. and J.F. All authors have read and agreed to the published version of the manuscript.

Funding: This work was supported by the National Key R & D Program of China [grant number 2022YFE0114100], the National Key R & D Program of China [grant number 2017YFE0101400], and the German Federal Ministry of Education and Research (BMBF) [grant number 02P17X000 ff].

Conflicts of Interest: The authors declare no conflict of interest.

References

1. Mehrabi, M.G.; Ulsoy, A.G.; Koren, Y.; Heytler, P. Trends and perspectives in flexible and reconfigurable manufacturing systems. *J. Intell. Manuf.* **2002**, *13*, 135–146. [CrossRef]
2. Bortolini, M.; Galizia, F.G.; Mora, C. Reconfigurable manufacturing systems: Literature review and research trend. *J. Manuf. Syst.* **2018**, *49*, 93–106. [CrossRef]
3. Tao, F.; Qi, Q.; Liu, A.; Kusiak, A. Data-driven smart manufacturing. *J. Manuf. Syst.* **2018**, *48*, 157–169. [CrossRef]
4. Gönnheimer, P.; Kimmig, A.; Mandel, C.; Stürmlinger, T.; Yang, S.; Schade, F.; Ehrmann, C.; Klee, B.; Behrendt, M.; Schlechtendahlc, J.; et al. Methodical approach for the development of a platform for the configuration and operation of turnkey production systems. *Procedia CIRP* **2019**, *84*, 880–885. [CrossRef]
5. *DIN SPEC 91345:2016-04*; Reference Architecture Model Industrie 4.0 (RAMI4.0). Beuth: Berlin, Germany, 2016.
6. Pisching, M.A.; Pessoa, M.A.; Junqueira, F.; dos Santos Filho, D.J.; Miyagi, P.E. An architecture based on RAMI 4.0 to discover equipment to process operations required by products. *Comput. Ind. Eng.* **2018**, *125*, 574–591. [CrossRef]
7. Fast Radius. The Biggest Revolution in Manufacturing since the Assembly Line. Available online: https://www.fastradius.com/capabilities/cloud-manufacturing/ (accessed on 27 September 2022).
8. Phoenixcontact. Digital Factory Now. Available online: https://www.phoenixcontact.com/zh-cn/industries/digital-factory (accessed on 27 September 2022).
9. Rold. Discover Smartfab. Available online: https://www.rold.com/rold-smartfab-bring-company-4-0-universe/ (accessed on 27 September 2022).
10. Instrumentation Technology and Economy Institute, PR. China. Intelligent Manufacturing Comprehensive Test Platform. Available online: http://www.itei.cn/imct.html (accessed on 27 September 2022).
11. Schneider Electric. EcoStruxure Platform. Available online: https://www.se.com/ww/en/work/campaign/innovation/platform.jsp (accessed on 27 September 2022).
12. Pal Singh, P.; Madan, J.; Singh, H. Composite performance metric for product flow configuration selection of reconfigurable manufacturing system (RMS). *Int. J. Prod. Res.* **2020**, *59*, 3996–4016. [CrossRef]
13. Sabioni, R.C.; Daaboul, J.; Le Duigou, J. Concurrent optimisation of modular product and Reconfigurable Manufacturing System configuration: A customer-oriented offer for mass customisation. *Int. J. Prod. Res.* **2022**, *60*, 2275–2291. [CrossRef]
14. Eynaud, A.B.D.; Klement, N.; Roucoules, L.; Gibaru, O.; Durville, L. Framework for the design and evaluation of a reconfigurable production system based on movable robot integration. *Int. J. Adv. Manuf. Technol.* **2022**, *118*, 2373–2389. [CrossRef]
15. Khan, A.S.; Homri, L.; Dantan, J.Y.; Siadat, A. An analysis of the theoretical and implementation aspects of process planning in a reconfigurable manufacturing system. *Int. J. Adv. Manuf. Technol.* **2022**, *119*, 5615–5646. [CrossRef]
16. Demir, L.; Koyuncuoğlu, M.U. The impact of the optimal buffer configuration on production line efficiency: A VNS-based solution approach. *Expert Syst. Appl.* **2021**, *172*, 114631. [CrossRef]
17. Guo, D.R.; Zhong, Y.; Ling, S.; Rong, Y.; Huang, G.Q. A roadmap for Assembly 4.0: Self-configuration of fixed-position assembly islands under Graduation Intelligent Manufacturing System. *Int. J. Prod. Res.* **2020**, *58*, 4631–4646. [CrossRef]
18. Hou, S.X.; Gao, J.; Wang, C. Design for mass customisation, design for manufacturing, and design for supply chain: A review of the literature. *IET Collab. Intell. Manuf.* **2022**, *4*, 1–16. [CrossRef]
19. Sabioni, R.C.; Daaboul, J.; le Duigou, J. An integrated approach to optimize the configuration of mass-customized products and reconfigurable manufacturing systems. *Int. J. Adv. Manuf. Technol.* **2021**, *115*, 141–163. [CrossRef]
20. Zhang, Y.; Tang, D.B.; Zhu, H.H.; Li, S.P.; Nie, Q.W. A flexible configuration method of distributed manufacturing resources in the context of social manufacturing. *Comput. Ind.* **2021**, *132*, 103511. [CrossRef]
21. Khettabi, I.; Benyoucef, L.; Boutiche, M.A. Sustainable reconfigurable manufacturing system design using adapted multi-objective evolutionary-based approaches. *Int. J. Adv. Manuf. Technol.* **2021**, *115*, 3741–3759. [CrossRef]

22. Liu, Q.; Leng, J.; Yan, D.; Zhang, D.; Wei, L.; Yu, A.; Zhao, R.; Zhang, H.; Chen, X. Digital twin-based designing of the configuration, motion, control, and optimization model of a flow-type smart manufacturing system. *J. Manuf. Syst.* **2021**, *58*, 52–64. [CrossRef]
23. Xia, K.; Sacco, C.; Kirkpatrick, M.; Saidy, C.; Nguyen, L.; Kircaliali, A.; Harik, R. A digital twin to train deep reinforcement learning agent for smart manufacturing plants: Environment, interfaces and intelligence. *J. Manuf. Syst.* **2021**, *58*, 210–230. [CrossRef]
24. Tao, F.; Zhang, Y.; Cheng, Y.; Ren, J.; Wang, D.; Qi, Q.; Li, P. Digital twin and blockchain enhanced smart manufacturing service collaboration and management. *J. Manuf. Syst.* **2022**, *62*, 903–914. [CrossRef]
25. Li, L.; Lei, B.; Mao, C. Digital twin in smart manufacturing. *J. Ind. Inf. Integr.* **2022**, *26*, 100289. [CrossRef]
26. Wang, P.; Luo, M. A digital twin-based big data virtual and real fusion learning reference framework supported by industrial internet towards smart manufacturing. *J. Manuf. Syst.* **2021**, *58*, 16–32. [CrossRef]
27. Luo, W.; Hu, T.; Ye, Y.; Zhang, C.; Wei, Y. A hybrid predictive maintenance approach for CNC machine tool driven by Digital Twin. *Robot. Comput. Integr. Manuf.* **2020**, *65*, 101974. [CrossRef]
28. Liu, C.; le Roux, L.; Korner, C.; Tabaste, O.; Lacan, F.; Bigot, S. Digital Twin-enabled Collaborative Data Management for Metal Additive Manufacturing Systems. *J. Manuf. Syst.* **2022**, *62*, 857–874. [CrossRef]
29. Ahola, T.; Laitinen, E.; Kujala, J.; Wikström, K. Purchasing strategies and value creation in industrial turnkey projects. *Int. J. Proj. Manag.* **2008**, *26*, 87–94. [CrossRef]
30. Brady, T.; Davies, A.; Gann, D.M. Creating value by delivering integrated solutions. *Int. J. Proj. Manag.* **2005**, *23*, 360–365. [CrossRef]
31. Gönnheimer, P.; Kimmig, A.; Ehrmann, C.; Schlechtendahl, J.; Güth, J.; Fleischer, J. Concept for the Configuration of Turnkey Production Systems. *Procedia CIRP* **2019**, *86*, 234–238. [CrossRef]
32. Adolphs, P.; Ulrich, E. *Reference Architecture Model Industrie 4.0 (RAMI4.0)*; VDI/VDE & ZVEI: Berlin, Germany, 2015.
33. Bangemann, T.; Bauer, C.; Bedenbender, H.; Diesner, M.; Epple, U.I.; Elmas, F.; Jens, F.; Grüner, S. *Industrie 4.0—Technical Assets—Basic Terminology Concepts, Life Cycles and Administration Models*; VDI/VDE & ZVEI: Berlin, Germany, 2016.
34. Dorst, W.; Glohr, C.; Hahn, T.; Knafla, F.; Loewen, U.; Rosen, R.; Schiemann, T.; Vollmar, F.; Winterhalter, C.; Diegner, B.; et al. *Implementation Strategy Industrie 4.0: Report on the Results of the Industrie 4.0 Platform*; Bitkom, VDMA & ZVEI: Berlin, Germany, 2016.
35. D'Agostino, N.; Annacondia, E.; Bentkus, A.; Bianchi, G.; Briant, J.; Heidel, R.; Hoffmeister, M.; Lamboley, P.; Lensi, R.; Rossi, G.; et al. *The Structure of the Administration Shell: Trilateral Perspectives from France, Italy and Germany*; Plattform Industrie 4.0 & Piano Nazionale Impresa 4.0; Ministry of Economy and Finances: Paris, France; Federal Ministry for Economic Affairs and Energy (BMWi): Berlin, Germany; Ministero dello Sviluppo Economico: Rome, Italy, 2018.
36. Bader, S.; Barnstedt, E.; Bedenbender, H.; Billmann, M.; Boss, B.; Braunmandl, A.; Clauer, E.; Deppe, T.; Diedrich, C.; Flubacher, B.; et al. *Details of the Asset Administration Shell: Part 1—The Exchange of Information between Partners in the Value Chain of Industrie 4.0 (Version 3.0RC01)*; Plattform Industrie 4.0; ZVEI: Berlin, Germany, 2020.
37. Bedenbender, H.; Billmann, M.; Epple, U.; Hadlich, T.; Hankel, M.; Heidel, R.; Hillermeier, O.; Hoffmeister, M.; Huhle, H.; Jochem, M.; et al. *White Paper: Examples of the Asset Administration Shell for Industrie 4.0 Components—Basic Part*; ZVEI: Berlin, Germany, 2017.
38. *DIN SPEC 16593-1*; Reference Model for Industrie 4.0 Service Architectures—Part 1: Basic Concepts of an Interaction-Based Architecture. Beuth: Berlin, Germany, 2018.
39. Bedenbender, H.; Bentkus, A.; Epple, U.; Hadlich, T.; Heidel, R.; Hillermeier, O.; Hoffmeister, M.; Huhle, H.; Kiele-Dunsche, M.; Koziolek, H.; et al. *Industrie 4.0 Plug-and-Produce for Adaptable Factories: Example Use Case Definition, Models, and Implementation*; Plattform Industrie 4.0; ZWEI: Berlin, Germany, 2017.
40. Billmann, M.; Boss, B.; Epple, U.I.; Garrels, K.; Hadlich, T.; Hankel, M.; Heidel, R.; Hillermeier, O.; Jochem, M.; Hoffmeister, M.; et al. *Which Criteria Do Industrie 4.0 Products Need to Fulfil? Guideline 2019*; Plattform Industrie 4.0; ZVEI: Berlin, Germany, 2019.
41. Angeli, C.; Boss, B.; Braunmandl, A.; Brost, G.; Claus, M.; Dressel, O.; Förder, T.; Garrels, K.; Hoffmeister, M.; Houdeau, D.; et al. *Plattform Industrie 4.0. Access Control for Industrie 4.0 Components for Application by Manufacturers, Operators and Integrators*; Federal Ministry for Economic Affairs and Energy (BMWi): Berlin, Germany, 2019.

machines

MDPI

Article

A Multi-Part Production Planning System for a Distributed Network of 3D Printers under the Context of Social Manufacturing

Inno Lorren Désir Makanda, Maolin Yang, Haoliang Shi, Wei Guo and Pingyu Jiang *

State Key Laboratory for Manufacturing Systems Engineering, Xi'an Jiaotong University, Xi'an 710049, China; innolorren@stu.xjtu.edu.cn (I.L.D.M.); maolin@xjtu.edu.cn (M.Y.); shihaoliang@stu.xjtu.edu.cn (H.S.); weiguo1128@xjtu.edu.cn (W.G.)
* Correspondence: pjiang@mail.xjtu.edu.cn

Abstract: Additive manufacturing (AM) systems are currently evolving into network-based models, where the distributed manufacturing resources from multiple enterprises are coordinated to complete product orders. The layer-by-layer approach of AM technologies gives manufacturers unprecedented freedom to create complex parts tailored to customer needs, but this comes at slow build rates. Consequently, for AM to become mainstream in the industry, challenges in production planning remain to be addressed to increase AM system productivity. This paper considers two practical problems encountered in AM systems, namely, production planning and part-to-printer assignment, and a series of heuristic algorithms are proposed to solve these problems. In particular, an approach for automatically determining part orientation, part-to-printer allocation, and nesting of multiple parts for a distributed network of fused filament fabrication three-dimensional printers is described to reduce the total production cost and time regarding the context of social manufacturing. The proposed method is implemented through a web application. The case study, using real-world parts and comparative analysis findings, indicated that the proposed method produces high-performance results.

Keywords: social manufacturing; additive manufacturing; production planning; distributed manufacturing systems; heuristic algorithms

Citation: Makanda, I.L.D.; Yang, M.; Shi, H.; Guo, W.; Jiang, P. A Multi-Part Production Planning System for a Distributed Network of 3D Printers under the Context of Social Manufacturing. *Machines* **2022**, *10*, 605. https://doi.org/10.3390/machines10080605

Academic Editor: Francisco J. G. Silva

Received: 11 June 2022
Accepted: 19 July 2022
Published: 24 July 2022

1. Introduction

Additive manufacturing (AM), regularly referred to as three-dimensional (3D) printing, is a set of technologies that manufacture objects from digital 3D models by layering material rather than subtracting it [1]. As a result, it enables the creation of customized and highly complex component geometries that can be difficult to manufacture using conventional processes. While the possibilities from a design perspective are endless, the production throughput speed of AM systems is slow, limiting their use for mass production [2]. In this regard, AM is still considered a niche technology for the rapid prototyping and production of small-batch products. Therefore, today's AM challenges revolve around enhancing productivity to meet industry demands for mass personalization.

Meanwhile, the latest advancements in AM combined with social networks are opening up a new world of possibilities for the manufacturing industry and fostering the development of new collaborative business models, such as social manufacturing (SocialMfg) [3,4]. A key factor behind the growth of SocialMfg is the ability of AM technologies to enable on-demand production and rapid innovation for small and medium-sized enterprises (SMEs), and even individual entrepreneurs [5,6]. In a nutshell, in SocialMfg, heterogeneous orders from multiple customers are coordinated to be completed by a distributed network of AM machines from several SMEs that interact and collaborate on a peer-to-peer (p2p) basis, as shown in Figure 1. In this setting, it becomes critical to investigate production planning approaches for a network of AM machines in which several 3D printers are shared to allow on-demand production for flexible and geographically distributed production units.

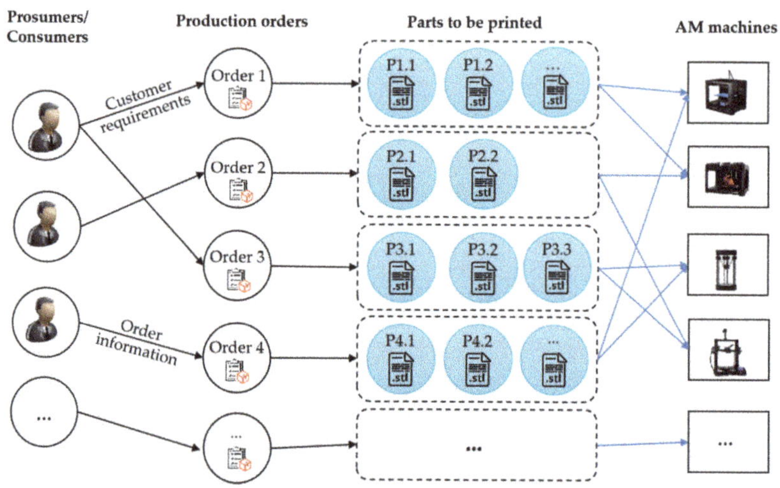

Figure 1. Concept model for AM production planning in SocialMfg.

In AM, production planning includes selecting part orientation, packing or nesting multiple parts together, setting process parameters, slicing, and assigning parts to machines. Each of these tasks requires a high degree of manual intervention, which affects the overall efficiency of the 3D printing process. Indeed, selecting the build orientation of a part remains a manual operation and relies on the knowledge and experience of the process planner. Furthermore, parts with similar production requirements are usually grouped by batch to meet order due dates [7]. Several parts must be nested in a build plate for each batch and printed simultaneously in a single run to reduce the production time and operating costs [8,9]. Manually arranging the placement of a group of parts with complex geometries is, however, time-consuming and can become a difficult task [10,11]. As a result, the automation of production planning looks highly promising to stimulate the productivity of AM systems, especially for a network-based organization involving shared 3D printers.

This study investigates the feasibility of an automated multi-part production planning system in the context of SocialMfg. The paper addresses, in particular, three research questions: how to optimally orient a part, how to allocate parts to AM machines, and how to optimally place multiple parts in build plates. A further question is whether this automated production planning system can quickly assign parts, using the few possible build plates, while ensuring part quality. Although these questions have been substantially investigated in the literature, no study has yielded an integrated solution. However, it is necessary to consider part orientation selection, part-to-printer assignment, and nesting, as a whole, to increase AM production rates and lower operating costs.

In this work, we focus on a prominent material extrusion technology called fused filament fabrication (FFF), one of the most widely used AM techniques. Initially adopted for rapid prototyping, FFF is gradually used for industrial applications to deliver rapid tooling and spare parts. This paper aims to enhance the FFF process via an industrial-like production logic by providing a realistic solution to production planning and scheduling problems. First, efficient heuristic algorithms are presented for optimizing part build orientation, part-to-printer assignment, and the nesting of parts. In fact, in the field of combinatorial optimization, heuristic algorithms have proven to be effective. Compared to other methods, they allow for satisfactory results with less computational effort. A web application (web app) is also developed in which the parts' stereolithography (STL) files are entered, and the packing result is returned. Lastly, the effectiveness and operability of the developed system are investigated using a case study with real-world components and comparative analysis.

The contribution of this work is threefold:

- As part of the SocialMfg, this is one of the first works in AM production planning and scheduling that includes part orientation selection, part-to-printer assignment, and two-dimensional nesting;
- Orientation-dependent factors that affect the part quality and printing time are considered, and the two-dimensional (2D) packing is solved using a rasterization technique and quadtree representation. Moreover, the proposed 2D nesting algorithm outperforms the top existing methods in terms of packing density and computational time;
- A real-world case study to evaluate the proposed method confirms the effectiveness of considering part orientation and nesting in AM machine scheduling.

The balance of this paper is arranged as follows: Section 2 summarizes the related work and indicates research gaps. Section 3 describes the scenario considered and details the proposed methodology. The prototype web app and a case study are reported in Section 4. In Section 5, the performance of the proposed method is discussed. Finally, Section 6 provides the conclusion to this investigation.

2. Literature Review

This section summarizes existing studies regarding orientation selection in AM and provides an overview of the most prevalent approaches for AM production planning and scheduling.

2.1. Part Orientation in AM

The problem of part orientation in AM has been thoroughly investigated [12]. Some researchers presented a one-step approach. Notably, they proposed a customized algorithm [13–15] or applied an existing optimization algorithm, most commonly an evolutionary algorithm, such as a genetic algorithm [16,17], to directly search an orientation, allowing several considered factors to be optimal from a large number of alternative orientations. The challenge with this one-step approach is to define the suitable rotation angle or step size. A large rotation angle reduces the number of computations and raises the risk of missing the real optimal build orientation. Contrarily, the risk is reduced when the rotation angle is slight, but the number of computations will increase considerably. This situation attracted several researchers to investigate the part orientation problem from other perspectives and propose a two-step approach. The initial phase is to generate some practical alternative orientations among all possible orientations. The second phase consists of identifying the best orientation among the generated options. A multi-criteria decision-making (MCDM) technique is usually used to select the best orientation [13,18,19].

Despite decades of research, automatic determination of optimal build orientation remains an open issue in AM. Compared to the vat photopolymerization (VPP) process, few attempts have been made to identify optimal part build orientation for the FFF process, where the staircase effect affects the surface quality more severely. Selecting the optimal build orientation should improve the print quality, regardless of the intended purpose of the part. Considering this background, this research proposed a two-step strategy and new orientation-dependent factors, inexpensive in computation, to address the optimal build orientation determination problem for the FFF technique.

2.2. Planning and Scheduling of AM Production

In the context of scheduling a set of unrelated parallel machines with non-identical job sizes, Li et al. [20] proposed heuristics based on best-fit and longest processing time (LPT) to reduce the makespan. Similarly, Arroyo and Leung [21] introduced several heuristics that allow the realization of varied-sized jobs with unrelated parallel machines, while minimizing the makespan and grouping jobs into batches. Ransikanbum et al. [22] adjusted the workload balance between FFF 3D printers to lower the parts' total cost and completion time. Moreover, Li et al. [23] investigated several techniques to manufacture multiple components in an industrial setting. They examined the costs and reaction times of centralized and decentralized manufacturing methods. Zhang et al. [24] addressed the part-to-printer assignment and part placement problems for the VPP process. They also developed a heuristic and genetic algorithm that can be used for the FFF technique. Cadiou et al. [25] recently suggested a framework based on reasoning algorithms to solve the on-demand production planning of non-identical AM parts in Fab labs made up of unrelated parallel FFF 3D printers.

In addition, several studies investigated the 3D nesting problem [8,26] to enhance AM machine utilization by reducing the overall build time and cost per machine. However, the literature review here focuses on prior research that has explored the 2D nesting problem. Indeed, this study focuses on the FFF process, where parts are placed in a single layer rather than stacked [10]. Approaches for nesting can be categorized as parallel or serial [27]. In parallel nesting, all parts are arranged simultaneously in the build plate, and the packing result is evaluated. Such approaches are described in [10,27]. However, in serial nesting, the parts are arranged one at a time, emulating how a human operator can place parts in a print bed. This strategy has been implemented in most earlier works [28–31] because it can decrease the computational cost while still providing satisfactory results.

A decisive step of the nesting process is the need to develop a geometric tool for handling the complexity of the parts to be packed. The simplest method uses the bounding box [28,29,32] of part projections on the printer bed. Using a bounding box to replace a part's actual shape can waste the workspaces of an AM machine when packing parts, although such an approach eases the problem of detecting collision between parts. There are several other solutions for part representation using polygons [10,27,30]. However, in the AM nesting process, the input is not polygons, but a set of triangular facets (i.e., STL files). The polygon union of all these facets can be computed using the no-fit-polygon (NFP) algorithm, or direct trigonometric method, to represent the shape of the parts. Nevertheless, as the number of facets in the STL model increases drastically, such an operation becomes computationally complex and time-consuming [30].

Furthermore, a critical problem with NFP is that as the number of parts to place increases, or as pieces rotate, the NFP computation becomes too large to be practical. Another possible solution for dealing with the geometry of parts in 2D nesting problems is the pixel/raster method [31]. This technique can generate a nearly exact part shape, regardless of its geometrical complexity, in a short time. Although the computation cost can increase rapidly with increasing resolutions to obtain a more accurate raster representation of the STL files, high resolution is not usually required [31,33].

Another methodological concern of nesting for AM is the rotation of parts in the print bed to find the best possible position. In many previous works, no rotation of the parts is allowed [26,29,30]. On the other hand, some studies considered the rotation of parts on the XY plane by 90 degrees increment [8,28]. In such an approach, the part is rotated before placement to find an optimal positioning. It should also be emphasized that most current nesting methods rely on genetic algorithms to optimize the part placement strategy. Although this algorithm can achieve reasonably good solutions, it has a high computational burden.

In synthesis, in the past, methods and techniques for planning and scheduling AM production primarily focused on the part-to-printer assignment, or the nesting of several components in a single printing run. Concerning the nesting problem, despite the considerable number of solutions that have been proposed, a closer inspection of the literature reveals several gaps and shortcomings. On the one hand, little attention has been given to 2D nesting for the FFF process. Previous studies have almost exclusively focused on the VPP and laser powder bed fusion (LPBF) processes. Without considering the characteristics and constraints of the FFF technique, generating part layout based on part geometry is ineffective.

On the other hand, approaches based on genetic algorithms used to determine part layout are practical, but they appear to be computationally expensive [31]. Moreover, many studies are simplified using bounding boxes or rectangular shapes in the packing process, resulting in a waste of workspace of AM machines. Consequently, this study proposes a new nesting approach based on pixel/raster representation of the STL models, which considers the FFF constraints.

3. Methods

In SocialMfg, orders from several customers can be allocated to a network of AM machines. Each order can include multiple parts. After receiving the orders, planners compose batches of parts to be assigned to AM machines, according to customer specifications and different planning objectives (e.g., reducing operating cost and processing time). Concerning a particular printer, multiple parts composing a batch need to be optimally placed to use the printer efficiently. However, when placing several parts in a print bed, the space utilization ratio must be maximized to lower the production time and cost, and the quality of the parts must be guaranteed [10,27]. In addition, with the FFF process, producing multiple parts in a single run does not necessarily improve productivity if part orientation is not optimal. Part orientation, nesting, and part-to-printer assignment are all significant considerations in this situation. As a result, the methodology presented here begins with a group of heterogeneous parts assigned to one or more FFF 3D printers. First, the optimal build orientation of each part is identified. Afterward, the optimally oriented parts are taken as an input set and are assigned to the FFF 3D printers. Finally, parts are optimally placed in build plates according to two objectives: to maximize the print bed utilization and ensure the quality of the parts. Figure 2 shows a graphic depiction of this approach.

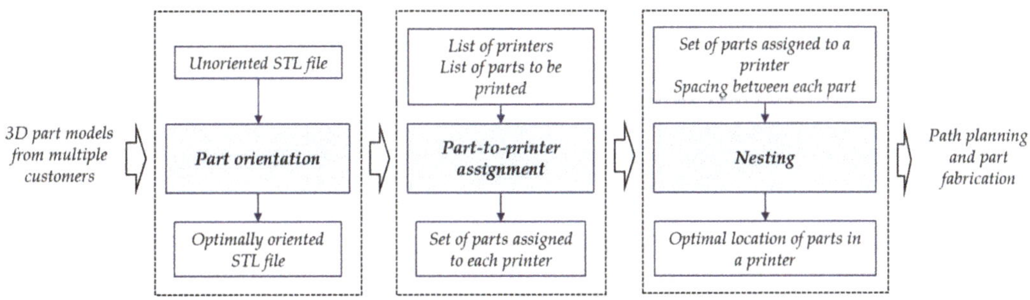

Figure 2. The framework of the proposed approach.

The following assumptions have been formulated to limit the production planning problems to our research objectives.

- The parts to be oriented and nested are assigned to a FFF 3D printer and have identical manufacturing requirements. In other words, the service provider and service demander agreed on the parts' materials, production deadline, and characteristics.
- The optimal build orientation of each part is selected before the nesting process. Nevertheless, the rotation of the parts along the vertical axis is allowed because of its relatively small impact on the quality of printed parts [22,30].

It is essential to clarify that AM service providers receive diverse printing tasks from multiple customers. Thus, some parts can be pre-oriented adequately by customers. Accordingly, this paper explores the optimal build orientation determination for a single part instead of orientation optimization for several parts in a build, as in the work of [10,27].

The optimization strategy adopted here can be segmented into three consecutive stages. The first stage involves optimizing each part's build orientation separately to improve the production quality. Then, the second stage aims to assign the parts to FFF 3D printers to minimize the makespan, ensure load balance among the printers, and efficiently use the manufacturing resources. The last stage involves placing the parts on build plates in the most compact way to achieve higher productivity with the FFF 3D printers. The following paragraphs of this section detail the heuristic algorithms proposed to solve these combinatorial optimization problems.

3.1. Heuristic Algorithm-Based Part Orientation Selection

In theory, a part can be built in many possible orientations. By orienting the part in different directions, there is a significant difference in part quality. Therefore, optimizing part orientation is crucial in AM. For many industrial applications, the surface quality of the part is critical. Good surface quality is more appealing and gives greater dimensional accuracy necessary for the correct functioning of assembly parts. In actual experiments, it is found that the key to achieving good quality printed parts with the FFF process is to prevent overhangs. Indeed, the overhanging surfaces of an STL model are difficult to print; they usually require additional support structures and print poorly.

Nevertheless, there are admissible overhangs. These are typically overhanging surfaces with angles between the normal vector and the build direction lower than 135°. Therefore, it is preferable to differentiate the support contact area [34,35] (i.e., the overhang that requires support structures) from other tolerable overhanging surfaces. Intuitively, minimizing the support contact area would reduce the need for support structures, thus leading to better surface quality and decreasing the post-processing time. As a result, build orientation optimization is the determination of a desirable orientation that minimizes the support contact area. It involves generating a set of alternative build orientations (ABOs) and determining the best build orientation from the ABOs generated.

Another way to enhance the surface quality of the parts is to maximize the area of the non-stepped surfaces [36,37]. It is possible to achieve this by maximizing the bottom surface area (i.e., the area of the part in contact with the build plate). There are at least two reasons to consider optimizing this criterion. First, surfaces horizontal to the build plate do not have a stair-stepping effect, guaranteeing a good surface quality. Second, using a large area as the bottom face can drastically decrease the number of layers and the build time. It is also necessary to consider the perimeter of this bottom face because for STL models with a spherical shape, for example, the bottom face could be a circle without any resulting area. Accordingly, in the present work, three orientation-dependent factors are considered when searching for the optimal build orientation: support contact area, bottom surface area, and bottom surface perimeter. These can be directly computed using geometrical analysis of the STL model [34–36]. These factors are weighted and summed to calculate the degree quality for each ABO.

The optimal build orientation determination procedure is formalized with pseudocode, as shown in Algorithm 1. The algorithm requires an STL file as input and produces an STL file optimally oriented as output. The appropriate weights to assign to each orientation-dependent factor must be found because the optimal build orientation depends strongly on the chosen weight values.

Algorithm 1: Pseudocode for the orientation optimization algorithm

```
Input: Unoriented STL file
Output: Optimally oriented STL file
Generate the alternative build orientations of the part (Oi)
Compute the orientation-dependent factors (F_j) of the i-th alternative orientation
for each value of the j-th factor of the i-th alternative orientation (x_ij) do
    Convert each orientation-dependent factor value into a number in [0, 1] as follows:
        α_{i,j} = x_ij / sqrt(Σ_{i=1}^{m} x_ij^2)
    Normalize the converted results using
        X_{i,j} = { α_{i,j}       if F_j is a positive factor
                  { 1 − α_{i,j}   if F_j is a positive factor
end for
Assign the weights to the considered factors
Compute a summary value of factors for each alternative orientation
Determine the optimal build orientation for the part
return the optimally oriented STL file
```

3.2. *Heuristic Algorithm-Based Part-to-Printer Assignment*

As stated in Section 1, this paper investigates the production planning of a network of dispersed AM machines to meet the low-quantity needs of multiple customers. The network of AM machines can be made up of unrelated parallel machines that can accomplish the same task but have different capacities. The entire available build volume determines the capacity of an AM machine. The problem is allocating parts from multiple customers to a network of AM machines, with varying operating costs and speed characteristics, while minimizing the total production cost and time.

According to the customer orders, there is a set of parts with various sizes, heights, and geometries. The orders are allocated to AM machines on a part-by-part basis, considering the part sizes and heights, and then divided into batches or printing jobs. In fact, due to the limited build volume of AM machines, some parts may not be able to be manufactured on all available machines. For example, a part higher than the maximum height supported by a machine cannot be assigned to that machine.

When given a set of parts, each with a different build time and several available AM machines, the part-to-printer assignment involves allocating all parts to machines with the goal of optimizing the workloads. As the SocialMfg can receive multiple print jobs from several customers and a network of various AM machines, production scheduling decisions must be relatively quick to effectively schedule these jobs for all the distributed 3D printers. Therefore, this study proposes an optimized LPT algorithm [38] to solve the part-to-printer assignment due to its simplicity and practical effectiveness. Indeed, LPT has good adaptability to a variety of scheduling tasks. This algorithm sorts parts in non-ascending order of their build time and then assigns them to the machine with the earliest end time. In particular, it entails attributing the parts one at a time to the FFF 3D printers to minimize the processing times of the machines. The LPT rule schedules the longest tasks first so that no large task overruns at the end of the schedule, significantly extending the execution time of the last job. The total manufacturing times of all the parts assigned to a machine is its completion time. The printing time is estimated after the parts are oriented in their optimal build direction. The heuristic procedure for part-to-printer assignment is realized through Algorithm 2.

Algorithm 2: Heuristic procedure for the part-to-printer assignment

Input: Set of AM machines M, set of oriented parts to be printed P
Output: Batch of parts allocated to each machine

Estimate the build time of all the parts
Sort the parts in decreasing order of build time
while P is not empty **do**
    Let P_k be the next part in the sorted order
    if P_k can be scheduled **then**
        Assign the part to the machine which has the minimum total processing time
        Remove the assigned part from the list of parts
    end if
end while
return parts allocated to AM machines

3.3. Heuristic Algorithm-Based Nesting

The FFF technique allows the printing of multiple parts simultaneously when they fit the build plate. Nesting optimizes the process of laying out parts on the print bed of an AM machine. The gain is the reduction in the total processing time to build and arrange the parts. Manually finding the position of multiple STL files on the print bed can take exponential time. However, the nesting or part packing in AM should be relatively fast because, unlike in subtractive manufacturing, the only gain is production time (i.e., no waste or material). Thus, the time spent nesting parts becomes proportional to the build time.

The proposed nesting method considers the rotation of the parts along the vertical axis (i.e., build direction), in 90° increments, to find a better positioning, which can improve the packing solution in terms of compactness. Figure 3 shows the flowchart of the nesting procedure. First, data are initialized to specify the spacing between each part and the printing space dimensions. Next, a rasterization algorithm transforms the STL models into raster images. The main iteration of the nesting strategy repeats until all parts have been placed. Within the space available in the print bed, parts are sequentially placed as far as feasible to the bottom, then, as far as possible to the left. The available space dimensions must be larger than part dimensions in order to arrange a part within the packing area. It follows that if this requirement is not met, the current packing area is abandoned, and another is tried until all available spaces have been explored. When the placement of the part is feasible, the selected slot is removed from the list of available space, and the part is removed from the part list. The next part is chosen once the previous one has been placed, and the same packing procedure is repeated. When all the parts in the input list are placed, the optimized part layout is generated as an STL file. The rasterization procedure and placement strategy are further discussed in the following subsections.

3.3.1. Raster Method

Here, the rasterization is the task of converting the STL file into a 2D bitmap. The simplest form of a bitmap, a black and white image, is used; all white pixels are represented as 0, meaning void space, and all black pixels are represented as 1, which means the presence of a part. Rasterization aims to divide the irregular shapes of STL files into discrete areas, which thus reduces the geometric information. The basic idea is that a blank scene whose size corresponds to the print bed is created at the beginning. The facets of the STL model are projected onto the scene as 2D triangles. Once a part has been placed in the scene, it is updated. In the first step of the rasterization process, each triangular facet in the STL model is transformed from 3D to 2D space using projection matrices. The second step is to fill all the pixels in the region covered by the triangles. These two steps are illustrated in Figure 4. Finding out which pixels are covered by the resulting projected profiles, i.e., which pixel should be 0 or 1, is realized using the D-function [33].

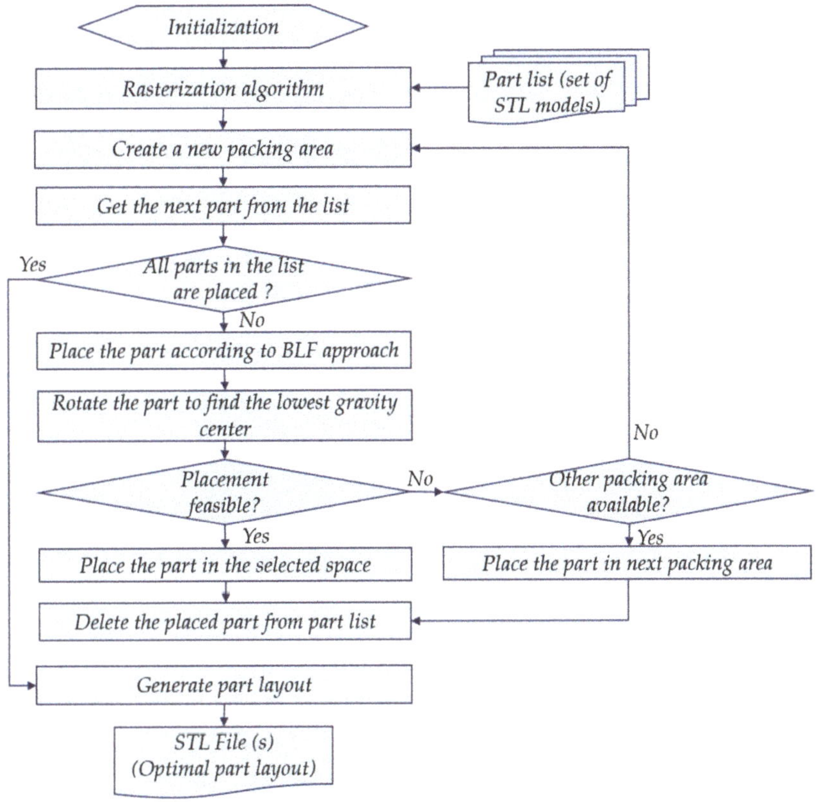

Figure 3. Flowchart of the 2D nesting algorithm.

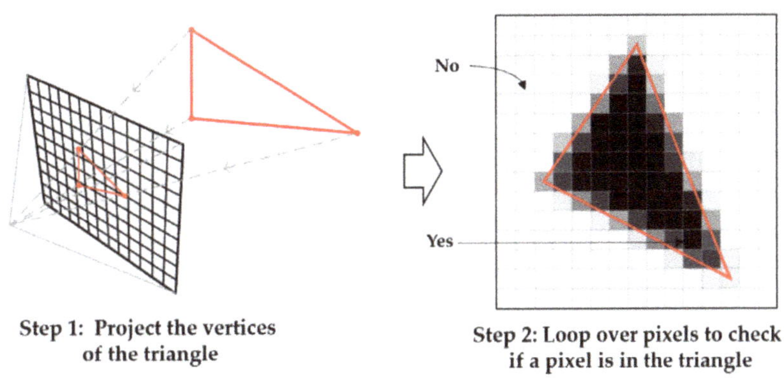

Figure 4. Two-step procedure of the raster method.

Part projection profiles are represented as region quadtrees [39] to reduce the complexity and computation time. Moreover, 2D pixel dilation is used to handle the part spacing. This is an operation to enlarge the boundaries of the pixels. The raster method provides a simplistic and relatively inexpensive approach for processing the geometric representation of the STL files. In addition, it makes the test of possible placement trivial because we must simply check that all the pixels are located where the part will be placed.

3.3.2. Improved Bottom-Left-Fill Algorithm

In this paper, the proposed placement heuristic, called improved bottom-left-fill (I-BLF), follows the BLF strategy [40], with a slight variation, using the lowest gravity center of the part. The algorithm will prefer the orientation where the center of mass of the part is as close as possible to the lower left of the build plate. For several parts, experiments have shown that placing the center of mass as close as possible to the lower-left leaves a maximum amount of space on the top right. The parts are placed using an exhaustive search, as the algorithm tries every possible orientation (relative to the vertical axis) for the parts. The center of mass of the part is calculated using the average of the coordinates of the pixels covered by the raster image. The placement algorithm is described in Algorithm 3. The sequence order of the parts is significant and can affect the solution obtained because the algorithm is greedy. Sorting parts from the largest to the smallest is the best approach; from the authors' knowledge, it always leads to better results than sorting the smallest to the largest area.

The packing strategy begins with a print bed of fixed dimensions and a set of STL files. The part is placed according to the I-BLF strategy for each STL file. The packing position of the chosen part will be checked to see if it is viable. The algorithm performs two checks: first, it ensures that the part does not penetrate the boundary of the print bed. Last, it is verified that the part does not overlap with previously packed parts. If part *i* fits in position *j*, the part position is set as *j*, and the void space of the layout is updated by deleting position *j*. If part *i* cannot be packed into all *j*-th positions of the current packing area, the next available packing area is considered. The packing operation ends once all the parts on the list have been placed, and an STL file representing the nesting layout is generated.

Algorithm 3: Pseudocode for the improved-BLF packing algorithm

```
Input: Set of STL files
Output: STL file (s) of the placement layout
Initialize print bed size and part spacing
repeat
    for each part (Pi) of the set do
        Place the part according to the BLF policy
        Rotate the part to find the lowest gravity center
        for the j-th position of the part Pi do
            if the part can be packed into j-th position then
                Set position j as the packed position of part Pi
                Delete part Pi from the part list
                Update the available space in the layout
            end if
        end for
    end for
until the part list is empty
return placement layout as STL file
```

4. Prototype Implementation and Case Study

In this section, the proposed approach is encapsulated into a web app to illustrate its effectiveness in solving the AM production planning and scheduling problems of multiple parts with a network of FFF 3D printers, and a case study is conducted.

4.1. Prototype Implementation

The web app is based on a browser/server architecture and consists of three main modules: a user interface embedded in a web browser, an autorotation module to find the best build orientation of the parts, and a nesting module to find out the optimal part layout. HTML, JavaScript, CSS, and other programming languages are used to build the web app. In particular, the orientation algorithm introduced in our previous work [41] is adopted, as its reasoning capabilities align with the technique described in Section 3.1. The LPT rule presented in Algorithm 2 has been implemented in Python language. Lastly, the 2D nesting algorithm detailed in Figure 3 has been coded in C++. All the algorithms were tested on a computer with a 3 GHz Intel Core i5-9500 processor and 8 GB RAM.

The web app operation is as follows. First, the operator/user imports a set of STL files. The user interface serves as a visualization environment for manipulating STL models and an interface to the production planning and scheduling services. Second, if necessary, the user can invoke the autorotation function to orient the parts in optimal build orientation. The optimally oriented STL model is returned to the user. Then, the STL files can be assigned to the available FFF 3D printers following the LPT rule. Finally, the user can invoke the nesting algorithm to generate an optimal layout from the STL files. For this last step, the user needs to input the quantity of each part and the print bed size. Each build plate generated by the algorithm is returned to the user as an STL file.

4.2. Case Study

The operability and overall workflow of the web app are illustrated through a case study. A set of 19 non-identical parts (named 1 to 19 in Figure 5), whose shapes reflect actual part geometries, are introduced. Table 1 sums up the information of these parts. The aim is to optimize the STL files in terms of print direction before assigning them to FFF 3D printers and to compactly place them in build plates for fabrication. We assume that all the parts have similar manufacturing requirements and priorities.

Regarding the FFF 3D printers, we used a JG Aurora Z-603S-C and a JG Aurora Z-603S (hereafter Machine 1 and Machine 2, respectively) available in our lab; the first machine has a build volume of 300 mm × 200 mm × 200 mm, while the second has a build area of 280 mm × 180 mm and 180 mm in height. However, in most desktop FFF 3D printers, the parts cannot occupy the entire print bed area. In practice, the extruder cannot travel to the edges of the build plate to precisely deposit a layer. Therefore, to ensure an efficient printing process, we have defined a packing area of 280 mm × 180 mm for Machine 1, and 260 mm × 160 mm for Machine 2, instead of the full print-bed size. Furthermore, the spacing between each part is fixed at 1.5 mm to avoid overlapping due to the accuracy error of the 3D printer and to prevent stringing caused by excess filament oozing from the nozzle.

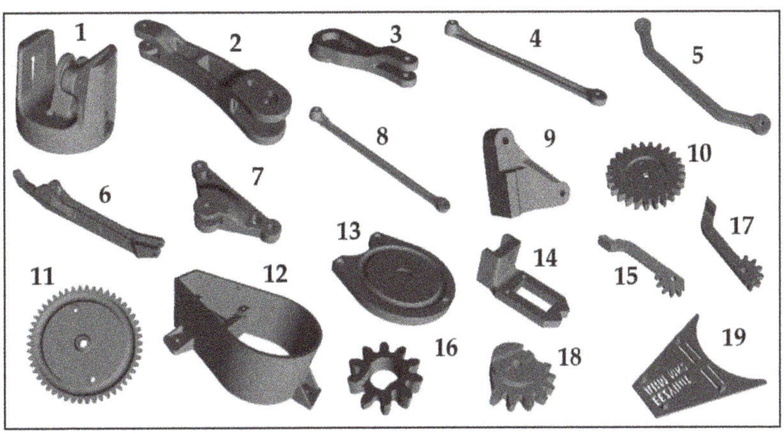

Figure 5. Group of parts considered for production planning. The numbers 1 to 19 represent the part IDs.

Table 1. Sample data related to the parts used for the case study.

ID	Part Name	Build Time (h)	Length (mm)	Width (mm)	Height (mm)	Projection Area (mm^2)
1	EBAmk_001_base	5.25	80	85.968	71.18	5726.149
2	EBAmk_002_mainarm	3.75	36.792	160.996	33	5171.117
3	EBAmk_003_varm	0.78	78	29.991	16	1673.323
4	EBAmk_004_link135	0.62	12.998	148	10.50	1351.975
5	EBAmk_005_link135angled	0.68	27.5	150	5.50	1623.272
6	EBAmk_006_horarm	3.63	28	214.224	40.854	4565.190
7	EBAmk_007_trialink	1.06	83.994	40.260	20	2025.812
8	EBAmk_008_link147_new	0.65	12.998	160	10.50	1459.943
9	EBAmk_009_trialinkfront	0.88	40.888	22	45.692	839.428
10	EBAmk_010_gearservo	0.43	40.474	40.394	5	1074.972
11	EBAmk_011_gearmast_full	1.77	77.910	77.991	11	4349.470
12	EBAmk_012_mainbase	6.82	160.187	89.525	55	3690.305
13	EBAmk_013_lower_base	2.1	85.8	81.567	8	5524.398
14	EBAmk_014_claw_base	0.62	54	25	25	1002.040
15	EBAmk_015_claw_finger_dx	0.18	57.576	14	9	417.015
16	EBAmk_016_claw_gear_drive	0.08	17.371	18	5	125.522
17	EBAmk_017_claw_finger_sx	0.2	58.271	21.511	9	418.462
18	EBAmk_018_claw_gear_driven	0.13	25.310	20.959	7	293.427
19	EBAmk_019_drive_cover	0.85	76.277	78.062	8	2609.353

The web app operation for the case study example is shown in Figure 6. First, the parts are uploaded to the web app user interface. The visualization window allows for the displaying of the selected STL file. Then, the parts are oriented in their optimal build orientation using the autorotation function. The web app provides an estimated build time for each STL file, which is necessary for the part-to-printer assignment step. The dimensions of the parts supplied in Table 1 show that none of the parts is higher than the maximum height capacity of Machine 1 and Machine 2. As a result, all the parts can be assigned to both machines. The parts are selected individually and assigned to AM machines following the LPT rule in Section 3.2. Table 2 lists the results of the part-to-printer assignment.

The packing area of each printer is specified to generate the optimal layout for printing the parts. The batch of parts allocated to Machines 1 and Machine 2 can be produced in a single run, as shown in Figure 7. In particular, the nesting result of Machine 1 (see Figure 7a) exhibits the advantage of the I-BLF strategy, which allows small parts to fill up the void area left by the previously placed large parts. In addition, the large parts are present in the two machines; this helps to balance the workload and reduce the total processing time. As all components have the same priority, printing operations must be completed concurrently to minimize production delays and maintain consistent operating costs.

Table 2. Allocation of the parts.

Machine	Scheduled Part	Total Build Time (h)
Machine 1	12, 2, 13, 7, 19, 5, 4, 10, 15	14.58
Machine 2	1, 6, 11, 9, 3, 8, 14, 17, 18, 16	14.25

Figure 6. A snapshot of the web app operation.

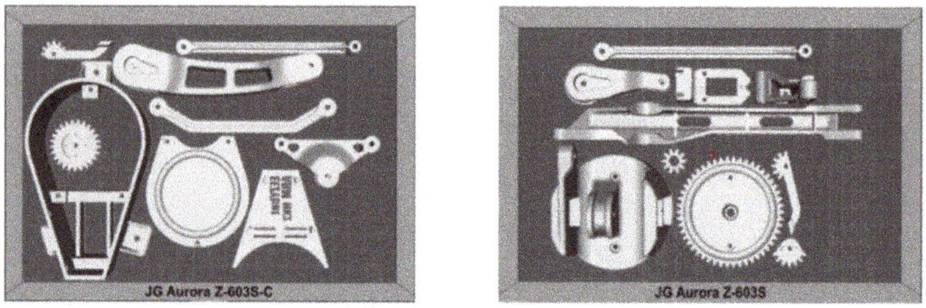

(a) (b)

Figure 7. Results of the part-to-printer assignment with nesting: (**a**) Machine 1; (**b**) Machine 2.

4.3. Performance Analysis

The ratio of the total projection area of all nested parts to the area of the print bed, called printing area covered (PAC) [30], must be as great as possible.

$$PAC = \frac{\sum_{i=1}^{n} \text{Projection\_Area\_of\_the\_ith\_part}}{\text{Printing\_Area}}$$

Since the parts in AM have irregular shapes and freeform surfaces, it is challenging to guarantee a high PAC value. The defined spacing between the parts also reduces the overall PAC value. Therefore, the performance of the nesting algorithm may vary in different situations and is heavily influenced by the geometry of the parts. As the objective of this paper is to investigate the feasibility of an automated production planning system, the benchmark criteria for evaluating the efficiency of the proposed method are the number of build plates generated and the computation time. The nesting results for the case study example are calculated in tens of seconds, as shown in Table 3, indicating that the proposed method exhibits satisfactory performance. The computation time includes rasterization and part placement.

Table 3. Computation results of the case study.

Machine	No. of Parts	Computation Time	Number of Build Plate
Machine 1	9	4.65 s	1
Machine 2	10	2.47 s	1

5. Discussion

5.1. Comparison of the Proposed Nesting Approach to Existing Methods

A comparative study is conducted to illustrate and validate the efficacy of the proposed nesting method with other existing methods reported by the authors of [30,31], and the results are listed in Tables 4 and 5, respectively.

Table 4. Comparison between the LBDB and NFP method in the study in [30] and the proposed approach.

	LBDB and NFP in [30]	Proposed Method
No. of parts packed	19	20
Printing area covered (%)	78	83.4
Total computation time (s)	2600	12.77

Table 5. Packing density comparison between the method in the study in [31] and the proposed method.

Printing Space	No. of Parts Packed in [31]	Printing Area Covered in [31]	No. of Parts Packed Using the Proposed Method	Printing Area Covered Using the Proposed Method
1	20	81.2%	25	97.5%
2	13	65.2%	8	46%

When the packing results of the approaches detailed in [30,31] and the proposed method are compared, the proposed method outperforms the others in terms of packing efficiency and computing time. Table 4 shows that better nesting results for the test case in [30] are obtained using the method proposed in this paper. All the models to be nested are present in the generated nesting layout (see Figure 8), unlike the layout arrangement obtained using the left-border-down-border (LBDB) heuristic and NFP placement strategy detailed in [30], where a part was not present. Moreover, as shown in Table 5, approximately 97.5% of the first print bed is used compared to 81.2% in the scene-driven method [31], resulting in a 16.3% increase in PAC value. The proposed method packed 25 parts (see Figure 9), while the approach reported in [31] only packed 20 parts in the same printing area. These performances are mainly due to the ability of the I-BLF algorithm to fill the void spaces surrounded by parts already placed, which allows for increased compactness.

Comparing this study with two existing methods has demonstrated that the proposed method can provide viable nesting solutions, with maximum print bed utilization ratio and high compactness, considerably quickly. Thus, the process planner can benefit from the time saved, to a certain extent. In the SocialMfg environment, this advantage makes it possible to process many parts quickly. In addition, with such a system, service providers can quickly and easily adjust production plans when additional parts are introduced into the initial batch.

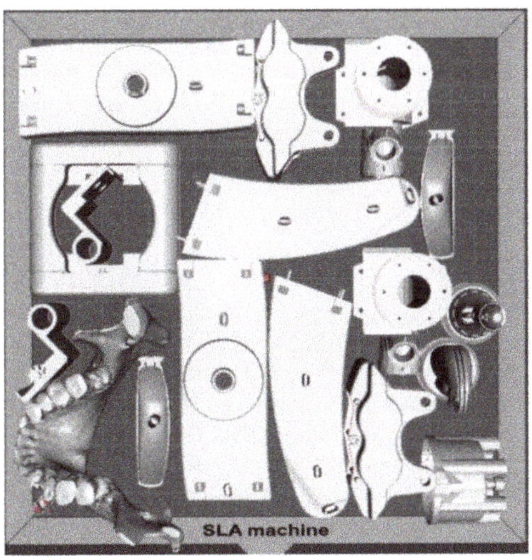

Figure 8. Nesting layouts for the test case data from [30] using the proposed method.

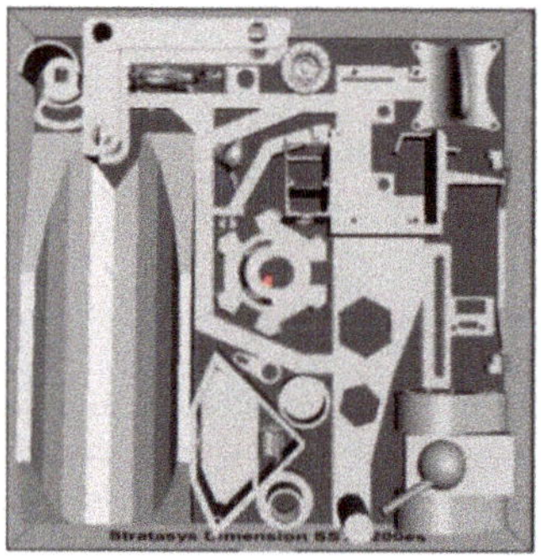

(a)

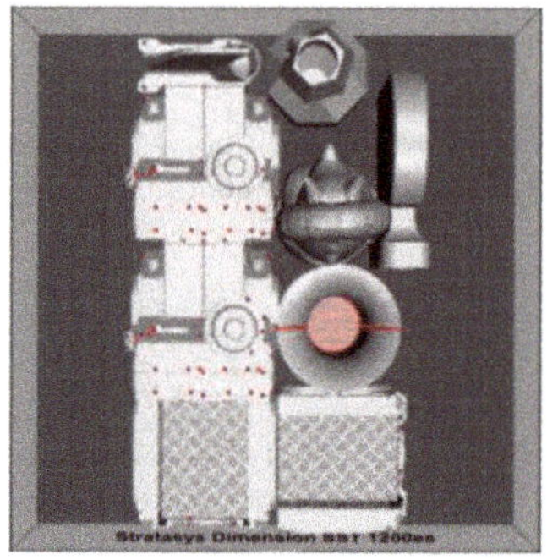

(b)

Figure 9. Nesting layouts for the dataset from [31] using the proposed method: (**a**) The first batch; (**b**) the second batch.

5.2. Multi-Parts Production in SocialMfg Scenario

Recently, SocialMfg has gained significant interest from researchers and manufacturers. It provides a network-based production approach in which distributed service consumers and providers can interact and share resources. In the SocialMfg environment, several print jobs from multiple customers may need to be assigned to a network of FFF 3D printers. In such a scenario, there is a need for software solutions to allocate parts to printers rapidly. The work presented in this paper corresponds in this direction. Indeed, the proposed approach can improve the production rates of FFF 3D printers by packing multiple parts into a single print bed to ensure the efficient use of AM resources.

Some of the assumptions made in this study must be discussed to provide new prospects. In particular, negotiation, order allocation, and resource sharing techniques between multiple service providers have not been studied. However, investigating these mechanisms will respond to major scientific concerns, which will help advance the SocialMfg paradigm. Indeed, in a network of FFF 3D printers, order splitting, and resource sharing might be considered to be a strategic measure to balance production and meet production deadlines. Transferring some workloads from one service provider to another may reduce the maximum processing time. Nevertheless, it should be noted that decomposing a print job into sub-jobs to balance production in AM may not lead to a considerable advantage. Indeed, splitting a job entails dividing the part into numerous bodies, each of which will require an assembly afterward.

6. Conclusions

This paper introduced a multi-part production planning system for a distributed network of FFF 3D printers. Part orientation selection, part-to-printer assignment, and nesting of the parts are carried out successively to minimize the total production time and cost while ensuring part quality. A comparative study verified the performance of the proposed method, and a real-world case study demonstrated its feasibility and added value. The findings of this study indicate that the proposed nesting algorithm, which is based on a rasterization technique and an I-BLF algorithm, exhibits good performance in most cases. Furthermore, with its robustness and relatively low computation time, the proposed method can be employed in a SocialMfg context where multiple parts must be efficiently assigned to distributed AM machines.

Despite its effectiveness, the proposed approach can still be extended in many ways. For instance, this study only considers 2D nesting because of the FFF process, while many AM technologies allow pieces to be stacked vertically. As a result, future investigations could be conducted to provide effective packing solutions for LPBF processes where parts can be stacked on top of each other. Besides, as heuristic algorithms for scheduling can lack generality, future studies could overcome this issue by training a reinforcement learning model or a neural network capable of generating highly efficient AM machine scheduling solutions.

Author Contributions: Conceptualization, I.L.D.M.; funding acquisition, P.J.; methodology, I.L.D.M.; resources, W.G.; supervision, P.J.; visualization, I.L.D.M. and H.S.; writing—original draft, I.L.D.M. and H.S.; writing—review and editing, M.Y. All authors have read and agreed to the published version of the manuscript.

Funding: The research work was funded by the National Natural Science Foundation of China, grant number 51975464.

Data Availability Statement: Not applicable.

Acknowledgments: The authors are grateful to Yuanbin Wang, from the State Key Laboratory of Fluid Power and Mechatronic Systems of Zhejiang University in China, and Vassilios Canellidis, from the Laboratory of Advanced Manufacturing Technologies and Testing of the University of Piraeus in Greece, for providing the STL files.

Conflicts of Interest: The authors declare no conflict of interest.

References

1. *F2792-12a;* Standard Terminology for Additive Manufacturing Technologies. ASTM International: West Conshohocken, PA, USA, 2012.
2. Thomas-Seale, L.E.J.; Kirkman-Brown, J.C.; Attallah, M.M.; Espino, D.M.; Shepherd, D.E.T. The barriers to the progression of additive manufacture: Perspectives from UK industry. *Int. J. Prod. Econ.* **2018**, *198*, 104–118. [CrossRef]
3. Jiang, P.; Ding, K.; Leng, J. Towards a cyber-physical-social-connected and service-oriented manufacturing paradigm: Social Manufacturing. *Manuf. Lett.* **2016**, *7*, 15–21. [CrossRef]
4. Guo, W.; Li, P.; Yang, M.; Liu, J.; Jiang, P. Social Manufacturing: What are its key fundamentals? *IFAC-Pap. OnLine* **2020**, *53*, 65–70. [CrossRef]
5. Hamalainen, M.; Karjalainen, J. Social manufacturing: When the maker movement meets interfirm production networks. *Bus Horiz.* **2017**, *60*, 795–805. [CrossRef]
6. Markillie, P. A Third Industrial Revolution: Collaborative Manufacturing. The Economist. 2012. Available online: http://web.mit.edu/pie/news/Economist.png (accessed on 19 April 2022).
7. Zhang, Y.C.; Bernard, A. Grouping parts for multiple parts production in Additive Manufacturing. Variety Management in Manufacturing. *Procedia CIRP* **2014**, *17*, 308–313. [CrossRef]
8. Gogate, A.S.; Pande, S.S. Intelligent layout planning for rapid prototyping. *Int. J. Prod. Res.* **2008**, *46*, 5607–5631. [CrossRef]
9. Manogharan, G.; Wysk, R.A.; Harrysson, O.L.A. Additive manufacturing-integrated hybrid manufacturing and subtractive processes: Economic model and analysis. *Int. J. Comput. Integr. Manuf.* **2016**, *29*, 473–488. [CrossRef]
10. Zhang, Y.C.; Bernard, A.; Harik, R.; Fadel, G. A new method for single-layer-part nesting in additive manufacturing. *Rapid Prototyp. J.* **2018**, *24*, 840–854. [CrossRef]
11. Zhang, X.; Zhou, B.; Zeng, Y.; Gu, P. Model layout optimization for solid ground curing rapid prototyping processes. *Robot. Comput.-Integr. Manuf.* **2002**, *18*, 41–51. [CrossRef]
12. Di Angelo, L.; Di Stefano, P.; Guardiani, E. Search for the Optimal Build Direction in Additive Manufacturing Technologies: A Review. *J. Manuf. Mater. Process.* **2020**, *4*, 71. [CrossRef]
13. Byun, H.-S.; Lee, K.H. Determination of the optimal build direction for different rapid prototyping processes using multi-criterion decision making. *Robot. Comput.-Integr. Manuf.* **2006**, *22*, 69–80. [CrossRef]
14. Lan, P.T.; Chou, S.Y.; Chen, L.L.; Gemmill, D. Determining fabrication orientations for rapid prototyping with stereolithography apparatus. *Comput.-Aided Des.* **1997**, *29*, 53–62. [CrossRef]
15. Cheng, W.; Fuh, J.Y.H.; Nee, A.Y.C.; Wong, Y.S.; Loh, H.T.; Miyazawa, T. Multi-objective optimization of part-building orientation in stereolithography. *Rapid Prototyp. J.* **1995**, *1*, 12–23. [CrossRef]
16. Masood, S.H.; Rattanawong, W.; Iovenitti, P. A generic algorithm for a best part orientation system for complex parts in rapid prototyping. *J. Mater. Process. Technol.* **2003**, *139*, 110–116. [CrossRef]
17. Zhang, J.; Li, Y. A unit sphere discretization and search approach to optimize building direction with minimized volumetric error for rapid prototyping. *Int. J. Adv. Manuf. Technol.* **2013**, *67*, 733–743. [CrossRef]
18. Qin, Y.C.; Qi, Q.F.; Shi, P.Z.; Scott, P.J.; Jiang, X.Q. Automatic determination of part build orientation for laser powder bed fusion. *Virtual Phys. Prototyp.* **2021**, *16*, 29–49. [CrossRef]
19. Zhang, Y.C.; Bernard, A.; Gupta, R.K.; Harik, R. Feature based building orientation optimization for additive manufacturing. *Rapid Prototyp. J.* **2016**, *22*, 358–376. [CrossRef]
20. Li, X.L.; Huang, Y.L.; Tan, Q.; Chen, H.P. Scheduling unrelated parallel batch processing machines with non-identical job sizes. *Comput. Oper. Res.* **2013**, *40*, 2983–2990. [CrossRef]
21. Arroyo, J.E.C.; Leung, J.Y.T. Scheduling unrelated parallel batch processing machines with non-identical job sizes and unequal ready times. *Comput. Oper. Res.* **2017**, *78*, 117–128. [CrossRef]

22. Ransikarbum, K.; Ha, S.; Ma, J.; Kim, N. Multi-objective optimization analysis for part-to-Printer assignment in a network of 3D fused deposition modeling. *J. Manuf. Syst.* **2017**, *43*, 35–46. [CrossRef]
23. Li, Y.; Cheng, Y.; Hu, Q.; Zhou, S.H.; Ma, L.; Lim, M.K. The influence of additive manufacturing on the configuration of make-to-order spare parts supply chain under heterogeneous demand. *Int. J. Prod. Res.* **2019**, *57*, 3622–3641. [CrossRef]
24. Zhang, J.M.; Yao, X.F.; Li, Y. Improved evolutionary algorithm for parallel batch processing machine scheduling in additive manufacturing. *Int. J. Prod. Res.* **2020**, *58*, 2263–2282. [CrossRef]
25. Cadiou, T.; Demoly, F.; Gomes, S. A Multi-Part Production Planning Framework for Additive Manufacturing of Unrelated Parallel Fused Filament Fabrication 3D Printers. *Designs* **2022**, *6*, 11. [CrossRef]
26. Wu, S.; Kay, M.G.; King, R.E.; Vila-Parrish, A.R.; Fitts, E.P.; Warsing, D.P. Multi-objective Optimization of 3D Packing Problem in Additive Manufacturing. In Proceedings of the IIE Annual Conference, Montréal, QC, Canada, 31 May–3 June 2014.
27. Zhang, Y.C.; Gupta, R.K.; Bernard, A. Two-dimensional placement optimization for multi-parts production in additive manufacturing. *Robot. Comput.-Integr. Manuf.* **2016**, *38*, 102–117. [CrossRef]
28. Wodziak, J.R.; Fadel, G.M.; Kirschman, C. A genetic algorithm for optimizing multiple part placement to reduce build time. In Proceedings of the Fifth International Conference on Rapid Prototyping, Dayton, OH, USA, 12–15 June 1994; pp. 201–210.
29. Canellidis, V.; Dedoussis, V.; Mantzouratos, N.; Sofianopoulou, S. Pre-processing methodology for optimizing stereolithography apparatus build performance. *Comput. Ind.* **2006**, *57*, 424–436. [CrossRef]
30. Canellidis, V.; Giannatsis, J.; Dedoussis, V. Efficient parts nesting schemes for improving stereolithography utilization. *Comput.-Aided Des.* **2013**, *45*, 875–886. [CrossRef]
31. Wang, Y.B.; Pai, Z.; Xun, X.; Yang, H.Y.; Jun, Z. Production planning for cloud-based additive manufacturing-A computer vision-based approach. *Robot. Comput.-Integr. Manuf.* **2019**, *58*, 145–157. [CrossRef]
32. Li, Q.; Kucukkoc, I.; Zhang, D.Z. Production planning in additive manufacturing and 3D printing. *Comput. Oper. Res.* **2017**, *83*, 157–172. [CrossRef]
33. Bennell, J.A.; Oliveira, J.F. The geometry of nesting problems: A tutorial. *Eur. J. Oper. Res.* **2008**, *184*, 397–415. [CrossRef]
34. Allaire, G.; Bihr, M.; Bogosel, B. Support optimization in additive manufacturing for geometric and thermo-mechanical constraints. *Struct. Multidiscip. Optim.* **2020**, *61*, 2377–2399. [CrossRef]
35. Das, P.; Mhapsekar, K.; Chowdhury, S.; Samant, R.; Anand, S. Selection of build orientation for optimal support structures and minimum part errors in additive manufacturing. *Comput.-Aided Des. Appl.* **2017**, *14*, 1–13. [CrossRef]
36. Pande, S.S.; Kumar, S. A generative process planning system for parts produced by rapid prototyping. *Int. J. Prod. Res.* **2008**, *46*, 6431–6460. [CrossRef]
37. Pandey, P.M.; Reddy, N.V.; Dhande, S.G. Part deposition orientation studies in layered manufacturing. *J. Mater. Process. Technol.* **2007**, *185*, 125–131. [CrossRef]
38. Della Croce, F.; Scatamacchia, R. The Longest Processing Time rule for identical parallel machines revisited. *J. Sched.* **2020**, *23*, 163–176. [CrossRef]
39. Samet, H.; Tamminen, M. Computing Geometric-Properties of Images Represented by Linear Quadtrees. *IEEE Trans. Pattern Anal. Mach. Intell.* **1985**, *7*, 229–240. [CrossRef] [PubMed]
40. Chazelle, B. The Bottom-Left Bin-Packing Heuristic—An Efficient Implementation. *IEEE Trans. Comput.* **1983**, *32*, 697–707. [CrossRef]
41. Makanda, I.L.D.; Jiang, P. A Web-based Generative Process Planning System for FDM-based Additive Manufacturing. *IFAC-Pap. Online* **2020**, *53*, 83–88. [CrossRef]

machines

Article

Data Acquisition Network Configuration and Real-Time Energy Consumption Characteristic Analysis in Intelligent Workshops for Social Manufacturing

Chaoyang Zhang [1,*], Juchen Zhang [2,3], Weixi Ji [1] and Wei Peng [1]

1 Jiangsu Key Laboratory of Advanced Food Manufacturing Equipment & Technology, Jiangnan University, Wuxi 214122, China
2 School of Mechanical Engineering, Hefei University of Technology, Hefei 230009, China
3 College of Mechanical and Electrical Engineering, Nanjing University of Aeronautics & Astronautics, Nanjing 210016, China
* Correspondence: cyzhang@jiangnan.edu.cn

Abstract: To achieve energy-saving production, one critical step is to calculate and analyze the energy consumption and energy efficiency of machining processes. However, considering the complexity and uncertainty of discrete manufacturing job shops, it is a significant challenge to conduct data acquisition and energy consumption data processing of manufacturing systems. Meanwhile, under the growing trend of personalization, social manufacturing is an emerging technical practice that allows prosumers to build individualized services with their partners, which produces new requirements for energy data processing. Thus, a real-time energy consumption characteristic analysis method in intelligent workshops for social manufacturing is established to realize data processing and energy efficiency evaluation automatically. First, an energy-conservation production architecture for intelligent manufacturing processes is introduced, and the configuration of a data acquisition network is described to create a ubiquitous manufacturing environment. Then, an energy consumption characteristic analysis method is proposed based on the process time window. Finally, a case study of coupling-part manufacturing verifies the feasibility and applicability of the proposed method. This method realizes a combination of social manufacturing and real-time energy characteristic analysis. Meanwhile, the energy consumption characteristics provide a decision basis for the energy-saving control of intelligent manufacturing workshops.

Keywords: data acquisition network; real-time energy consumption; characteristic analysis; intelligent workshops; social manufacturing

Citation: Zhang, C.; Zhang, J.; Ji, W.; Peng, W. Data Acquisition Network Configuration and Real-Time Energy Consumption Characteristic Analysis in Intelligent Workshops for Social Manufacturing. *Machines* **2022**, *10*, 923. https://doi.org/10.3390/machines10100923

Academic Editor: Jose L. Martinez Lastra

Received: 19 August 2022
Accepted: 7 October 2022
Published: 10 October 2022

1. Introduction

Nowadays, the increasing emissions of carbon dioxide have made a crucial contribution to global warming, especially in industrial fields. With the improvements made around the awareness of saving energy and the enhancement of environmental concerns, energy-efficient and low-carbon emissions should be considered as key factors in many fields, such as transportation facilities [1], manufacturing management [2] and disassembly sequencing [3]. Owing to the explosive demand for energy worldwide, energy efficiency in manufacturing processes has become a challenging goal. In the last few years, the reduction in energy consumption has attracted many researchers. A survey on electric energy consumption showed that up to 54% of electric energy is used in production processes, or more accurately, production machines. Therefore, the energy conservation of machine tools can significantly reduce carbon emissions in manufacturing. To realize the energy conservation of machine tools and environmentally friendly production, researchers have conducted many theoretical and practical studies, which mainly include machining parameter optimization [4], the state control of machine tools [5], process planning [6] and production scheduling [7].

Meanwhile, under the growing trend of personalization and socialization, social manufacturing (SM) is an emerging technical and business practice that allows prosumers to build personalized products and individualized services with their partners [8]. SM can realize a customer's requirements "from mind to products", and fulfill tangible and intangible needs of a prosumer, i.e., a producer and consumer at the same time [9]. By establishing a cyber–physical–social connection via decentralized social media, various communities can be formed as complex, dynamic, autonomous systems to cocreate customized and personalized products and services [10]. From the principle of SM, the consumers are acting as prosumers and participate in the energy production and consumption process over the Internet [11]. For a manufacturing workshop, its operation state will be focused on by several prosumers, such as equipment providers, manufacturers and consumers [12]. How to solve these different demands is a new problem. With the development of the Internet of Things (IoT) technology, many kinds of data are generated in a manufacturing workshop, and these data present some characteristics of big data, such as large volume, high variety and velocity. Thus, it becomes a research focus to acquire and process these data, and various data analysis methods have been proposed, including fog computing [13], deep-learning approaches [14] and big-data analytics [2]. Meanwhile, many energy-conservation models and algorithms have been proposed for single-machine tools [15] or closed-loop flow-shop plants [16]. However, there are still two research gaps in the recent research. First, many models and algorithms for data acquisition and analysis have been proposed to mine knowledge and guide the production process. Most of these studies neglect the adaptability of manufacturing data network on the shop floor. Since discrete manufacturing is a manufacturing process that does not follow sequential steps or formulas to develop products, the production process in discrete manufacturing job shops has stochasticity. The data of energy and processes that need to be acquired are dynamic and diverse, and so is the energy data acquisition network. Second, although there are many studies on energy data analysis and evaluation from different perspectives, most of them focus on the study of theoretical models. Scant sustainability research has been conducted on SM [17]. It is still an application difficulty to combine this production information with real-time energy consumption data in an intelligent workshop for SM.

Considering these two research gaps, a real-time energy consumption characteristic analysis method for intelligent manufacturing (IM) workshops is established, which involves the manufacturing data acquisition network configuration and energy consumption characteristic analysis of machining processes. The contributions of the paper include two aspects. First, the configuration of a data acquisition network is described to create a ubiquitous manufacturing environment, which can deal with the complexity of discrete manufacturing processes. Second, an energy consumption characteristic analysis method is proposed based on the process time window. This method realizes the combination of production information with real-time energy consumption data. The remainder of this paper is organized as follows. A literature survey on the data acquisition of IM workshops and energy consumption data evaluation is reviewed in Section 2. Section 3 introduces an energy-conservation production architecture for IM processes, and the configuration of a data acquisition network is described to create a ubiquitous manufacturing environment. Then, an energy consumption characteristic analysis method is established based on the process time window in Section 4. A case study on coupling-part manufacturing is presented to verify the feasibility and applicability of the proposed methods in Section 5. We conclude with the main contributions and future research directions in Section 6.

2. Literature Review

2.1. Data Acquisition and Analysis of IM Workshops

With the development of sensor network technology, the industry is increasingly moving towards digitally enabled "smart factories" that utilize the IoT to realize IM [18]. A large amount of data is gathered in a manufacturing workshop, and these data, including on energy consumption, present some characteristics of big data, such as large volume, high variety and velocity [19]. Thus, the acquisition and processing of these data are research focuses. With the beginning of the era of big data, an enormous amount of real-time data was used for the risk analysis of various industrial applications, and Ding et al. proposed a real-time big-data-gathering algorithm based on an indoor wireless sensor network for the risk analysis of industrial operations [20]. In this algorithm, sensor nodes can screen the data collected from the environment and equipment according to the requirements of the risk analysis. As managing industrial big data has become a challenging task for factories, designing a generic architecture for implementing cyber–physical systems in manufacturing is necessary. Thus, Lee et al. proposed a systematic architecture for applying cyber–physical systems in manufacturing to automate and centralize data processing, health assessments and prognostics [21].

In addition to data acquisition, an increasing number of researchers have focused on data analysis and data mining in various applications. Industrial big-data integration and sharing (IBDIS) is of great significance in managing and providing data for big-data analysis in manufacturing systems; thus, Wang et al. proposed a novel fog-computing-based IBDIS approach to integrate and share industrial big data with high raw-data security and low network-traffic loads by moving the integration task from the cloud to the edge of networks [13]. A deep-learning approach for anomaly detection with industrial time-series data in a refrigerator manufacturing enterprise was proposed, and was designed to be deployed in a decision support system to assist human operators [14]. Zhong et al. extended the physical Internet concept to manufacturing shop floors where typical logistics resources were converted into smart manufacturing objects by using the IoT and wireless technologies [22]. This study introduced big-data analytics for radio frequency identification device (RFID) logistics data by defining different behaviors of smart manufacturing objects. Since current task scheduling is mainly concerned with the availability of machining resources rather than the potential errors after scheduling, Ji and Wang presented a fault prediction approach based on big-data analytics for shop-floor scheduling to minimize such errors in advance [23]. An innovative, big-data-enabled, intelligent immune system has been developed to monitor, analyze and optimize machining processes over lifecycles in order to achieve energy-efficient manufacturing [2]. The novelty of this study is that big-data analytics and intelligent immune mechanisms have been integrated systematically to achieve condition monitoring, analysis and energy-efficient optimization over manufacturing execution lifecycles. According to the literature, big-data analytics and smart manufacturing have been individually researched in academia and industry [24]. To provide theoretical foundations for the research community to further develop scientific insights into applying big-data analytics to smart manufacturing, a comprehensive overview of big data in smart manufacturing was conducted, and a conceptual framework was proposed from the perspective of the product lifecycle. A review of the literature suggests that production research enabled by data has shifted from analytical models to data-driven models [25]. Ghahramani et al. proposed a dynamic algorithm for gaining useful insights about semiconductor manufacturing processes and to address various challenges [26]. White et al. developed a fault diagnosis tool, which can robustly detect, locate and isolate occurred faults in an Industry 4.0 context [27]. Ding and Jiang provided an RFID-based production data analysis method for production control in IoT-enabled smart job shops [28]. Yuan et al. established an integrated, deep-learning, continuous time network structure that consists of a sequential encoder, a state decoder and a derivative module to learn the deterministic state-space model from thickening systems [29].

Based on the current research, data acquisition and data analysis have become research hotspots, and many models and algorithms have been proposed to mine knowledge and guide production processes in turn. However, this research mainly focused on data processing and data analysis, especially big-data analysis, and data acquisition network configuration was rarely studied. Meanwhile, the theory research about data mining in an ideal manufacturing environment was very deep, which neglected the complexity and dynamics of an actual production environment. Considering the complexity of manufacturing processes, it is still difficult for operators to configure the data acquisition network, especially when the production processes experience frequent adjustments. Thus, a configuration method of the sensor network is proposed for dynamic manufacturing tasks.

2.2. Energy consumption Data Processing and Evaluation

After the production data acquisition, energy data processing and evaluation needs to be conducted to obtain the characteristics and energy efficiency of manufacturing processes. Many manufacturing energy data processing and evaluation methods have been proposed. For example, considering that the energy consumption evaluation and analysis of a product's entire life cycle is a key issue for realizing green and sustainable manufacturing, an IoT and cloud-based novel approach for product energy consumption evaluation and analysis was proposed, in which the IoT technologies were employed for real-time and dynamic collection of energy consumption-related data [30]. For a machine process, its energy consumption can be decomposed into two parts: energy consumption of the steady state and energy consumption of the transient state. Jia et al. proposed a finite-state, machine-based energy consumption modeling method for the machining transient state [31]. Cai et al. proposed the use of energy benchmarking to strengthen the evaluation of energy demand and achieve efficiency improvements for machining systems, in which drivers for energy benchmarking and their characteristics were analyzed first [32]. Finkbeiner et al. explored the current status of a life-cycle sustainability assessment for products and processes [33]. Saxena et al. considered the sustainability metrics in tandem with other traditional manufacturing metrics such as time, flexibility and quality, and presented a novel framework that integrates information and requirements from computer-aided technology systems [34]. Swarnakar et al. identified and prioritized experts' consensus on the structured set of triple-bottom-line indicators through an open-ended questionnaire [35]. However, these evaluation methods cannot realize the real-time evaluation of manufacturing processes, and they neglect data acquisition and the processing process. Wang et al. presented a real-time energy efficiency optimization method for energy-intensive manufacturing enterprises [36]. In this study, a multilevel event model and complex event processing were used to obtain real-time, energy-related, key performance indicators that extend the production performance indicators to the energy efficiency area. Owing to the complicated energy flow and dynamic energy changes of the machining workshop, Chen et al. proposed an energy efficiency monitoring and management system with the support of the newly emerging IoT technology, in which the energy characteristics and energy efficiency indicators of the machining workshop were analyzed and defined [37]. To improve the generalization ability, Xiao et al. combined the machining parameters and configuration parameters into energy efficiency models, for which machine-learning algorithms were used to consider the lack of theoretical formulas [38]. Considering the complexity of discrete manufacturing workshops, a big-data-analysis approach for the real-time carbon efficiency evaluation of discrete manufacturing workshops was proposed in an IoT-enabled ubiquitous environment [39]. Based on advanced technologies such as cloud manufacturing, IoT, and cyber–physical systems, an energy–cyber–physical-system-enabled green manufacturing model for future smart factories was proposed, in which qualitative and quantitative synergetic models based on an energy–cyber–physical system were developed for cleaner manufacturing [40]. Along with the advent of globalization in the industrial sector, the distributed manufacturing systems became an important production process since they enable the efficient collaboration of multiple factories; thus, the stochastic, multiobjective modeling and optimization of an energy-conscious distributed-permutation flow-shop scheduling problem was proposed [41]. To ensure the fastest production and least energy consumption of steelmaking-continuous casting, a mixed-integer mathematical programming model was presented with the objectives of minimizing the maximum completion time, idle time penalties, and energy consumption penalties related to waiting time [42].

Li et al. proposed a modified dynamic programming algorithm for the optimization of total energy consumption in flexible manufacturing systems [43]. In addition, Diaz et al. outlined and discussed most of the resent research regarding the technologies and strategies to improve energy efficiency in manufacturing systems [44].

From these studies, it can be seen that there are many studies on energy data analysis and evaluation from different perspectives. However, most of them mainly focused on the study of theoretical models, and neglected the correlation between energy consumption data and manufacturing processes. Since the ultimate realization of energy savings and emission reductions is closely related to manufacturing processes, the practical application value of the current studies is limited. It is still difficult to combine production information with real-time energy consumption data in IM workshops.

3. Energy-Conservation Production Architecture and Data Acquisition Network Configuration

3.1. Energy-Conservation Production Architecture for IM Processes

To implement energy-aware production, an energy-conservation production architecture in an intelligent workshop for SM is established, as shown in Figure 1. The architecture provides a step-by-step guide for controlling computer numerical control (CNC) machine tools and discrete manufacturing processes. The architecture is established mainly from the perspective of data processing, that is, data acquisition, data filtering, data integration, data conversion, data reduction and knowledge analysis. It consists of three modules: the data acquisition network module, energy characteristic analysis and energy-conservation control module, and energy consumption service module. Since the data of different manufacturing processes are disparate, the data acquisition network module is carried out on the basis of discrete manufacturing processes. Then, the gathered data from the data sensor network can be analyzed through the data mining and analysis module, which cannot be divorced from manufacturing processes. The energy-saving strategies generated by the energy-conservation control module must be finally implemented in discrete manufacturing processes.

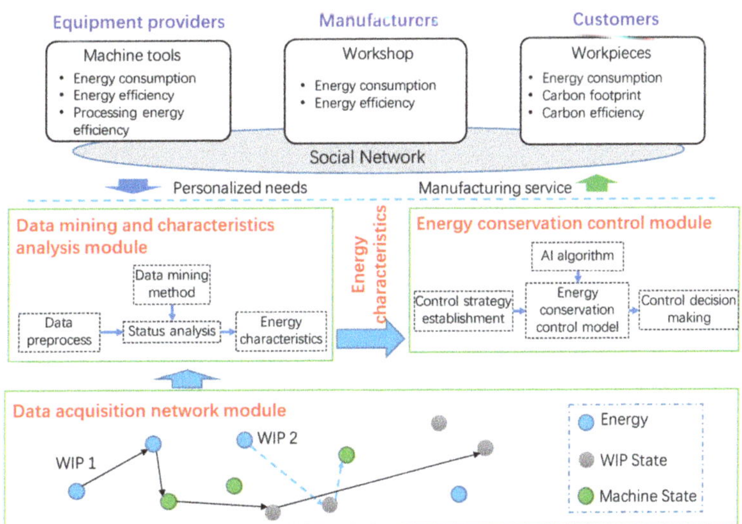

Figure 1. Energy-conservation production architecture in intelligent workshops for SM.

3.1.1. Data Acquisition Network Module

As the basic layer of the architecture, discrete manufacturing processes reflect the physical resource configuration of a workshop, including facility layout, selection of machine and tools, setting of processing parameters and process planning. In the process flow, there are three types of flows: energy flow, raw material flow and operation flow. The inputs of the process flow are raw materials, energy and supporting materials. The outputs are semifinished or end products, and waste materials or effluent. In each step of the process flow, the emissions from the energy and supporting materials occur at all times. Based on manufacturing processes, an energy-acquisition sensor network is established to gather the relative data for realizing energy-saving production, for example, on the energy consumption of machine tools, the state of the work-in-process (WIP) and machine tools. Initially, each machine may be configured with different sensors that contain energy sensors, position sensors and testing sensors. However, for different production tasks, disparate sensors are integrated into a sensor network involved in real production. Therefore, the data acquisition network varies constantly from time to time, as discussed in Section 3.2.

3.1.2. Energy Characteristic Analysis and Energy-Conservation Control Module

For the gathered data from the data sensor network, data processing needs to be implemented first, which includes data filtering, data integration, data conversion and data interaction [45]. As the first step, the data filter needs to be adopted to remove useless data because many data are continuous and abundant. Then, data integration needs to be used to deal with each single data point for comparison and analysis. Data conversion and interaction must be adopted for the entire manufacturing system. Then, the data need to be converted into information on the state of machine tools and WIPs. Using the above information, the energy characteristics can be analyzed, which will be used to support the energy-conservation control of CNC machine tools.

To realize energy-aware production, the energy-conservation control model is based on an artificial intelligence algorithm. The input of this model is the above energy characteristic values (that can reflect the machining state and energy efficiency), and the output is the selection of energy-conservation strategies.

3.1.3. Energy Consumption Service Module

Through the above energy consumption analysis, energy characteristics of different layers in a workshop can be obtained, which include the machine tool layer, workshop layer and workpiece layer. Different SM participants can put forward the personalized manufacturing and service needs through the social network. Meanwhile, different data and services will be provided to them.

3.2. *Configuration of Manufacturing Data Acquisition Network*

Here, the manufacturing data acquisition network is aimed at energy-conservation manufacturing, and the acquired data include energy consumption, state of the machine tools, and WIPs. Considering the real production processes, the configuration of the sensor network contains two parts: the static and the dynamic network constructions. The static network configuration is realized after the design of a manufacturing system, which is a part of the physical configuration process of the manufacturing system. On the other hand, the dynamic sensor network is mainly applied to one or several certain production tasks after production planning and scheduling. After the dynamic network construction, the required data can be acquired and the usage effectiveness of the sensors will be improved.

3.2.1. The Static Sensor Network Configuration

This configuration process uses a rule-based inference engine to realize the intelligent configuration of the data acquisition network of machine tools, as shown in Figure 2. The configuration process consists of three parts: establishment of a configuration knowledge base, construction of rules and the case base, and realization of the reasoning engine.

The configuration knowledge base is the extraction and description of the related knowledge based on ontology, including feature description information, precision information, key parameters, etc. The rule and instance base is a configuration case that contains expert experience and the actual case, and the ontology rule base is generated based on this case. The reasoning engine is a combination of case-based reasoning and ontology-based reasoning approaches. First, case-based reasoning is performed based on similarity matching. If the configuration does not meet the requirements, ontology rules are used to determine a new configuration scheme. The process of ontology rule matching is an iterative process that may not meet the requirements one time. The new data acquisition network configuration plan will be stored in the instance library for later use. The input of the configuration process includes the demand information of the machining features, and the output is a viable data acquisition network for the machine tool.

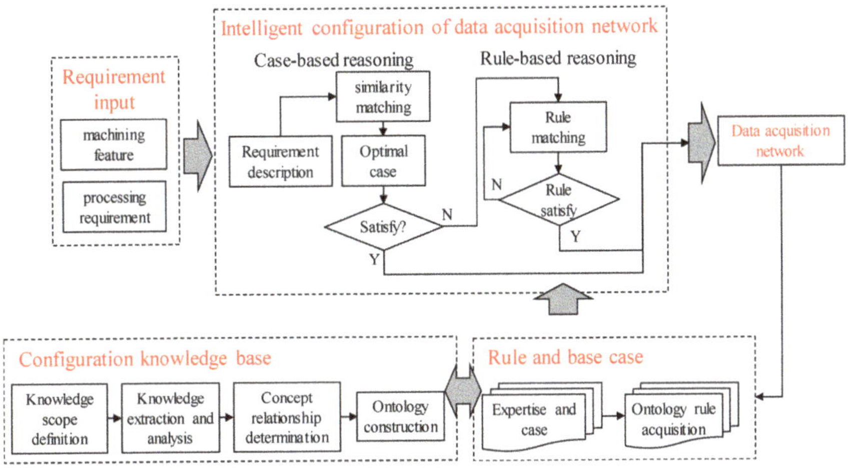

Figure 2. Intelligent configuration of manufacturing data acquisition network.

3.2.2. Dynamic Network Construction for a Certain Production Task

To improve the efficiency of configuration reasoning, case-based reasoning (CBR) is used to configure the data acquisition subnetwork, which is used to search for similar cases based on the matching of the above-mentioned processing-task attributes. The matching steps of CBR are described as follows: (1) define the processing task, including its characteristics or attributes; (2) search the instance database and find the highest similarity instance in the sample database according to the processing task's characteristic data; (3) constitute a data acquisition network configuration scheme for the processing tasks to serve as a new paradigm; and (4) save the valuable new configuration examples obtained in the instance database for future energy data acquisition subnetwork configuration reference. Owing to the different attribute categories of machining tasks, the machining type is text data, the machining size is continuous numerical data and the machining accuracy is discrete numerical data. Therefore, the similarity ($SIM_{i,j}$) calculation is an integration of the similarities of different types of data as follows:

$$SIM_{i,j} = \prod_a Sim_{i,j}^a \cdot \sum_b w_b \cdot Sim_{i,j}^b \tag{1}$$

$$Sim_{i,j}^a = \begin{cases} 1, & if\ mta_i^a\ is\ the\ same\ to\ mta_j^a \\ 0, & otherwise \end{cases} \tag{2}$$

$$Sim_{i,j}^b = 1 - \frac{\sqrt{\left(mta_i^b - mta_j^b\right)^2}}{\sqrt{\left(\max_i mta_i^b - \min_j mta_j^b\right)^2}} \tag{3}$$

$$\sum_b w_b = 1 \quad (4)$$

where $SIM_{i,j}$ denotes the total similarity of the ith and jth manufacturing tasks. $Sim^a_{i,j}$ and $Sim^b_{i,j}$ represent the similarity of the text-type property and the value-type property, respectively, and a and b represent the indexes of these two similarities. mta^a_i and mta^a_j represent the ath text-type property of the ith and jth manufacturing tasks, respectively. mta^b_i and mta^b_j represent the bth value-type property of the ith and jth manufacturing tasks, respectively. w_b denotes the weight of the bth value-type property.

4. Energy Consumption Characteristic Analysis Based on Process Time Window

4.1. Data Modeling of Discrete Manufacturing Processes

Different kinds of data can be obtained after the manufacturing data acquisition network configuration. To realize data processing and data storage, date modeling is conducted first. The discrete manufacturing processes mainly include three types of data related to energy conservation: process data, energy consumption data and supporting-material data.

Process data model:

The process data involve the information of each machining process, for example, the process name, machine tool and process time. Because all the manufacturing activities are carried out according to process planning, process data are the core data for the discrete manufacturing workshop. The process data are modeled in Equation (5), and these data can be obtained from process planning and the WIP state.

$$PD = < PDID,\ WID,\ PID, PName,\ MID, STime,\ ETime > \quad (5)$$

where $PDID$ is the index of the process data, and WID, PID and MID are the indexes of the workpiece, process and machine tool, respectively. $PName$ denotes the process name. $STime$ and $ETime$ represent the starting and ending time of the process, respectively.

Energy consumption data model:

For a manufacturing workshop, the energy consumption mainly comes from machine tools; thus, the energy consumption data can be modeled by Equation (6).

$$ECD = < ECDID,\ MID, EData, T > \quad (6)$$

where $ECDID$ is the index of the energy consumption data, and $EData$ represents the real-time power of a machine tool at time T.

Supporting-material data model

In addition to the energy consumption, a machining process also consumes some supporting materials that may be related to energy consumption, such as compressed air and cooling liquid. Thus, the supporting-material data can be expressed by Equation (7).

$$SMD = < SMDID, MID, SType,\ SMData,\ T > \quad (7)$$

where $SMDID$ is the index of the supporting-material data, and $SType$ represents the type of supporting material. $SMData$ is the real-time usage amount of the supporting material.

4.2. Energy Consumption Data-Partition Method Based on Process Time Window

For a manufacturing process, the process data are discrete, whereas the data on energy consumption are continuous. To relate the energy consumption data with process data, the former needs to be divided according to machining processes or feeds. Thus, an energy consumption data-partition method is proposed based on the process time window. The time window method can divide time series into segments, and has been used in many applications [46]. Because the traditional time window moves forward in succession over a fixed or variable time, it is not suitable for discrete manufacturing processes. The concept of the process time window is defined, which means that the interval time of the time window is determined on the basis of the processing stages, as shown in Figure 3.

For a single process, the time window can be obtained from the enterprise resource planning system, that is, the starting and ending times of the process. However, this process may contain several steps. For example, a turning process may include cylindrical turning, face cutting and internal cylindrical turning. Meanwhile, a step contains different states, such as standby, idle, air cutting and cutting, as shown in Figure 3. The power of different processes and states shows a high variation due to different cutting parameters [47,48].

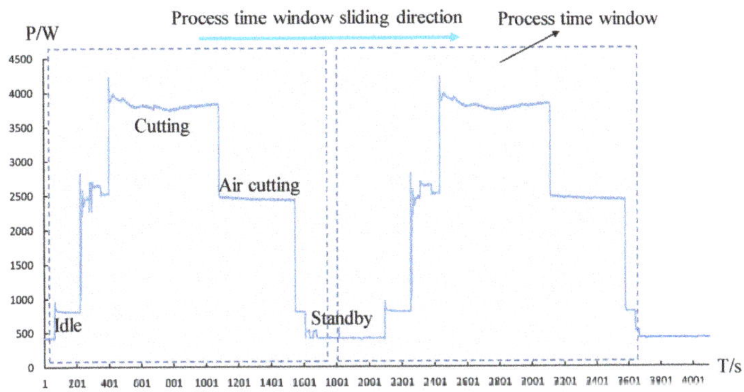

Figure 3. Energy consumption data partition based on process time window.

To analyze the energy characteristics of different steps or cutting states, the continuous energy data belonging to different steps are divided based on the process time window. The power of the air-cutting state is used to judge whether a machine tool is in a cutting state, and the power of the cutting state is for identifying different processes or cutting steps. Here, ten real-time energy data are analyzed every time, that is, $P^t = \{p_i^t\}, i = 1\ldots10$. Then, the maximum and minimum values of these data are deleted to exclude the influence of chance factors. The average value of the remaining data $P^{t'} = \{p_i^{t'}\}, i = 1\ldots8$ is used as the judgment standard, as denoted in Equation (8).

$$MeanCE_t = \frac{1}{8}\sum_{i=1}^{8} p_i^{t'} \tag{8}$$

The algorithm flow of the entire energy data-partition method based on the process time window is shown in Algorithm 1.

4.3. Real-Time Energy Consumption Characteristic Analysis

For a machining process, the original energy consumption data cannot be directly used to evaluate or optimize the machining processes. Important characteristic values must be derived. In this study, these characteristic values mainly fall into two categories: instantaneous and cumulative.

The instantaneous characteristics mainly reflect the variation law of cutting power, such as the maximum, minimum and mean power of a process or feed, as follows:

$$MaxP_i = \max_{t\in[s_i,e_i]} p_t \tag{9}$$

$$MinP_i = \min_{t\in[s_i,e_i]} p_t \tag{10}$$

$$MeanP_i = \frac{1}{N_i}\sum_{t\in[s_i,e_i]} p_t \tag{11}$$

where p_t denotes the power at time t, and $MaxP_i$, $MinP_i$ and $MeanP_i$ represent the maximum, minimum and mean of the energy consumption values, respectively. s_i and e_i denote the starting and ending times of a process time window, respectively. N_i is the number of cutting-power data points during the process time window.

Algorithm 1: Energy Data Partition Method

Input: the power of air cutting $P_{aricutting}$
Output: the time window node of each steps $s_{i,j}$ and $e_{i,j}$
Algorithm flow:
According to process planning, obtain the starting time s_i and ending time e_i
For each process
  For each energy data of this process
    The starting time of the first step is s_i, obtain the mean value $MeanCE_t = \frac{1}{8}\sum_{i=1}^{8} p_i^{t'}$
    If $\frac{MeanCE_t - P_{aricutting}}{P_{aricutting}} > 0.05$
      The step is in cutting state, and obtain the mean power value $P^1_{cutting}$
    Else
      Obtain the time point $e_{i,1}$
    End if
    If $\frac{\left|P^2_{cutting} - P^1_{cutting}\right|}{P^1_{cutting}} > 0.05$
      This is the next step, and the above $e_{i,1}$ is the starting time of the next step $s_{i,2}$
    End if
  End For
End For
Return $s_{i,j}$ and $e_{i,j}$

These instantaneous characteristics can reflect the operating state of a manufacturing system and have been used to detect recessive production anomalies [49].

On the other hand, the cumulative characteristics mainly show the overall energy consumption and energy efficiency of a process. These characteristics can be obtained as follows:

$$TotalE_i = \sum\nolimits_{t\in[s_i,e_i]} p_t * \Delta t \tag{12}$$

$$EnergyEff_i = \frac{\sum_{t\in[s_i,e_i]\cap[ms_i,me_i]} p_t * \Delta t}{TotalE_i} \tag{13}$$

$$ProcessEff_i = \frac{V_i}{TotalE_i} \tag{14}$$

where $TotalE_i$ denotes the total energy consumption during the process time window. Δt is the energy consumption data-sampling interval. $EnergyEff_i$ denotes the energy efficiency, which is the ratio of the material-removal energy consumption to the total energy consumption during the process time window. $ProcessEff_i$ is the material-removal volume of each energy consumption. V_i denotes the material-removal volume during this process time window. ms_i and me_i mean the starting time and ending time of the material-removal process.

The cumulative characteristic values can reveal the energy consumption feature at the process level, which can support process improvement and parameter optimization.

5. Case Study

5.1. Case Description

To verify the proposed method, a case of a discrete manufacturing workshop that mainly produces coupling parts for different customers is studied. This workshop contains three types of machine tools from different equipment providers: a CNC lathe, milling center and drilling machine. There are four machining processes for a coupling part: cylindrical lathe cutting, milling flat, drilling hole and tapping. The relationship between the machining processes and machine tools is presented in Table 1. The power of the air-cutting state of each machine tool is also listed in the table.

Table 1. The machining processes of a coupling part.

No.	Processes	Machine Tool	Machine No.	Power of Air Cutting (W)
1	Cylindrical lathe cutting	CNC lathe	M1	2411
2	Milling flat	Milling center	M2	2941
3	Drilling hole			
4	Tapping	Drilling machine	M3	2182

To realize the monitoring of machining states, a static data acquisition network is first configured for each machining system in the manufacturing workshop, as shown in Figure 4a. The processing characteristics, accuracy, optional sensing-equipment set and measuring-equipment set of the machining system are input to build the rule-reasoning library. A feasible sensing-equipment recommended list is obtained according to the specific process system information. Then, the appropriate configuration instances for a specific processing task are retrieved from the instance database through similarity calculation. According to a specific instance, the operator selects and modifies sensing devices from the static data acquisition network, and then a dynamic data acquisition subnetwork is formed for the current task, as shown in Figure 4b.

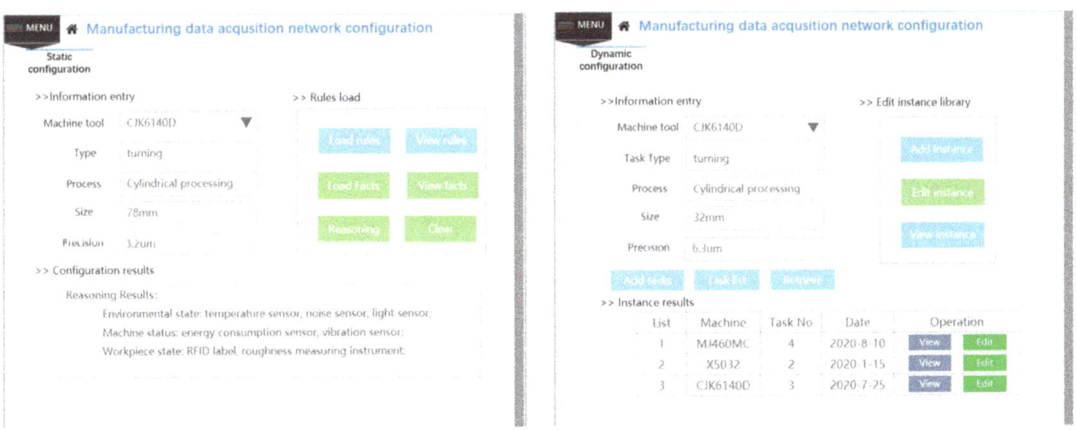

(a) **(b)**

Figure 4. Configuration of manufacturing data acquisition network; (**a**) static data acquisition network configuration; (**b**) dynamic data acquisition network configuration.

5.2. Energy Consumption Monitoring and Characteristic Analysis

Based on the above data acquisition network, the machining state and energy consumption data are obtained. Meanwhile, a prototype system of energy consumption monitoring and characteristic analysis is developed, as shown in Figure 5. This system is designed based on the browser/server (B/S) architecture. On the server side, the Java web-programming language is adopted, whereas HTML5/CSS/JavaScript is used to develop the browser side. This system can be conveniently visited by networked computers or remote handheld terminals. The prototype system mainly contains two functions: energy consumption monitoring and energy characteristic analysis. Figure 5a shows the real-time energy consumption monitoring module, which contains the machine tool, the current process and real-time energy. Figure 5b shows the energy consumption characteristic analysis module, which includes the mean energy consumption value, energy efficiency and material-removal volume of each energy consumption.

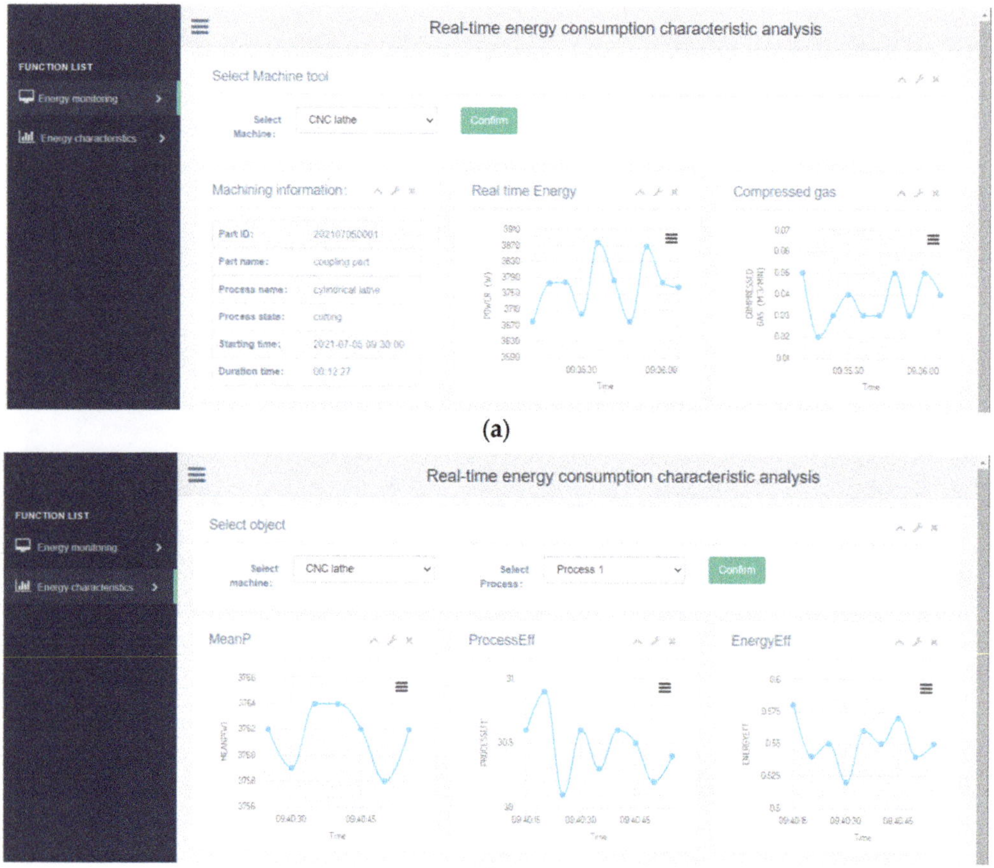

(a)

(b)

Figure 5. A prototype system of energy consumption monitoring and characteristic analysis; (**a**) real-time energy consumption monitoring; (**b**) energy consumption characteristic analysis.

To analyze the accuracy of the proposed energy data-partition method, three tests with different machining times and sample sizes are chosen for comparison, as shown in Table 2. The sample sizes are 2700, 5400 and 8100. The partition accuracy is calculated by the ratio of properly segmented data and total energy consumption data. The results show that the partition accuracies of these three tests are all more than 98%, and it reaches 99.5% for Test No. 1. Moreover, clustering approaches were often used in the data-partition process [50]. In order to evaluate the validity of the proposed method, the clustering approach is used as a contrast in Table 2. The results show that the accuracy of the clustering approach for Test No. 1 is 98.7%, and the results for the other two tests are 97.8% and 97.1%. In summary, the proposed method has a high partition precision for manufacturing energy consumption data. The partitioned energy consumption data can then be used to analyze the characteristics of energy efficiency.

Table 2. The accuracy of the energy data-partition method.

Test No.	1	2	3
Machining time (min)	15	30	45
Sample size	2700	5400	8100
Partition accuracy (%)	99.5	98.6	98.4
Accuracy of clustering approach (%)	98.7	97.8	97.1

According to the proposed energy analysis method, some energy characteristic values of each process and machine tool can be obtained. The instantaneous energy characteristics of different processes are shown in Figure 6. It can be seen that the total energy of the four processes is descending, and the cylindrical lathe-cutting process consumes the most energy, which reaches 2.41 kWh. Therefore, this lathe process needs more attention, and some adjustment strategies can be implemented by equipment providers to realize energy conservation, such as process route modification or process parameter optimization. Moreover, the maximum power and mean power of the cutting state of the different processes are consistent. The power of the drilling hole on M2 is the highest, which is 4097 kW. The differentiation of mean power over time can be used to detect abnormal production conditions, for example, machine tool performance degradation and cutting tool wear. Additionally, the whole energy consumption of coupling parts can be obtained, which can be used by customers to optimize their product design.

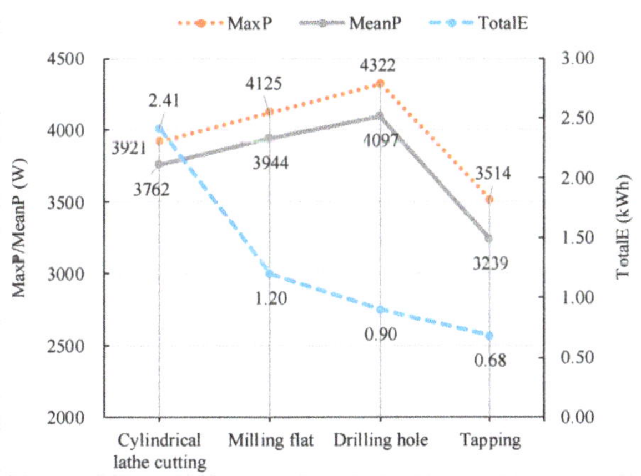

Figure 6. The energy characteristic analysis of different processes.

The energy characteristic values of different machine tools are shown in Figure 7. This shows that the energy efficiency of M3 is the highest, reaching 0.62, which means that most of the energy consumption of M3 is used to conduct the material-removal process. That of M2 is the lowest, and more energy is wasted in standby or air-cutting states. For this problem, the NC code on M2 can be improved to increase its energy efficiency. For the process efficiency, M1 is the best, whereas M3 has the lowest process efficiency. This characteristic value can be used by manufacturers to select the appropriate machine tool with the highest processing efficiency, especially for roughing processes.

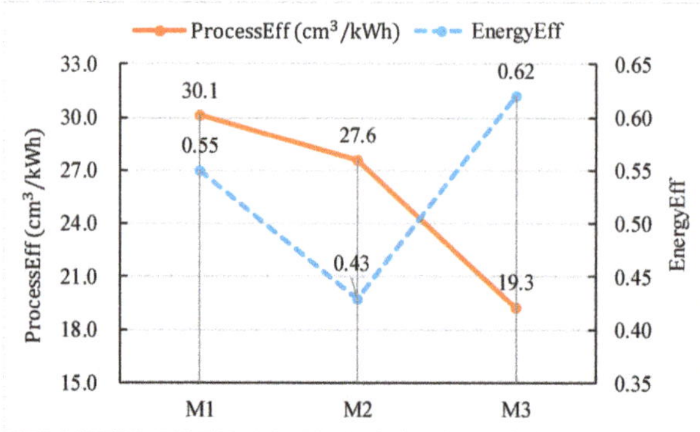

Figure 7. The energy characteristic analysis of different machine tools.

6. Conclusions and Future Work

To realize energy conservation and carbon emission reduction of manufacturing processes, one important step is to calculate and analyze the energy consumption and energy efficiency of machining processes. In this paper, a real-time energy consumption characteristic analysis method for IM workshops is established. First, an energy-conservation production architecture for IM processes is introduced, and the configuration of the data acquisition network is described to create a ubiquitous manufacturing environment. After the dynamic network construction, the required data can be acquired and the usage effectiveness of the sensor network can be improved. Then, an energy consumption data-partition and characteristic analysis method is proposed based on the process time window. The results show that the partition accuracies of these three tests are all more than 98%. Thus, the proposed method has a high partition precision for manufacturing energy consumption data. The energy characteristic values of different processes and machine tools can be obtained, which can be used by manufacturers to select the appropriate machine tool with the highest processing efficiency. The obtained energy characteristics can be used by different SM participants. This method realizes a combination of SM and real-time energy characteristic analysis.

In addition, there are also some limitations in the proposed methods which need to be researched in future works. First, in this study, only the energy consumption data of machine tools are analyzed, and many other important production data are neglected, such as energy consumption data of logistic processes and the workshop environment, the production capacity, and the utilization rate of equipment. Thus, more aggregative indicators are necessary for the overall assessment of manufacturing processes. Second, the actual manufacturing processes may be more complex, and include machine fault and cutting tool wear. Manufacturing data are dynamic and present uncertainties, and an intelligent processing method for complex data is required. Third, an energy consumption characteristic analysis cannot directly reduce the energy consumption of manufacturing systems. It is necessary to combine the energy characteristic analysis with some energy-saving strategies, such as multicriteria decision making about the selection of appropriate machine tools, process planning and production scheduling.

Author Contributions: Methodology, C.Z.; software, C.Z. and W.J.; validation, J.Z.; formal analysis, W.P.; writing—original draft preparation, C.Z.; writing—review and editing, C.Z. and J.Z.; visualization, W.J. All authors have read and agreed to the published version of the manuscript.

Funding: This research received no external funding.

Data Availability Statement: Not applicable.

Acknowledgments: This work is partially supported by the National Natural Science Foundation of China under Grant 51805213, the Natural Science Foundation of Jiangsu Province under Grant BK20170190 and BK20190419, and the Major Scientific and Technological Innovation Project of Shandong Province under Grant 2019JZZY020111.

Conflicts of Interest: The authors declare no conflict of interest.

References

1. Tian, G.; Liu, Y. Energy-Efficient Models of Sustainable Location for a Vehicle Inspection Station With Emission Constraints. *IEEE Trans. Autom. Sci. Eng.* **2015**, *12*, 238–243. [CrossRef]
2. Wang, S.; Liang, Y.; Li, W.; Cai, X. Big Data enabled Intelligent Immune System for energy efficient manufacturing management. *J. Clean. Prod.* **2018**, *195*, 507–520. [CrossRef]
3. Tian, G.; Ren, Y.; Feng, Y.; Zhou, M.; Zhang, H.; Tan, J. Modeling and Planning for Dual-Objective Selective Disassembly Using and/or Graph and Discrete Artificial Bee Colony. *IEEE Trans. Ind. Inform.* **2019**, *15*, 2456–2468. [CrossRef]
4. Zhang, C.Y.; Li, W.D.; Jiang, P.Y.; Gu, P.H. Experimental Investigation and Multi-objective Optimization Approach for Low-Carbon Milling Operation of Aluminum. *Proc. Inst. Mech. Eng. Part C J. Mech. Eng. Sci.* **2017**, *231*, 2753–2772. [CrossRef]
5. Frigerio, N.; Matta, A. Energy-Efficient Control Strategies for Machine Tools With Stochastic Arrivals. *IEEE Trans. Autom. Sci. Eng.* **2015**, *12*, 50–61. [CrossRef]
6. Newman, S.; Nassehi, A.; Imani-Asrai, R.; Dhokia, V. Energy efficient process planning for CNC machining. *CIRP J. Manuf. Sci. Technol.* **2012**, *5*, 127–136. [CrossRef]
7. Zhang, C.; Gu, P.; Jiang, P. Low-carbon scheduling and estimating for a flexible job shop based on carbon footprint and carbon efficiency of multi-job processing. *Proc. Inst. Mech. Eng. Part B J. Eng. Manuf.* **2015**, *229*, 328–342. [CrossRef]
8. Jiang, P.; Leng, J.; Ding, K. Social manufacturing: A survey of the state-of-the-art and future challenges. In Proceedings of the 2016 IEEE International Conference on Service Operations and Logistics, and Informatics (SOLI), Beijing, China, 10–12 July 2016; pp. 12–17.
9. Xiong, G.; Wang, F.-Y.; Nyberg, T.R.; Shang, X.; Zhou, M.; Shen, Z.; Li, S.; Guo, C. From mind to products: Towards social manufacturing and service. *IEEE/CAA J. Autom. Sin.* **2018**, *5*, 47–57. [CrossRef]
10. Jiang, P.; Leng, J.; Ding, K.; Gu, P.; Koren, Y. Social manufacturing as a sustainable paradigm for mass individualization. *Proc. Inst. Mech. Eng. Part B: J. Eng. Manuf.* **2016**, *230*, 1961–1968. [CrossRef]
11. Xiong, G.; Tamir, T.S.; Shen, Z.; Shang, X.; Wu, H.; Wang, F.-Y. A Survey on Social Manufacturing: A Paradigm Shift for Smart Prosumers. *IEEE Trans. Comput. Soc. Syst.* **2022**, 1–19. [CrossRef]
12. Ding, K.; Zhang, Y.; Chan, F.T.; Zhang, C.; Lv, J.; Liu, Q.; Leng, J.; Fu, H. A cyber-physical production monitoring service system for energy-aware collaborative production monitoring in a smart shop floor. *J. Clean. Prod.* **2021**, *297*, 126599. [CrossRef]
13. Wang, J.; Zheng, P.; Lv, Y.; Bao, J.; Zhang, J. Fog-IBDIS: Industrial Big Data Integration and Sharing with Fog Computing for Manufacturing Systems. *Engineering* **2019**, *5*, 662–670. [CrossRef]
14. Carletti, M.; Masiero, C.; Beghi, A.; Susto, G.A. A deep learning approach for anomaly detection with industrial time series data: A refrigerators manufacturing case study. *Procedia Manuf.* **2019**, *38*, 233–240. [CrossRef]
15. Mouzon, G.C.; Yildirim, M.B.; Twomey, J.M. Operational methods for minimization of energy consumption of manufacturing equipment. *Int. J. Prod. Res.* **2007**, *45*, 4247–4271. [CrossRef]
16. Mashaei, M.; Lennartson, B. Energy Reduction in a Pallet-Constrained Flow Shop Through On–Off Control of Idle Machines. *IEEE Trans. Autom. Sci. Eng.* **2013**, *10*, 45–56. [CrossRef]
17. Hamalainen, M.; Mohajeri, B.; Nyberg, T. Removing barriers to sustainability research on personal fabrication and social manufacturing. *J. Clean. Prod.* **2018**, *180*, 666–681. [CrossRef]
18. Egger, J.; Masood, T. Augmented reality in support of intelligent manufacturing—A systematic literature review. *Comput. Ind. Eng.* **2020**, *140*, 106195. [CrossRef]
19. Wang, K.; Li, H.; Feng, Y.; Tian, G. Big Data Analytics for System Stability Evaluation Strategy in the Energy Internet. *IEEE Trans. Ind. Inform.* **2017**, *13*, 1969–1978. [CrossRef]
20. Ding, X.; Tian, Y.; Yu, Y. A Real-Time Big Data Gathering Algorithm Based on Indoor Wireless Sensor Networks for Risk Analysis of Industrial Operations. *IEEE Trans. Ind. Inform.* **2016**, *12*, 1232–1242. [CrossRef]
21. Lee, J.; Ardakani, H.D.; Yang, S.; Bagheri, B. Industrial Big Data Analytics and Cyber-physical Systems for Future Maintenance & Service Innovation. *Procedia CIRP* **2015**, *38*, 3–7. [CrossRef]
22. Zhong, R.Y.; Xu, C.; Chen, C.; Huang, G.Q. Big Data Analytics for Physical Internet-based intelligent manufacturing shop floors. *Int. J. Prod. Res.* **2017**, *55*, 2610–2621. [CrossRef]
23. Ji, W.; Wang, L. Big data analytics based fault prediction for shop floor scheduling. *J. Manuf. Syst.* **2017**, *43*, 187–194. [CrossRef]
24. Ren, S.; Zhang, Y.; Liu, Y.; Sakao, T.; Huisingh, D.; Almeida, C.M.V.B. A comprehensive review of big data analytics throughout product lifecycle to support sustainable smart manufacturing: A framework, challenges and future research directions. *J. Clean. Prod.* **2019**, *210*, 1343–1365. [CrossRef]
25. Kuo, Y.-H.; Kusiak, A. From data to big data in production research: The past and future trends. *Int. J. Prod. Res.* **2018**, *57*, 4828–4853. [CrossRef]

26. Ghahramani, M.; Qiao, Y.; Zhou, M.C.; O'Hagan, A.; Sweeney, J. AI-based modeling and data-driven evaluation for smart manufacturing processes. *IEEE/CAA J. Autom. Sin.* **2020**, *7*, 1026–1037. [CrossRef]
27. White, A.; Karimoddini, A.; Karimadini, M. Resilient fault diagnosis under imperfect observations—A need for Industry 4.0 era. *IEEE/CAA J. Autom. Sin.* **2020**, *7*, 1279–1288. [CrossRef]
28. Ding, K.; Jiang, P. RFID-based production data analysis in an IoT-enabled smart job-shop. *IEEE/CAA J. Autom. Sin.* **2018**, *5*, 128–138. [CrossRef]
29. Yuan, Z.; Li, X.; Wu, D.; Ban, X.; Wu, N.-Q.; Dai, H.-N.; Wang, H. Continuous-Time Prediction of Industrial Paste Thickener System With Differential ODE-Net. *IEEE/CAA J. Autom. Sin.* **2022**, *9*, 686–698. [CrossRef]
30. Zuo, Y.; Tao, F.; Nee, A. An Internet of things and cloud-based approach for energy consumption evaluation and analysis for a product. *Int. J. Comput. Integr. Manuf.* **2018**, *31*, 337–348. [CrossRef]
31. Jia, S.; Tang, R.; Lv, J.; Yuan, Q.; Peng, T. Energy consumption modeling of machining transient states based on finite state machine. *Int. J. Adv. Manuf. Technol.* **2017**, *88*, 2305–2320. [CrossRef]
32. Cai, W.; Liu, F.; Xie, J.; Liu, P.; Tuo, J. A tool for assessing the energy demand and efficiency of machining systems: Energy benchmarking. *Energy* **2017**, *138*, 332–347. [CrossRef]
33. Finkbeiner, M.; Schau, E.M.; Lehmann, A.; Traverso, M. Towards Life Cycle Sustainability Assessment. *Sustainability* **2010**, *2*, 3309–3322. [CrossRef]
34. Saxena, P.; Stavropoulos, P.; Kechagias, J.; Salonitis, K. Sustainability Assessment for Manufacturing Operations. *Energies* **2020**, *13*, 2730. [CrossRef]
35. Swarnakar, V.; Singh, A.R.; Antony, J.; Jayaraman, R.; Tiwari, A.K.; Rathi, R.; Cudney, E. Prioritizing Indicators for Sustainability Assessment in Manufacturing Process: An Integrated Approach. *Sustainability* **2022**, *14*, 3264. [CrossRef]
36. Wang, W.; Yang, H.; Zhang, Y.; Xu, J. IoT-enabled real-time energy efficiency optimisation method for energy-intensive manufacturing enterprises. *Int. J. Comput. Integr. Manuf.* **2018**, *31*, 362–379. [CrossRef]
37. Chen, X.; Li, C.; Tang, Y.; Xiao, Q. An Internet of Things based energy efficiency monitoring and management system for machining workshop. *J. Clean. Prod.* **2018**, *199*, 957–968. [CrossRef]
38. Xiao, Q.; Li, C.; Tang, Y.; Chen, X. Energy Efficiency Modeling for Configuration-Dependent Machining via Machine Learning: A Comparative Study. *IEEE Trans. Autom. Sci. Eng.* **2020**, *18*, 717–730. [CrossRef]
39. Zhang, C.; Ji, W. Big Data Analysis Approach for Real-Time Carbon Efficiency Evaluation of Discrete Manufacturing Workshops. *IEEE Access* **2019**, *7*, 107730–107743. [CrossRef]
40. Ma, S.; Zhang, Y.; Lv, J.; Yang, H.; Wu, J. Energy-cyber-physical system enabled management for energy-intensive manufacturing industries. *J. Clean. Prod.* **2019**, *226*, 892–903. [CrossRef]
41. Fu, Y.; Tian, G.; Fathollahi-Fard, A.M.; Ahmadi, A.; Zhang, C. Stochastic multi-objective modelling and optimization of an energy-conscious distributed permutation flow shop scheduling problem with the total tardiness constraint. *J. Clean. Prod.* **2019**, *226*, 515–525. [CrossRef]
42. Tan, Y.; Zhou, M.; Zhang, Y.; Guo, X.; Qi, L.; Wang, Y. Hybrid Scatter Search Algorithm for Optimal and Energy-Efficient Steelmaking-Continuous Casting. *IEEE Trans. Autom. Sci. Eng.* **2020**, *17*, 1814–1828. [CrossRef]
43. Li, X.; Xing, K.; Zhou, M.; Wang, X.; Wu, Y. Modified Dynamic Programming Algorithm for Optimization of Total Energy Consumption in Flexible Manufacturing Systems. *IEEE Trans. Autom. Sci. Eng.* **2019**, *16*, 691–705. [CrossRef]
44. Jenny, L.D.C.; Ocampo-Martinez, C. Energy efficiency in discrete-manufacturing systems: Insights, trends, and control strategies. *J. Manuf. Syst.* **2019**, *52*, 131–145.
45. Zhang, C.; Jiang, P.; Cheng, K.; Xu, X.W.; Ma, Y. Configuration Design of the Add-on Cyber-physical System with CNC Machine Tools and its Application Perspectives. *Procedia CIRP* **2016**, *56*, 360–365. [CrossRef]
46. Ren, L.; Wei, Y.; Cui, J.; Du, Y. A sliding window-based multi-stage clustering and probabilistic forecasting approach for large multivariate time series data. *J. Stat. Comput. Simul.* **2017**, *87*, 2494–2508. [CrossRef]
47. Zhang, C.; Jiang, P. Sustainability Evaluation of Process Planning for Single CNC Machine Tool under the Consideration of Energy-Efficient Control Strategies Using Random Forests. *Sustainability* **2019**, *11*, 3060. [CrossRef]
48. Wang, W.; Tian, G.; Chen, M.; Tao, F.; Zhang, C.; Ai-Ahmari, A.; Li, Z.; Jiang, Z. Dual-objective program and improved artificial bee colony for the optimization of energy-conscious milling parameters subject to multiple constraints. *J. Clean. Prod.* **2020**, *245*, 118714. [CrossRef]
49. Zhang, C.; Wang, Z.; Ding, K.; Chan, F.T.; Ji, W. An energy-aware cyber physical system for energy Big data analysis and recessive production anomalies detection in discrete manufacturing workshops. *Int. J. Prod. Res.* **2020**, *58*, 7059–7077. [CrossRef]
50. Tabassian, M.; Verbeke, R.; Tourwe, T.; Tsiporkova, E. Data-Driven Divide-and-Conquer for Estimating Build Times of 3D Objects. In Proceedings of the 2021 International Conference on Data Mining Workshops (ICDMW), Auckland, New Zealand, 7–10 December 2021; pp. 268–277. [CrossRef]

machines

Article

An Efficient IIoT Gateway for Cloud–Edge Collaboration in Cloud Manufacturing

Yi Zhang, Dunbing Tang *, Haihua Zhu, Shihui Zhou and Zhen Zhao

College of Mechanical and Electrical Engineering, Nanjing University of Aeronautics and Astronautics, Nanjing 210016, China
* Correspondence: d.tang@nuaa.edu.cn

Abstract: The cloud manufacturing system can provide consumers with on-demand manufacturing services, which significantly improve the utilization rate of distributed manufacturing resources and the response speed of personalized product needs. In the cloud manufacturing platform, the successful implementation of various industrial applications relies on the uploading and streaming of related field-level manufacturing data. For example, the realization of manufacturing service composition application should match the manufacturing tasks with distributed manufacturing resources according to their working state data and performance measurement data. Therefore, this paper proposes a data integration and analysis framework of a cloud manufacturing system based on cloud–edge collaboration and the Industrial Internet of Things (IIoT). A service-oriented information model is established to uniformly describe the related operational data and functional attributes of heterogeneous manufacturing resources. Secondly, a real-time transmission and integration method of high-volume operational field and sensor data based on message middleware is proposed to realize the remote monitoring of distributed manufacturing resources and efficient distribution of related data. Finally, a cloud–edge collaboration mechanism is put forward to train and update the parameters of various artificial intelligence models deployed at edge gateways. In the experiment, taking the computer numerical control (CNC) lathe as an example, the effectiveness of the proposed manufacturing resource access method is verified. Taking the fault diagnosis model of the CNC lathe as an example, the efficiency of the proposed cloud–edge collaboration mechanism for model updating is verified.

Keywords: cloud manufacturing; Industrial Internet of Things; data acquisition; cloud–edge collaboration; resource virtualization; gateway

Citation: Zhang, Y.; Tang, D.; Zhu, H.; Zhou, S.; Zhao, Z. An Efficient IIoT Gateway for Cloud–Edge Collaboration in Cloud Manufacturing. *Machines* **2022**, *10*, 850. https://doi.org/10.3390/machines10100850

Academic Editors: Pingyu Jiang, Gang Xiong, Timo R. Nyberg, Zhen Shen, Maolin Yang and Guangyu Xiong

Received: 30 August 2022
Accepted: 20 September 2022
Published: 23 September 2022

1. Introduction

Cloud manufacturing is a service-oriented manufacturing paradigm in which all of the distributed manufacturing resources involved in the whole life cycle of product production are encapsulated as various manufacturing services [1]. It then provides on-demand manufacturing services for personalized manufacturing tasks submitted by consumers, so as to finally improve the response speed to meet the personalized requirements for customers and the utilization rate of manufacturing resources for manufacturers. At present, most researchers focus on these scientific issues, including virtualization and servitization encapsulation of manufacturing resources, manufacturing service selection, matching and combination. The successful implementation of service matching and composition needs to determine the current production capacity of manufacturing resources according to their real-time running status and load conditions, and then select and optimize the service combination based on this. However, the working state and workload of manufacturing resources can only be obtained by analyzing and processing the corresponding field operational data of manufacturing equipment. The operational and status data of manufacturing resources provide basic informational support for the industrial applications of the cloud platform (such as matching, selection and combination of manufacturing services, equipment fault diagnosis), and are used to drive the stable operation of business. That is to say, the efficient acquisition, transmission and integration of the working state data and related sensor data of distributed manufacturing resources have become the key research issue in cloud manufacturing. However, there is little research on how to upload and transmit the operational data and production status of distributed manufacturing resources to the cloud platform.

Up to now, the existing literature studies have fulfilled the cloud-based access operation of physical equipment data through supervisory control and data acquisition (SCADA), open platform communications unified architecture (OPC UA), ISA-95, etc. All the above methods are used to store, transmit and query the equipment data from the physical workshop to the cloud in a centralized process. It is not only difficult to guarantee the real-time data query requirements of cloud applications (i.e., order completion time prediction [2] and equipment fault diagnosis [3]), but also to avoid the serious overload of the cloud platform computing workload. This will lead to the slow response speeds of cloud applications in order to deal with unpredictable events, such as poor performance of task scheduling and difficulty in quickly rescheduling manufacturing resources when the pre-established service composition is abnormal. Therefore, the efficient and low-delay network transmission of field equipment data is the basic and key technology of all networked and service-oriented manufacturing systems.

With the continuous development of edge computing theory, the edge gateway [4] has become a new tool to establish information communication and data transmission between physical manufacturing resources and the cloud platform. It undertakes a part of the computing workload of the cloud platform to ensure the real-time performance of data analysis applications. The edge gateway is a development platform that integrates the core capabilities of network, computing, storage, and application at the edge of the network close to the data source, and provides edge intelligence services to meet the key requirements of digital factories in agile connections, real-time business intelligence, data analysis, and intelligent decision making. With the help of the edge gateway, it is not necessary to upload all of the generated data of the workshop to the cloud directly. The edge gateway preprocesses the operational data and sensor data of physical manufacturing resources, and analyzes the local data to a certain extent. Then, the preprocessed data and key data are uploaded to the cloud, and the cloud platform mainly focuses on big data processing and the execution of industrial applications with large amounts of computation. To sum up, cloud–edge collaboration not only greatly improves the real-time response speed of cloud applications, but also reduces the computing load of the cloud platform, which shows great potential for the development and application of the cloud manufacturing system.

The structure of the paper is organized as follows: Firstly, the background for, and key issues of, cloud manufacturing and cloud–edge collaboration is presented in the "Introduction". An overview of the relevant literature in terms of the service-oriented manufacturing paradigm and the Industrial Internet of Things is discussed in the "Literature Review". The section entitled "Research Methodology" elaborates the framework of the cloud manufacturing system, service-oriented informatization model, data transmission and integration method and the cloud–edge collaboration mechanism for training the artificial intelligence (AI) model. A prototype cloud manufacturing system with the functions of distributed manufacturing resources for cloud access and AI model training is designed, and two types of experiments are applied to verify the effectiveness and superiority of the above-proposed method in the "Case study". Finally, the "Conclusions" section summarizes the key technologies and achievements demonstrated in this paper.

2. Literature Review

2.1. Service-Oriented and Networked Manufacturing Paradigm

In order to effectively adapt to the fierce market competition and rapidly changing business environment, several service-oriented and networked manufacturing paradigms have been put forward by scholars, including cloud manufacturing [5], social manufacturing [6] and shared manufacturing [7]. The common purpose of the above-mentioned manufacturing modes is to integrate and gather massive manufacturing resources in different regions and different enterprises to form a manufacturing service pool, and to quickly respond to personalized product requirements through the efficient configuration and scheduling of manufacturing services. In the industrial and academic fields, cloud manufacturing is one of the most popular manufacturing modes, and its main research contents include the virtualization and servitization of heterogeneous manufacturing resources, and manufacturing service matching and composition.

The virtualization and servitization of manufacturing resources is the premise and foundation for the efficient access of distributed manufacturing resources to the cloud platform. Manufacturing resources are encapsulated into various manufacturing services with different functions, and aggregated into a service pool of the cloud platform, so as to facilitate the efficient usage and management of manufacturing resources. Zhang [8] proposed a general modeling method for heterogeneous manufacturing resources, and used ontology technology to establish a service-oriented packaging model for additive manufacturing and subtractive manufacturing resources, realizing that multiple types of manufacturing resources are connected to the collaborative manufacturing platform in a unified form, and facilitate the access and invocation of distributed manufacturing resources.

Manufacturing service matching and composition is the most popular research issue in cloud manufacturing. The matching process of manufacturing services is to select the appropriate services from the manufacturing service pool to establish a candidate set of services for multi-granularity manufacturing tasks through ontology technology. The manufacturing service composition selects high-quality service aggregates from the candidate sets of manufacturing resources based on logistics and time factors, and provides a complete manufacturing service scheduling scheme for personalized product requirements. Li [9] established a universal framework to describe the machine tools and their requirements in cloud manufacturing in detail, and proposed a machine tool matching method based on Markov decision processes and cross-entropy for dealing with a single manufacturing task with complex machine tool application requirements. Yang [10] proposed an event-driven and IoT-enabled dynamic service selection paradigm, using a service reservation mechanism based on AHP and simple additive weighting to evaluate whether the expected manufacturing service should be retained, so as to avoid the situation where there are no qualified services available and ensure that the high-QoS manufacturing service is provided to customers in a dynamic business environment. Zhang [11] put forward a flexible configuration method of distributed manufacturing resources, and adopted the NSGA-III algorithm to fulfill the manufacturing service composition that meets multi-objective optimization. Secondly, a manufacturing community with specific capability was detected and established from the manufacturing resource collaborative network based on the Louvain algorithm, and was applied to quickly respond to personalized order requirements and abnormal production situations.

2.2. Industrial Internet of Things

With the development of the new generation of information and communication technology, the concept of the Industrial Internet of Things (IIoT) is rapidly heating up in the industrial field, and was first articulated by General Electric (GE) [12]. The emergence of the IIoT allows the ubiquitous manufacturing resources to achieve effective connectivity, which enables information sharing and communication between manufacturing equipment from different geographical locations and enterprises, as well as between manufacturing equipment and the cloud platform, breaking the limitations and boundaries of physical space. With the help of IIoT, the field-level operational data generated by manufacturing equipment can be transmitted or pushed to the cloud platform or other edge nodes for analysis in real time. The collected industrial big data provide informational support for various value-added microservices deployed in the cloud platform [13], enabling the cloud platform to monitor, optimize, and control the distributed manufacturing process [14,15].

Liu [16] proposed an IIoT-supported general system architecture for the cloud manufacturing system, and developed a service-oriented data schema and the plug-and-play IIoT gateway solution, which realized efficient storage and querying of field-level manufacturing equipment data for the usage of cloud applications. Zhang [17] put forward a data collection and integration architecture of distributed manufacturing resources based on the IIoT, and designed a real-time manufacturing information integration service, which realized seamless two-way connectivity and interoperability between heterogeneous enterprise information systems. Wang [18] proposed an interoperable and flat IIoT architecture for low-latency data collection based on OPC UA and software-defined networking (SDN). The proposed edge gateway realized the adaptation of heterogeneous devices by providing a unified semantic information model and a standard data access interface. Lu [19] developed a general system architecture of cloud-based manufacturing equipment based on cyber-physical systems and big data analysis, which allows manufacturing equipment to connect to the cloud, and can be used to provide on-demand manufacturing services. It successfully handles the issue of Internet unavailability in machine monitoring and unpredictable big data processing challenges. Gonzalez [20] proposed an innovative six-layer IIoT architecture to organize the heterogeneous elements into microgrids for energy automation, information sharing and monitoring with the aid of the Modbus TCP network. Samer [21] introduced two scenarios to build the IIoT environment. In the first scenario, Modbus TCP is used for synchronous polling communication. This solution conforms to most industrial control systems and SCADA-like applications. In the second scenario, message queuing telemetry transport (MQTT) was applied to supplement the function of Modbus TCP to meet the requirements of asynchronous event-based communications. Through the structural analysis of the existing IIoT reference architectures, their classifications and target concerns, Bader [22] provided the consistency of shared concepts, and presented the structural mapping of concerns in each part of their respective reference architectures. Wollschlaeger [23] summarized the impact of the Internet of Things and cyber-physical systems on industrial automation from the perspective of Industry 4.0, investigated the working status of time-sensitive networking (TSN), and clarified the role of fifth generation (5G) technology in automation.

The IIoT technology has brought the possibility of cloud–edge collaboration. The focus of the edge node is to realize data analysis applications with high real-time performance, while that of the cloud platform is to realize data analysis applications with high computational load. Network and communication connections could be established between the edge node and cloud platform through message middleware, and business cooperations between them (i.e., AI model training and complex task scheduling) can be performed and realized based on their data analysis results. For example, the manufacture of a complex product needs the cooperation of multiple manufacturing enterprises in cloud manufacturing. The edge node is responsible for analyzing and monitoring the production progress of the corresponding manufacturing enterprise, and regularly reporting the analysis results to the cloud platform with the help of IIoT. Then, the cloud platform configures and schedules resources according to the production status of the manufacturing enterprise.

Zhang [24] proposed a multi-agent manufacturing system integrating a self-organization mechanism and self-learning strategy. Based on the historical scheduling data and cloud–edge collaboration mechanism, the production scheduling model inside the machine agent was periodically trained and optimized by the proximal policy optimization algorithm. Finally, the self-organization production for personalized order and adaptive decision under dynamic disturbance was realized. Wang [25] proposed a knowledge sharing mechanism between manufacturing resources based on cloud–edge collaboration. In this mechanism, the ontology model is applied to encode the manufacturing process knowledge, and the existing primitive knowledge contained in the knowledge base deployed on the cloud platform and edge nodes is effectively screened, matched, and combined to realize the learning process between manufacturing resources. Yang [26] proposed a microservice-enabled condition monitoring platform for smart manufacturing systems, which effectively integrates the powerful computing and storage capabilities of the cloud layer and the real-time data analysis of the edge layer, and introduces the cloud–edge collaborative mechanism to realize real-time equipment fault diagnosis and improve its prediction accuracy. In addition, the microservice management system is applied to construct lightweight edge and cloud services, realizing the flexible deployment, and upgrading of manufacturing services.

2.3. Research Gaps

From the literature review, most of the studies focus on the virtualization and servitization of heterogeneous manufacturing resources, manufacturing service matching and composition, and the architecture design of the IIoT-enabled cloud manufacturing system. However, there are still the following issues to be studied to promote the further application of the cloud manufacturing system in the practical industrial field.

(1) There is a lack of an efficient method to transmit and distribute the generated massive operational data and sensor data from the physical resource to the cloud platform or edge nodes, and the corresponding storage mechanism of massive data. The collection, transmission, storage, and distribution of field-level equipment data is fundamental for the stable operation of various value-added manufacturing services and cloud applications.

(2) There is a lack of flexible deployment and remote upgrade technology in AI models. Most studies in the literature focus on the network structure design of AI models and their parameter optimization. In addition, the execution of manufacturing services depends on the high-level decision making ability of the cloud platform, which leads to its lack of real-time data and significantly increases the workload of the cloud platform. In this paper, a cloud–edge collaboration mechanism is proposed to realize the flexible deployment of various AI models (i.e., diagnostic and prediction models) to the edges in a plug-and-play manner, and the MQTT-supported remote updating of model parameters after training in the cloud platform.

3. Research Methodology

3.1. The Overall Framework of the Cloud Manufacturing System Based on Edge Intelligence

Inspired by the edge computing theory, a three-layer IoT framework (including physical resources, edge computing node, and the cloud platform) is developed to meet the requirements of real-time local data analysis and application, as well as the efficient training of AI models and big data analysis. The physical resources are the executors of production activities and the sources of mass manufacturing data. The edge computing node is responsible for real-time processing, analysis, and application of the collected local data. The cloud platform is responsible for analyzing the stored mass manufacturing data, driving the stable operation of industrial applications and the efficient training of the AI model.

The overall framework of the cloud manufacturing system consists of three parts: the cloud manufacturing platform layer, edge gateway layer, and physical resource layer. The cloud manufacturing platform is equipped with powerful computing, data storage, data processing, and data analysis capabilities. Industrial applications with various functional types are deployed and ran in the cloud platform, including task decomposition, service matching, selection and composition, model training and remote monitoring of workshop execution processes. Among them, service matching and composition are the key components, which organize and configure manufacturing services efficiently to meet changing product requirements. In addition, with the support of strong computing power, machine learning models, including the equipment fault diagnosis model, remaining life prediction model, order completion time prediction model, etc., could be trained and optimized periodically by the cloud platform so that the optimized machine learning model can be downloaded and deployed in the edge gateway. The concept of "Edge" refers to the physical workshop, which is composed of a great deal of manufacturing equipment, and has the ability to provide the part-level production services.

The edge gateway is a bridge for information communication and business cooperation between the physical manufacturing resources and the cloud platform. It is connected to the manufacturing resources through Ethernet, serial port, and radio frequency identification (RFID), and is connected to the cloud platform based on MQTT protocol. The edge gateway can not only capture equipment operational data and sensor data, but also send control instructions to equipment or workshop systems based on the data analysis results of its edge application. The physical resource layer refers to all manufacturing entities involved in the whole life cycle of product manufacturing, and is the core executor of production activities, which mainly include CNC machine tools, machining centers, 3D printers, robots, and logistics equipment, etc. In addition, programmable logic controllers (PLCs), as the most popular type of controller in the real-world industry, should also be placed in the physical resource layer.

The running logic of the cloud manufacturing system is shown in Figure 1. Firstly, heterogeneous physical manufacturing resources are encapsulated into various manufacturing services according to their functions, characteristics, and performance parameters, and are shared and configured in the cloud platform. Secondly, the connection between the cloud platform and physical workshop is established through the edge gateway. The edge gateway captures the real-time data of equipment workload, spindle vibration, motor temperature, operational process, and workpiece processing progress from the physical equipment. The collected data are subsequently preprocessed, including data standardization, data cleaning and feature extraction. The preprocessed data are input into various edge applications for data analysis, such as equipment fault diagnosis and remaining life prediction based on spindle vibration information, and production scheduling based on the workload of equipment and process attributes of work in progress (WIP). Based on the data analysis results, control instructions will be released to physical manufacturing resources to guide them to complete the corresponding production actions.

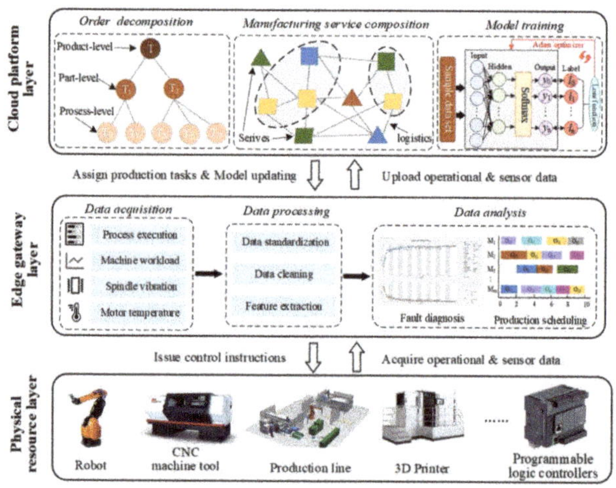

Figure 1. The framework of the cloud manufacturing system based on cloud–edge collaboration.

In addition, the equipment-related, sensor-related, and product-related operational process data will be uploaded to the cloud platform after preprocessing, and stored in the cloud database. These data are distributed to various industrial application interfaces of the cloud platform in a publish–subscribe system. On the one hand, the collected industrial big data are analyzed and processed by the cloud platform, and are streamed to the industrial application data items they intend to stream to the cloud (i.e., remote monitoring, task scheduling, manufacturing service matching and composition). For example, the orders submitted by consumers are decomposed into the part-level manufacturing tasks, and then various manufacturing services are matched and combined based on the working status and performance indicators of shared manufacturing resources in the resource pool of the cloud platform to form a service scheduling scheme. Finally, part-level manufacturing tasks are distributed to the edge server of the corresponding workshop based on the scheduling scheme. On the other hand, the cloud intelligence application will periodically read and query the historical data of the execution process of the equipment/workshop stored in the database, which will be applied to train and optimize the parameters of the machine learning models deployed on the edge gateway (i.e., the equipment fault diagnosis model, the remaining completion time prediction model, and the production scheduling model). Finally, the updated machine learning model is deployed and downloaded to the edge gateway in a cloud–edge collaborative process. It is beneficial to enhance and upgrade the analysis ability of edge applications. Finally, the physical manufacturing resources carry out production activities according to the control instructions issued by the corresponding edge gateway.

3.2. Service-Oriented Informatization Model for Manufacturing Equipment

In the cloud platform, manufacturing resources (i.e., CNC machine tools, 3D printers, and robots) from different regions and different enterprises are encapsulated into various types of manufacturing services, aiming at providing service demanders with on-demand service compositions. How to describe heterogeneous manufacturing services has become a research issue. In this paper, a service-oriented manufacturing resource informatization model is established to describe manufacturing resources with different structures and functions in a unified way, and provide informational support for subsequent service matching, composition, and scheduling, and other applications.

As shown in Figure 2, the information model of manufacturing services consists of the following parts: basic information, service capability, service quality, and working status. The basic information shows the ID, name, geographical location, granularity, and working hours of manufacturing services. Service granularity is divided into three levels, the process-level, part-level, and product-level, which correspond to the granularity of manufacturing tasks when performing service matching operations. Geographical location refers to the geographical location of the manufacturing services, which is used to determine the logistics costs of products transferred between manufacturing resources when performing service composition operations. Working hours refers to the daily working hours of the manufacturing service, and the unit of its value is min/d. Service capabilities mainly describes the static function and dynamic performance of manufacturing resources, including service type, available material information, production efficiency, service cost per unit time, and service performance. Service types include machining, 3D printing, etc. The available material information indicates the materials that can be processed by the manufacturing service. Production efficiency refers to the speed at which manufacturing services complete each task, and the range of its values provided by service providers according to the experience of operators. In the case of machining service, the service performance indicates the machining size of the part, the machining accuracy of the process and the achievable surface roughness. In the case of 3D printing service, the service performance indicates the printing size, printing accuracy, and printing shape. Service quality is determined by the following aspects: product qualification rate, delivery cycle, cost performance, response speed, and service reputation. The qualification rate indicates the comprehensive qualification rate of the manufacturing service to complete all manufacturing tasks to date. The response speed indicates the time from receiving the task request sent by the platform to confirming the receipt of the manufacturing task. Service reputation reflects the evaluation of service demanders to service providers after the completion of manufacturing tasks, and is periodically maintained and updated by the platform manger. Service performance and service quality are used as indicators to guide the matching of manufacturing services with manufacturing tasks. Working status shows the current workload status, task progress, running state, and health condition of manufacturing services. Among them, workload status consists of idle, light load, and full load. Running state indicates the operation status of key components or the whole instrument, including surface temperature, vibration, working current, spindle speed, etc.

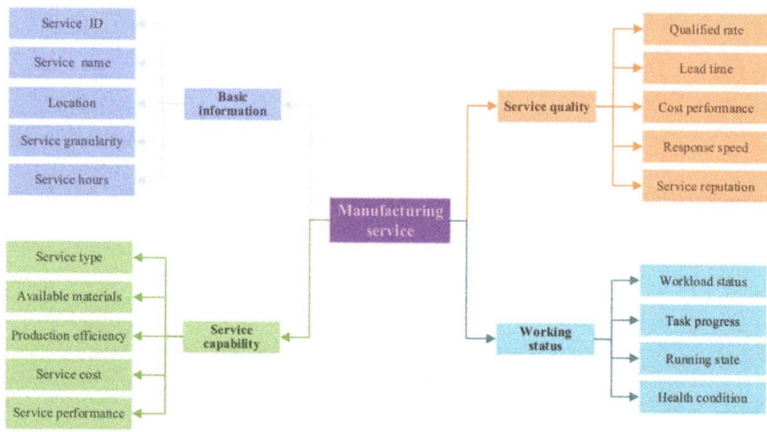

Figure 2. Service-oriented informatization model for manufacturing equipment.

3.3. *Data Transmission and Integration Method Based on Message Middleware*

The operational data and sensor data of field-level manufacturing equipment are the bases supporting the stable operation of various industrial applications in the cloud platform. Therefore, how to acquire, transmit, and store the real-time data of massive and heterogeneous manufacturing equipment from different regions and different enterprises have become some of the key technological factors for the successful implementation of the cloud manufacturing system. Various types of manufacturing equipment are connected with edge gateways (i.e., Raspberry Pi and industrial personal computers) through Ethernet ports. The acquisition of operational data and the release of control instructions are realized with the support of the connection between the edge gateway and manufacturing equipment. The overall data transmission and integration process is shown in Figure 3.

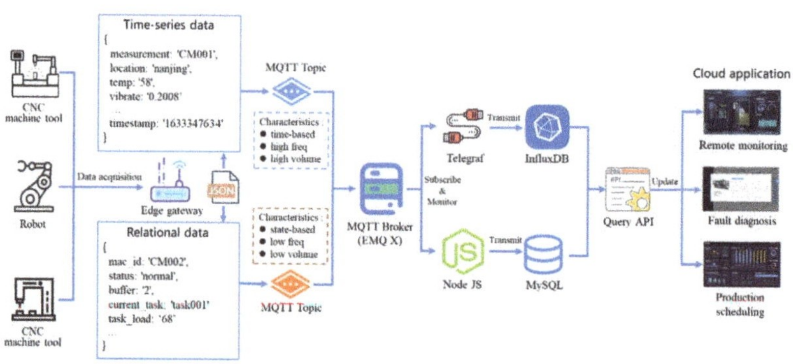

Figure 3. Data acquisition and transmission method for distributed manufacturing resources.

Firstly, the intelligent gateway deployed at the edge of the workshop is applied to acquire the operational data and sensor data of manufacturing equipment with different communication interfaces and industrial protocols (i.e., Modbus, Modbus TCP, OPC UA, and Focas) through its protocol analysis module. The acquired data mainly include the operational data related to manufacturing equipment (i.e., machining progress, working status, and remaining workload), the relevant sensor data (i.e., spindle vibration, motor temperature, and working current), and the execution process data of the workshop. The operational data of manufacturing equipment include spindle speed, operation status of machine, remaining tasks to be processed in the buffer, operation information of current jobs, machining progress of production tasks, total number of processed workpieces, working status, and alarm information. The execution process data of the workshop include job scheduling scheme, product qualification rate, machine utilization rate, and workload distribution of various machines. The sensor data mainly include the vibration signal and acoustic emission signal of the spindle motor, the working current of the equipment, and the surface temperature of its key components.

According to the characteristics of the data, the collected equipment data are divided into time series data and relational data. The time series data mainly refer to the sensor data collected from the equipment, and exhibit the characteristics of time dependence, high frequency, and high volume. The time series data are conducive to the deep learning model to evaluate the state and performance of manufacturing equipment. For example, the fault diagnosis model analyzes the fault cause of the machine tool based on the vibration data of the spindle motor. The relational data mainly refer to the working status of manufacturing equipment or workshop system, and possess the characteristics of state dependence, low frequency, and low volume. The relational data are used to assist the service matching and composition module application in the cloud platform to make the task allocation decisions and manufacturing resource configurations, and are also applied to support the model training application in the cloud platform to optimize the scheduling strategy of workshop production activities.

Secondly, both of the above two types of equipment data are encapsulated in a JavaScript Object Notation (JSON) format. The JSON format has the characteristics of simple structure, easy reading and writing capacity, low bandwidth occupation, and is compatible with various programming languages (i.e., Python, Java, and JavaScript). The JSON file format of time series data is shown in Table 1, which consists of various necessary attributes such as measurement, tag, field, and timestamp. The measurement indicates the identification of the shared equipment. The tag data indicate the basic information of the shared equipment with index (i.e., ownership, location), and are stored in the form of key–value pairs. The field data indicate the real-time sensor data without indices (i.e., temperature, voltage, vibration, and spindle speed), and are stored in the form of key–value pairs. The timestamp indicates the time point of data collection. Among them, the attribute names of measurement, timestamp, and tag need to be specified in a data collector called Telegraf. However, it is not necessary to specify the attribute names of the field data. The JSON file of relational data contain attributes such as equipment identification, device working status, and task execution progress. The transmission and storage procedures of the time series data are shown in the Figure 4.

Table 1. The JSON format of time series data uploaded to the cloud platform.

Data Format	Attributes	Description
{		
"measurement": "CM001",	Measurement	Identification of the equipment
"host": "NUAA",	Tag: host	The stored basic information with indices (i.e., ownership, location)
"location": "Beijing",	Tag: location	
"surface temperature": 50,	Field: temperature	The stored sensor data without indices (i.e., temperature, voltage, vibration, and spindle speed)
"working voltage": 220	Field: voltage	
"spindle vibration": 3	Field: vibration	
"spindle speed": 1500	Field: spindle_speed	
...	Field	
"timestamp": 1633590165557,	Timestamp	Time point of data collection
}		

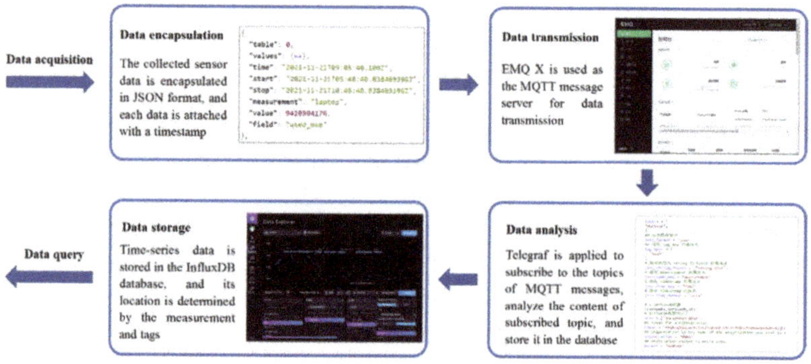

Figure 4. Transmission and storage procedures of time series data.

The JSON-format data are then published via MQTT Broker (i.e., EMQ X) in the form of MQTT topics through MQTT client on the edge gateway. MQTT Broker is a message server oriented to MQTT protocol, which acts as a message middleware system, and is applied to transmit and distribute the collected data related to physical equipment instantly. Telegraf is a data collection tool for InfluxDB, developed by InfluxData Company. It is applied to monitor and subscribe the topics of MQTT messages, analyze the content of subscribed topics, and store these data corresponding to the topics in the time series database InfluxDB. The storage location of the data is determined by the measurement and tags in the JSON data schema. The topics of the relational data are subscribed to the message middleware through NodeJS service, and the corresponding data are stored in the relational database MySQL.

Finally, industrial applications (i.e., equipment status monitoring, predictive maintenance, service composition, and equipment fault diagnosis) exchange information with InfluxDB and MySQL databases through the predesigned query interfaces to obtain real-time running data of distributed equipment, thus establishing a technical foundation for driving the equipment remote monitoring module, real-time detection of processing progress, and fault diagnosis of high-end equipment.

3.4. Cloud–Edge Collaboration Mechanism

Edge intelligence is a new technical scheme combining edge computing with artificial intelligence. On the one hand, with the development of the Internet of Things, the generated data from physical manufacturing equipment need to be analyzed by intelligent algorithms to realize intelligent data application and provide higher data usability. The intelligent algorithms represented by deep learning can extract implicit and effective information from equipment data to improve the efficiency and accuracy of decision making. On the other hand, edge computing can provide more real-time data and application scenarios for intelligent algorithms. Generally speaking, artificial intelligence (AI) and big data analysis applications are deployed and ran in cloud servers and big data centers, while edge computing nodes can directly obtain massive operational data from physical equipment and realize intelligent calculation and analysis, which will strongly promote the further popularization and development of AI applications, especially IIoT-enabled applications. The proposed cloud–edge collaboration mechanism is shown in Figure 5.

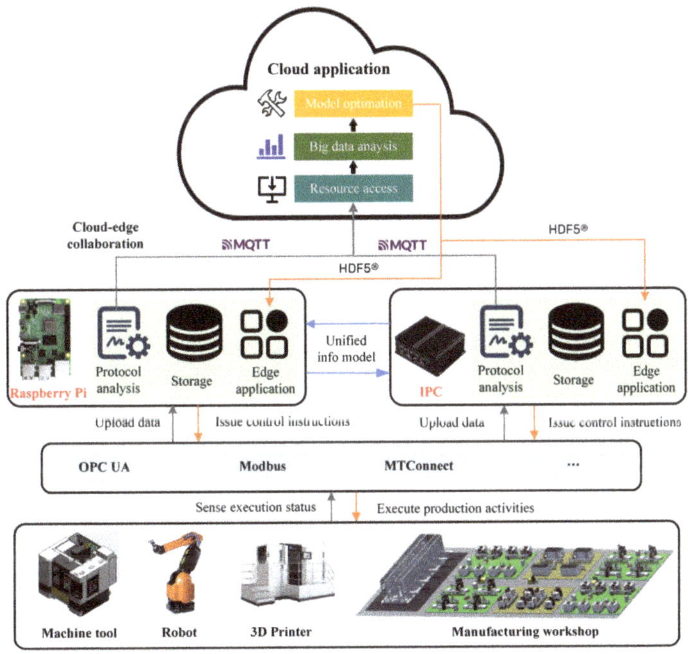

Figure 5. Cloud–edge collaboration mechanism.

In the workshop, the edge gateway establishes communication connection with manufacturing equipment, robots, and logistics equipment. The functional modules of gateway are composed of protocol analysis, data storage, and edge application. The industrial protocols carried out by different equipment are inconsistent (i.e., OPC UA, OPC DA, Modbus, Modbus RTU, and MT Connect). The edge gateway carries out information extraction, data format conversion, and data preprocessing on real-time and heterogeneous equipment data collected from the bottom layer through the protocol analysis module. The preprocessed industrial process data are then stored in the edge server in a standardized format, and the data access interface is provided to the edge application. The edge gateway is responsible for the localized analysis and processing of real-time equipment data and workshop operational data, which improves the security of industrial process data analysis and real-time data processing. For example, the equipment fault diagnosis application deployed on the edge gateway can analyze and judge the fault causes of machine tools through the collected spindle vibration data and motor surface temperature. In addition, the production scheduling application schedules tasks and resources according to the working state of equipment, and the process attribute information of WIP. Based on the analysis results of edge applications, the edge gateway sends task requirements and control instructions to the corresponding equipment to guide them to complete the production activities. Information communication between gateways is based on a unified information model, which can realize task division, cooperation, and information sharing.

In addition, the artificial intelligence model of edge application is constantly being updated and optimized with the help of the cloud–edge collaboration mechanism. The resource access module in the cloud platform subscribes to related topics of messages through MQTT protocol, and receives the preprocessed industrial process data from the edge. These data will be stored and managed in the cloud database for other industrial applications to call. For example, the big data analysis module calls historical data about equipment operation or workshop operation, and extracts useful characteristic information behind the data. With the powerful computing performance of the cloud platform, the model optimization module trains and optimizes the artificial intelligence models (i.e., equipment fault diagnosis model, production scheduling model, logistics scheduling model) by feeding into the platform labeled historical sample data, the extracted feature information, and various optimizers (i.e., Adam, RMSProp, and SGD). Then, the neural network parameters of the trained artificial intelligence models are generated into Hierarchical Data Format Version 5 (HDF5) files. These HDF5 files are then transmitted, deployed, and updated at the edge gateway through TCP/IP protocols, which are applied to upgrade the parameters of machine learning models in edge applications.

The efficient training and flexible deployment mechanism of the artificial intelligence model based on cloud–edge collaboration will provide a large number of modular, plug-and-play, intelligent data analysis tools, and edge applications for manufacturers in the discrete manufacturing field, and continuously improve the intelligent manufacturing capacity of factories. At the same time, it takes advantage of the real-time performance of edge intelligence and the powerful computing and analysis capabilities of the cloud platform.

4. Case Study

4.1. Cloud Access and Remote Monitoring of Distributed Manufacturing Resources

In this paper, the computer numerical control (CNC) lathe is selected as the experimental object to verify the effectiveness of the cloud access and remote monitoring methods proposed in this paper. As shown in Figure 6, the temperature sensor, vibration sensor, and hall current sensor are installed to sense the surface temperature, motor vibration, and working current of the spindle motor of the CNC lathe. The data acquisition card is used to collect the output analog signals (i.e., 5–20 mA) of the sensors, and then convert them into digital signals (i.e., binary value). Then, the digital signal data are transmitted by the data acquisition card to the industrial personal computer (IPC) via RS485 protocol. The operational data of the equipment (i.e., task progress, cumulative workload, and job schedule) can be obtained from the equipment interface through Ethernet. The working status of PLC is acquired by the IPC through the equipment interface. The IPC is applied as the edge gateway to upload and transmit the collected operational data and sensor data of the CNC lathe. The uploaded data are stored in the cloud database and provide the query application program interfaces (APIs) for the usage of cloud applications.

Tables 2 and 3 show the relational data structure and time series data schema of the CNC lathe. In Table 2, the basic information, manufacturing capacity, and service quality of the CNC lathe are elaborated in detail based on the service-oriented information model proposed in Section 3.2. Among them, basic information and manufacturing capability information are mostly static data, which are stored in the system by manual input. Part of the information on manufacturing capability is in the form of dynamic data, which can be obtained from the edge gateway and CNC system. The service quality information is calculated by the cloud platform according to the historical operational data. In Table 3, the basic information, working status, and task progress of the CNC lathe are elaborated in detail. Its basic information is in the form of static data, which is stored in the system by manual input. Work status and task progress information are dynamic data, which is mainly obtained from the data acquisition card and CNC system.

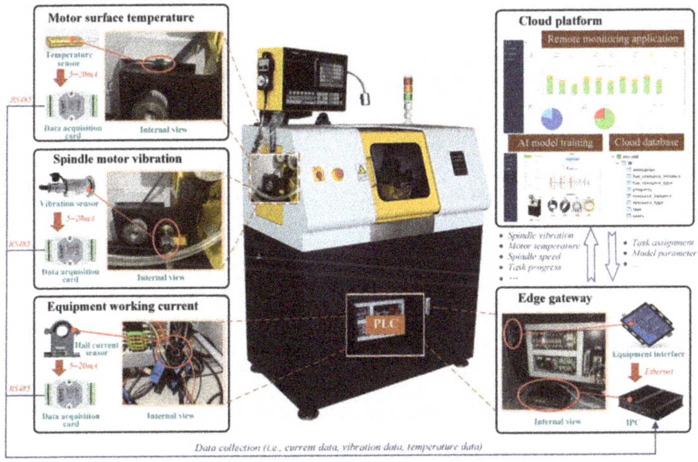

Figure 6. Digital transformation of the CNC lathe.

Table 2. Relational data schema of the CNC lathe.

Category	Column	Value	Data Source
Basic information	ID (Primary key)	CM001	Manual input
	Host	NUAA	Manual input
	Location	Nanjing	Manual input
	Granularity	Part-level	Manual input
	Service hours	8 (h)	Manual input
Capability	Service type	Machining	Manual input
	Available materials	Steel	Manual input
	Production efficiency	50 (min/PC)	Manual input
	Service cost	11 ($/h)	Manual input
	Service performance	Machining size: 80 × 80 × 300 (mm) Repeat accuracy: 0.02 (mm)	Manual input
	Current workload	68 (min)	CNC system
	Job schedule	Task queue	Edge gateway
	Health condition	Normal	Edge gateway
Quality	Qualified rate	0.96	Cloud platform
	Lead time	2 (day)	Cloud platform
	Cost performance	Good, normal, or bad	Cloud platform
	Response speed	<1 (h)	Cloud platform
	Service reputation	85	Cloud platform

The process of data transmission and application is demonstrated in Figure 7. Firstly, the field sensor data and operational data of CNC lathe are acquired and collected by RS485 and Modbus TCP. Secondly, the edge gateway preprocesses the collected data (i.e., outlier elimination and value conversion) and converts this data into JSON format, and the JSON files are sent to the message middleware through MQTT client. Then, the related MQTT topics are monitored and received by the cloud platform, and these data are stored in the cloud database. Finally, cloud applications use flux and SQL language to query data.

Table 3. Time series data schema of CNC lathe.

Measurement	Tag and Field	Value	Data Source
Basic information	Equipment ID	CM001	Manual input
	Host	NUAA	Manual input
	Location	Nanjing	Manual input
Working status	Equipment ID	CM001	Manual input
	Motor surface temperature (°C)	36	Data acquisition card
	Spindle vibration (Hz)	4.5	Data acquisition card
	Spindle speed (r/min)	1500	CNC system
	Feed speed (mm/min)	70	CNC system
	X, Y, and Z axis coordinates	(110.8, 22.6, 10.0)	CNC system
	Working current (A)	4.3	Data acquisition card
	Working voltage (V)	220	Data acquisition card
Task progress	Equipment ID	CM001	Manual input
	Task progress (%)	70	CNC system
	Current machining time (min)	21	CNC system
	Remaining machining time (min)	9	CNC system

As shown in Figure 8, the prototype system of the cloud platform is divided into homepage module, manufacturing resource module, manufacturing task module, equipment monitoring, and health management module. The manufacturing resource provider can log into the system and view the relevant information (i.e., name, type, and description) and current status (i.e., normal operation, idle, and fault) of the shared manufacturing resources in the cloud manufacturing system. In the homepage module, manufacturers can view the overall working status of manufacturing resources, including product qualification rate, number of tasks, revenue, enterprise reputation, equipment workload, and health status. The manufacturing resource provider submits the attribute information of the manufacturing resources to be shared in the manufacturing resource module, and checks the task progress and accepts new tasks in the manufacturing task module. The equipment monitoring and health management module queries the values of relevant parameters (i.e., spindle speed, task completion rate, processing progress, and motor surface temperature) by accessing the database API, and dynamically displays them on the monitoring page in the form of data curves and tables.

The traditional method of industrial data transmission and integration generally adopts the data storage scheme based on MySQL, while this paper proposes a hybrid data storage scheme based on InfluxDB and MySQL. In this scheme, MySQL is used to store relational data, and InfluxDB is used to store time series data. In order to verify the superiority of the proposed method, the above two schemes are used to write and store the collected data, and their performance is compared in terms of write time and storage space. The results of the data writing experiments of the two schemes are shown in Figure 9. Under the condition that the proposed data storage method is limited by the hardware of the test computer, the data write speed can reach up to 90,000 row/s. If the hardware configuration of the test computer is upgraded, the write speed can be further improved. However, the writing speed of the traditional data storage method is maintained at about 40,000 row/s. The experiment result proves that the writing speed of the proposed method is more than twice that of the traditional method. The storage space experiment results of the two data integration schemes are shown in Figure 10. The proposed method can compress the collected time series data, but the traditional method cannot. When the amount of data transmitted is 1.0*108 rows, the storage space occupied by the proposed method is 235 MB, while the storage space occupied by the traditional method is 562 MB. In other words, the storage space occupied by the proposed method is obviously less than that of the traditional method, which is nearly two times larger. In summary, the proposed data acquisition and transmission method is significantly better than the traditional method in terms of writing speed and storage space.

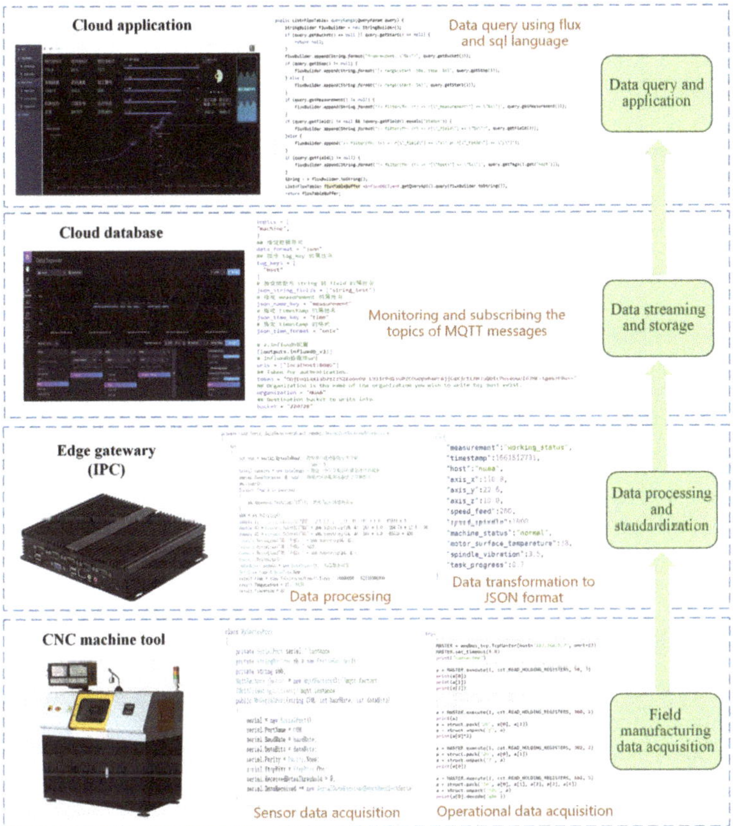

Figure 7. Data transmission and application process of the edge gateway for the CNC lathe.

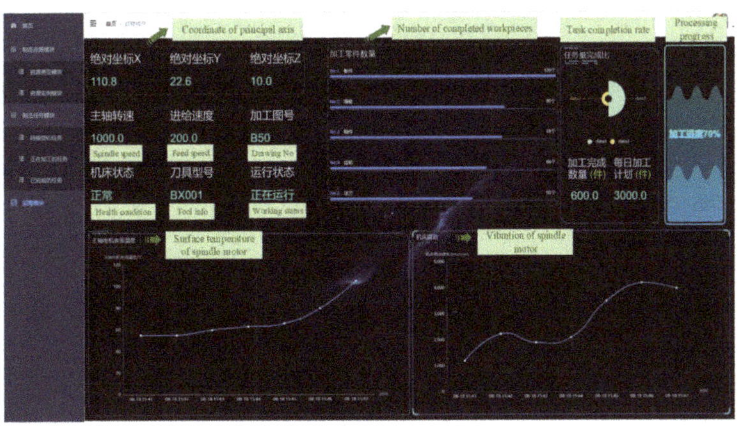

Figure 8. Remote monitoring of equipment working status.

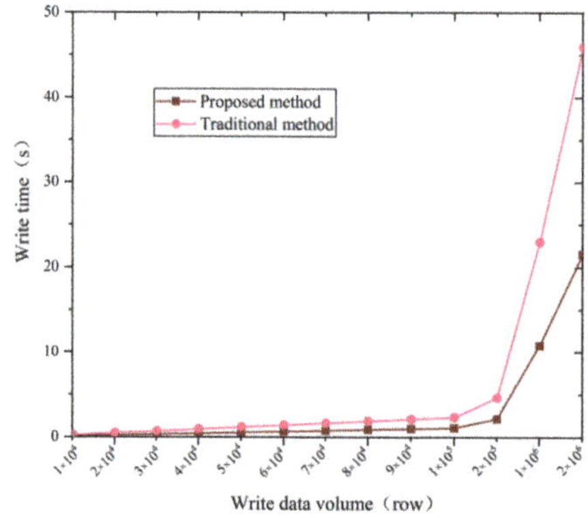

Figure 9. Write speed comparison.

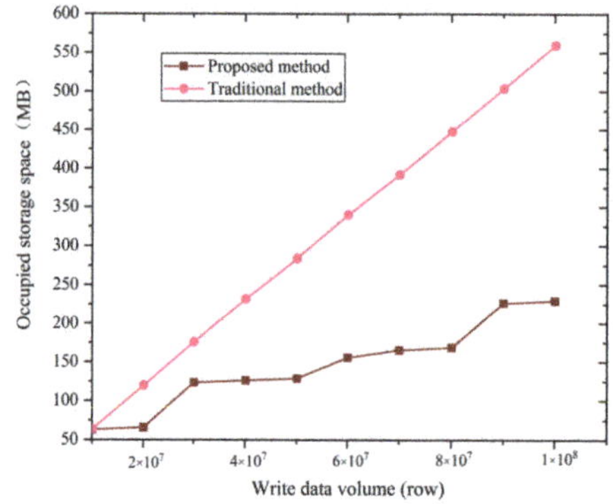

Figure 10. Storage performance comparison.

4.2. Cloud–Edge Collaborative Updating of Artificial Intelligence Model

In Figure 11, the user interface of the equipment monitoring and health management module in the cloud platform is shown in detail. For this interface, the results of the time–frequency domain analysis of vibration data are shown in Figure 11a. The AI model updating, parameter settings, and structure of the convolutional neural network, equipment identification, and fault cause analysis are depicted in Figure 11b.

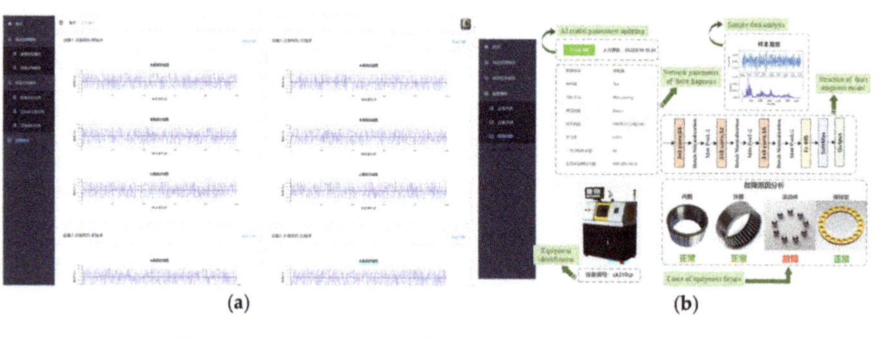

Figure 11. Remote upgrade of fault diagnosis model: (**a**) preliminary analysis of vibration data; (**b**) application and upgrade of fault diagnosis model.

The equipment monitoring and health management interface of the cloud platform adopts the frontend and backend separation development mode. The Springboot, Mybatis, and Springcloud frameworks are applied to develop the backend of the cloud platform. The InfluxDB and MySQL databases are used for the persistent operation of data storage. We also apply the progressive JavaScript framework (i.e., Vue) to design and build the view layer of the monitoring interface, and use an open-source JavaScript visualization library (i.e., Apache Echarts) to dynamically display temperature and vibration changes in the form of charts. The edge gateway is an industrial personal computer with 4G memory and i5-6200U CPU. The cloud platform is equipped with a cloud server with 16G memory and NVIDIA V100 GPU. Python 3.6 is used to write the code of the fault diagnosis model, and TensorFlow 1.14 is applied to build the neural network structure of fault diagnosis model.

The convolutional neural networks were applied to build the fault diagnosis model in this experiment, and this fault diagnosis model includes three convolution layers and three full connection layers. Among them, the first convolution layer adopts a convolution kernel with a size of 3×3, the moving step of the convolution kernel is 1, the shape of the input data is 64×64, the padding mode is the same, and maximum pooling is applied to compress the data. The second convolution layer adopts a convolution kernel with a size of 3×3, the moving step of the convolution kernel is 1, the shape of the input data is 32×32, the padding mode is the same, and maximum pooling is applied to compress the data. The third convolution layer adopts the same convolution kernel, moving step, padding mode, and pooling mode. The shape of its input data is 16×16. The fully connected layer is a three-layer neural network with 485,300,100 neurons in each layer, and the activation function is ReLU. Finally, the fault diagnosis model outputs the prediction results of five labels through the Softmax layer. The training parameters of the model are set as follows: learning rate is 0.001, batch size is 64, and the loss function is root mean square error (RMSE). The parameter updating of the fault diagnosis model will be completed after the button "model update" is clicked. The fault diagnosis model analyzes the vibration signal of the spindle motor to identify the cause of the equipment fault. In this paper, the factors leading to machine failure mainly include the inner race, outer race, ball, and holder.

The underlying feature engineering process includes two parts: outlier elimination and data enhancement. In order to prevent abnormal data from causing error in the fault diagnosis results, the feature engineering process shall first eliminate abnormal data points. The standard deviation of the sample data is calculated, taking three times the standard deviation as the limit, and then the sample data points that exceed the limit are removed and classified as abnormal data. In the actual industry, the sample data of machine failures are very precious and rare. Therefore, data enhancement is often used to enrich the number of samples for machine failures. In this paper, the technique of overlapping sampling is applied to increase the number of samples for machine failures. As shown in Figure 12, when the training samples are collected from the original signal, there is an overlapping area between a signal segment and the subsequent signal segment, and the sampling step is set to 1/3 of the sample length (i.e., 1024). Using this method, the number of training samples can be increased to nearly three times that of the original.

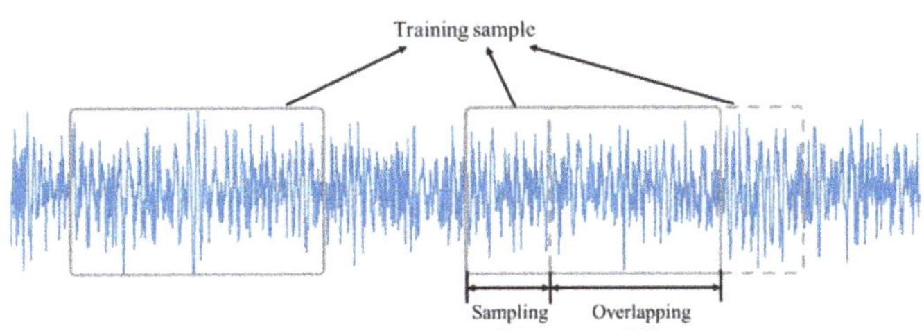

Figure 12. Overlapping sampling method.

In order to better demonstrate the effectiveness of our findings, the proposed cloud–edge collaborative fault diagnosis method is compared with the edge-based method and cloud-based method in terms of three indicators (i.e., execution time, training time, and prediction accuracy). The edge-based fault diagnosis method processes and analyzes the acquired vibration data locally, and the training and updating of the fault diagnosis model are also carried out on the local server. The cloud-based fault diagnosis method uploads all of the collected vibration data to the cloud platform, then processes and analyzes these sample data through cloud applications, as well as conducts model training, and finally feeds back the prediction results of the sample data to the workshop. The proposed cloud–edge collaborative fault diagnosis method directly processes and analyzes the collected vibration data through edge applications, and uploads the preprocessed sample data to the cloud platform. The saved sample data are applied to train the fault diagnosis model and optimize the parameters by the model optimization application in the cloud platform. The parameters of the trained fault diagnosis model are periodically transmitted and sent to the edge gateway in the form of HDF5 files through the cloud–edge collaboration mechanism. The edge gateway updates and deploys the parameters of the fault diagnosis model based on HDF5 files. In this experiment, a total of 100 sample data points (including 20% fault samples and 80% normal samples) were tested for fault diagnosis, and the overall accuracies of the predicted results were taken as the evaluation indicator of model performance. In addition, the training time of the fault diagnosis model is used as another evaluation indicator.

The traditional AI model training method (i.e., edge-based) optimizes and updates the model through the terminal equipment at the edge, while the proposed method is to train the model on the cloud platform, and deploy and apply the trained model at the edge. The experimental results are shown in Table 4. Due to the powerful GPU computing capability of the cloud platform, the proposed and cloud-based methods only take 11 min to complete the model training process. However, the traditional method (i.e., edge-based) is limited by the computing capacity of the edge server, and takes 56 min. In other words, the model training speed of the proposed and cloud-based methods is much faster than that of the traditional method. In addition, the fault prediction accuracy of the proposed method and the cloud-based method is 96%, while that of the traditional method is only 91%. The proposed and cloud-based methods only take a short time to complete the training of the AI model; therefore, the fault prediction accuracy of the AI model optimized by the proposed and cloud-based methods is higher than that of the traditional method in the same training period. The execution time represents the time consumed by the fault diagnosis model from data generation to fault prediction completion. The shorter the execution time, the better the real-time analysis performance of industrial applications. In Table 4, the execution time of the edge-based or the proposed method is 0.8 s, while the cloud-based method consumes 4.6 s. The reason for this result is that the cloud-based method needs to transmit the generated vibration data from the equipment to the cloud platform, and then generate the prediction results through the cloud application system and feed them back to the workshop. This data transmission process consumes a lot of time.

Table 4. Performance comparison of fault diagnosis models after upgrading.

Method	Execution Time (s)	Training Time (min)	Prediction Accuracy
Only edge	0.8	56	91%
Cloud-based	4.6	11	96%
Cloud–edge collaborative	0.8	11	96%

5. Conclusions

In cloud manufacturing, the data acquisition, data transmission, and cloud access of distributed manufacturing resources are the crucial research factors that provide support for the stable operation of various machine control software and industrial applications (i.e., service matching and composition, AI model training, and remote monitoring) in the cloud platform.

With the support of the IIoT, this paper proposes a cloud–edge framework of the cloud manufacturing system, which realizes the cloud access of distributed manufacturing resources, the efficient acquisition of equipment data available at any time, and real-time data analysis and processing. Firstly, a service-oriented information model for heterogeneous manufacturing equipment is proposed, which can unify the collected heterogeneous equipment operational data and sensor data. Secondly, based on the message middleware, the running status data and related sensor data of physical equipment are transmitted and pushed from the workshop to the cloud platform, which realizes the efficient storage and ordered distribution of time series data and relational data. Finally, the training, upgrading, and deployment of the edge AI model is proposed based on cloud–edge collaboration mechanism. With the support of powerful computing ability of the cloud platform, the training speed of the AI model is greatly accelerated, and its prediction accuracy increases subsequently. With the help of the real-time data processing capability of multiple edge nodes, the response speed of the AI model deployed on the edge getaway is greatly improved. Taking a numerical control machine tool as an example, the experimental results verify the effectiveness of cloud access of distributed manufacturing resources proposed in this paper. Taking the updating of fault diagnosis AI model as an example, it is verified that the cloud–edge collaboration mechanism proposed in this paper realizes the high efficiency of the AI model training and updating methods.

In this paper, the management of edge applications lacks some flexibility, and it is impossible to directly download the complete industrial intelligence applications from the cloud, or uninstall edge applications. In the future research, the edge application management method based on microservices deserves more attention

Author Contributions: Conceptualization, Y.Z.; methodology, Y.Z.; software, Y.Z.; validation, Y.Z., D.T. and H.Z.; formal analysis, S.Z.; investigation, Y.Z.; writing—original draft preparation, Y.Z.; writing—review and editing, Y.Z. and D.T.; visualization, Z.Z. and S.Z.; funding acquisition, D.T. All authors have read and agreed to the published version of the manuscript.

Funding: This research was funded by National Key Research and Development Program of China, grant number 2018YFE0117000, National Natural Science Foundation of China, grant number 52075257, Key Research and Development Program of Jiangsu Province, grant number BE2021091, and Chengfei-Nanhang "Zhi Hui Lan Tian" school-enterprise collaborative education platform project, grant number 2022QYGCSJ38.

Data Availability Statement: Not applicable.

Conflicts of Interest: The authors declare no conflict of interest.

References

1. Okwudire, C.E.; Madhyastha, H.V. Distributed manufacturing for and by the masses. *Science* **2021**, *372*, 341–342. [CrossRef]
2. Wang, C.; Jiang, P. Deep neural networks based order completion time prediction by using real-time job shop RFID data. *J. Intell. Manuf.* **2019**, *30*, 1303–1318. [CrossRef]
3. Liu, C.; Tang, D.; Zhu, H.; Nie, Q. A Novel Predictive Maintenance Method Based on Deep Adversarial Learning in the Intelligent Manufacturing System. *IEEE Access* **2021**, *9*, 49557–49575. [CrossRef]
4. El Kaed, C.; Khan, I.; Van Den Berg, A.; Hossayni, H.; Saint-Marcel, C. SRE: Semantic Rules Engine for the Industrial Internet-Of-Things Gateways. *IEEE T. Ind. Inform.* **2018**, *14*, 715–724. [CrossRef]
5. Xu, X. From cloud computing to cloud manufacturing. *Robot. Comput. Integr. Manuf.* **2012**, *28*, 75–86. [CrossRef]
6. Jiang, P.; Ding, K.; Leng, J. Towards a cyber-physical-social-connected and service-oriented manufacturing paradigm: Social manufacturing. *Manuf. Lett.* **2016**, *7*, 15–21. [CrossRef]
7. Yu, C.; Xu, X.; Yu, S.; Sang, Z.; Yang, C.; Jiang, X. Shared manufacturing in the sharing economy: Concept, definition and service operations. *Comput. Ind. Eng.* **2020**, *146*, 106602. [CrossRef]
8. Qian, C.; Zhang, Y.; Liu, Y.; Wang, Z. A cloud service platform integrating additive and subtractive manufacturing with high resource efficiency. *J. Clean. Prod.* **2019**, *241*, 118379. [CrossRef]
9. Li, X.; Fang, Z.; Yin, C. A machine tool matching method in cloud manufacturing using Markov Decision Process and cross-entropy. *Robot. Comput. Integr. Manuf.* **2020**, *65*, 101968. [CrossRef]

10. Yang, C.; Peng, T.; Lan, S.; Shen, W.; Wang, L. Towards IoT-enabled dynamic service optimal selection in multiple manufacturing clouds. *J. Manuf. Syst.* **2020**, *56*, 213–226. [CrossRef]
11. Zhang, Y.; Tang, D.; Zhu, H.; Li, S.; Nie, Q. A flexible configuration method of distributed manufacturing resources in the context of social manufacturing. *Comput. Ind.* **2021**, *132*, 103511. [CrossRef]
12. Boyes, H.; Hallaq, B.; Cunningham, J.; Watson, T. The industrial internet of things (IIoT): An analysis framework. *Comput. Ind.* **2018**, *101*, 1–12. [CrossRef]
13. Erasmus, J.; Grefen, P.; Vanderfeesten, I.; Traganos, K. Smart Hybrid Manufacturing Control Using Cloud Computing and the Internet-of-Things. *Machines* **2018**, *6*, 62. [CrossRef]
14. Wang, J.; Xu, C.; Zhang, J.; Bao, J.; Zhong, R. A collaborative architecture of the industrial internet platform for manufacturing systems. *Robot. Comput. Integr. Manuf.* **2020**, *61*, 101854. [CrossRef]
15. Li, Q.; Yang, Y.; Jiang, P. An Industry 4.0 Platform for Equipment Monitoring and Maintaining in Carbon Anode Production. *IFAC-Pap. OnLine* **2022**, *55*, 37–41. [CrossRef]
16. Liu, C.; Su, Z.; Xu, X.; Lu, Y. Service-oriented industrial internet of things gateway for cloud manufacturing. *Robot. Comput. Integr. Manuf.* **2022**, *73*, 102217. [CrossRef]
17. Zhang, Y.; Zhang, G.; Wang, J.; Sun, S.; Si, S.; Yang, T. Real-time information capturing and integration framework of the internet of manufacturing things. *Int. J. Comput. Integr. Manuf.* **2014**, *28*, 811–822. [CrossRef]
18. Wang, R.; Gu, C.; He, S.; Shi, Z.; Meng, W. An interoperable and flat Industrial Internet of Things architecture for low latency data collection in manufacturing systems. *J. Syst. Archit.* **2022**, *129*, 102631. [CrossRef]
19. Lu, Y.; Xu, X. Cloud-based manufacturing equipment and big data analytics to enable on-demand manufacturing services. *Robot. Comput. Integr. Manuf.* **2019**, *57*, 92–102. [CrossRef]
20. González, I.; Calderón, A.J.; Portalo, J.M. Innovative Multi-Layered Architecture for Heterogeneous Automation and Monitoring Systems: Application Case of a Photovoltaic Smart Microgrid. *Sustainability* **2021**, *13*, 2234. [CrossRef]
21. Jaloudi, S. Communication Protocols of an Industrial Internet of Things Environment: A Comparative Study. *Future Internet* **2019**, *11*, 66. [CrossRef]
22. Bader, S.R.; Maleshkova, M.; Lohmann, S. Structuring Reference Architectures for the Industrial Internet of Things. *Future Internet* **2019**, *11*, 151. [CrossRef]
23. Wollschlaeger, M.; Sauter, T.; Jasperneite, J. The Future of Industrial Communication: Automation Networks in the Era of the Internet of Things and Industry 4.0. *IEEE Ind. Electron. Mag.* **2017**, *11*, 17–27. [CrossRef]
24. Zhang, Y.; Zhu, H.; Tang, D.; Zhou, T.; Gui, Y. Dynamic job shop scheduling based on deep reinforcement learning for multi-agent manufacturing systems. *Robot. Comput. Integr. Manuf.* **2022**, *78*, 102412. [CrossRef]
25. Wang, X.; Wan, J. Cloud-Edge Collaboration-Based Knowledge Sharing Mechanism for Manufacturing Resources. *Appl. Sci.* **2021**, *11*, 3188. [CrossRef]
26. Yang, H.; Ong, S.K.; Nee, A.Y.C.; Jiang, G.; Mei, X. Microservices-based cloud-edge collaborative condition monitoring platform for smart manufacturing systems. *Int. J. Prod. Res.* **2022**, 1–10. [CrossRef]

machines

Article

Application of Industrial Internet for Equipment Asset Management in Social Digitalization Platform Based on System Engineering Using Fuzzy DEMATEL-TOPSIS

Yuguang Bao [1,2], Xianyu Zhang [1,2], Tongtong Zhou [1,2], Zhihua Chen [1,2] and Xinguo Ming [1,2,*]

1 Institute of Industrial Engineering and Management, School of Mechanical Engineering, Shanghai Jiao Tong University, Shanghai 200240, China
2 SJTUSME-COSMOPlat Joint Research Center for New Generation Industrial Intelligent Technology, School of Mechanical Engineering, Shanghai Jiao Tong University, Shanghai 200240, China
* Correspondence: xgming@sjtu.edu.cn; Tel.: +86-13918034401

Abstract: In any industry, Equipment Asset Management (EAM) is at the core of the production activities. With the rapid development of Industrial Internet technologies and platforms, the EAM based on the Industrial Internet has become an important development trend. Meanwhile, the paradigm of EAM is changing, from a single machine to integrated systems, from the phase of using them to the end of their lifecycle, from breakdown maintenance to predictive maintenance, and from local decision-making to collaborative optimization. However, because of the lack of a unified understanding of the Industrial Internet platforms (IIPs) and the lack of a comprehensive reference architecture and detailed implementation framework, the implementation of EAM projects will face greater risks according to special needs in different industries. Based on the method of system engineering, this study proposes a general reference model and a reference architecture of implementation for the Industrial Internet Solution for Industrial Equipment Asset Management (I3EAM). Further, to help enterprise to evaluate and select their best-fit I3EAM scheme and platform partner, we proposed a set of performance indicators of I3EAM schemes and a quantitative decision-making method based on fuzzy DEMATEL-TOPSIS. Finally, a case study for an I3EAM in automated container terminals was conducted. In the multi-criteria decision environment with complex uncertainty, the project group identified the I3EAM metrics priorities and social digitalization platforms that were more in line with the actual needs of the automated container terminal and firms. The complexity and time of the decision-making process were dramatically reduced. In terms of feasibility and validity, the decision result was positively verified by the feedback from the enterprise implementation. The given model, architecture, and method in this study can create a certain reference value for various industrial enterprises to carry out the analysis and top-level planning of their I3EAM needs and choose the partner for co-implementation. In addition, the research results of this study have the potential to support the construction of standard systems and the planning and optimization of the cross-domain social platform, etc.

Keywords: industrial internet; equipment asset management; digitalization; social digitalization platform; system engineering; fuzzy DEMATEL-TOPSIS

Citation: Bao, Y.; Zhang, X.; Zhou, T.; Chen, Z.; Ming, X. Application of Industrial Internet for Equipment Asset Management in Social Digitalization Platform Based on System Engineering Using Fuzzy DEMATEL-TOPSIS. *Machines* **2022**, *10*, 1137. https://doi.org/10.3390/machines10121137

Academic Editor: Kai Cheng

Received: 29 September 2022
Accepted: 20 November 2022
Published: 29 November 2022

1. Introduction

Smart manufacturing represents the development trend of the global industry, and the integration, construction, and application of smart factories based on cyber-physical systems are leading the change in the manufacturing paradigm. Equipment Asset Management (EAM) plays an important role in a smart factory, and it becomes a necessary guarantee of achieving reliable and efficient production [1]. With the development of Industrial Internet technologies, equipment operation and maintenance (O and M) businesses have shifted from traditional regular inspection and contingent maintenance to data-driven equipment lifecycle management and predictive maintenance [2–4]. In China, nearly 40% of the Industrial Internet platforms focus on applications and solutions for equipment management. Through the interconnection of machines, processes, and people, the efficiency of equipment maintenance has been improved, and the networked collaborative evolution of various production factors and business processes closely related to equipment assets has been further realized [5,6]. Since complex equipment usually has high added value, the initial ecosystem for the operation and maintenance services will gradually form around the value chain of equipment [7,8], involving the equipment owners (factories/enterprises), equipment manufacturers, module suppliers, material suppliers, service providers, etc. In addition, based on Industrial Internet platforms and technologies, many outstanding solution providers for equipment operation and maintenance have emerged, such as Siemens MindSphere, GE Predix, SANY ROOTCLOUD, etc. Third-party Industrial Internet Platforms (IIPs) have become a common option for enterprises to achieve digitalization. Industrial enterprises leverage the platform's technical capabilities and ecosystem to accelerate business–IT alignment in different scenarios, including the EAM scenario.

However, the industry still lacks a unified understanding of the IIP [9], and only a few studies have focused on the detailed implementation framework and path for special industrial needs [10]. It is difficult for industrial enterprises to comprehensively and precisely identify the implementation needs and choose their best-fit IIP. All of the above factors will lead to them incurring the risk of EAM projects. In addition, according to our previous work [11], a systematic model and reference architecture are necessary for the collaboration platforms with various industrial enterprises across different fields. This study also conducted a comprehensive status review of the studies on the Industrial Internet-based EAM (Section 2). To our best knowledge and industrial investigations across a wide range, the current research mainly focuses on the explicit technologies of EAM schemes, while relatively few studies have focused on the comprehensive top-level framework and implementation path. It hinders the application and synergy of Industrial Internet technologies and platforms in the field of industrial EAM.

The research motivation and objective of this paper is to improve the capability of complex system awareness and decision-making for EAM projects with Industrial Internet technologies and to provide a reference for the top-level planning and effective implementation of different industrial enterprises. This study proposes a general reference model and a reference architecture of implementation for the Industrial Internet Solution for Industrial Equipment Asset Management (I3EAM). In addition, this study also develops a set of performance indicators of I3EAM and a scheme selection method by using fuzzy the DEMATEL-TOPSIS approach. A case study of its practice in an automated container terminal project demonstrates the feasibility and efficiency of the proposed model, architecture, and approach. In addition, the research results of this study have the potential to support the construction of standard systems, and the planning and optimization of cross-domain social platforms, etc.

The organization structure of this paper is arranged as follows. Section 2 conducts a status review of the studies on Industrial Internet for EAM. Section 3 introduces the main research methodology for a system analysis of I3EAM. In Sections 4 and 5, a general reference model and a reference architecture of I3EAM in the social digitalization platform are proposed and clarified, respectively. Section 6 introduces an effective method for the evaluation and selection of the I3EAM schemes. In Section 7, a real industry case study is conducted to validate the proposed model, architecture, and method. Finally, the discussion, conclusions and future works of this study are given in Section 8.

2. Status Review for Industrial Internet and EAM

2.1. Related Concepts

2.1.1. Industrial Internet

In 2012, GE first introduced the concept of "Industrial Internet". The Industrial Internet was defined as the third wave in the development of the technological revolution.

They defined the Industrial Internet as "the convergence of the global industrial system with the power of advanced computing, analytics, low-cost sensing and new levels of connectivity permitted by the internet". Subsequently, GE, along with Cisco, Intel, AT&T, IBM, and other companies, formed the Industrial Internet Consortium (IIC) [9]. The IIC proposed the Industrial Internet of Things (IIoT) system and defined it as "a system that connects and integrates industrial control systems with enterprise systems, business processes and analytics" [12]. From the IIC's reference architecture [12], the Industrial Internet will serve multiple industrial sectors, covering the manufacturing, transportation, energy, health care ones, and more. In China, in 2010, the Internet Society of China proposed that they were "to take into account both consumer and production-oriented Internet". The Alliance of Industrial Internet (AII) regards the Industrial Internet as a new industrial ecology, a key infrastructure, and a new application model. The AII proposed that the core connotation of Industrial Internet is to realize the comprehensive connection of the whole element, the whole industrial chain and the whole value chain through the comprehensive interconnection of people, machines, and things. Industrial Internet System Architecture 2.0 (China) [13], together with IIRA (USA) [12], RAMI4.0 (Germany) [14] and IVRA (Japan) [15] have become the main reference architecture worldwide that are used to guide the exploration and practice of Industrial Internet applications.

2.1.2. Equipment Asset Management (EAM)

Manufacturing companies are highly dependent on the proper operation of their equipment assets to ensure good business operations [1,5,8]. The purpose of Equipment Asset Management is to reduce the asset's maintenance and repair costs, extend the asset's life, and improve the asset's utilization [1]. Asset management is defined as "a set of strategic, integrated and comprehensive processes (finance, management, engineering, operations and maintenance) to obtain maximum lifecycle efficiency, utilization and return from physical assets (production and operating equipment and structures)" [16]. An AM system can be defined as "a system that plans and controls asset-related activities and their relationships to ensure that asset performance meets the organization's intended competitive strategy" [17]. In practice, engineering asset management systems are increasingly focused on the full lifecycle of the engineering assets [8,18–20]. The Asset Life Cycle Management (ALCM) approach was proposed, which originated as a design philosophy and management principle in the U.S. Department of Defense's weapons and equipment acquisition process. The ALCM is used to address the tension between reducing the lifecycle cost of the assets and improving the asset's utilization [19]. In general, the whole lifecycle of the equipment refers to the entire process of designing and developing, producing, using, and safeguarding the equipment until decommissioning it, and the material recovery stage [4,8,18,20]. El-Akruti et al. developed a typical representation of the major asset-related activities in an organization, which provides a comprehensive framework for fully defining the asset management system activities and the relationships between the activities [1,17].

2.1.3. Smart Product Service System (SPSS)

Product service system (PSS) is a system of products, services, collaborative networks of the stakeholders, and supporting infrastructure, which are jointly capable of creating diverse functional and non-functional value economically and sustainably for the customers (based on [21–23]). Tukker categorizes PSSs into three modes, i.e., product-oriented PSS, use-oriented PSS, and result-oriented PSS [22]. After three industrial technology upgrades have taken place: mechanization, electrification, and information technology, the era of intelligence will soon come [24,25]. The products become Smart Connected Products (SCPs) through advanced sensing and IoT technologies [25], such as smart machines. Based on the concept of SCPs, Valencia et al. first proposed and defined the Smart Product Service System, which is "a bundle of SCP and the generated smart services by leveraging SCP as the media and tool, which is delivered to market for satisfying personalized needs of customers as well as providing more environmental and social benefits" [26]. With a data and knowledge-driven approach, the SPSS is able to provide more powerful and flexible capabilities than a traditional PSS can, including smart sensing, smart connectivity, smart analytics, and smart delivery [27,28]. In response to the trend of the convergence of smart machines and digital platforms, the German Academy of Engineering has proposed the concept of "smart services and its ecologization" as one of the key strategic directions of Industry 4.0 [29]. Zheng defined the Smart Product Service Ecosystem (SPSE), from the system of systems perspective, whereby the SPSE consists of six components, including relationship, user experience, smart technologies, marketing, business models, and connections [30].

2.1.4. Social Digitalization Platform (SDP)

A digitalization platform is often used to describe a centralized system to help enterprises to become digital and improve the way organizations deliver value to the customers and remain competitive [31]. Technology architecture upgrades are making digitalization platforms a standalone business model that helps application companies to achieve business–IT alignment by providing IaaS (infrastructure as a service), PaaS (platform as a service), and SaaS (software as a service) solutions [31,32]. Based on the definition of platform leverage logic (PLL) defined by Tian et al. [33] and the study of the typology of "platform" in the literature [34–36], we extended the leverage rationale of digitalization platforms to include production rationale, transaction rationale, customization rationale, and innovation rationale. The difference between the customization rationale and the innovation rationale is whether the platform is licensed for data sharing and utilization when one is providing digitalization solutions [32,37]. In this paper, we focus on those large, cross-industry, third-party digitalization platforms, and we refer to them as social digitalization platforms (SDPs) [38]. SDPs are third-party Industrial Internet platforms that can develop and deliver enterprise digitalization solutions for different industries or domains [32]. SDPs have stronger technical resource pools, actor connectivity, and system integration capabilities [36,37,39], while the PLL of the SDPs can be defined as being "focused on building platform ecosystem based on the innovation leverage rationale, bringing together solution suppliers, developers and users around digital servitization". For example, the "cross-industry and cross-region" Industrial Internet platforms in China [40], e.g., Siemens MindSphere, GE Predix, etc., are typical SDPs.

2.1.5. The Relationship between Industrial Internet, EAM, SPSS, and SDP

(1) The relationship between industrial internet and EAM

Industrial Internet technologies can support various digitalization needs, including Equipment Asset Management (EAM) [41]. EAM is a typical application scenario of Industrial Internet that is used by manufacturing companies to improve their organizational performance by establishing synergy links with the reliability of the equipment assets, key business, and stakeholders [5]. Digitalization solutions based on Industrial Internet can facilitate enterprise integration, industry chain collaboration and value chain extension around complex equipment objects and EAM-related activities [4,5,41,42]. In this paper, "Industrial Internet solution for industrial equipment asset management" is the core comprehensive concept which is used to represent a digital service solution for the EAM needs of a specific enterprise (corresponding to the "digitalization service instance" in Figure 1). For convenience, it is referred to as "I3EAM".

(2) The relationship between Industrial Internet, EAM, and SPSS

A Smart Product Service System (SPSS) is regarded as a bundle of the business model and a delivery system of I3EAM [30,43]. Diverse SPSSs will be developed for specific EAM-related application scenarios, such as remote maintenance [43], service network configuration [44], and design optimization [42], etc. The SPSS will include smart equipment, EAM-related digital services, stakeholder collaboration networks, and supporting infrastructure [23,45]. Multiple SPSSs will gradually form a smart product service ecosystem and create organizational benefits by generating complex relationships and synergies [46].

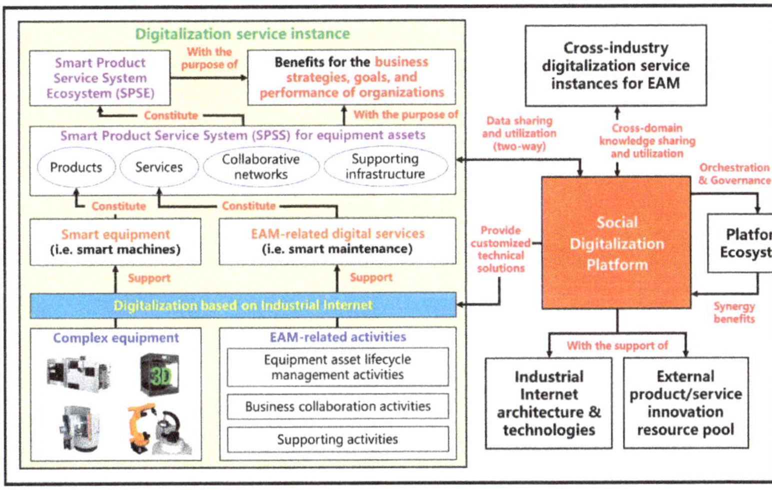

Figure 1. The relationship between Industrial Internet, EAM, SPSS, and SDP.

(3) The relationship between Industrial Internet, EAM, SPSS, and SDP

The digitalization platform uses Industrial Internet technologies to help companies to realize their vision of digital servitization [36,47]. The Industrial Internet technology system provides an open architecture for social digital platforms to build their platform ecosystem [34], and it expands the organizational boundaries of platform-based enterprises [38]. Social digitalization platforms (SDPs) focus on building their own Industrial Internet technology platforms and provide customized EAM technical solutions and SPSSs for enterprises in different industries through the use of business models such as IaaS, PaaS, and SaaS. At the same time, SDPs will utilize interaction mechanisms (such as data sharing), external innovation resource pools, strong system integration capabilities, and platform ecosystem orchestration and governance to achieve better product/service innovation [32,37,46,47].

(4) What are the benefits of SDP for EAM?

The value of SDPs is reflected in four aspects, (1) reducing the cost of enterprise digitalization and reducing repetitive and inefficient R&D; (2) facilitating data sharing and utilization and enhancing cross-domain knowledge transfer to cope with a wide variety of complex equipment and EAM business scenarios; (3) accelerating manufacturing enterprise, industry chain and value chain integration and collaboration on the basis of reliability of equipment assets and systems; (4) focusing on gathering solution providers, developers and users to activate the innovation capability of the platform ecosystem.

2.2. Status of Key Technologies for EAM

As shown in Table 1, the advanced Industrial Internet technologies for EAM include ten main aspects, including the key technologies for PHM such as data acquisition, signal processing, fault diagnosis, health assessment, maintenance decision-making, and remote monitoring, and co-maintenance. In data acquisition, sensing technology usually involves 13 types including the vibration, temperature, current, pressure, infrared, gas, sound, strain, position, torque, image, video, and laser types, etc. Wireless transmission, large-scale low-cost access, and intelligent edge acquisition are the main technical difficulties of the current data acquisition processes. For signal processing, its main work is to improve the signal-to-noise ratio through conducting data pre-processing, thus facilitating the extraction and highlighting of fault characteristics. Around the vibration signal, a more comprehensive vibration feature parameter library has been formed, including the time domain, frequency domain, and time-frequency domain features of the vibration signal.

The interpretable physical linkage between the signal processing and the feature extraction is the key technology that is used to breaking through the machine condition monitoring. For fault diagnosis, the solutions include historical data-based decision making, physical model-based decision making, and data-driven decision making. On the one hand, the threshold and physical model approaches are suitable for edge measurement time-sensitive intelligent diagnostic operations. On the other hand, a wide variety of deep and shallow neural networks are used for online fault diagnoses to explore more robust learning methods for nonlinear mapping relationships between the data and the anomalous patterns. Thus, the physics, the data, and the expert experience are moving toward convergence, together serving the accurate identification of an early fault diagnosis and generalized migration across the devices. The core objectives in health assessment mainly include the performance degradation assessment, health index (HI) construction, and residual useful life (RUL) prediction of the machines. Monitoring the initial failure time and constructing health indices by multi-sensing signal fusion are the main research directions. In addition, the difficulty of obtaining full lifecycle data of complex equipment is a key challenge in practice, and so scaling models and migration learning will be used in some application scenarios. For maintenance decision making, fixed and empirical maintenance resource management strategies need to be optimized in conjunction with new equipment condition monitoring tools, such as optimized maintenance strategies based on equipment health indices, reliability-based preventive maintenance strategies, etc. Other maintenance resource allocation strategies should also be considered, including the scientific spare parts management, multi-skilled staffing, and overhaul of critical equipment allocation. For remote monitoring and co-maintenance, new Industrial Internet technologies have been applied to equipment operation and maintenance, such as cloud-edge collaboration, digital twins, and AR, and VR.

Table 1. Literature review of advanced Industrial Internet technologies for EAM.

Category	The State-of-the-Art Technologies	Future Direction
Data acquisition	• Sensing technologies (13 types usually): vibration, temperature, current, pressure, infrared, gas, sound, strain, position, torque, image, video, laser, etc., [48]; • Informative sensor identification technologies [49]; • Multi-sensor fusion technologies [50,51].	Wireless transmission, large-scale low-cost access, intelligent edge acquisition, and multi-modal information fusion
Signal processing	• Feature extraction methods: many signal features have been proposed, especially for the vibration signal for critical structural components (such as gear and bearing) [52], including the time domain, frequency domain, and time-frequency domain features [53,54]; • Main technologies: including methods for data of redundancy [55], sparsity [56,57], and deficiency [58].	Interpretable physical linkage of signal processing and feature extraction [56]
Fault diagnosis	• Main research flow: fault diagnosis based on the historical data, fault diagnosis based on the physical model, and data-driven fault diagnosis; • Main technologies: the threshold and physical model approaches for time-sensitive scenarios [56,57], and a wide variety of deep and shallow neural networks for more robust learning and higher accuracy [59,60], including the CNN [61], DNN [62], GAN [63], transfer learning [64], deep echo state network [65], and GNN ones [66], etc.	Convergence of physics, data, and expert experience for higher accuracy of pattern recognition [60,66]

Table 1. *Cont.*

Category	The State-of-the-Art Technologies	Future Direction
Health assessment	• Main research flow: performance degradation assessment, health index (HI) construction, and residual useful life (RUL) prediction of machines; • Main technologies: monitoring initial failure time [56] and constructing health indices by multi-sensing signal fusion [50,67], residual useful life (RUL) prediction based on physical models [68], machine learning, deep learning (DBN [69], GRU [70], TCN [71], and GNN [72], etc.), and fusion models [73], and transfer learning for RUL prediction with data from similar objects [74], simulation [75], or scaling models.	Powerful public data sets [76], cross-scenario adaptive learning methods, and health assessment with consideration of multiple maintenance influences [77]
Maintenance decision-making	• Main maintenance modes: condition-based maintenance, preventive maintenance, predictive maintenance, cognitive maintenance; • Main maintenance decision-making problems: the main research problems include ageing probabilistic risk/safety assessment [78,79], cost estimation [80], maintenance policy selection and optimization [81–83], maintenance outsourcing and leased manufacturing [84,85], spare parts management [86], multi-skilled staffing and scheduling (resource dynamic allocation) [87], overhaul and retrofitting decision-making for critical, complex equipment [88], etc.; • Main methods and technologies: stochastic model, Markov process modelling, Bayesian network, probabilistic relational model, multi-criteria decision-making (MCDM) methods, mathematical programming model, D-S theory, evolutionary algorithms for NP-hard problems.	Joint optimization for effective equipment maintenance in the integrated environment with multi-component systems [86], multi-state systems, and multi-layer systems [78]
Remote monitoring and co-maintenance	• Edge-cloud collaboration technologies: task offloading for time-sensitive PHM scenarios [89], online PHM with incremental learning [90]; • Digital-twin-based technologies: digital-twin-driven PHM approach [91], digital twin-driven cooperative awareness and interconnection framework for total factors of maintenance decision-making environment [6], DT-based equipment lifecycle management [4], DT-based maintenance cost joint optimization [92]; • AR/VR-based technologies: smart asset management functions (monitoring, recognition, positioning, classification, instruction, prognosis, optimization, and their combinations) and human–machine collaboration [93,94].	Real-time synchronization, multi-discipline, multi-scale, faithful mapping, and high-fidelity modelling for complex equipment and maintenance decision-making environment [6]
Service network setting and optimization	• Maintenance network design and planning [95]; • Service network modelling [44]; • Robustness of maintenance service networks and adaptive mechanisms [96].	Technology-driven business model innovation for data sharing and utilization [97]
Knowledge sharing and privacy preservation	• Maintenance decision making oriented cross-organization knowledge sharing blockchain network [46]; • Privacy-preserving PHM framework [97].	

Table 1. *Cont.*

Category	The State-of-the-Art Technologies	Future Direction
Equipment optimization driven by reliability data	• Reliability data for equipment lifecycle [42], equipment design [98,99], and others [100].	Joint optimization model and algorithm design which is driven by various collaborative scenarios, missions and utility goals [101]
Business optimization driven by reliability data	• Uncertainty modelling and business optimization [102]; • Mission reliability-driven functional healthy state modelling [101]; • Social network of collaborating industrial assets [103]; • Collaborative prognostics in social asset networks [104].	

As shown in Table 1, there are also some other important research streams in key technologies, including service network setting and optimization, knowledge sharing and privacy preservation, equipment optimization that is driven by reliability data, and business optimization that is driven by reliability data.

2.3. Status of Industrial Application for EAM

In the terms of the application object, Industrial Internet technologies have been widely used in numerous industrial fields and industrial equipment types, including the bearings and gears of general rotating machinery [6,62,63,71,72], CNC machining centers [84,89], aircraft gas turbine engines [51,105], aircraft turbofan engine [68,70,73], sea port equipment [77], complex electromechanical systems [79], train bogies [80], and wind turbines [91], etc. Aiming at the functional requirements of EAM, different industries and equipment objects show large differences, including differences in data acquisition, signal processing, fault diagnosis, health management, maintenance decisions, and collaborative operations. Therefore, it is essential to customize the technical solution according to the actual needs.

In the term of the decision support system for PHM (Prognostics and Health Management), the typical EAM technical solution suppliers include MindSphere, Predix, ROOTCLOUD, Thingswise iDOS, and so on. Currently, most Industrial Internet platforms have established their EAM-related services and solutions. Table 2 provides a brief overview of the typical Industrial Internet platforms, as well as their EAM technical solutions and industrial applications. In addition, several scholars have provided systematic insights for the development of decision support systems for PHM. Lee et al. proposed a PHM system design and implementation methodology that includes the Streamline, Smart Processing, Synchronize and See, Standardize and Sustain stages [2]. Meanwhile, the data, predictive techniques, information management with visualization tools, standards, and closed-loop management systems are considered to be key elements of the decision support systems for equipment operation and maintenance [2]. Tao et al. introduced digital twin (DT) technology for the PHM of complex equipment [91]. Based on five dimensions: physical entities, virtual models, service activities, data, and connectivity, a componentized PHM system has been designed and performed for unified management and on-demand use [91].

Table 2. A brief overview of typical Industrial Internet solutions and applications for EAM.

Name of the Industrial Internet Platform	Technical Framework and Its Characters	Industrial Application
Siemens MindSphere	• Cloud-based Open IoT Architecture: including three main layers: MindConnect, MindCloud, and MindApps; • Edge-cloud collaboration, digital twin, and low-code industrial service development capabilities; • Abundant common application components for Equipment Asset Management.	Aerospace and defense, automotive and transportation, electronics and semiconductors, heavy equipment, and industrial machinery, etc.
GE Predix	• Asset Performance Management (APM) Software: featuring Digital Twin analytics, work process automation and built-in GE industry expertise.	Energy, paper and pulp, chemical, mining and metallurgy, and fertilizer, etc.
Microsoft Azure	• Secured reliable computing and industrial IoT solutions (from chip to cloud); • Powerful machine learning capabilities, edge intelligence solutions, and software ecosystems.	Transportation and logistics, industrial robot, casting, elevator, and food and drink industry, etc.
SANY ROOTCLOUD	• Powerful connectivity (supporting 1100+ common industrial protocols and protocols customization); • Powerful modelling capacity (multi-level nested models for equipment: abstract thing model, thing model, basic elements of thing model, thing instances); • Mature business model of product service system.	Industrial machinery, discrete manufacturing, steel and non-ferrous metal, chemical, and mining industry, etc.
CASICCloud INDICS	• INDCIS + CMSS: Industrial Internet social platform + Cloud manufacturing support systems; • Supporting 21 types of scenarios of equipment data collection, including machining, welding, and simulation, etc.; • Safe, autonomous and controllable.	Aerospace, electronics, machinery, and automotive industry, etc.
You-Ye Thingswise iDOS	• A unified and loosely coupled system architecture and reuse of general technology components; • Powerful industrial mechanism modeling capacity.	Process manufacturing industries including energy, metallurgy, and material processing, etc.

3. Research Methodology

3.1. Method of System Engineering Analysis

In this paper, each particular Industrial Internet Solution for Industrial Equipment Asset Management (I3EAM) is considered as a complex artificial system that helps the companies to achieve business–IT alignment in the EAM domain. The development of a generic system model and customized implementation path for I3EAM is the research goal of this paper. The I3EAM functions for a specific enterprise's needs will be the functional subset of the generic system model, while the social digitalization platform will provide the Industrial Internet technology system support for the customized I3EAM. The system engineering analysis method can deal with the complexity and integrity of the system, involving the system's elements, input, output, hierarchical structure, information interaction, and self-feedback mechanism, etc. To provide a multi-faceted and comprehensive description of the I3EAM system, three dimensions is defined from the Industrial Internet element (II), Industry (I), and EAM activities. The system engineering analysis was used to explore the customization-oriented implementation path.

The general steps of the systems engineering analysis approach include clearing the questions, selecting the objectives, system synthesis, system analysis, scheme establishment, making a decision, and putting it into practice. Complex systems can be built successfully by using the system engineering processes of definition, planning, assessment, implementation, and feedback in accordance with the fictitious visions and actual needs. In this paper, the seven specific steps of system engineering analysis are shown below.

(1) Step 1: Clear the questions (from trends and visions to demands)

EAM has become an important application scenario for Industrial Internet applications and the digitalization of manufacturing companies. From a technical perspective, new decision theories and technologies around Equipment Asset Management are constantly being developed. However, there is still a lack of system framework models around EAM in the existing literature, which hinders the precise matching of the academic research results to the industrial application needs. In terms of the service scenarios, the value of asset management extends to the full lifecycle of the equipment and business decisions of the enterprise organization. Equipment interconnection, deeper multi-level system integration, and business model innovation are promoting the creation of diverse scenarios and product service systems. From an industrial application perspective, the demand for EAM in SMEs is very active, but the backward digitalization level of the SMEs cannot support the realization of the business–IT alignment. Common models and implementation paths need to develop for supporting the customization of solution, and they will facilitate the functional improvement and synergy of the Industrial Internet platforms.

(2) Step 2: Select the objectives (from demands to goals and measurement criteria)

The objective of this paper is to develop a generic reference model and implementation architecture for the customization of I3EAM in social digitalization platforms by using the technical architecture of Industrial Internet. Our goals are focused on exploring a systematic technical solution that meets the current state and application needs of EAM, involving break throughs in the connection of heterogeneous equipment, industrial cloud platform, industrial big data engine, industrial modelling and simulation methods, information models for collaborative, and service system planning and development.

(3) Step 3: System synthesis (from goals to investigations and plans in detail)

In order to achieve the selected objective above, we needed to investigate, review, and refine the target system's status quo and plan in a comprehensive manner. This process is similar to the classical approach of Business Process Reengineering, where the investigation of the current situation is considered as the process of acquiring the "as-is" of the system, while the development of the detailed plan is considered as the process of acquiring the "to-be" of the system. Our research path is presented as a combination of scientific research, industrial practice, and a case study, and it consists of four phases.

1. The first stage was to summarize and analyze the existing literature on Industrial Internet-based technologies and solutions for Equipment Asset Management (Sections 2.2 and 2.3) and deduce the application status of I3EAM in different industries.
2. The second stage was to perform a detailed on-site investigation of the digitalization platforms and enterprises in different industries and analyze the industrial demand of I3EAM to obtain its constituent elements, functional modules, and their interaction relations.
3. The third stage was to plan and design the general model and the reference architecture of co-implementation path, which leverage the research results of the system architecture of Industrial Internet. This stage is the most crucial stage, which is divided into six sub-steps as follows:
    - Sub-step 1: Planning for the resource and edge layer of I3EAM;
    - Sub-step 2: Planning for the infrastructure and platform layer of I3EAM;
    - Sub-step 3: Planning for the big data engine layer of I3EAM;
    - Sub-step 4: Planning of the modelling and simulation layer of I3EAM;

- Sub-step 5: Planning of the information model for integration and collaboration layer of I3EAM;
- Sub-step 6: Planning of the organization, business, and service layer of I3EAM.

4. The fourth stage was to select the representative enterprises in typical areas as a case study and verify the reference model and detailed implementation steps of I3EAM, while helping those enterprises with the digitalization solution to achieve a top-level design and platform partner selection. Finally, the customized I3EAM usually can achieve industrial profit from operation cost reductions or the expansion of service capacities.

(4) Step 4: System analysis (from designed system to performance metrics)

The planning and design work derives a sophisticated system and a methodology for system implementation, including (1) a general reference model (Section 4) and (2) a reference architecture for the implementation path (Section 5). The system and methodology provide a comprehensive, multi-perspective overview of and guidance on the implementation of I3EAM in different industries. In addition, a set of performance metrics were established in order to evaluate the performance of the system implementation (Section 6.1).

(5) Step 5: Scheme collection (to collect and compare the I3EAM based on different SDPs according to the needs of the special enterprise)

I3EAMs based on specific enterprise needs are the subsets of the function and instances of the general system model. SDPs differ significantly in terms of technology, service, ecosystem, and industrial application (Section 2.3). Therefore, there are multiple potential I3EAM schemes that can meet the needs of the enterprises, but they differ in terms of their performance, cost, etc. This step focused on collecting feasible solutions from these potential digital service providers and evaluating the solutions according to the metrics framework that is presented earlier in this paper.

(6) Step 6: Make a decision (to select an SDP partner for co-implementation with a group decision-making process)

In the set of feasible solutions, the optimal feasible solution is selected based on the special needs, decision preferences, and decision constraints of the enterprise. A group decision-making process was performed for solution evaluation and partner selection, and as well as this, the methods of fuzzy DEMATEL and TOPSIS were used to analyze the correlation between the performance metrics and to find the ideal approximate solution (Section 6).

(7) Step 7: Put it into practice (to customize and co-implement the I3EAM, and to build feedback mechanisms between the industrial practices and the reference model/architecture)
According to the results of the above decision, the plan was implemented, including:

1. A reference model of I3EAM in the marine engineering equipment industry (Section 7.1);
2. A reference architecture of implementation path for I3EAM in marine engineering (Section 7.2);
3. The selection of the SDP partner for I3EAM co-implementation in marine engineering (Section 7.3).

3.2. The System Model for the Research of I3EAM

As shown in Figure 2, a system model is put forward, and it was used to conduct research of I3EAM with the support of the social digitalization platform. According to the method of system engineering above, the system model consists of the input, the system environment, the system structure, the output, and the system function.

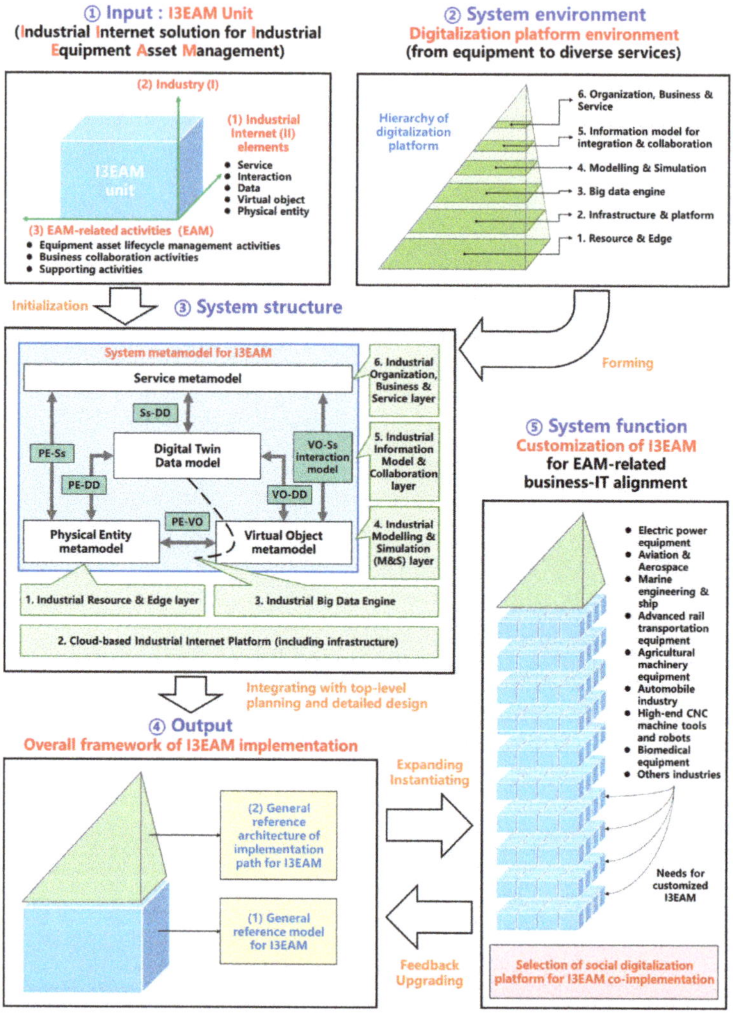

Figure 2. The system model for the research of I3EAM.

Firstly, as the input of the whole system, the I3EAM unit is proposed with three dimensions, which are basic and independent, namely, the Industrial Internet element (II), Industry (I), and the EAM-related activities (EAM), respectively. On the basis of the Industrial Internet elements, I3EAM is designed for achieving the digitalization of the EAM-related activities following the special needs in different industries.

Secondly, the system environment I3EAM is the digitalization platform because Industrial Internet platform (IIP) provides the implementation foundation for I3EAM. Referring to Industrial Internet System Architecture 2.0 [13], the hierarchy of the IIP includes the resource and edge layer, the infrastructure and platform layer, the big data engine layer, the modelling and simulation layer, the information model layer, and the organization, business, and service layer.

Thirdly, the I3EAM unit and system environment jointly induce the system structure of I3EAM, which is mediated by a digital twin-based metamodel for complex equipment. The system metamodel is comprised of the physical entity, the virtual object, the digital twin data, the interaction, and the service. Meanwhile, the hierarchy of the Industrial

Internet platform embodies the system support for five types of elements (corresponding to the system metamodel) and a unified cloud platform environment.

Next, as the output of the whole system, an overall framework of I3EAM implementation is given in a detailed manner after the system synthesis. The framework involves two parts, a general reference model for I3EAM and a general reference architecture of the implementation path for I3EAM.

Lastly, the overall framework above produces the targeted system function, to achieve the growth of the capacity of the social digitalization platforms for the customization of I3EAM and to accelerate the business–IT alignment for the special EAM-related needs in various industries. Additionally, based on a set of performance metrics, the system model can help different enterprises to select their digitalization platform partner for the co-implementation of I3EAM.t

4. A General Reference Model of I3EAM in Social Digitalization Platform

Based on the I3EAM unit, the general system model of I3EAM is defined in three dimensions: the Industrial Internet element (II), the Industry (I), and the EAM-related activities (EAM). The three dimensions have good independence and avoid the ambiguity of description. Meanwhile, the three dimensions cover the core concerns of I3EAM, including the core content of the model (corresponding to the Industrial Internet element (II) dimension), the object served by the model (corresponding to the Industry (I) dimension), and the scenario-based value of the model (corresponding to the EAM-related activities (EAM) dimension). As shown in Figure 3, we made full use of the existing architectural research results to refine and elaborate the three dimensions.

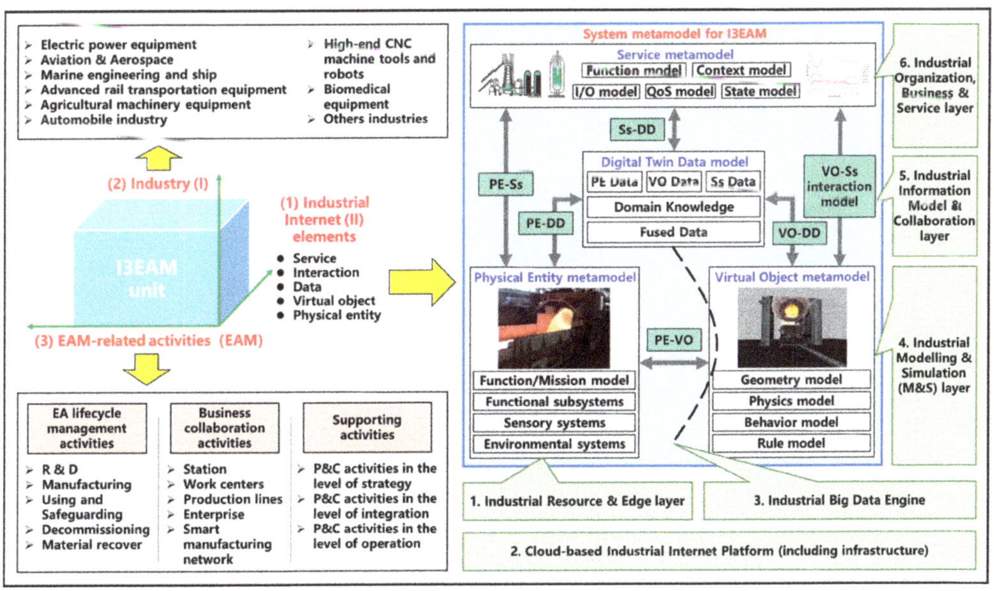

Figure 3. A general reference model of I3EAM in social digitalization platform.

(1) Dimension 1: Industrial Internet element (II) dimension

The Industrial Internet element (II) dimension defines the system metamodel for I3EAM, including the physical entity, the virtual object, the digital twin data, the interaction, and the service. This dimension is a reference to the digital twin general model of complex equipment [91], while we expand on the details of the metamodel. The essence of digitalization is to create parallel control systems for specific systems, including sensing, modelling, and managing physical entities, information flows, and social relationships [106]. The physical entity metamodel consists of the function/mission model, the functional subsystems, the sensory systems, and the environmental systems. The virtual object metamodel consists of the geometry model, the physics model, the behavior model, and the rule model. The digital twin data model includes the data from the physical entities, the virtual objects, and the services, while the domain knowledge and fused data are also the parts of the data model. The service metamodel comprises the function model, the context model, the I/O model, the QoS (quality of service) model, and the state model, etc. Finally, the interaction model builds the bridges among the different models. As described in Section 5, this dimension helps the general reference model to establish a direct mapping to the Industrial Internet architecture [13].

(2) Dimension 2: Industry (I) dimension

The research objective of this paper is to develop a general reference model that used is across various industry sectors for I3EAM customization in social digitalization platforms. The Industry (I) dimension includes, and is not restricted to, electric power equipment, aviation and aerospace, marine engineering and ship, advanced rail transportation equipment, agricultural machinery equipment, automobile, high-end CNC machine tools and robots, biomedical equipment, and other industries.

(3) Dimension 3: EAM-related activities (EAM) dimension

Digitalization needs to rely on explicit business activities in order to generate a scenario-based value. In this paper, we propose a comprehensive EAM-related activity model for the system structure analysis of I3EAM to support the modular design of model and service functions. The EAM functional requirements of a specific enterprise are a functional subset of the EAM-related activity model. The EAM-related activities (EAM) dimension is divided into three types of activity, including the equipment asset lifecycle management activities [8], the business collaboration activities [5,6], and the supporting activities [1,17]. The equipment asset lifecycle management activities are further split into the activities in the phases of R&D, manufacturing, using, safeguarding, decommissioning, and material recovery [4,8,18,20,42]. Business collaboration activities are derived from the hierarchy of system integration [5,14], including the station, the work center, the production line, the enterprise, and the connected collaborative network [78,103,104]. According to the literature [17], the supporting activities for EAM consist of various EAM-related planning and control activities in the level of strategy, integration, and operation. The analysis of the activity will help to effectively locate and identify the tangible and intangible system elements with interpretable links to organizational performance and business goals [5].

5. An Implementation Path of I3EAM in Social Digitalization Platform

5.1. Overall Planning: A General Reference Architecture of Implementation Path

Six layers of I3EAM were derived from the metamodel, corresponding to the five dimensions of the DT model and the generic cloud computing architecture. As illustrated in Figure 4, each layer is considered to be necessary and harmonious, serving to provide common solutions for building digital twins and parallel management systems for complex equipment. The industrial edge layer enables the sensing and control of the physical entities of complex equipment and their environments. The industrial cloud platform will facilitate the flexible matching of heterogeneous computing resources and needs, as well as establish connectivity between the edge devices and the industrial services. Industrial big data and prior knowledge will support the creation of "life-like virtual objects or digital models" of complex engineering systems. Meanwhile, complex coupling optimization mechanisms for the equipment and its environment will be better identified and presented in the cycle of data collection, modeling, simulation, and collaborative interaction. In the context of evolving business goals and organizational performance, data-driven services will support the generation of better planning and control strategies and solutions, dynamically. Each layer will be clarified in detail in Section 5.2 through Section 5.6.

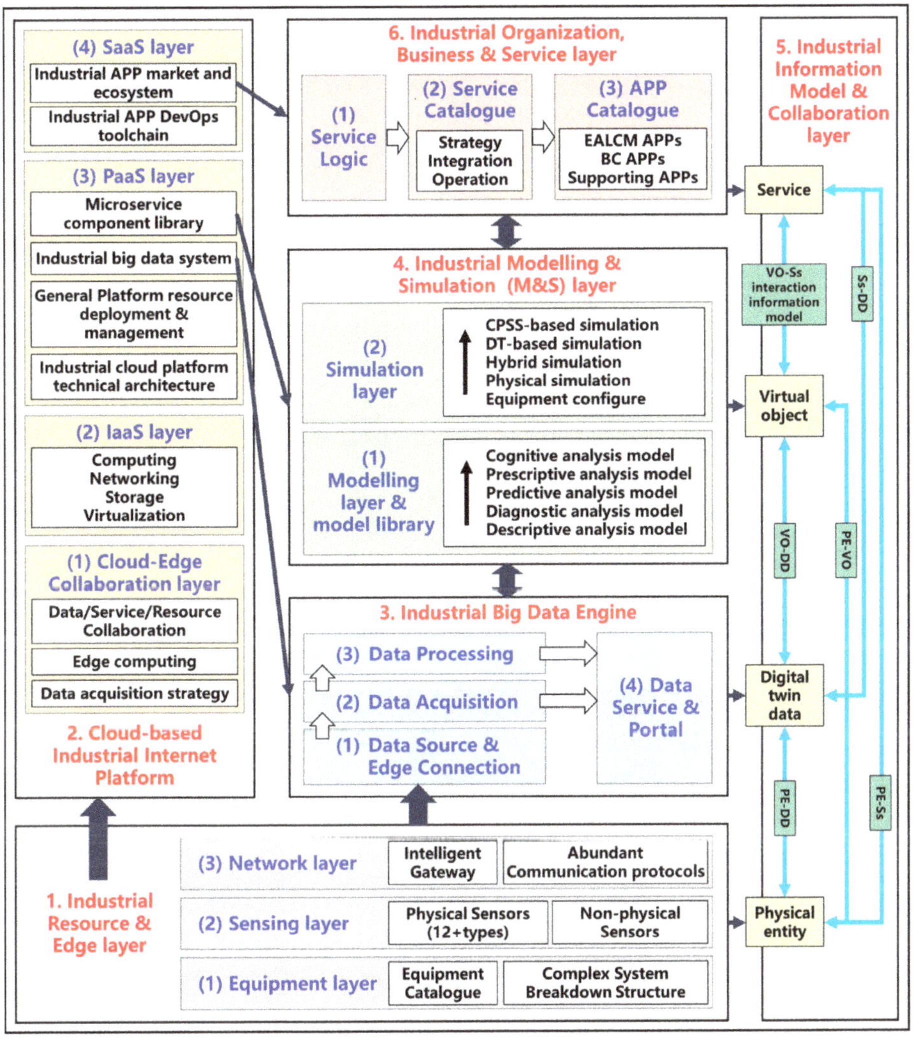

Figure 4. A general reference architecture of the implementation path for I3EAM in SPD.

5.2. Step 1: Planning and Designing of Industrial Resource and Edge Layer

As shown in Figure 5, industrial equipment usually consists of three main categories based on predefined functions or missions: product equipment, production equipment, and non-production equipment. The equipment needs to be maintained in a reasonable interval of physical state during its operation to ensure the reliability of mission execution. The monitoring of the equipment status needs to be based on an understanding of the functional subsystems and a reasonable arrangement of sensing control schemes. For specific scenarios, non-physical signals as well as time-sensitive requirements are incorporated into edge solutions. There are tens of thousands of industrial buses or protocols with private controllers for automation devices around the world, with less than 15% of the general purpose PLCs and more than 60% of the "dumb" devices. Therefore, the edge solutions should provide "plug-and-play" capabilities with technology options for sensing and networking. The industrial edge layer includes low-cost sensor solutions for different system levels and signal types, as well as smart gateways that support a variety of industrial protocols. The industrial protocols include three categories: (1) General communication protocols: 4G, 5G, TCP/IP, Wi-Fi, NB-IoT, ZigBee, Bluetooth, Wired Network, etc.; (2) Mainstream industrial protocols: OPC-UA, IEC10x, IEC61850, Modbus, LT645, DNP3.0, CDT, Open Protocol, MQTT, CoAP, DataExchange, etc.; (3) Other specialized or custom protocols. In addition, edge computing solutions are introduced in the cloud-edge collaboration solution of the industrial cloud platform in the next section.

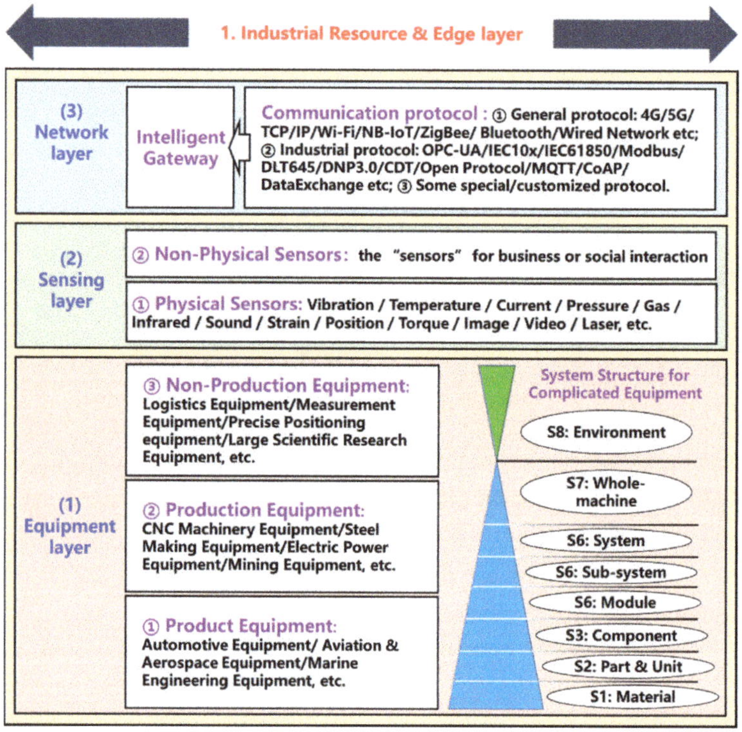

Figure 5. Industrial resource and edge layer.

The industrial cloud platform system provides all of the necessary computer system resources for I3EAM. Due to the scalability of cloud-based computing resources, the platform that is here is not a single software system. The industrial cloud platform may involve multiple subjects and multiple layers of computer resources, including the edge computing systems, the enterprises' local OT/IT systems, the enterprise-level Industrial Internet platforms, and the cross-industry third-party service providers. The industrial cloud platform consists of a cloud-edge collaboration layer, an IaaS (Infrastructure as a service) layer, a PaaS (Platform as a service) layer, and a SaaS (Software as a service) layer.

The cloud-edge collaboration layer will utilize the time-sensitive response capability of the edge side (e.g., real-time equipment monitoring) and the powerful back-end computing (e.g., big data analysis), and unified configuration capability (e.g., computer resource allocation) is measured by the cloud platform. The core of cloud-edge collaboration lies in selecting reasonable distributed resource and function configurations according to the application scenarios and establishing a collaborative link among the cloud, edge, and end points. The cloud-edge collaboration includes resource collaboration, service collaboration, and application collaboration. Through the abstraction of hardware resources such as underlying devices, it realizes the convenient deployment, communication, operation, and whole lifecycle management of the infrastructure resources, service components, and application software on both sides of the cloud and the edge. As shown in Figure 6, the cloud-edge collaboration layer includes edge data collection strategies and edge computing schema.

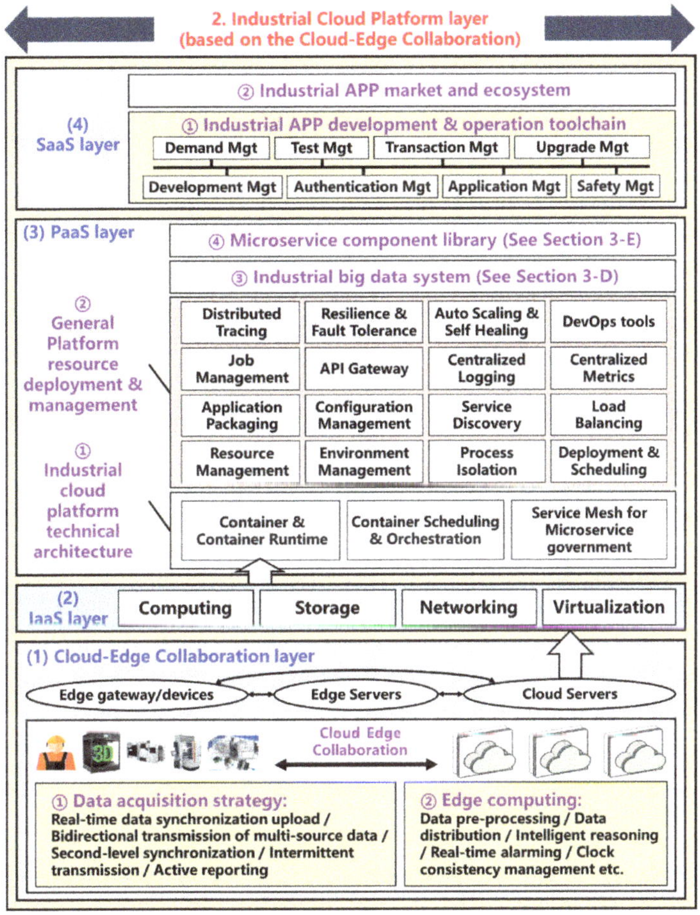

Figure 6. Industrial cloud platform layer.

5.3. Step 2: Planning and Designing of Industrial Cloud Platform Layer

The IaaS layer provides infrastructure involving computing, storage, networking, and virtualization. The PaaS layer offers the technical architecture of the industrial cloud platform and the common capabilities for platform resource deployment and management, and it supports the integration of industrial big data systems and microservice component libraries. To establish the digital platform for the group-type company or cross-industry service provider, a strong PaaS layer is required to provide infrastructure services for wider access to the edge devices and industrial services. A flexible service-oriented technology architecture is created by combining containers and microservices. Simultaneously, in order to better handle the diverse computer system resources and microservice requirements, the platform offers a number of common services such as resource management, environment management, configuration management, load balancing, and an API gateway. In a larger range of application scenarios, service grids are used to control the service communication crisis that is caused by the increased business complexity. Finally, as described subsequently in Sections 5.2 and 5.7, the SaaS components or industrial APPs will create direct value for business scenarios and users. The industrial cloud platform will promote the formation of industrial APP ecosystems with the toolchain of industrial APP development and operation and business model innovation.

5.4. Step 3: Planning and Designing of Industrial Big Data Engine Layer

The remote sensing data of equipment through I3EAM has the characteristics of multi-source, massive and heterogeneous. On the one hand, the platform must construct a data management and governance system, to realize the effective and real-time management of the equipment data and to provide unified, complete, standardized, and visible data resources for various data application projects and data analysts. On the other hand, the platform must provide rich data service functions to improve the efficiency of the users' access to data and the level of the data that they are using. As illustrated in Figure 7, the industrial big data system provides data sources, edge connections, data collection, data processing, data services, and portals. Different data sources should use appropriate edge access solutions, data collection strategies, and data quality standards in accordance with the data volume, data characteristics, and real-time requirements. Different databases are interconnected through the middleware, and ETL (Extract–Transform–Load) formation is developed to support data processing and integration. Based on microservices and industrial APPs, the platform should provide the users with self-service and convenient data services and support the users' secondary development activities based on the data resources.

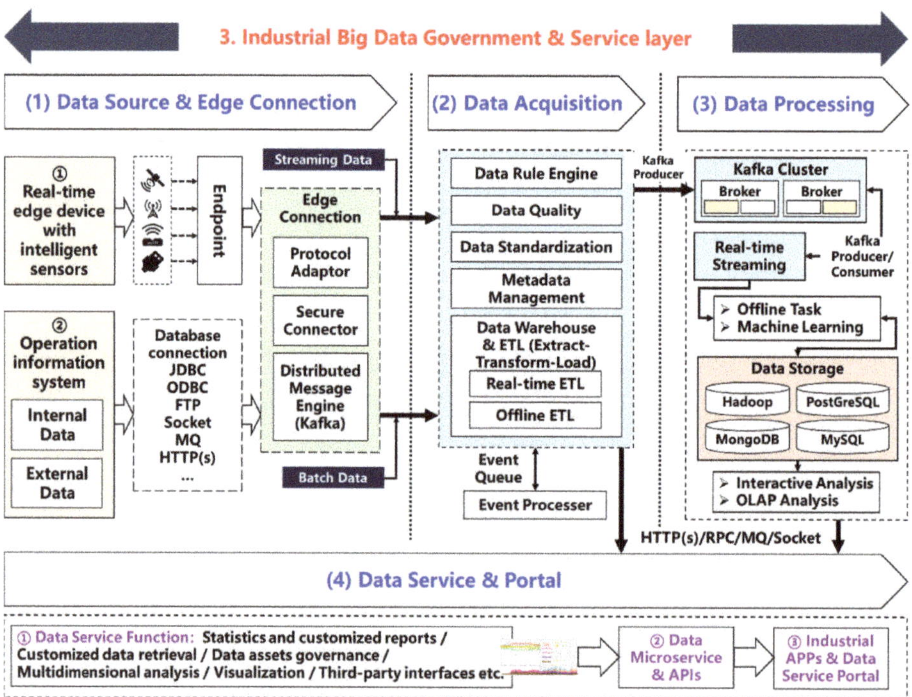

Figure 7. Industrial big data government and service layer.

5.5. Step 4: Planning and Designing of Industrial Modeling and Simulation Layer

Constructing a digital twin and parallel management system for complex equipment requires data insights and prior knowledge. The mechanistic models and common methods that have already been developed in the PHM field serve as the foundation for model encapsulation and the development of microservice libraries. As shown in Figure 8, I3EAM should build modeling and simulation systems related to Equipment Asset Management (EAM) in the Industrial Internet platform environment to reduce the time and cost of building the models for complex equipment with different properties and behaviors. The proposed modeling layer contains five types of analytical models: the descriptive, diagnostic, predictive, prescriptive, and cognitive analysis models. From the functional perspective of the models, I3EAM needs to consider other decision issues related to human factors and reliability of integrated systems in addition to the models in the PHM domain.

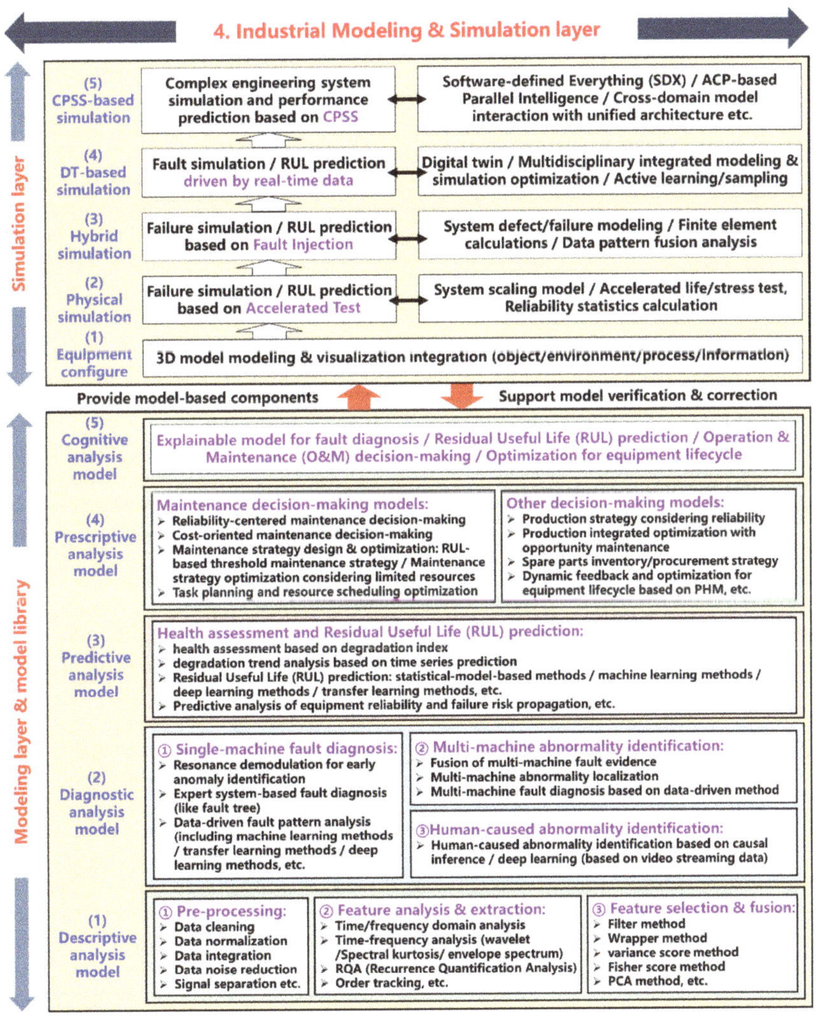

Figure 8. Industrial modeling and simulation layer.

Simulation has great potential for application in the area of equipment operation and maintenance, but it is limited by the cost of constructing high-fidelity digital models and the acquisition of rare fault data. Digital twin technology is facilitating the convergence of physical models, data models, and prior knowledge. The interaction between the physical entities and the virtual mirrors will be rolled up and optimized through generalized autoregressive models, and this will improve the effectiveness of equipment operation and maintenance decisions based on collaborative computing and consistency checks. The I3EAM simulation layer contains the equipment configuration, the physical simulation, the hybrid simulation, the digital twin-based simulation, and the CPSS-based simulation. Social factors such as the business' processes, the human factors, and the organization, likewise, need to be defined by the software and integrated with the physical dimension of the model. For example, the impact of the manual operation of maintenance on the condition and residual useful life (RUL) of the equipment is ignored in most of the equipment PHM work.

5.6. Step 5: Planning and Designing of Industrial Information Model Layer

Here is an example from the steel industry to illustrate the objective and function of the industrial information model and collaboration layer. The inability to cooperatively optimize the equipment and accessories is caused by the lack of equipment operation and maintenance (O and M) data for the key steel production equipment that is owned by the steel mill (the equipment owner) and the outsourced equipment O and M vendors. The equipment manufacturers and lubricant companies want to obtain these data, but the equipment owners lack the motivation to collect and share it. Likewise, due to the intellectual property, it might be challenging to exchange the equipment design data and lubricant formulation data with the manufacturers or other stakeholders to improve the equipment maintenance models. The main purpose of the information model layer is to build safe and interoperative industrial information integration models, technology solutions, and collaborative systems in order to lower the barriers of knowledge transfer and collaboration to meet the challenges of new equipment, conditions, and scenarios. Federal learning and the blockchain are considered to be potential integration technology directions. In addition, information model mapping for heterogeneous devices and information systems in the cloud-edge collaboration framework is necessary. For example, the OPC-UA's information modeling framework supports the integration of information models and protocols, allowing the direct modeling and transfer of semantic data through user-specified data formats. Based on the correspondence between the ontology model and the information model, the information is obtained from the device/edge node by semantic reasoning, and then, the device information model is exposed to the network through the OPC-UA server.

5.7. Step 6: Planning and Designing of Industrial Organization, Business, and Service Layer

Equipment assets are particularly important in industrial companies, and they are closely linked to production activities and profit generation. Industrial Internet technologies support Equipment Asset Management solutions for an increasing number of industrial enterprises. As shown in Figure 9, the service logic of I3EAM based on the Industrial Internet platform is first expressed using a simple concept diagram. The equipment reliability is associated with business goals and organizational performance in different dimensions and at different levels. Therefore, the service function requirements and the development of industrial APPs will be closely focused on the planning and control activities related to Equipment Asset Management. In the previous studies of decision support systems, the meta-model of service scenarios is missing or one-sided. In this paper, a service catalogue is given to cover the planning and control activities related to Equipment Asset Management at the strategic, integration, and operational levels based on the proposed AM framework in the literature [17]. Meanwhile, the EAM APP catalog of I3EAM is expected to be consistent with the service catalogue to provide a systematic solution for Equipment Asset Management in industrial enterprises.

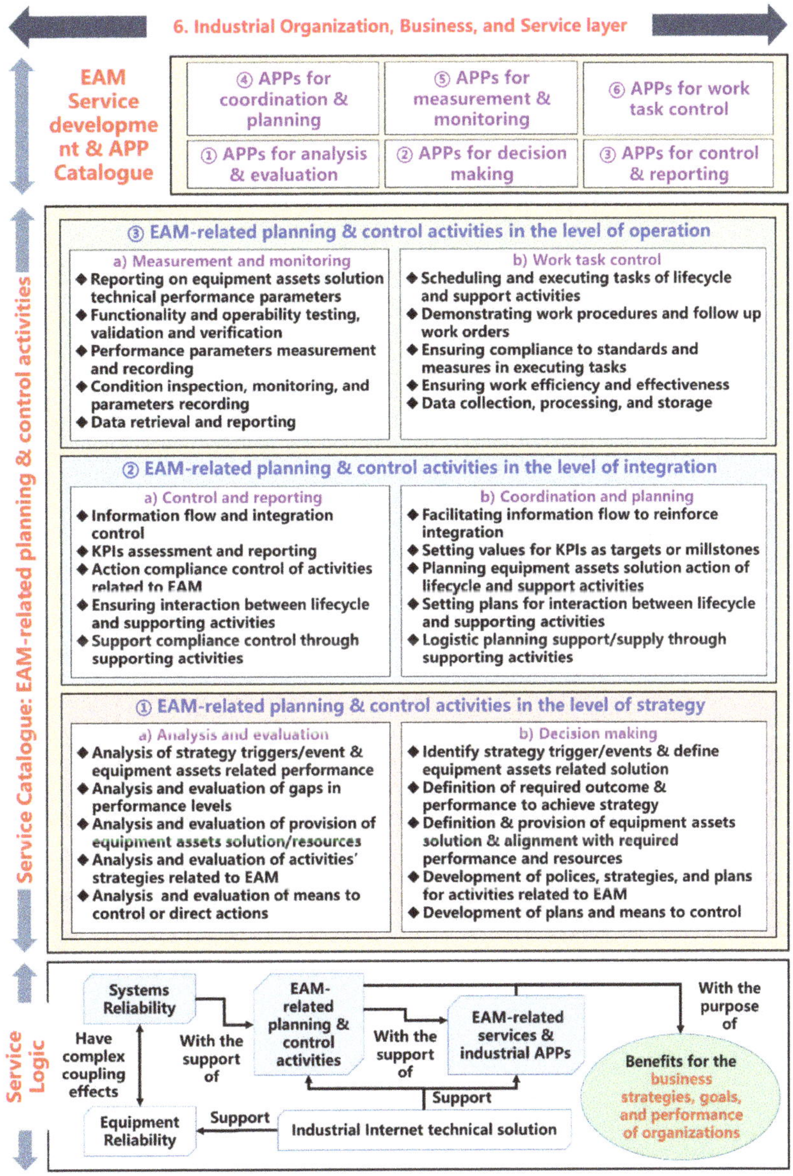

Figure 9. Industrial organization, business, and service layer.

6. Evaluation and Selection of I3EAM Schemes Based on Fuzzy DEMATEL-TOPSIS

6.1. The Establishment of Performance Metric Set for I3EAM Scheme

The core objective of this paper is to help enterprises to customize I3EAM using social digital platforms. The generic model and reference architecture provide effective guidance for identifying the requirements of I3EAM and planning the implementation path of I3EAM. However, there may be multiple social digitalization platforms to choose from around the specific needs of enterprises, and the I3EAM solutions that they offer have various characteristics. Therefore, based on the systematic synthesis, a set of performance metrics need to be constructed to help the enterprises to evaluate different solutions according to their preferences. The company will finally obtain an optimal solution, and the selected social digital platform will become the digital partner of the company to implement I3EAM together. The I3EAM evaluation system is crucial for EAM business–IT alignment for all of the types of enterprises, but it has not yet been studied and proposed.

The construction of I3EAM is a complex system-engineering project. The generic model and reference architecture focus on the identification of functional instances and architectural design, but they cannot completely reflect the performance of I3EAM in terms of its feasibility, capability, and ecology. Establishing an evaluation system for I3EAM will help identify the weak connections between the requirements, the solutions, and the platform partners more accurately and comprehensively, and increase the value created by the joint implementation of I3EAM. When designing and building the performance metric set for an I3EAM scheme, multiple factors and platform leverage logics need to be taken into account, such as the scheme's feasibility, the requirement coverage level, the technical capability, the application benefits, and the potential value in the medium and long term. Based on the research of the system model and enterprise requirements, this study proposes a performance metric assessment framework for the I3EAM scheme and its corresponding social digitalization platform. The framework was designed from three dimensions, including the basic platform service attribute, the technical capacity for I3EAM, and the value of the platform application and ecology synergy, as well as fifteen specific second grade indicators which are presented, as shown in Table 3.

Table 3. Performance metric set for I3EAM scheme.

First-Grade Indicator	Second-Grade Indicator	Explanation of the Indicator
Basic platform service attribute (B)	B1: Talent guarantee	Whether the platform constructor has the necessary industry technology and IT technology capabilities; The number of professionals engaged in the platform construction and operation; The coverage of the professionals' fields (knowledge and skills for ICT, specific industrial fields, and platform ecology construction &and operation).
	B2: System Security guarantee	Whether a management system of information security has been built, and whether the hidden dangers can be controlled; Whether effective countermeasures of information security have been constructed; The adoption of technologies for networking security, data security, and industrial access security, etc.
	B3: Cost of digitalization service	The level of the total cost for the implementation of I3EAM.
Technical capability for I3EAM (T)	T1: IT resource management capability	The comprehensive performance of IT infrastructure resources (including computing, storage, and network), and the capabilities of cloud-based management and scheduling of IT resources.
	T2: Industrial equipment adaptation capability	The capability of connection of heterogenous equipment asset resources (including type, sensing scheme, gateway, protocol, and cloud-edge collaboration scheme).
	T3: Industrial big data engine performance	The comprehensive performance of industrial big data engine (including data acquisition, data processing, and data service and portal).
	T4: Industrial modeling and simulation service capability	The capability of modeling and simulation for Equipment Asset Management (including intelligent algorithms, industrial mechanism models, industrial domain knowledge, digital twins schemes, and human–machine interactions).

Table 3. *Cont.*

First-Grade Indicator	Second-Grade Indicator	Explanation of the Indicator
	T5: Microservice development and government capability	The application level of microservice technical architecture, including capability maturity for microservice development and governance, the number and type of the EAM-related microservice, and the coverage of the needs of I3EAM.
	T6: Development environment supporting	The potential for custom development, secondary development and ongoing development of the microservices and industrial APPs (including cost of learning, low-code scheme, the friendliness level of user interfaces, and development toolchain).
	T7: Scheme technical architecture performance	The comprehensive performance of the technical architecture for I3EAM, namely the combination and integration scheme of the platform basic services, EAM-related microservices and industrial APPs.
	T8: System integration and collaboration capability	The level of compatibility and interoperability with enterprise information systems, the degree of standardization of information models, and the capability of middleware services.
Value of platform application and ecology synergy (A)	A1: Platform scale and activity	The platform scale and activity, including the user scale, user activity, developer scale, developer activity, service provider scale, service provider category, and service provider capability level.
	A2: Platform application value	The application level of the platform (including the number and type of long-term served industries, and the number and type of long-term served industrial scenarios related to EAM).
	A3. Open and sharing mechanism	Whether the platform has the mechanisms and tools for data and knowledge sharing and utilization across the various users and fields.
	A4: Platform ecology value	The ecology synergy effect of the platform (including the synergy effect in group company, global company, value chain, industry chain, and platform ecosystem).

The dimension of the basic platform service attribute includes three aspects, namely, (1) the talent guarantee, (2) the system security guarantee, and (3) the cost of the digitalization service, which are used in order to evaluate the feasibility of the implementation scheme and the basic service capability of the digitalization platform. The dimension of the technical capability for I3EAM is comprised of eight aspects, including (1) the IT resource management capability, (2) the industrial equipment adaptation capability, (3) the industrial big data engine capability, (4) the industrial modeling and simulation service capability, (5) the microservice development and government capability, (6) the development environment supporting, (7) the scheme technical architecture performance, and (8) the system integration and collaboration capability. The performance of technical capability is the core part of the assessment metric, and its objective is to provide a quantitative evaluation of I3EAM on the basis of the previously mentioned general model and reference architecture. The dimension of value of platform application and ecology synergy involves four aspects, namely, (1) the platform scale and activity, (2) the platform application value, (3) the open and sharing mechanism, and (4) the platform ecology value. This dimension is proposed for assessing the application benefits and potential value of the chosen digitalization platform in the medium-to-long term.

In addition, for the sake of conciseness in the subsequent expressions, we assigned tags to each indicator, i.e., B1–B3 for the dimension of a basic platform service attribute, T1–T8 for the dimension of a technical capability for I3EAM, and A1–A4 for the dimension of a value of platform application and ecology synergy.

6.2. The Approach to Evaluate and Select I3EAM Schemes Based on Different SDPs

In this section, a hybrid approach combining fuzzy DEMATEL and fuzzy TOPSIS is proposed for evaluating the I3EAM schemes based on different social digitalization platforms and the selection of the best scheme and co-implementation partner (the SDP which provides the best I3EAM scheme). The computational steps of our hybrid approach are clarified as follows:

(1) Step 1. Establishing fuzzy number and determining the linguistic variables

According to the fuzzy set theory, it is indicated that the elements have a degree of membership in a fuzzy set. A triangular fuzzy number (TFN) is defined as a fuzzy number whose membership function $f_{\tilde{a}}(x) : \mathbb{R} \to [0,1]$ is equal to following Equation (1). Additionally, the TFN can be denoted as $\tilde{a} = (l, m, u)$, where $l \leq m \leq u$.

$$f_{\tilde{a}}(x) = \begin{cases} 0, & x < l \text{ or } x > u \\ \frac{x-l}{m-l}, & l \leq x \leq m \\ \frac{u-x}{u-m}, & m \leq x \leq u \end{cases} \quad (1)$$

In order to design the questionnaire for collecting the decision makers' judgements, the linguistic terms and its variables were defined. In this paper, the computational technique is based on the following membership function of linguistic scale which has been defined in the literature [107] (as shown in Table 4). We used this kind of expression to evaluate the dependent influence between each two performance indicators, which are given in Section 6.1. The linguistic terms include: "No influence (N)", "Very low influence (VL)", "Low influence (L)", "High influence (H)", "Very high influence (VH)", and "Extremely high influence (EH)", as well as their corresponding TFNs separately, which are (0,0,0), (0.5,0.5,1), (0.5,1,1.5), (1.5,2,2.5), (2.5,3,3.5), and (3.5,4,4), respectively.

(2) Step 2. Constructing dependence matrices of performance indicators and group fuzzy initial direct-relation matrix

Table 4. Membership function of linguistic scale.

Fuzzy Number	Linguistic (Abbreviation)	Scale of Fuzzy Number
4	Extremely high influence (EH)	(3.5,4,4)
3	Very high influence (VH)	(2.5,3,3.5)
2	High influence (H)	(1.5,2,2.5)
1	Low influence (L)	(0.5,1,1.5)
0.5	Very low influence (VL)	(0.5,0.5,1)
0	No influence (N)	(0,0,0)

According to the previous research and industry investigations, we believe that the interaction between the performance indicators will produce a positive influence on the decision making about the importance of them (the weight of each indicator). This study used the method of group fuzzy DEMATEL for revealing the interaction relationships between every two indicator sets. For the I3EAM, such a complex system, the method is applicable to analyze and segment complex factors by group decision in an uncertain environment.

The experts were invited to make pair-wise comparisons in terms of crisp scores for the performance indicator interactions (No influence (N) = 0, Very low influence (VL) = 0.5, Low influence (L) = 1, High influence (H) = 2, Very high influence (VH) = 3, and Extremely high influence (EH) = 4). The experts were from multiple disciplines, including CIOs, procurement managers, project managers, technical staff, and financial experts. Based on the scoring system, the group judgements on the relationships between n performance

indicators were collected from q decision-makers. The kth decision maker's initial direct-relation matrix M_k is as follows:

$$M_k = \begin{bmatrix} 0 & s_{12}^k & \cdots & s_{1n}^k \\ s_{21}^k & 0 & \cdots & s_{2n}^k \\ \vdots & \vdots & \ddots & \vdots \\ s_{n1}^k & s_{n2}^k & \cdots & 0 \end{bmatrix} \quad k = 1, 2, \ldots, q \tag{2}$$

where s_{ij}^k is the kth decision maker's judgement with crisp score for the ith I3EAM performance indicator's influence on the jth indicator, q is the number of experts, and n is the number of I3EAM performance indicators.

Then, through the use of Formula (1) and the membership function in Table 4, the M_k can be fuzzified, and it was converted to a triangular fuzzy direct-relation matrix $\widetilde{M}_k$. The elements of $\widetilde{M}_k$ are triangular fuzzy numbers $\widetilde{s}_{ij}^k = (l_{ij}^k, m_{ij}^k, u_{ij}^k)$. Thus, the group fuzzy initial direct-relation matrix $\widetilde{S}$ can be obtained as follows:

$$\widetilde{S} = \begin{bmatrix} \hat{0} & \hat{s}_{12} & \cdots & \hat{s}_{1n} \\ \hat{s}_{21} & \hat{0} & \cdots & \hat{s}_{2n} \\ \vdots & \vdots & \ddots & \vdots \\ \hat{s}_{n1} & \hat{s}_{n2} & \cdots & \hat{0} \end{bmatrix} \tag{3}$$

where $\hat{s}_{ij} = (\hat{l}_{ij}, \hat{m}_{ij}, \hat{u}_{ij})$, $\hat{l}_{ij} = \left\{l_{ij}^1, \ldots, l_{ij}^k \ldots, l_{ij}^q\right\}$, $\hat{m}_{ij} = \left\{m_{ij}^1, \ldots, m_{ij}^k \ldots, m_{ij}^q\right\}$, and $\hat{u}_{ij} = \left\{u_{ij}^1, \ldots, u_{ij}^k \ldots, u_{ij}^q\right\}$. Then, by calculating the average value of each element in $\hat{s}_{ij}$, the group fuzzy direct-relation matrix $\widetilde{S}^A$, which was aggregated, could be obtained.

(3) Step 3. Acquiring the fuzzy normalized direct relation matrix and fuzzy total-relation matrix

After the process of aggregation, the fuzzy normalized direct-relation matrix $\widetilde{NS}^A = \left[\hat{ns}_{ij}^A\right]_{n\times n}$ can be acquired through dividing each element in $\widetilde{S}^A$ by a crisp number $NF(u)$. The $NF(u)$ can be calculated by using Equation (4). Through Equation (5), the matrix $\widetilde{S}^A$ was converted to the fuzzy normalized direct-relation matrix $\widetilde{NS}^A = \left[\hat{ns}_{ij}^A\right]_{n\times n}$, which is denoted as Equation (6). The element in $\widetilde{NS}^A$ is denoted as Equation (7).

$$NF(u) = max \sum_{j=1}^{n} u_{ij}^A \tag{4}$$

where $u_{ij}^A = \frac{1}{q}\sum_{k=1}^{q} u_{ij}^k$.

$$\hat{ns}_{ij}^A = \frac{\hat{s}_{ij}^A}{NF(u)} = \frac{\hat{s}_{ij}^A}{max \sum_{j=1}^{n} u_{ij}^A} \tag{5}$$

where $\hat{s}_{ij}^A = (l_{ij}^A, m_{ij}^A, u_{ij}^A) = (\frac{1}{q}\sum_{k=1}^{q} l_{ij}^k, \frac{1}{q}\sum_{k=1}^{q} m_{ij}^k, \frac{1}{q}\sum_{k=1}^{q} u_{ij}^k)$.

$$\widetilde{NS}^A = \left[\hat{ns}_{ij}^A\right]_{n\times n} = \begin{bmatrix} \hat{0} & \hat{ns}_{12}^A & \cdots & \hat{ns}_{1n}^A \\ \hat{ns}_{21}^A & \hat{0} & \cdots & \hat{ns}_{2n}^A \\ \vdots & \vdots & \ddots & \vdots \\ \hat{ns}_{n1}^A & \hat{ns}_{n2}^A & \cdots & \hat{0} \end{bmatrix} \tag{6}$$

$$\hat{ns}_{ij}^A = \left[\frac{l_{ij}^A}{max \sum_{j=1}^{n} u_{ij}^A}, \frac{m_{ij}^A}{max \sum_{j=1}^{n} u_{ij}^A}, \frac{u_{ij}^A}{max \sum_{j=1}^{n} u_{ij}^A}\right] \tag{7}$$

Next, the fuzzy total-relation matrix $\widetilde{TS}$ was acquired as follows:

$$\widetilde{TS} = \lim_{k\to\infty}((\widetilde{NS}^A)^1 + (\widetilde{NS}^A)^2 + \cdots + (\widetilde{NS}^A)^k) = \widetilde{NS}^A \times (E - \widetilde{NS}^A)^{-1} \tag{8}$$

$$\widetilde{TS} = \left[\hat{fs}_{ij}\right]_{n\times n} = \begin{bmatrix} \hat{0} & \hat{fs}_{12} & \cdots & \hat{fs}_{1n} \\ \hat{fs}_{21} & \hat{0} & \cdots & \hat{fs}_{2n} \\ \vdots & \vdots & \ddots & \vdots \\ \hat{fs}_{n1} & \hat{fs}_{n2} & \cdots & \hat{0} \end{bmatrix}, \ \hat{fs}_{ij} = (\hat{fs}^l_{ij}, \hat{fs}^m_{ij}, \hat{fs}^u_{ij}) \tag{9}$$

$$\begin{cases} \left[\hat{fs}^l_{ij}\right] = \hat{ns}^l_{ij} \times (E - \hat{ns}^l_{ij})^{-1} \\ \left[\hat{fs}^m_{ij}\right] = \hat{ns}^m_{ij} \times (E - \hat{ns}^m_{ij})^{-1} \\ \left[\hat{fs}^u_{ij}\right] = \hat{ns}^u_{ij} \times (E - \hat{ns}^u_{ij})^{-1} \end{cases} \tag{10}$$

(4) Step 4. Defuzzifying the fuzzy total-relation matrix

In order to acquire a crisp total-relation matrix, it was necessary to defuzzify the element of the fuzzy total-relation matrix above. In this study, the calculation operator for the process of defuzzifying used the operator in the literature [108].

$$\hat{fs}_{ij} = \frac{\hat{fs}^l_{ij} + 4\hat{fs}^m_{ij} + \hat{fs}^u_{ij}}{6} \tag{11}$$

(5) Step 5. Determining the weight of performance indicators of I3EAM

After obtaining the crisp total-relation matrix, the sum of the rows and the sum of the columns are separately denoted as X_i and Y_j, respectively, within the crisp total-relation matrix through the following formulas:

$$X_i = [x_i] = \left[\sum_{j=1}^{n} \hat{fs}_{ij}\right], \ Y_j = [y_j] = \left[\sum_{i=1}^{n} \hat{fs}_{ij}\right] \tag{12}$$

Next, based on the crisp total-relation matrix, x_i and y_j were used to calculate the "Prominence" and "Relation", respectively. In the approach of DEMATEL, the "Prominence" represents the strength of the influences that were both dispatched and received, and the "Relation" was used to determine the difference between the dispatched ones and the received ones. The vector p_i named "Prominence" was obtained by adding x_i to y_j. Similarly, the vector r_i named "Relation" was obtained by subtracting x_i to y_j. The vector p_i reveals how much importance the criterion has, and the vector r_i divides the criteria into groups of the cause and the effect.

$$p_i = x_i + y_j, \ r_i = x_i - y_j, \ i = j \tag{13}$$

Then, the importance of the ith I3EAM performance indicator ω_i was calculated using Equation (14) as follows.

$$\omega_i = \sqrt{{p_i}^2 + {r_i}^2} \tag{14}$$

The importance of any I3EAM performance indicator can be normalized with Equation (15) as follows:

$$W_i = \frac{\omega_i}{\sum_{1\le i\le n} \omega_i} \tag{15}$$

(6) Step 6. Constructing the group fuzzy initial decision matrix with determining the appropriate linguistic variables

Assuming that there are p social digitalization platforms (SDPs) being evaluated by q decision makers (DMs) with respect to n performance indicators, the kth decision maker's linguistic initial decision matrix N_k is as follows:

$$N_k = \begin{matrix} SDP_1 \\ SDP_2 \\ \vdots \\ SDP_p \end{matrix} \begin{bmatrix} d_{11}^k & d_{12}^k & \cdots & d_{1n}^k \\ d_{21}^k & d_{22}^k & \cdots & d_{2n}^k \\ \vdots & \vdots & \ddots & \vdots \\ d_{p1}^k & d_{p2}^k & \cdots & d_{pn}^k \end{bmatrix} k = 1, 2, \ldots, q \quad (16)$$

where d_{ij}^k is the kth decision maker's judgement with a linguistic variable for the ith SDP under the jth I3EAM performance indicator, q is the number of experts, and n is the number of I3EAM performance indicator.

Aiming to evaluate the alternatives concerning the performance indicators, the linguistic terms include: "Very poor (VP)", "Poor (P)", "Fair (F)", "Good (G)", and "Very good (VG)", as well as their corresponding TFNs, which are (0,0,0.25), (0,0.25,0.5), (0.25,0.5,0.75), (0.5,0.75,1), and (0.75,1,1), respectively. Similarly, the group fuzzy initial decision matrix $\widetilde{D}$ can be obtained as follows:

$$\widetilde{D} = \begin{bmatrix} \hat{d}_{11} & \hat{d}_{12} & \cdots & \hat{d}_{1n} \\ \hat{d}_{21} & \hat{d}_{22} & \cdots & \hat{d}_{2n} \\ \vdots & \vdots & \ddots & \vdots \\ \hat{d}_{p1} & \hat{d}_{p2} & \cdots & \hat{d}_{pn} \end{bmatrix} \quad (17)$$

where $\hat{d}_{ij} = (\hat{l}_{ij}^d, \hat{m}_{ij}^d, \hat{u}_{ij}^d)$, $\hat{l}_{ij}^d = \{l_{ij}^{d1}, \ldots, l_{ij}^{dk} \ldots, l_{ij}^{dq}\}$, $\hat{m}_{ij}^d = \{m_{ij}^{d1}, \ldots, m_{ij}^{dk} \ldots, m_{ij}^{dq}\}$, and $\hat{u}_{ij}^d = \{u_{ij}^{d1}, \ldots, u_{ij}^{dk} \ldots, u_{ij}^{dq}\}$. Then, by calculating the average value of each element in $\hat{d}_{ij}$, the group fuzzy decision matrix $\widetilde{D}^A$, which was aggregated, can be obtained.

(7) Step 7. Acquiring the weighted fuzzy decision matrix

The weighted fuzzy decision matrix $\widetilde{D}^*$ was obtained as follows:

$$\widetilde{D}^* = \left[\hat{d}_{ij}^*\right]_{n \times n}, \ i = 1, 2, \ldots, p; \ j = 1, 2, \ldots, n \quad (18)$$

where $\hat{d}_{ij}^* = \hat{W}_j \otimes \hat{d}_{ij}^A$, $\hat{W}_j$ is the weight of the jth performance indicator calculated by Equation (15), and $\hat{d}_{ij}^A = (l_{ij}^{dA}, m_{ij}^{dA}, u_{ij}^{dA}) = (\frac{1}{q}\sum_{k=1}^q l_{ij}^{dk}, \frac{1}{q}\sum_{k=1}^q m_{ij}^{dk}, \frac{1}{q}\sum_{k=1}^q u_{ij}^{dk})$.

(8) Step 8. Determining the fuzzy positive-ideal solution (FPIS) and fuzzy negative-ideal solution (FNIS)

The FPIS and FNIS were identified from the weighted fuzzy decision matrix by calculating Equations (19) and (20), respectively. The $\hat{d}_j^{*+}$ and $\hat{d}_j^{*-}$ separately represent the FPIS and FNIS, respectively.

$$\hat{d}_j^{*+} = \max_i(\hat{d}_{ij}^*) \ where \ \hat{d}_j^{*+u} = \max_i(\hat{d}_{ij}^{*u}) \quad (19)$$

$$\hat{d}_j^{*-} = \min_i(\hat{d}_{ij}^*) \ where \ \hat{d}_j^{*-l} = \min_i(\hat{d}_{ij}^{*l}) \quad (20)$$

(9) Step 9. Calculating the distance of each SDP alternative from FPIS and FNIS

The distances (D_i^+ and D_i^-) of each alternative from $\hat{d}_j^{*+}$ and $\hat{d}_j^{*-}$ can calculated by using the following equations.

$$D_i^+ = \sqrt{\frac{1}{3}\sum_{j=1}^{n}\left[(\hat{d}_{ij}^{*l} - \hat{d}_{ij}^{*+l})^2 + (\hat{d}_{ij}^{*m} - \hat{d}_{ij}^{*+m})^2 + (\hat{d}_{ij}^{*u} - \hat{d}_{ij}^{*+u})^2\right]} \tag{21}$$

$$D_i^- = \sqrt{\frac{1}{3}\sum_{j=1}^{n}\left[(\hat{d}_{ij}^{*l} - \hat{d}_{ij}^{*-l})^2 + (\hat{d}_{ij}^{*m} - \hat{d}_{ij}^{*-m})^2 + (\hat{d}_{ij}^{*u} - \hat{d}_{ij}^{*-u})^2\right]} \tag{22}$$

(10) Step 10. Acquiring the closeness coefficients and the priority of alternatives

The closeness coefficients CC_i represents the relative gaps–degree of each SPD alternative, and CC_i can be obtained through calculating Equation (23). On the basis of CC_i, we can also acquire the priority of all of the SDP alternatives (digitalization service provider for the co-implementation of I3EAM).

$$CC_i = \frac{D_i^-}{D_i^+ + D_i^-} \tag{23}$$

7. Case Study

7.1. Reference Model of I3EAM Schemes in Automated Container Terminals

In this section, a case study of building an I3EAM solution in an Automated Container Terminal (ACT) application scenario will be presented. The case was also taken as an example to demonstrate the application of the proposed general model and reference architecture. The context of the study is the digital transformation needs of the ports, which require the use of Industrial Internet technologies and platforms to build a parallel management system for container terminal equipment assets. We invited several industry experts, marine equipment companies, and port terminal companies to discuss and use the general reference model to collect and qualitatively analyze the requirements for the I3EAM solution for automated container terminals. Figure 10 shows the reference model of I3EAM schemes in automated container terminals (ACTs).

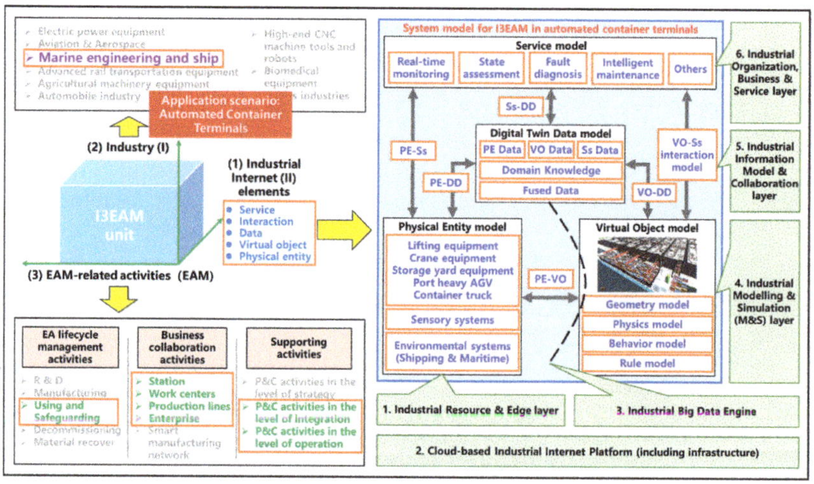

Figure 10. Reference model of I3EAM schemes in automated container terminals.

(1) Dimension 1: Industrial Internet element (II) dimension

The Equipment Asset Management of automated container terminals presents comprehensive requirements for the Industrial Internet elements, including the physical entity, the virtual object, the digital twin data, the interaction, and the service. In this study, the physical entity models in the ACTs include lifting equipment, crane equipment, storage yard equipment, port heavy AGV, the container truck and its sensory systems and environmental systems. To build the ACT simulation and emulation systems, the construction and management of virtual object models is considered to be necessary to realize the sensing, modeling, and management of the physical entities, the information flows, and the social relationships. Multiple sources of heterogeneous data were required to be integrated and managed, including the data from the physical entities, virtual objects, and services, the domain knowledge, and the fused data. The service model is comprised of real-time monitoring, state assessment, fault diagnosis, intelligent maintenance, and intelligent decision making (i.e., production scheduling, etc.). The interaction models built the connections among the different models.

(2) Dimension 2: Industry (I) dimension

Container terminals are closely related to marine engineering and shipping, transportation and other industries. From the perspective of equipment assets, they should belong to the marine engineering and shipping industry, which will inherit the characteristics of the industry and adapt to the development needs of the industry. Due to various reasons such as efficiency, costs, and epidemics, the container terminals and unmanned terminals have become the trend of industry development, and there is an urgent need to improve the level of terminal digitalization, networking, and intelligence.

(3) Dimension 3: EAM-related activities (EAM) dimension

In Section 4, the comprehensive EAM-related activity model is introduced, and it was used for analyzing and recognizing the scenario-based value in automated container terminals. From the perspective of the equipment's lifecycle, using and safeguarding them is regarded as the main application scenario in automated container terminals. From the perspective of business collaboration, automated container terminals constitute a complex organic system, which includes the level of equipment health that will affect different levels of the integrated systems (including the station, the work center, the production line, and the enterprise). From the perspective of system support, the entire ACT system will also be affected by random factors such as the arrival of the cargo vessel, and so the system usually requires dynamic response and scheduling. Therefore, the necessary planning and control activities need to be provided at the levels of both integration and operation.

7.2. *Reference Architecture for the Implementation of I3EAM in Automated Container Terminals*

Referring to the architecture that is proposed in Section 5, the implementation path of I3EAM in the automated container terminals is composed of six common steps and special elements about the automated container terminal corresponding to the enterprise needs. Figure 11 shows the whole picture of the reference architecture for the I3EAM implementation in automated container terminals.

(1) Step 1: Industrial resource and edge layer in ACTs

For automated container terminals, the main equipment includes the lifting equipment, the crane equipment, the storage yard equipment, the port heavy AGV, and the container truck. The focused sub-systems in the EAM business include the reduction gearbox, the pressure pump, the steel structure, the rotating mechanical parts, the electrical control system, the AGV battery, the AGV motor, and the motor synergy system. For connection and communication, the sensory systems should provide the data perception capacities about the various physical signals involving the vibration, temperature, current, pressure, position, torque, image, video, and laser ones. Additionally, many RFID-based systems and smart terminals are used for non-physical signals. With the consideration of the heterogenous equipment and edge sensors, the intelligent gateway and configurable protocol pool were planned and developed.

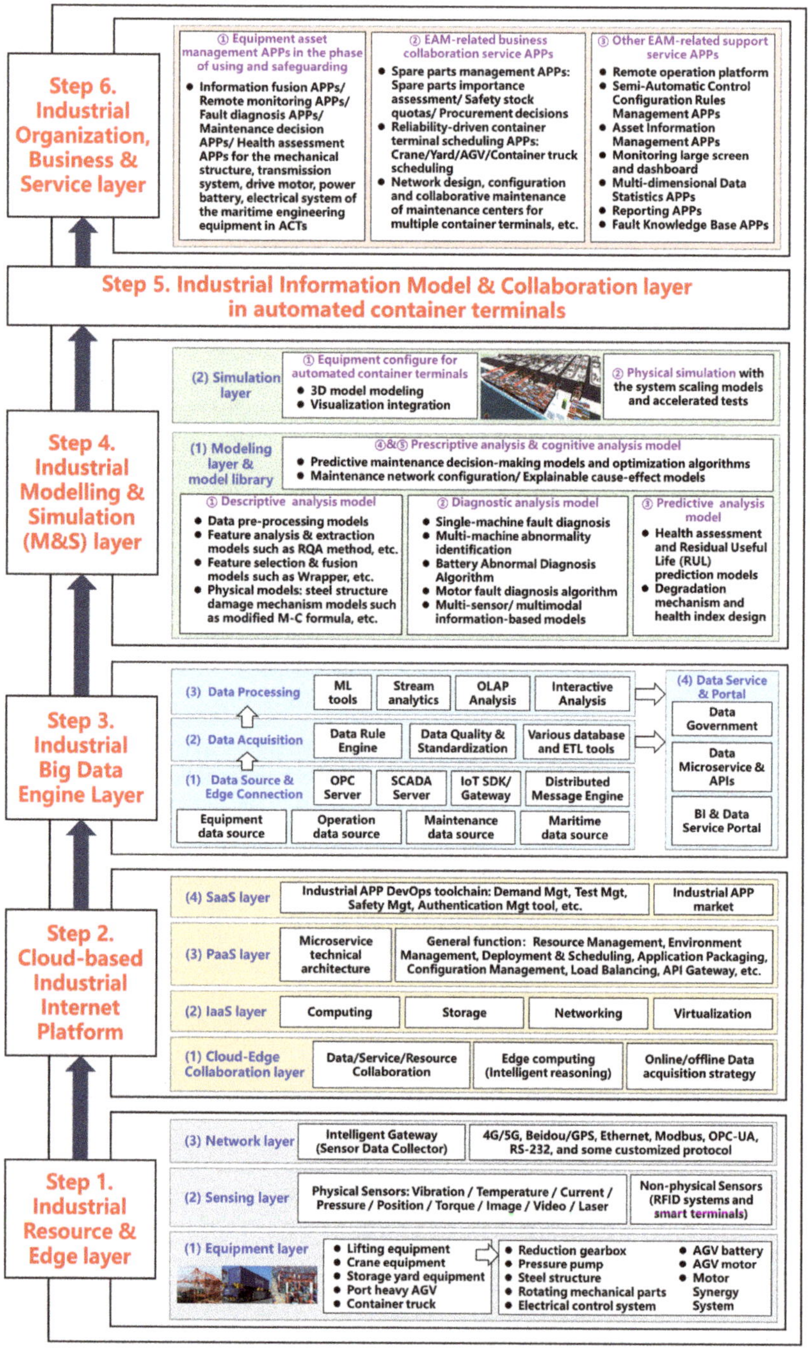

Figure 11. Reference architecture for the I3EAM implementation in automated container terminals.

(2) Step 2: Cloud-based Industrial Internet platform layer in ACTs

The cloud environment is necessary for the open, flexible, and dynamic business regarding automated container terminals. The designed platform of I3EAM in the ACTs is divided into four basic layers, including the cloud-edge collaboration layer, the IaaS layer, the PaaS layer, and the SaaS layer. Between the heterogenous integrated equipment and the remote control platform, it was necessary to establish a time-sensitive pipeline for the data/service/resource collaboration. In addition, intelligent reasoning for the edge sides and flexible data acquisition strategies were also required. The platform should be based on the microservice technical architecture. Meanwhile, the general enabling functions for resource virtualization, resource management, and service development and deployment were selected from the reference architecture, including the Resource Management, Environment Management, Deployment and Scheduling, and Configuration Management ones, etc. The APP DevOps toolchain and ecosystem are also important for the whole I3EAM scheme. Based on the protocol adaptor, secure connector, and distributed message engine, the various source data were collected in the form of streaming data or batch data.

(3) Step 3: Industrial big data engine layer in ACTs

The third step was to plan and construct the big data engine for the common requirements in the perspective of the data source and the edge connection, the data acquisition, the data processing, and the data service and portal. In the data warehouse, the subject data which were to be collected for the automated container terminals included the equipment source data, the operation source data, the maintenance source data, and the maritime source data. The system should have a data extraction capability and it should use ETL (Extract-Transform-Load) tools to realize data import and fusion into the database. Through the data rule engine, the data quality management and its standardization, and the standard data, which are scattered, fragmented, and non-uniform, should be loaded and integrated into the data warehouse. In addition, data processing functions in the online and offline modes were needed, including machine learning (ML) tools, stream analytics, an OLAP analysis, and an interactive analysis. To satisfy different data service requirements, the tools for the full data lifecycle will be encapsulated as microservices and invoked through the APIs to support various business intelligence (BI) processes and to provide access to the users' data.

(4) Step 4: Industrial modelling and simulation (M and S) layer in ACTs

Through the demand analysis of predictive maintenance for automated terminals, the system decision strategy will use a combination of online and offline, stand-alone and system, equipment and process, local diagnosis, local diagnosis and remote diagnosis, wired communication and wireless communication systems. The I3EAM solution requires multi-level and multi-faceted condition monitoring, fault diagnosis, and intelligent maintenance for the automated terminals to fully ensure the stable and reliable operation of the terminal production and operation. Modeling and model libraries play an important role in the process of the intellectualization of the container terminals, including intelligent perception, intelligent control, and intelligent decision making. In the environment of rich data sources, multiple layers of analyses need to be developed, including descriptive, diagnostic, predictive, prescriptive, and cognitive analyses.

The descriptive analysis model contains the models of data pre-processing, feature analysis and extraction (such as the RQA method, etc.), feature selection and fusion (such as Wrapper, etc.), and the special physical models such as the steel structure damage mechanism models such as the modified M-C formula, etc. The diagnostic analysis model contains the single-machine fault diagnosis, the multi-machine abnormality identification, the battery abnormal diagnosis, the motor fault diagnosis, and the multi-sensor or multimodal information-based models. The predictive analysis model contains the health assessment and residual useful life (RUL) prediction models, and the degradation mechanisms and health index for the various ACT equipment and the key components. Additionally, the prescriptive and cognitive analyses need to pay more attention to causal-based decision making, including the predictive maintenance decision-making models and the optimization algorithms, the maintenance network configuration, and the explainable cause–effect models.

In addition, the simulation layer was needed, and as well as this, the equipment configuration and physical simulations were considered. Combining the domain knowledge of complex equipment and equipment performance, the visual integration of the automated container terminals was realized based on 3D models. Through equipment configuration management and imitation learning based on scaling models and accelerated tests, a more realistic simulation was carried out to optimize and improve it, and better targeting was achieved from multiple perspectives, such as single machine equipment performance and system control.

(5) Step 5: Industrial information model and collaboration layer in ACTs

The industrial information model is an essential agreement that is shared by the participants and components in any networked and integrated system, facilitating compatibility and interoperability. Under the application scenario of the ACTs, different Industrial Internet elements (the physical entity, the virtual object, the data, and the service) in I3EAM are evolving and generating new interaction and collaboration requirements. The system needs to support the high level of interoperability, customized development, and the flexible configuration of the industrial information models. During the implementation, the different elements and their interactions were delicately and unambiguously named and assigned according to the conceptual, declarative, and programmatic representations.

(6) Step 6: Industrial organization, business, and service layer in ACTs

Based on the EAM dimensions in the reference model and the service activity metamodel given in Section 5.7, the scenario-based service requirements of the automated container terminals were identified. The automated container terminals are the application scenario of the offshore equipment such as cranes, which mainly consider the use and protection phase of the equipment. According to the level of system integration, the status performance of single machine equipment needed to be linked with the parts (spare parts), workstations, terminal areas, enterprises, and service networks. The identified service scenarios will be transformed into industrial APP development requirements.

In the I3EAM of the targeted automated container terminal, three types of industrial APPs needed to be developed as follows:

- The Equipment Asset Management APPs of using and safeguarding: information fusion APPs, remote monitoring APPs, fault diagnosis APPs, maintenance decision APPs, and health assessment APPs for the mechanical structure, transmission system, drive motor, power battery, electrical systems of the marine engineering equipment in the ACTs.
- The EAM-related business collaboration service APPs: spare parts management APPs (such as spare parts importance assessment, safety stock quotas, and procurement decision-making), reliability-driven container terminal scheduling APPs (such as crane/yard/AGV/container truck scheduling), maintenance network design and configuration, and collaborative maintenance of maintenance centers for multiple container terminals, etc.
- The EAM-related support service APPs: remote operation platform, semi-automatic control configuration rules management APPs, asset information management APPs, monitoring large screen and dashboard, multi-dimensional data statistics APPs, reporting APPs, and fault knowledge-based APPs.

7.3. Evaluation and Selection of I3EAM Schemes in Automated Container Terminals

(1) I3EAM scheme alternative profile

The case background is a new automated container terminal construction project in a coastal city in China. In order to enhance the digitalization and intelligence of the container terminal, there was an urgent need to develop a new Equipment Asset Management solution to help the terminal improve its operational capabilities. Social digitalization platforms have become the enabler and the accelerator for transformation in traditional industries. Aiming at the I3EAM in the automated container terminal, there were four SDP alternatives for the co-implementation of I3EAM, including MindSphere (Siemens), Azure (Microsoft), ROOTCLOUD (SANY), INDICS (CASICCloud), at the start of the project. According to the implementation requirements that were analyzed previously, the four SDPs submitted their technical solutions. The project team applied the proposed framework and approach to decide for the best scheme and platform partner. Due to confidentiality reasons, the schemes are randomly named as P_1–P_4.

(2) Expert-based evaluation data collection

The project team consists of eight decision makers, including industry experts, marine equipment companies, and port terminal companies, all of whom have an average of more than 10 years of experience in terminal construction or digital transformation. Similarly, the decision makers' profiles and the order of the decision making are not declared here, but they are randomly labelled as DM_1–DM_8.

The performance metric set and the evaluation and selection approach in Section 6 were used to help select the best alternative of I3EAM scheme and its social digitalization platform for the automated container terminals. To determine the weight of the performance indicators and the priority of the alternatives, each member of the decision group provided two initial evaluation matrices, including the dependence matrix of performance indicators, and the decision matrix of I3EAM schemes with its SDPs. The original data are provided in Appendix A (Tables A1–A3). There is one table containing the brief information of the decision makers, and there are two tables for the two groups of fuzzy initial matrices.

(3) Performance indicator weight analysis by fuzzy DEMATEL

In this study, the computational tool and environment for the fuzzy DEMATEL method was Matlab2022a. The process of the weighting analysis is outlined as follows. Through calculating Formula (1) and using the membership function in Table 4, the initial dependence matrix of performance indicators was converted to the triangular fuzzy direct-relation matrix. Next, by using Equations (4)–(10), the fuzzy normalized direct-relation matrix and fuzzy total-relation matrix were obtained. Then, the fuzzy total-relation matrix was defuzzified through Equation (11). Finally, based on the crisp total-relation matrix, the normalized weights can be obtained by calculating Equations (12)–(15). As shown in Table 5, the computation results were obtained.

Table 5. The importance and normalized weights of performance indicators of I3EAM.

Indicators	ω_i	W_i
B1	2.4254	0.0746
B2	1.0255	0.0315
B3	2.8756	0.0884
T1	1.8564	0.0571
T2	1.8314	0.0563
T3	3.5128	0.1080
T4	3.8550	0.1185
T5	1.5719	0.0483
T6	2.4328	0.0748
T7	2.5513	0.0784
T8	2.9460	0.0906
A1	1.6440	0.0505
A2	1.9866	0.0611
A3	0.9139	0.0281
A4	1.1009	0.0338

In our case, the T4 (Industrial modeling and simulation service capability), T3 (Industrial big data engine performance), T8 (System integration and collaboration capability), B3 (Cost of digitalization service), T7 (Scheme technical architecture performance), T6 (Development environment supporting), and B1 (Talent guarantee) emerge as the top six performance indicators of I3EAM in the automated container terminals.

(4) I3EAM scheme sorting and selection by fuzzy TOPSIS

Similarly, the computational tool and environment for the fuzzy TOPSIS method was Matlab2022a. By applying Equation (18) and combining the normalized weights, the weighted fuzzy decision matrix was obtained. Then, through Equations (19) and (20), both the fuzzy positive-ideal solution (FPIS) and the fuzzy negative-ideal solution (FNIS) were calculated, and they are listed in Table 6.

Table 6. The fuzzy positive-ideal solution (FPIS) and fuzzy negative-ideal solution (FNIS).

Indicators	FPIS	FNIS
B1	(0.0443, 0.0629, 0.0722)	(0.0117, 0.0280, 0.0466)
B2	(0.0099, 0.0177, 0.0246)	(0.0039, 0.0118, 0.0197)
B3	(0.0497, 0.0718, 0.0884)	(0.0000, 0.0221, 0.0442)
T1	(0.0428, 0.0571, 0.0571)	(0.0161, 0.0303, 0.0446)
T2	(0.0281, 0.0422, 0.0563)	(0.0088, 0.0229, 0.0369)
T3	(0.0776, 0.1046, 0.1080)	(0.0337, 0.0607, 0.0877)
T4	(0.0741, 0.1037, 0.1185)	(0.0222, 0.0518, 0.0815)
T5	(0.0076, 0.0196, 0.0317)	(0.0015, 0.0060, 0.0181)
T6	(0.0374, 0.0561, 0.0678)	(0.0000, 0.0187, 0.0374)
T7	(0.0515, 0.0711, 0.0784)	(0.0147, 0.0343, 0.0539)
T8	(0.0566, 0.0792, 0.0906)	(0.0255, 0.0481, 0.0708)
A1	(0.0221, 0.0347, 0.0474)	(0.0063, 0.0190, 0.0316)
A2	(0.0382, 0.0534, 0.0611)	(0.0057, 0.0210, 0.0363)
A3	(0.0114, 0.0184, 0.0255)	(0.0000, 0.0061, 0.0132)
A4	(0.0212, 0.0296, 0.0338)	(0.0000, 0.0042, 0.0127)

Using Equations (21)–(23), the distance to the ideal solutions, closeness coefficient, and priority of each alternative can be calculated, as shown in Table 7. Meanwhile, the sorting of I3EAM scheme alternatives was obtained as follows: $P_3\ (CC_3 = 0.6802) > P_2\ (CC_2 = 0.4574) > P_4\ (CC_4 = 0.4553) > P_1\ (CC_1 = 0.3376)$. Therefore, the P_3 was the selected alternative.

Table 7. The closeness coefficients and the priority of alternatives.

Alternatives	D_i^+	D_i^-	CC_i	Sorting
P_1	0.0494	0.0969	0.3376	4
P_2	0.0668	0.0792	0.4574	2
P_3	0.0941	0.0442	0.6802	1
P_4	0.0700	0.0837	0.4553	3

7.4. The Results

In the case study above, the I3EAM's system model and reference architecture provided the knowledge structure and systematic decision-making approach for the digitalization of the targeted global port. In our case, the applicational results of our model will be tested and examined in two main aspects: (1) whether it supports the correct implementation of the automated container terminal EAM system, and (2) whether it reduces the complexity and uncertainty of decision making.

For the first question, we examined the deviations of the implementation results of the ACT from the planning. The results show that the requirements and system architecture planning that were analyzed by I3EAM cover the list of functions of the ACT in the review phase. Around the three core equipment assets of the ACT, the project finally established capabilities such as structure resolution, edge connectivity, platform aggregation, big data engine, the model repository and ACT simulation tools, and the APP service pool. The operational status of the equipment assets can be transmitted back to the remote monitoring cloud platform through multi-channel and multi-type sensors in real time, and the joint diagnosis can be performed based on remote experts and online models. The cargo vessel demands and equipment status were incorporated into a joint scheduling model, while the swarm intelligence approaches are used to coordinate multiple equipment clusters. The overall port throughput performance meets the expected design target of 4 million containers per year. At the same time, the ACT's flexible architecture enables a scalability up to a forward throughput of 6.3 million containers per year.

For the second question, we went back to five of the decision makers of the project team and several other project members. They represented the port group, the equipment manufacturer, the terminal operator, and the external experts. The performance of the proposed decision-making model and process was validated via 12 semi-structure interviews and the backtracking of documents. All of the respondents indicated that our knowledge structure had a significant impact during the project planning phase, especially as the Industrial Internet standards were being developed at that time. In addition, they believe that the reference architecture generalizes the multi-view knowledge well and provides analytical tools to help implement EAM systems in different domains. The respondents agreed that the research result provided a logical, understandable, and step-by-step guide for the EAM projects. The model facilitates the project teams' systematic understanding of the Industrial Internet and EAM, collaboration across organizations, and the identification of the best platform partner.

8. Discussion

8.1. Caparisions among the Existing Research and the Research in This Paper for EAM

In the part of the status review (see Section 2), a comprehensive survey and comparison of the existing studies about the Industrial Internet-based EAM were made, including the related concepts, key technologies, application fields, decision support systems, and technical solutions, etc. The existing research mainly focus on the explicit technologies of EAM and their local applications, and only a few works give a comprehensive view of EAM implementation under the Industrial Internet-based environment. In China, the Alliance of Industrial Internet (AII) has released "Industrial Internet System Architecture (IISA) version 2.0" [13], which means the reference architecture of the Industrial Internet Platform (IIP) has been standardized. From the perspective of the future consideration of the IIP, China's IIP needs to emphasize the scenario-driven architecture design and implementation to adapt to our national conditions [9]. However, as an important applicational scenario, the EAM domain still lacks a clear implementation path based on Industrial Internet for the actual projects. In this paper, the authors proposed a general system model, a reference architecture, and an evaluation system with the metric set and decision-making approach in order to establish the mapping with EAM application requirements and the IIP system architecture. The model of I3EAM is based on the IISA 2.0 and digital twin metamodel, and it provides more powerful high-level abstractions, and as well as this, the structured model can be developed and integrated quickly. The paradigm of Design Science Research (DSR) has been introduced into the study along fusion path between EAM and Industrial Internet. This is a valuable addition to existing research. Meanwhile, the system architecture research provides support for the horizontal collaboration of the existing research.

8.2. Caparisions between Traditional Information System and Social Digitalization Platform

Platform-based digitalization is emphasized here, particularly to leverage the abilities of social digitalization platforms. The traditional process of business–IT alignment focuses on the implementation of information systems, such as SCADA, ERP, MES, etc. The information system is seen as a system of records to assist staff in the local monitoring and managing of the operational status of the equipment assets. Social digitalization platforms based on the Industrial Internet are more focused on how to support the wider interaction and collaboration in heterogeneous environments. For example, two types of information systems have been developed by separate entities, and they support the operation of container terminals. One type is the product service system that is provided by the equipment manufacturers, i.e., the condition monitoring software for structures, motors, and transmissions. One type is the shipping data system of the port group company, which collects the vessel and cargo information. In the traditional model, there is no integration between the two information systems. However, the EAM model recognizes that automated container terminals require the integration of demanding information, equipment reliability, and O and M planning to enable rapid resource scheduling. Therefore, when one is evaluating different I3EAM solutions using a structured model, the modelling of heterogeneous data integration and collaborative scheduling issues were highlighted to be the highest priority. Meanwhile, the eventual winner (P_3 in Table 7), which shows the accumulation of its shipping data analysis capabilities and general technologies, obtained the favor of the decision makers. This implicit phenomenon reflects the unique ecosystem advantage of the social digitalization platforms over the traditional IT service providers.

8.3. Caparisions between Traditional Process and Proposed Process for Decision Making

In the past, information systems development has been implemented during the phase of assets operation. However, digitalization is merging asset development, business innovation, and IT support. In our case, the Yangshan automated container terminal was regarded as a whole complex system, and the project emphasized the importance of overall planning in system implementation. The digital capabilities for EAM are the basic needs of system planning and design. The results in this paper offer the foundation and system engineering approaches to find the whole picture of the ACT's EAM system. Complex engineered systems usually involve many stakeholders. Traditional decision-making processes adopt the strategy of face-to-face meetings and step-by-step implementation. Nonetheless, the stakeholders involved often have inconsistent knowledge structures and value propositions about the target system, which leads to inefficiencies in their decision making and negotiation processes. Our research provides decision support for both of the two-aspect challenges. First, the multi-layer architecture illustrates the different view and structured knowledge of the I3EAM. Next, the performance indicators define the core potential value propositions from different perspectives. Except for the models, the proposed evaluation method based on fuzzy DEMATEL-TOPSIS simplifies the decision-making processes and makes them clearer. On the basis of the model, work such as more targeted information gathering and consultation were carried out. At the same time, the value propositions of multiple relevant stakeholders are coordinated with a scientific decision-making approach. The fuzzy method alleviates the uncertainty in the decision-making process. As a rough estimate, the project time for early decision making and negotiations were reduced by 30%. The project implementation met the expected system functional requirements and retained extensible interfaces for high-level functions.

To sum up, our research results are feasible and effective. The given model, the architecture, and the method in this study can create a certain reference value for both the EAM-related business–IT alignment of various enterprises (especially SMEs) and the development of IIPs (especially SDPs).

9. Conclusions

The research motivation and the objective of this paper are to improve the capability of complex system awareness and decision-making for I3EAM projects and to provide reference knowledge for the top-level planning and the effective implementation of different industrial enterprises. This study proposes a general reference model and a reference architecture for the implementation of Industrial Internet Solution for Industrial Equipment Asset Management (I3EAM). In addition, this study also developed a set of performance indicators for I3EAM and a scheme selection method by using the fuzzy DEMATEL-TOPSIS approach. A case study from the practice in an automated container terminal project demonstrates the feasibility and efficiency of the proposed model, architecture, and approach. Various industrial enterprises can carry out the analysis and top-level planning of their I3EAM needs, and the evaluation and selection of a co-implementation platform partner with our research results. Industrial Internet Platforms (IIPs) can carry out self-evaluation and self-diagnosis continuously for their EAM technical solutions based on the general model and reference architecture.

The scientific contribution of our study is remarkable, especially the exploratory work of DSR (Design Science Research) for the combination of EAM and the Industrial Internet. The structured knowledge and reference architecture models will reduce the complexity and uncertainty of system development, directly. IISA 2.0 is the standardized reference architecture model for the Industrial Internet in China, and this work is also the first to establish a detailed IISA-based scenario application model. Our research provides a useful and valuable addition to the existing academic literature and community such as EAM, II, information system, and digital transformation, etc., and it promotes collaborative evolution among these fields.

Due to the limited space, there are some follow-up works that need to be studied to supplement these findings. The future works of this study mainly contain three aspects.

- Firstly, more detailed content should be studied and analyzed for the comprehensive I3EAM reference model and technical architecture, including key technologies, information models, governance mechanisms, and programmatic components.
- Secondly, more precise and refined performance metrics and group decision methods should be used for the design of complex systems and system architecture.
- Thirdly, they should be further applied and verified in various industries and fields.

Author Contributions: Conceptualization, Y.B. and X.M.; methodology, Y.B., X.Z., T.Z. and Z.C.; validation, Y.B.; investigation, Y.B. and Z.C.; data curation, Y.B.; writing—original draft preparation, Y.B.; writing—review and editing, Y.B., X.Z., T.Z., Z.C. and X.M.; supervision, X.M.; funding acquisition, X.M. All authors have read and agreed to the published version of the manuscript.

Funding: This research was funded by National Natural Science Foundation of China [Grant No. 71632008]; Transformation and Upgrading of Industry in 2017 (China Manufacturing 2025) [Grant No. ZL35060009002]; Special Development Fund of China (Shanghai) Pilot Free Trade Zone [No. 725 (Title of funding project: 'Integrated equipment R&D and demonstration application in Yangshan automation container terminal')].

Data Availability Statement: Not applicable.

Acknowledgments: The authors would like to thank Shanghai Key Laboratory of Advanced Manufacturing Environment, Shanghai Research Center for industrial Informatics (SRCI2), and SJTU Innovation Center of Producer Service Development (SICPSD) for the funding support to this research.

Conflicts of Interest: The authors declare no conflict of interest.

Appendix A

Table A1. The detail information for the decision-makers.

Number of DM	Type of DM	Working Years of DM
1	Vice General Manager and Project manager	15
2	Group IT Director	13
3	Group Chief Engineer	25
4	Vice Director of Manufacturer Software Inc.	16
5	General manager of ACT	13
6	Director of Strategic Planning	15
7	Group Financial Director	20
8	External Expert (Professor)	23

Table A2. The group fuzzy initial direct-relation matrix.

DM_1	B1	B2	B3	T1	T2	T3	T4	T5	T6	T7	T8	A1	A2	A3	A4
B1	0	2	3	1	1	0	4	3	0	3	3	0	0	0	0
B2	0	0	2	0	2	1	0	0	0	2	2	0	2	0	0
B3	0	0	0	0	0	2	2	0	0	0	0	0	0	0	0
T1	0	0	0.5	0	0.5	3	4	3	0.5	4	0.5	2	0	0	0
T2	2	0	1	0	0	2	1	0	0	0	3	3	0	2	1
T3	2	0	3	0	1	0	4	2	3	3	2	2	0	2	1
T4	3	0	4	1	0	3	0	1	2	1	4	2	1	0	2
T5	0	0	2	3	0	2	2	0	2	2	2	1	0	0	0
T6	2	0	3	0	0	2	2	2	0	1	4	1	2	0	1
T7	0	0	2	0	0	2	2	0	0	0	2	0	0	0	0
T8	0	0	1	0	0	1	3	0	0	2	0	0	0	0	0
A1	2	0	0.5	0	1	2	2	0.5	1	0	1	0	0	0	3
A2	4	0	1	0	0	0	3	1	0	2	3	1	0	1	2
A3	0	2	0	0	1	0	1	0	0	0	3	0	0	0	3
A4	0	1	1	0	0	0	0	0	0	0	0	0	0	0	0
⋮	⋮	⋮	⋮	⋮	⋮	⋮	⋮	⋮	⋮	⋮	⋮	⋮	⋮	⋮	⋮
DM_8	B1	B2	B3	T1	T2	T3	T4	T5	T6	T7	T8	A1	A2	A3	A4
B1	0	0	2	1	1	2	3	0	1	3	2	1	0	0	1
B2	0	0	2	0	1	3	1	0	0	0	1	0	0	2	0
B3	0	0	0	0	0	1	2	0	0.5	0	0	0	0	0	0
T1	0	0	3	0	1	2	2	0.5	0	3	2	1	0	0	0
T2	0	0	2	2	0	2	1	0	0	2	3	0	0	1	2
T3	2	0	4	0.5	3	0	4	1	1	3	3	2	0	1	0
T4	4	0	4	1	0.5	4	0	1	1	2	2	1	0	1	1
T5	0	0	1	1	0	1	1	0	1	1	2	0	0	0	0
T6	2	0	2	1	1	3	3	2	0	2	3	2	2	0	1
T7	0	0	3	0	2	3	3	1	1	0	3	1	0	0	0
T8	0	0	2	1	1	2	3	2	2	3	0	0	0	0	2
A1	2	0	1	0	1	2	2	0	2	0	1	0	1	1	2
A2	3	0	2	0	2	2	4	1	0	3	2	1	0	1	3
A3	0	1	1	0	0	0.5	2	0	0	0	2	1	0	0	2
A4	0	0	2	0	0	0	0	0	0	0	0	1	0	0	0

Table A3. The group fuzzy initial direct decision matrix.

		B1	B2	B3	T1	T2	T3	T4	T5	T6	T7	T8	A1	A2	A3	A4
DM_1	P_1	0.75	0.75	0.75	0.5	0.25	0.5	0.5	0.25	0.25	0.5	0.5	0.75	0.25	0.25	0.25
	P_2	0.5	0.25	0.25	0.5	1	0.5	0.5	0.25	0.75	1	0.75	0.5	0.5	0.5	0.5
	P_3	0.5	0.5	0.5	1	0.5	1	1	0.25	0.75	1	0.5	0.75	0.5	0.75	1
	P_4	1	0.5	1	0.25	0.75	0.5	0.25	0	0.25	0.75	0.5	0.5	1	0.25	0.5
DM_2	P_1	0.5	1	0.75	0.25	0.25	0.5	0.5	0.25	0.25	0.25	0.75	0.75	0.25	0	0
	P_2	0.5	0.5	0.25	0.5	1	0.75	0.5	0.5	0.5	0.75	1	0.75	0.5	0.25	0.75
	P_3	0.5	0.5	0.75	1	0.75	1	0.75	0.75	0.5	1	0.5	0.75	0.25	0.5	1
	P_4	0.75	0.5	0.75	0.25	0.75	0.5	0.5	0	0.25	0.5	0.5	0.25	0.75	0.25	0.5
DM_3	P_1	0.75	0.75	0.75	0.5	0.5	0.5	0.5	0.25	0.25	0.5	0.5	0.5	0.25	0.25	0.25
	P_2	0.75	0.25	0.25	0.75	1	0.75	0.25	0.25	0.75	0.75	0.75	0.75	0.5	0.5	0.5
	P_3	0.5	0.5	0.5	1	0.5	1	1	0.25	1	0.75	0.5	0.75	0.5	0.75	0.75
	P_4	1	0.5	0.75	0.5	0.75	0.75	0.5	0.25	0.25	0.25	0.5	0.25	0.75	0.25	0.75
⋮	⋮	⋮	⋮	⋮	⋮	⋮	⋮	⋮	⋮	⋮	⋮	⋮	⋮	⋮	⋮	⋮
DM_8	P_1	0.5	0.5	0.75	0.5	0.5	0.5	0.25	0.25	0.25	0.75	0.75	0.25	0.25	0	0
	P_2	0.5	0.25	0.25	0.75	1	0.5	0.5	0.25	0.5	0.75	1	0.75	0.75	0.5	0.5
	P_3	0.5	0.25	0.5	1	0.75	1	1	0.25	0.5	1	0.75	0.75	0.5	0.75	0.75
	P_4	1	0.5	0.75	0.75	0.75	0.5	0.25	0.25	0.25	0.75	0.75	0.25	1	0.25	0.75

References

1. El-Akruti, K.; Dwight, R.; Zhang, T. The strategic role of Engineering Asset Management. *Int. J. Prod. Econ.* **2013**, *146*, 227–239. [CrossRef]
2. Lee, J.; Wu, F.; Zhao, W.; Ghaffari, M.; Liao, L.; Siegel, D. Prognostics and health management design for rotary machinery systems—Reviews, methodology and applications. *Mech. Syst. Signal Process.* **2014**, *42*, 314–334. [CrossRef]
3. Zhang, P.; Gao, Z.; Cao, L.; Dong, F.; Zou, Y.; Wang, K.; Zhang, Y.; Sun, P. Marine Systems and Equipment Prognostics and Health Management: A Systematic Review from Health Condition Monitoring to Maintenance Strategy. *Machines* **2022**, *10*, 72. [CrossRef]
4. Ren, Z.; Wan, J.; Deng, P. Machine-Learning-Driven Digital Twin for Lifecycle Management of Complex Equipment. *IEEE Trans. Emerg. Top. Comput.* **2022**, *10*, 9–22. [CrossRef]
5. Hoffmann Souza, M.L.; da Costa, C.A.; de Oliveira Ramos, G.; da Rosa Righi, R. A survey on decision-making based on system reliability in the context of Industry 4.0. *J. Manuf. Syst.* **2020**, *56*, 133–156. [CrossRef]
6. Mi, S.; Feng, Y.; Zheng, H.; Wang, Y.; Gao, Y.; Tan, J. Prediction maintenance integrated decision-making approach supported by digital twin-driven cooperative awareness and interconnection framework. *J. Manuf. Syst.* **2020**, *58*, 329–345. [CrossRef]
7. Erguido, A.; Marquez, A.C.; Castellano, E.; Parlikad, A.K.; Izquierdo, J. Asset Management Framework and Tools for Facing Challenges in the Adoption of Product-Service Systems. *IEEE Trans. Eng. Manag.* **2019**, *69*, 2693–2706. [CrossRef]
8. Kiritsis, D.; Emmanouilidis, C.; Koronios, A.; Mathew, J. *Engineering Asset Lifecycle Management*; Springer-Verlag: London, UK, 2010.
9. Li, J.; Qiu, J.-J.; Zhou, Y.; Wen, S.; Dou, K.-Q.; Li, Q. Study on the Reference Architecture and Assessment Framework of Industrial Internet Platform. *IEEE Access* **2020**, *8*, 164950–164971. [CrossRef]
10. Zhang, X.; Ming, X. A comprehensive industrial practice for Industrial Internet Platform (IIP): General model, reference architecture, and industrial verification. *Comput. Ind. Eng.* **2021**, *158*, 107426. [CrossRef]
11. Bao, Y.; Sun, Z.; Liao, X.; Chang, Y.; Qiu, S.; Ming, X. An industrial equipment maintenance support system for networked collaborative evolution in the smart product service ecosystem. In Proceedings of the 2022 IEEE Conference on Digital Twins & Parallel Intelligence, Ningbo, China, 26–28 October 2022.
12. *The Industrial Internet of Things Volume G1: Reference Architecture*; Industrial Internet Consortium: Needham Heights, MA, USA, 2019; pp. 1–7.
13. AII. Industrial Internet System Architecture 2.0. Alliance of Industrial Internet (AII): 2020. Available online: http://www.aii-alliance.org/index/c315/n45.html (accessed on 20 April 2020).

14. DIN SPEC. *Reference Architecture Model Industrie 4.0 (RAMI4. 0)*; DIN SPEC: Berlin, Germany, 2016; p. 10.31030.
15. Industrial Value Chain Initiative. *Industrial Value Chain Reference Architecture (IVRA)*; Technical Report; Industrial Value Chain Initiative: Tokyo, Japan, 2016.
16. Amadi-Echendu, J.E.; Willett, R.; Brown, K.; Hope, T.; Lee, J.; Mathew, J.; Vyas, N.; Yang, B.S. What is engineering asset management? In *Definitions, Concepts and Scope of Engineering Asset Management*; Springer: Berlin/Heidelberg, Germany, 2010; pp. 3–16.
17. El-Akruti, K.; Dwight, R. A framework for the engineering asset management system. *J. Qual. Maint. Eng.* **2013**, *19*, 398–412. [CrossRef]
18. Ouertani, M.Z.; Parlikad, A.K.; McFarlane, D. Asset information management: Research challenges. In Proceedings of the 2008 Second International Conference on Research Challenges in Information Science, Marrakech, Morocco, 3–6 June 2008; IEEE: Manhattan, NY, USA, 2008.
19. Schuman, C.A.; Brent, A.C. Asset life cycle management: Towards improving physical asset performance in the process industry. *Int. J. Oper. Prod. Manag.* **2005**, *25*, 566–579. [CrossRef]
20. Macchi, M.; Roda, I.; Negri, E.; Fumagalli, L. Exploring the role of Digital Twin for Asset Lifecycle Management. *IFAC-PapersOnLine* **2018**, *51*, 790–795. [CrossRef]
21. Mont, O.K. Clarifying the concept of product–service system. *J. Clean. Prod.* **2002**, *10*, 237–245. [CrossRef]
22. Tukker, A. Eight types of product–service system: Eight ways to sustainability? Experiences from SusProNet. *Bus. Strat. Environ.* **2004**, *13*, 246–260. [CrossRef]
23. Song, W.; Sakao, T. A customization-oriented framework for design of sustainable product/service system. *J. Clean. Prod.* **2017**, *140*, 1672–1685. [CrossRef]
24. Tomiyama, T.; Lutters, E.; Stark, R.; Abramovici, M. Development capabilities for smart products. *CIRP Ann.* **2019**, *68*, 727–750. [CrossRef]
25. Porter, M.E.; Heppelmann, J.E. How smart, connected products are transforming competition. *Harv. Bus. Rev.* **2014**, *92*, 64–88.
26. Valencia, A.; Mugge, R.; Schoormans, J.; Schifferstein, H. The design of smart product-service systems (PSSs): An exploration of design characteristics. *Int. J. Des.* **2015**, *9*, 13–28.
27. Zheng, P.; Wang, Z.; Chen, C.-H.; Khoo, L.P. A survey of smart product-service systems: Key aspects, challenges and future perspectives. *Adv. Eng. Inform.* **2019**, *42*, 100973. [CrossRef]
28. Chen, Z.; Ming, X.; Vareilles, E.; Battaia, O. Modularization of smart product service: A framework integrating smart product service blueprint and weighted complex network. *Comput. Ind.* **2020**, *123*, 103302. [CrossRef]
29. Kagermann, H.; Riemensperger, F.; Hoke, D.; Helbig, J.; Stocksmeier, D.; Wahlster, W.; Schweer, D. *Smart Service Welt: Recommendations for the Strategic Initiative Web-Based Services for Businesses*; Acatech-National Academy of Science and Engineering: Berlin, Germany, 2014.
30. Zheng, M.; Ming, X.; Wang, L.; Yin, D.; Zhang, X. Status Review and Future Perspectives on the Framework of Smart Product Service Ecosystem. *Procedia CIRP* **2017**, *64*, 181–186. [CrossRef]
31. Cenamor, J.; Sjödin, D.R.; Parida, V. Adopting a platform approach in servitization: Leveraging the value of digitalization. *Int. J. Prod. Econ.* **2017**, *192*, 54–65. [CrossRef]
32. Paiola, M.; Gebauer, H. Internet of things technologies, digital servitization and business model innovation in BtoB manufacturing firms. *Ind. Mark. Manag.* **2020**, *89*, 245–264. [CrossRef]
33. Tian, J.M.; Coreynen, W.; Matthyssens, P.; Shen, L. Platform-based servitization and business model adaptation by established manufacturers. *Technovation* **2021**, *118*, 102222. [CrossRef]
34. Thomas, L.D.W.; Autio, E.; Gann, D.M. Architectural Leverage: Putting Platforms in Context. *Acad. Manag. Perspect.* **2014**, *28*, 198–219. [CrossRef]
35. Fehrer, J.A.; Woratschek, H.; Brodie, R.J. A systemic logic for platform business models. *J. Serv. Manag.* **2018**, *29*, 546–568. [CrossRef]
36. Zhou, T.; Ming, X.; Chen, Z.; Miao, R. Selecting industrial IoT Platform for digital servitisation: A framework integrating platform leverage practices and cloud HBWM-TOPSIS approach. *Int. J. Prod. Res.* **2021**, 1–23. [CrossRef]
37. Eloranta, V.; Turunen, T. Platforms in service-driven manufacturing: Leveraging complexity by connecting, sharing, and integrating. *Ind. Mark. Manag.* **2016**, *55*, 178–186. [CrossRef]
38. Gawer, A. Digital platforms' boundaries: The interplay of firm scope, platform sides, and digital interfaces. *Long Range Plan.* **2021**, *54*, 102045. [CrossRef]
39. Kretschmer, T.; Leiponen, A.; Schilling, M.; Vasudeva, G. Platform ecosystems as meta-organizations: Implications for platform strategies. *Strateg. Manag. J.* **2021**, 1–20. Available online: https://onlinelibrary.wiley.com/doi/10.1002/smj.3250 (accessed on 1 January 2022). [CrossRef]
40. Office of Ministry of Industry and Information Technology of the People's Republic of China. List of Cross-Industry and Cross-Domain Industrial Internet Platforms in 2022 (China). Available online: https://www.miit.gov.cn/zwgk/zcwj/wjfb/tz/art/2022/art_56f15b21b6014c2b889e9fa19e371257.html (accessed on 24 May 2022).
41. Zhang, X.; Ming, X.; Bao, Y.; Liao, X. Industrial Internet Platform (IIP) enabled Smart Product Lifecycle-Service System (SPLSS) for manufacturing model transformation: From an industrial practice survey. *Adv. Eng. Inform.* **2022**, *52*, 101633. [CrossRef]
42. Wang, J.; Meng, Z.; Gao, D.; Feng, L. Construction of an Integrated Framework for Complex Product Design Manufacturing and Service Based on Reliability Data. *Machines* **2022**, *10*, 555. [CrossRef]

43. Mourtzis, D.; Vlachou, A.; Zogopoulos, V. Cloud-Based Augmented Reality Remote Maintenance Through Shop-Floor Monitoring: A Product-Service System Approach. *J. Manuf. Sci. Eng.* **2017**, *139*, 061011. [CrossRef]
44. Chang, F.; Zhou, G.; Xiao, X.; Tian, C.; Zhang, C. A function availability-based integrated product-service network model for high-end manufacturing equipment. *Comput. Ind. Eng.* **2018**, *126*, 302–316. [CrossRef]
45. Zhang, X.; Ming, X.; Bao, Y.; Liao, X.; Miao, R. Networking-enabled product service system (N-PSS) in collaborative manufacturing platform for mass personalization model. *Comput. Ind. Eng.* **2021**, *163*, 107805. [CrossRef]
46. Chang, F.; Zhou, G.; Zhang, C.; Ding, K.; Cheng, W.; Chang, F. A maintenance decision-making oriented collaborative cross-organization knowledge sharing blockchain network for complex multi-component systems. *J. Clean. Prod.* **2021**, *282*, 124541. [CrossRef]
47. Kohtamäki, M.; Parida, V.; Oghazi, P.; Gebauer, H.; Baines, T. Digital servitization business models in ecosystems: A theory of the firm. *J. Bus. Res.* **2019**, *104*, 380–392. [CrossRef]
48. Cheng, S.; Azarian, M.H.; Pecht, M.G. Sensor Systems for Prognostics and Health Management. *Sensors* **2010**, *10*, 5774–5797. [CrossRef]
49. Li, T.; Si, X.; Pei, H.; Sun, L. Data-model interactive prognosis for multi-sensor monitored stochastic degrading devices. *Mech. Syst. Signal Process.* **2021**, *167*, 108526. [CrossRef]
50. Liu, K.; Huang, S. Integration of Data Fusion Methodology and Degradation Modeling Process to Improve Prognostics. *IEEE Trans. Autom. Sci. Eng.* **2014**, *13*, 344–354. [CrossRef]
51. Fang, X.; Paynabar, K.; Gebraeel, N. Multistream sensor fusion-based prognostics model for systems with single failure modes. *Reliab. Eng. Syst. Saf.* **2017**, *159*, 322–331. [CrossRef]
52. Sim, J.; Kim, S.; Park, H.J.; Choi, J.-H. A Tutorial for Feature Engineering in the Prognostics and Health Management of Gears and Bearings. *Appl. Sci.* **2020**, *10*, 5639. [CrossRef]
53. Tian, J.; Morillo, C.; Azarian, M.H.; Pecht, M. Motor Bearing Fault Detection Using Spectral Kurtosis-Based Feature Extraction Coupled With K-Nearest Neighbor Distance Analysis. *IEEE Trans. Ind. Electron.* **2015**, *63*, 1793–1803. [CrossRef]
54. Oh, H.; Jung, J.H.; Jeon, B.C.; Youn, B.D. Scalable and Unsupervised Feature Engineering Using Vibration-Imaging and Deep Learning for Rotor System Diagnosis. *IEEE Trans. Ind. Electron.* **2018**, *65*, 3539–3549. [CrossRef]
55. Zhu, Y.; Wu, J.; Wu, J.; Liu, S. Dimensionality reduce-based for remaining useful life prediction of machining tools with multisensor fusion. *Reliab. Eng. Syst. Saf.* **2021**, *218*, 108179. [CrossRef]
56. Wang, D.; Peng, Z.; Xi, L. The sum of weighted normalized square envelope: A unified framework for kurtosis, negative entropy, Gini index and smoothness index for machine health monitoring. *Mech. Syst. Signal Process.* **2020**, *140*, 106725. [CrossRef]
57. Hou, B.; Wang, D.; Wang, Y.; Yan, T.; Peng, Z.; Tsui, K.-L. Adaptive Weighted Signal Preprocessing Technique for Machine Health Monitoring. *IEEE Trans. Instrum. Meas.* **2020**, *70*, 1–11. [CrossRef]
58. Fang, X.; Zhou, R.; Gebraeel, N. An adaptive functional regression-based prognostic model for applications with missing data. *Reliab. Eng. Syst. Saf.* **2015**, *133*, 266–274. [CrossRef]
59. Liu, R.; Yang, B.; Zio, E.; Chen, X. Artificial intelligence for fault diagnosis of rotating machinery: A review. *Mech. Syst. Signal Process.* **2018**, *108*, 33–47. [CrossRef]
60. Wang, D.; Chen, Y.; Shen, C.; Zhong, J.; Peng, Z.; Li, C. Fully interpretable neural network for locating resonance frequency bands for machine condition monitoring. *Mech. Syst. Signal Process.* **2021**, *168*, 108673. [CrossRef]
61. Huang, T.; Zhang, Q.; Tang, X.; Zhao, S.; Lu, X. A novel fault diagnosis method based on CNN and LSTM and its application in fault diagnosis for complex systems. *Artif. Intell. Rev.* **2021**, *55*, 1289–1315. [CrossRef]
62. Jia, F.; Lei, Y.; Lin, J.; Zhou, X.; Lu, N. Deep neural networks: A promising tool for fault characteristic mining and intelligent diagnosis of rotating machinery with massive data. *Mech. Syst. Signal Process.* **2016**, *72–73*, 303–315. [CrossRef]
63. Li, Q.; Chen, L.; Shen, C.; Yang, B.; Zhu, Z. Enhanced generative adversarial networks for fault diagnosis of rotating machinery with imbalanced data. *Meas. Sci. Technol.* **2019**, *30*, 115005. [CrossRef]
64. Wang, J.; Mo, Z.; Zhang, H.; Miao, Q. Ensemble diagnosis method based on transfer learning and incremental learning towards mechanical big data. *Measurement* **2020**, *155*, 107517. [CrossRef]
65. Zhang, S.; Sun, Z.; Wang, M.; Long, J.; Bai, Y.; Li, C. Deep Fuzzy Echo State Networks for Machinery Fault Diagnosis. *IEEE Trans. Fuzzy Syst.* **2019**, *28*, 1205–1218. [CrossRef]
66. Chen, Z.; Xu, J.; Peng, T.; Yang, C. Graph Convolutional Network-Based Method for Fault Diagnosis Using a Hybrid of Measurement and Prior Knowledge. *IEEE Trans. Cybern.* **2021**, *52*, 1–13. [CrossRef]
67. Zheng, R.; Najafi, S.; Zhang, Y. A recursive method for the health assessment of systems using the proportional hazards model. *Reliab. Eng. Syst. Saf.* **2022**, *221*, 108379. [CrossRef]
68. Aremu, O.O.; Hyland-Wood, D.; McAree, P.R. A Relative Entropy Weibull-SAX framework for health indices construction and health stage division in degradation modeling of multivariate time series asset data. *Adv. Eng. Inform.* **2019**, *40*, 121–134. [CrossRef]
69. Hu, C.-H.; Pei, H.; Si, X.-S.; Du, D.-B.; Pang, Z.-N.; Wang, X. A Prognostic Model Based on DBN and Diffusion Process for Degrading Bearing. *IEEE Trans. Ind. Electron.* **2019**, *67*, 8767–8777. [CrossRef]
70. Zhang, Y.; Xin, Y.; Liu, Z.-W.; Chi, M.; Ma, G. Health status assessment and remaining useful life prediction of aero-engine based on BiGRU and MMoE. *Reliab. Eng. Syst. Saf.* **2022**, *220*, 108263. [CrossRef]

71. Chang, Y.; Chen, J.; Liu, Y.; Xu, E.; He, S. Temporal convolution-based sorting feature repeat-explore network combining with multi-band information for remaining useful life estimation of equipment. *Knowl.-Based Syst.* **2022**, *249*, 108958. [CrossRef]
72. Yang, X.; Zheng, Y.; Zhang, Y.; Wong, D.S.-H.; Yang, W. Bearing Remaining Useful Life Prediction Based on Regression Shapalet and Graph Neural Network. *IEEE Trans. Instrum. Meas.* **2022**, *71*, 1–12. [CrossRef]
73. Chao, M.A.; Kulkarni, C.; Goebel, K.; Fink, O. Fusing physics-based and deep learning models for prognostics. *Reliab. Eng. Syst. Saf.* **2021**, *217*, 107961. [CrossRef]
74. Zhang, W.; Li, X.; Ma, H.; Luo, Z.; Li, X. Transfer learning using deep representation regularization in remaining useful life prediction across operating conditions. *Reliab. Eng. Syst. Saf.* **2021**, *211*, 107556. [CrossRef]
75. Fan, Y.; Nowaczyk, S.; Rögnvaldsson, T. Transfer learning for remaining useful life prediction based on consensus self-organizing models. *Reliab. Eng. Syst. Saf.* **2020**, *203*, 107098. [CrossRef]
76. Mauthe, F.; Hagmeyer, S.; Zeiler, P. Creation of Publicly Available Data Sets for Prognostics and Diagnostics Addressing Data Scenarios Relevant to Industrial Applications. *Int. J. Progn. Health Manag.* **2021**, *12*. [CrossRef]
77. Qu, Y.; Hou, Z. Degradation principle of machines influenced by maintenance. *J. Intell. Manuf.* **2021**, *33*, 1521–1530. [CrossRef]
78. Luo, C.; Shen, L.; Xu, A. Modelling and estimation of system reliability under dynamic operating environments and lifetime ordering constraints. *Reliab. Eng. Syst. Saf.* **2021**, *218*, 108136. [CrossRef]
79. Song, Y.; Mi, J.; Cheng, Y.; Bai, L.; Chen, K. A dependency bounds analysis method for reliability assessment of complex system with hybrid uncertainty. *Reliab. Eng. Syst. Saf.* **2020**, *204*, 107119. [CrossRef]
80. Compare, M.; Antonello, F.; Pinciroli, L.; Zio, E. A general model for life-cycle cost analysis of Condition-Based Maintenance enabled by PHM capabilities. *Reliab. Eng. Syst. Saf.* **2022**, *224*, 108499. [CrossRef]
81. Ding, S.-H.; Kamaruddin, S. Maintenance policy optimization—Literature review and directions. *Int. J. Adv. Manuf. Technol.* **2014**, *76*, 1263–1283. [CrossRef]
82. Cavalcante, C.A.; Lopes, R.S.; Scarf, P.A. Inspection and replacement policy with a fixed periodic schedule. *Reliab. Eng. Syst. Saf.* **2020**, *208*, 107402. [CrossRef]
83. Xia, T.; Si, G.; Wang, D.; Pan, E.; Xi, L. Progressive Opportunistic Maintenance Policies for Service-Outsourcing Network With Prognostic Updating and Dynamical Optimization. *IEEE Trans. Reliab.* **2021**, *71*, 1340–1354. [CrossRef]
84. Xia, T.; Sun, B.; Chen, Z.; Pan, E.; Wang, H.; Xi, L. Opportunistic maintenance policy integrating leasing profit and capacity balancing for serial-parallel leased systems. *Reliab. Eng. Syst. Saf.* **2021**, *205*, 107233. [CrossRef]
85. Jafar-Zanjani, H.; Zandieh, M.; Sharifi, M. Robust and resilient joint periodic maintenance planning and scheduling in a multi-factory network under uncertainty: A case study. *Reliab. Eng. Syst. Saf.* **2021**, *217*, 108113. [CrossRef]
86. Nguyen, K.-A.; Do, P.; Grall, A. Joint predictive maintenance and inventory strategy for multi-component systems using Birnbaum's structural importance. *Reliab. Eng. Syst. Saf.* **2017**, *168*, 249–261. [CrossRef]
87. Qin, Y.; Zhang, J.; Chan, F.T.; Chung, S.; Niu, B.; Qu, T. A two-stage optimization approach for aircraft hangar maintenance planning and staff assignment problems under MRO outsourcing mode. *Comput. Ind. Eng.* **2020**, *146*, 106607. [CrossRef]
88. Sanchez-Londono, D.; Barbieri, G.; Fumagalli, L. Smart retrofitting in maintenance: A systematic literature review. *J. Intell. Manuf.* **2022**, 1–19. [CrossRef]
89. Wang, C.; Guo, R.; Yu, H.; Hu, Y.; Liu, C.; Deng, C. Task offloading in cloud-edge collaboration-based cyber physical machine tool. *Robot. Comput. Manuf.* **2023**, *79*, 102439. [CrossRef]
90. Calabrese, F.; Regattieri, A.; Bortolini, M.; Gamberi, M.; Pilati, F. Predictive Maintenance: A Novel Framework for a Data-Driven, Semi-Supervised, and Partially Online Prognostic Health Management Application in Industries. *Appl. Sci.* **2021**, *11*, 3380. [CrossRef]
91. Tao, F.; Zhang, M.; Liu, Y.; Nee, A.Y.C. Digital twin driven prognostics and health management for complex equipment. *CIRP Ann.* **2018**, *67*, 169–172. [CrossRef]
92. Savolainen, J.; Urbani, M. Maintenance optimization for a multi-unit system with digital twin simulation. *J. Intell. Manuf.* **2021**, *32*, 1953–1973. [CrossRef]
93. Vichare, N.; Pecht, M. Prognostics and health management of electronics. *IEEE Trans. Compon. Packag. Technol.* **2006**, *29*, 222–229. [CrossRef]
94. Baroroh, D.K.; Chu, C.-H.; Wang, L. Systematic literature review on augmented reality in smart manufacturing: Collaboration between human and computational intelligence. *J. Manuf. Syst.* **2020**, *61*, 696–711. [CrossRef]
95. Kafiabad, S.T.; Zanjani, M.K.; Nourelfath, M. Robust collaborative maintenance logistics network design and planning. *Int. J. Prod. Econ.* **2021**, *244*, 108370. [CrossRef]
96. Fan, D.; Lin, J.; Cai, B.; Liu, B. Robustness of maintenance support service networks: Attributes, evaluation and improvement. *Reliab. Eng. Syst. Saf.* **2021**, *210*, 107526. [CrossRef]
97. Zhang, K.; Xia, T.; Wang, D.; Chen, G.; Pan, E.; Xi, L. Privacy-preserving and sensor-fused framework for prognostic & health management in leased manufacturing system. *Mech. Syst. Signal Process.* **2023**, *184*, 109666. [CrossRef]
98. Martowicz, A.; Uhl, T. Reliability- and performance-based robust design optimization of MEMS structures considering technological uncertainties. *Mech. Syst. Signal Process.* **2012**, *32*, 44–58. [CrossRef]
99. Mourtzis, D.; Angelopoulos, J.; Panopoulos, N. Equipment Design Optimization Based on Digital Twin Under the Framework of Zero-Defect Manufacturing. *Procedia Comput. Sci.* **2021**, *180*, 525–533. [CrossRef]

100. Tong, X.; Liu, Q.; Pi, S.; Xiao, Y. Real-time machining data application and service based on IMT digital twin. *J. Intell. Manuf.* **2019**, *31*, 1113–1132. [CrossRef]
101. Wang, W.; He, Y.; Liao, R.; Cai, Y.; Zheng, X.; Zhao, Y. Mission reliability driven functional healthy state modeling approach considering production rhythm and workpiece quality for manufacturing systems. *Reliab. Eng. Syst. Saf.* **2022**, *226*, 108682. [CrossRef]
102. Colledani, M.; Yemane, A. Impact of Machine Reliability Data Uncertainty on the Design and Operation of Manufacturing Systems. *Procedia CIRP* **2013**, *7*, 557–562. [CrossRef]
103. Li, H.; Palau, A.S.; Parlikad, A.K. A social network of collaborating industrial assets. *Proc. Inst. Mech. Eng. Part O J. Risk Reliab.* **2018**, *232*, 389–400. [CrossRef]
104. Palau, A.S.; Liang, Z.; Lütgehetmann, D.; Parlikad, A.K. Collaborative prognostics in Social Asset Networks. *Future Gener. Comput. Syst.* **2019**, *92*, 987–995. [CrossRef]
105. Zhu, S.-P.; Huang, H.-Z.; Peng, W.; Wang, H.-K.; Mahadevan, S. Probabilistic Physics of Failure-based framework for fatigue life prediction of aircraft gas turbine discs under uncertainty. *Reliab. Eng. Syst. Saf.* **2016**, *146*, 1–12. [CrossRef]
106. Wang, F.Y.; Wang, X.; Li, L.; Li, L. Steps toward parallel intelligence. *IEEE/CAA J. Autom. Sin.* **2016**, *3*, 345–348.
107. Chen, Z.; Lu, M.; Ming, X.; Zhang, X.; Zhou, T. Explore and evaluate innovative value propositions for smart product service system: A novel graphics-based rough-fuzzy DEMATEL method. *J. Clean. Prod.* **2019**, *243*, 118672. [CrossRef]
108. Kutlu, A.C.; Ekmekçioğlu, M. Fuzzy failure modes and effects analysis by using fuzzy TOPSIS-based fuzzy AHP. *Expert Syst. Appl.* **2012**, *39*, 61–67. [CrossRef]

machines

MDPI

Review

Remote Monitoring and Maintenance for Equipment and Production Lines on Industrial Internet: A Literature Review

Qingzong Li, Yuqian Yang and Pingyu Jiang *

State Key Laboratory for Manufacturing Systems Engineering, Xi'an Jiaotong University, Xi'an 710054, China
* Correspondence: pjiang@mail.xjtu.edu.cn

Abstract: Monitoring and maintaining equipment and production lines ensure stable production by detecting and resolving abnormalities immediately. In the Industrial Internet, operational technology and advanced information technology are fused to improve the digitalization and intelligence of monitoring and maintenance. This paper provides a comprehensive survey of monitoring and maintenance of equipment and production lines on the Industrial Internet. Firstly, a brief review of its architecture is given, and a reference architecture is summarized accordingly, clarifying the key enabling technologies involved. These key technologies are data collection technologies, edge computing, advanced communication technologies, fog computing, big data, artificial intelligence, and digital twins. For each of the key technologies, we provide a detailed literature review of their state-of-the-art advances. Finally, we discuss the challenges that it currently faces and give some suggestions for future research directions.

Keywords: remote monitoring; maintenance; Industrial Internet

Citation: Li, Q.; Yang, Y.; Jiang, P. Remote Monitoring and Maintenance for Equipment and Production Lines on Industrial Internet: A Literature Review. *Machines* **2023**, *11*, 12. https://doi.org/10.3390/machines11010012

Academic Editor: Zhuming Bi

Received: 22 November 2022
Revised: 17 December 2022
Accepted: 19 December 2022
Published: 22 December 2022

1. Introduction

1.1. Background

Well-maintained equipment and production lines are the basis for regular production in a factory. It is necessary to monitor and maintain the equipment and production lines during the operation process effectively to avoid failures [1–3]. Traditionally, equipment and production lines are manually inspected [4] and maintained after failures arise [5]. However, this strategy cannot avoid the negative impact of equipment downtime on quality and capacity, which entails high costs [6]. With the development of wireless sensor networks [7], advanced communication technologies [8–13], big data [14–16], artificial intelligence [17,18], and digital twins [19], the Industrial Internet emerged, which brings fresh impetus to the monitoring and maintenance of equipment and production lines.

There are several vital issues that should be addressed in the monitoring and maintenance of equipment and production lines on the Industrial Internet. It is a challenge to acquire various types of data from different devices from diverse manufacturers in a factory, as they have different communication protocols that are not compatible with each other. After collecting a vast volume of raw data, it is also a difficult task to store, transfer, and process this data. It is also essential to extract valuable information from these data to determine the health of the equipment and display it to humans to facilitate proper decision-making. Therefore, we present a comprehensive survey of the monitoring and maintenance of equipment and production lines on the Industrial Internet.

1.2. Research Methodology for Literature Review

This paper provides a comprehensive investigation of some studies completed on the monitoring and maintenance of equipment and production lines on the Industrial Internet and discusses some research questions and challenges. We started with a brief review of the Industrial Internet's architecture, sorting out the associated key enabling technologies.

Addressing these key technologies, we discuss their latest advances—including data collection technologies, edge computing, communication technologies, fog computing, big data, artificial intelligence, digital twins, data analytics, operation and maintenance (O&M) optimization, and sustainability—enabling researchers to keep up with the pioneers quickly. Then some industrial application examples were demonstrated. We also discussed some of the technical challenges and provided a few suggestions for future research directions.

2. Architecture

Architecture is a higher level of abstraction description that helps identify issues and challenges for the monitoring and maintenance of equipment and production lines on the Industrial Internet. There various architectures have been proposed in the recent literature.

Wang et al. [20] proposed a cloud-assisted platform for large-scale continuous condition monitoring based on the Industrial Internet of Things. The platform is a three-tier architecture comprising an edge layer, a platform layer, and an enterprise layer. The edge tier is where data are collected, aggregated, and transmitted. The data are transmitted through the edge gateway to the platform tier for data storage, workflow processing, and other applications. At the enterprise tier, data analysis and mining are applied to support enterprise planning and decision-making. Yang et al. [21] proposed a monitoring platform with a three-layer architecture based on cloud manufacturing, comprising an edge layer, a fog layer, and a cloud layer. At the edge layer, raw data are acquired and preprocessed. The fog layer is devoted to the interconnection of devices and the transmission of data on the one hand and the deployment of trained models on the other. In the cloud, engineers monitor production status, make decisions remotely via screens, and train models for diagnosis and prediction. Li et al. [22] proposed a two-layer Industry 4.0 platform for equipment monitoring and maintaining, consisting of a machine layer and an application layer. The data are collected at the machine layer and then used at the application layer to monitor the equipment conditions, production processes, and product quality. Yang et al. [23] designed an integrated monitoring and maintenance framework for the grinding and polishing robot. The framework is divided into four layers: a physical layer, a key enabling technology layer, a business logic layer, and a data collection and processing layer. The physical layer contains all the devices and sensors. The key enabling technology layer describes the models and algorithms for monitoring and maintenance. The entire business logic is described in the business logic layer. The data collection and processing layer depicts the devices and their corresponding interaction logic for collecting and processing the working condition data in the production line.

From the above literature review, we summarize a reference architecture for monitoring and maintenance of equipment and production lines on the Industrial Internet, as shown in Figure 1. This architecture consists of three layers: the physical layer, the transport layer, and the application layer. The physical layer includes the equipment, sensors, actuators, controller, and data acquisition unit. This layer focuses on data collection and preprocessing, and the key technologies involved in this layer are data collection technology and edge computing. The transport layer contains network transport devices, computing devices, and databases. The main task of the transport layer is data transmission, aggregation, and forwarding, which also involves some data processing. The key enabling technologies in this layer are communication technologies and fog computing. The application layer consists of computing servers, databases, and application servers, whose main functions are data storage, model training, algorithm operation, etc. The key enabling technologies in this layer are artificial intelligence, big data, digital twins, data analytics, O&M optimization, and sustainability. A comprehensive overview of the key enabling technologies in each layer will be presented in the next section.

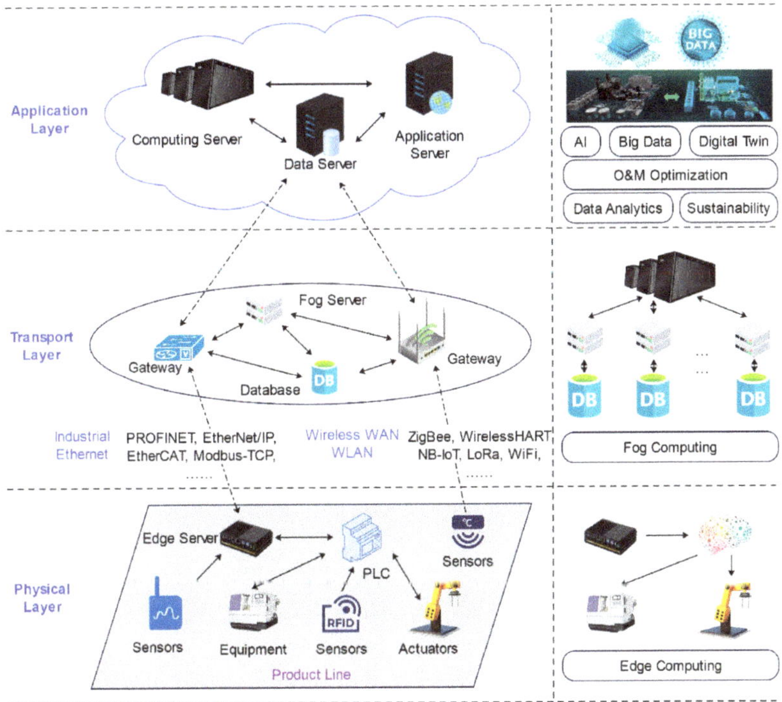

Figure 1. Reference architecture.

3. Key Enable Technologies

In the age of the Industrial Internet, traditional remote monitoring and maintenance technologies for equipment and production lines are becoming digital and intelligent as they are combined with rapidly evolving information technologies. This section focuses on the key enabling technologies in the above three layers: data collection technologies, edge computing, communication technologies, fog computing, big data, artificial intelligence, digital twins, data analytics, O&M optimization, and sustainability.

3.1. Physical Layer

The physical layer includes equipment and data acquisition devices. Its function mainly collects data from the actuators and sensors and processes the acquired data. Therefore, this subsection provides an overview of data acquisition technologies and edge computing.

3.1.1. Data Acquisition Technologies

The data acquisition unit collects data during the production process, including the data collected by sensors and the control data of actuators. From the perspective of physical connectivity, there are two data acquisition modes: wired and wireless.

A. Wired Data Acquisition Technologies

The wired data acquisition method is widely used in the industry. Short and Twiddle [24] developed a real-time condition monitoring and fault diagnosis system for large-scale rotating equipment in the water industry. The data acquisition unit of the system contains several temperatures and speed sensors. An ADC converter (AD7856) is used to convert the analog signals from the sensors to digital signals, a C167 microcontroller is used to process the digital signals, and a non-volatile memory chip (EEPROM) is used for data storage. It communicates with the outside through the RS-232 standard interface. Xia et al. [25] presented an intelligent fault diagnosis system for industrial robot bearings under varying conditions. The vibration from a sensor was collected by a signal acquisition board card (PXIe-4497) and sent to a PXI controller (NI PXIe-8840) together with the joint angle data from the robot controller.

As is shown in Figure 2, wired data acquisition devices usually consist of a data acquisition card, processor, memory, and transceiver. The control data of the equipment can be obtained directly from the controller. Low-frequency sensing data can generally be collected by the controller, but high-frequency sensor data needs to be collected with a dedicated data acquisition card. In general, the data will be processed at the edge, and the processed data will be transmitted to the next layer through the transceiver. The data acquisition card, processor, memory, and transceiver are always integrated into a single device as a data acquisition unit.

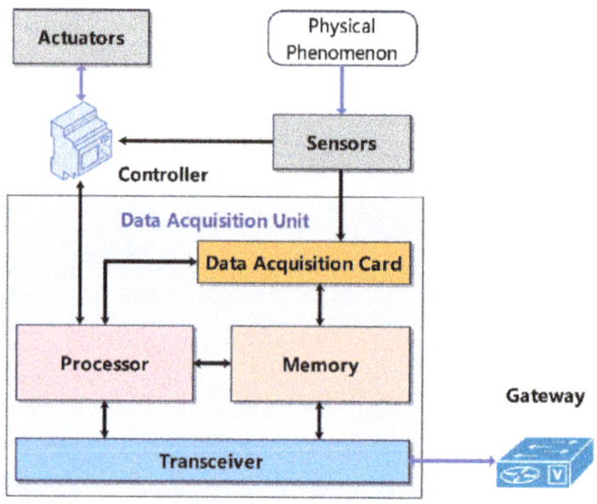

Figure 2. Wired data acquisition devices.

B. Wireless Data Acquisition Technologies

At present, wireless data collection methods are developing rapidly [26]. Li et al. [27] proposed a flexible strain sensor based on an aluminum nitride film. Its flexible substrate can stick well to bearings to detect micro-strain, therefore making it suitable for condition monitoring and failure prevention. Sancho et al. [28] proposed a wireless LC sensor to monitor the continuous wear in abradable blades in a paper mill, addressing the issue of wear monitoring in a distributed harsh industrial environment. Ahmed et al. [29] developed an optical camera communication-enabled wireless sensor network to monitor the state of industrial valves. The device consists of an AM2302 sensor to collect the temperature, an ATMEL 1430 TINY85 20SU microcontroller to process the sensor data and modulate the LED, and a transmitter to transfer the data. Walker et al. [30] proposed a method for real-time in-process monitoring of core motion in metal castings. A group of wireless Bluetooth inertial measurement sensors was integrated into the additive manufacturing sand cores to measure the acceleration and rotation during the casting. Lei and Wu [31] designed a wireless device to acquire mechanical vibration signals. The device consists of a high-precision MEMS acceleration sensor, a 16-bit resolution ADC acquisition chip, a high-performance control center (STM32), and a wireless transceiver core (Si4463), which enables high-frequency, high-precision acquisition of vibration signals. Patil et al. [32] proposed an architecture for wireless sensor nodes, arguing that the basic components of a wireless sensor node are a sensor, process unit, memory, transceiver, and battery.

The architecture of the wireless data acquisition devices can be summarized from the above literature, as shown in Figure 3, including sensors, a processor, memory, a wireless transceiver, and a battery. These components are integrated into a single device, commonly referred to as a wireless sensor node.

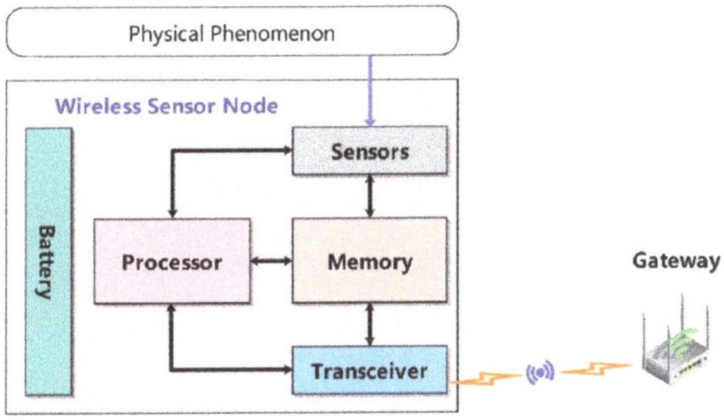

Figure 3. Wireless data acquisition devices.

There are numerous advantages of the wireless data acquisition device. It is easy to deploy without any pre-existing infrastructure [33]; the wireless sensor network is capable of covering a pervasive area [34]; the wireless sensor node is convenient to move [7]. Most importantly, wireless sensors can help us collect data in environments where wiring is impossible [35]. Therefore, the application of wireless data acquisition devices in equipment monitoring and maintenance has become a research hotspot.

However, there are several shortcomings in applying wireless sensor data collection methods in equipment and production lines. The wireless sensor node usually requires batteries for power and therefore requires periodic battery replacement, resulting in monitoring interruptions [36]. Even though some self-powered sensors have been developed, they can only be used in specific environments [37]. The real-time performance and efficiency of data transmission remain to be improved [38]. The data are susceptible to interference from the external environment during wireless transmission [39]. These deficiencies limit the application of wireless data acquisition methods in production lines, which require a high degree of reliability and stability of data in real time. It will be the emphasis of future research to address these issues.

3.1.2. Edge Computing

A large body of data is collected in production lines. The response time will be too long if all the data are sent to the cloud for processing. Hence, data can be processed at the edge, which reduces response times, increases processing efficiency, and reduces network stress [40–42]. Edge refers to resources and devices near the endpoint along the path between data sources and cloud data centers [43].

Edge computing has already been applied in the remote monitoring and maintenance of equipment and production lines [44]. Zhang et al. [45] developed a cyber-physical machine tool based on edge computing techniques to realize real-time monitoring of the machine tool. The edge devices were deployed on various manufacturing units to process the collected data. The processed data are graphed in real-time with digital twin technology to monitor the process and status of the machine. Edge computing techniques improve the accuracy and capability of virtual machine tools and reduce the mapping latency between physical and digital models. Wen et al. [46] designed a remote monitoring and intelligent maintenance platform for a sewage treatment plant based on edge computing instrumentation. The edge is composed of intelligent instruments and edge servers. Intelligent algorithms and processing units are integrated into intelligent instruments for real-time data collection. The data are sent to an edge server for data quality control, preprocessing, data aggregation, real-time data analysis, decision-making, and data cloud upload. The preprocessed data are used for digital twin modeling and predictive maintenance in the cloud.

The convergence of edge computing and artificial intelligence—called edge intelligence, is becoming a top research priority [18,47]—but there are still many challenges [48].

In the above reviews, the edge was only used for preprocessing of the raw data, and the training and inference of the model were conducted in the cloud. However, with limited bandwidth connectivity between the edge and the cloud, it is almost impossible to achieve real-time local decision-making [49], which is fatal in some hazardous production facilities [50]. Deploying machine learning at the edge is also significantly challenged by the limited computing capacity of edge devices. We discovered that some studies are working on these issues.

Lee et al. [51] presented a predictive maintenance system based on edge computing to maintain and manage motor equipment. The system collects audio data from the motor with an embedded acoustic recognition sensor and then preprocesses the raw data with a preprocessor server. The preprocessed data are transferred to an inference server, which classifies the health of the motor by using a trained model. Both the preprocessing server and the inference server are deployed at the edge. Bowden et al. [52] proposed a hybrid cloud-edge computing-based framework called SERENA for predictive maintenance. A machine-learning model was built, trained in the cloud, and then pushed to the edge device. In the edge device, the statistical characteristics of the raw data from the sensors are calculated first, and then machine learning models are used to diagnose the failures of machines.

From the above studies, a general edge intelligence architecture can be summarized, as shown in Figure 4. The raw data are preprocessed at the edge, and the results are sent to the cloud. In the cloud, the preprocessed data can be applied directly to remote monitoring and maintenance of equipment, as it can also be used to train machine learning models. The trained model can be deployed at the edge for inference of anomalies and consequently signaling an alarm.

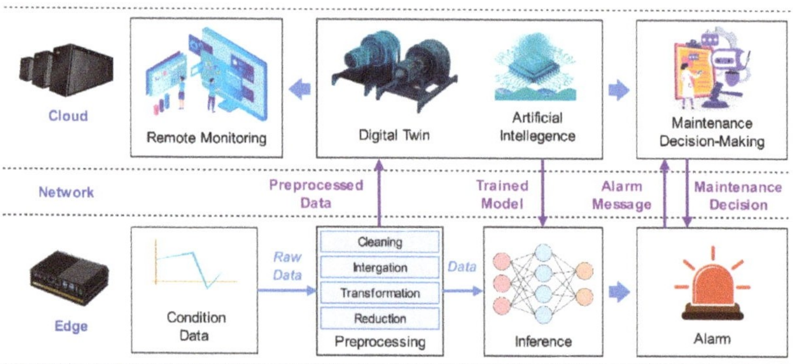

Figure 4. General edge intelligence architecture.

3.2. *Transport Layer*

The transport layer is mainly responsible for data transmission, aggregation, and forwarding. Hence, a review of communication technologies is necessary. As the volume of data increases, there is an additional need to store and compute data at this layer. Thus, an overview of fog computing is provided in the following paragraphs.

3.2.1. Communication Technologies

There are also two modes of data transmission: wired and wireless. This section reviews both wired and wireless communication protocols. In addition, because the Open Platform Communication Unified Architecture (OPC UA) enables the interconnection of devices under different protocols, it is also reviewed in this section

A. Wired Communication Technologies

Wired communication is the most universal and widely used mode. Wired communication protocols comprise Fieldbus and industrial Ethernet. Figure 5 illustrates the classification of the wired communication protocols.

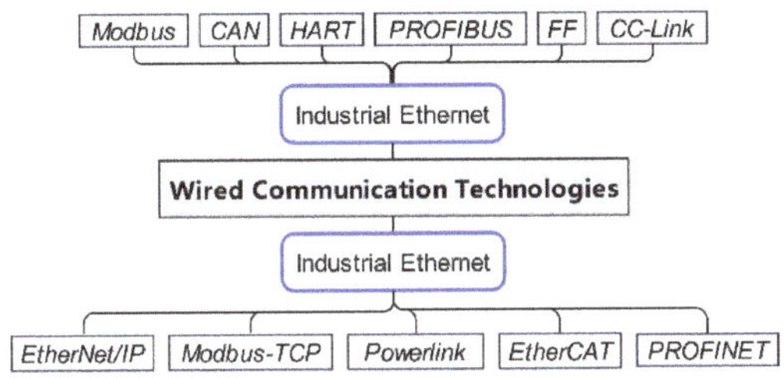

Figure 5. Wired communication technologies.

Fieldbus technology has been applied in industrial automation since the 1970s with a long history [53]. Because different companies developed their own products, numerous standards were created. The most frequently implemented Fieldbus protocols are Modbus [54], Controller Area Network (CAN) [55], Highway Addressable Remote Transducer (HART) [56], INTERBUS [57], PROFIBUS [58], Foundation Fieldbus (FF) [59], Control & Communication Link (CC-Link) [60], etc. Fieldbus technology is being phased out in the Industrial Internet age because devices with different protocols cannot communicate with each other.

With the development of Ethernet technology, Industrial Ethernet is gradually replacing the Fieldbus as the solution for the interconnection of equipment [61]. Protocols for Industrial Ethernet include EtherNet/IP [62], Modbus-TCP [63], Powerlink [64], EtherCAT [65], PROFINET [66], et al. Compared to Fieldbus technology, Industrial Ethernet has the advantages of fast transmission rates, long transmission distances, better interoperability, flexible topology, and easy integration [67]. However, the Industrial Ethernet does not solve the problem of interconnection between devices of different protocols either.

B. Wireless Communication Technologies

Although wired communication has the advantage of high reliability and low latency in the industry, there is still a place for wireless communication methods. Wireless communication methods are often used in environments where wiring is extremely hard, such as hazardous areas, moving equipment, etc. [68]. Figure 6 demonstrates the popular wireless communication protocols.

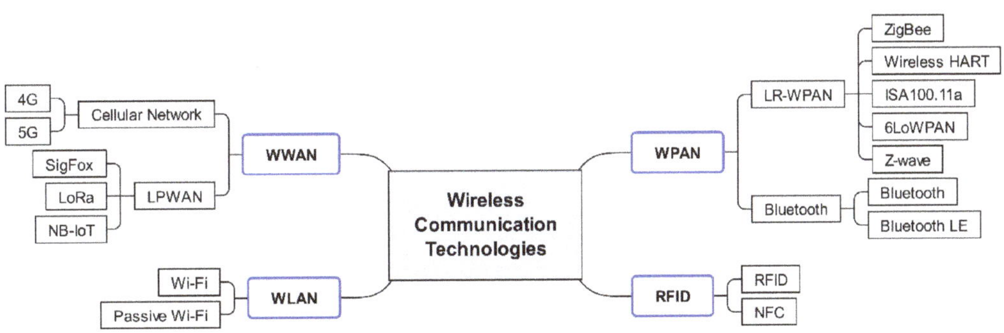

Figure 6. Wireless communication technologies.

A wireless wide area network (WWAN) is a telecommunications network that extends over a large geographic area. One approach to implementing WWAN is mobile telecommunication cellular network technologies, sometimes called Mobile Broadband. Technologies that can be used in the Industrial Internet include 4G [69] and 5G [70]. One of the drawbacks of the cellular network is its high power consumption. Hence the low-power wide area network (LPWAN) was proposed, which is a promising solution for remote and low-power Internet of Things [71]. The major LPWAN technology solutions include SigFox [13], LoRa [72], and Narrowband IoT (NB-IoT) [9]. A wireless local area network (WLAN) is a telecommunications network that links two or more devices within a limited area, such as a workshop or a production line [73]. The most widely used WLAN in the Industrial Internet is often known as Wi-Fi, based on the IEEE 802.11 standard [74]. To solve the problem of high power consumption of Wi-Fi, a low-power Wi-Fi called 'passive Wi-Fi' has been proposed recently [75]. A wireless personal area network (WPAN) is a telecommunications network within an individual's workspace [76]. The most widely used WPAN technology is Bluetooth [77] and Bluetooth Low Energy (Bluetooth LE) [78], based on the IEEE 802.15.1 standard, which has been used in the industry for a long time. However, the energy consumption of Bluetooth technology is relatively high, so the low-power, low-cost has been proposed, called low-rate wireless personal area networks (LR-WPAN), is also commonly used in industry [79]. The common LR-WPAN technologies are ZigBee [80], WirelesHART [81], ISA100.11a [82], 6LoWPAN [83], and Z-Wave [84]. Radio-frequency identification (RFID) [85] is a technology that automatically reads the information on a tag with electromagnetic fields, and Near Field Communication (NFC) [86] technology is developed based on it.

A qualitative and quantitative comparison between the wireless communication technologies is given in Table 1. In terms of the requirements of industrial applications, we have compared the performance of various wireless communication technologies from three perspectives: coverage range, power, and data rate. Cellular networks are suitable for the transmission of large amounts of data over long distances but require a stable energy supply. LPWAN is ideal for transmitting small amounts of data over long distances and benefits from low energy consumption. Wi-Fi is appropriate for transferring large amounts of data over short distances but consumes more energy. While passive Wi-Fi reduces energy consumption, it also reduces data transfer rates. Bluetooth and LP-WPAN are used for low-rate data transmission over short distances. Bluetooth Classic is slightly faster but consumes more energy; Bluetooth LE has lower power consumption but also lower speed. RFID and NFC require reading information from a tag at a short range, so they are commonly used for identifying and tracking objects. Every wireless communication technology has its own characteristics, and it is necessary to choose the right technology for application according to real industrial scenarios [87].

C. Open Platform Communication Unified Architecture

As can be seen from the previous review, there are different communication protocols in industrial applications, and devices with different protocols cannot interconnect with each other. However, in real production scenarios, a company's equipment always comes from various manufacturers and supports different communication protocols. The interconnection and interoperability of different devices must be achieved in the Industrial Internet [88]. The advent of Open Platform Communication Unified Architecture (OPC UA) provides a solution to this problem [89].

OPC UA is a cross-platform, open-source IEC 62,541 standard developed by the OPC Foundation that is used for the reliable, secure, and interoperable transfer of data [90]. The IEC 62,541 standard consists of the following parts: part 1—Overview and Concepts; part 2—Security Model; part 3—Address Space Model; part 4—Services; part 5—Information Model; part 6—Mappings; part 7—Profiles; part 8—Data Access; part 9—Alarms and Conditions; part 10—Programs; part 11—Historical Access; part 12—Discovery and Global Services; part 13—Aggregates; part 14—PubSub. Parts 1 to 7 specify the core functions of

OPC UA that define the modeling approach in the address space and the services associated with it. Parts 8 to 13 are the access type specifications of OPC UA. Part 14 enables OPC UA to support the Publish/Subscribe communication mode, improving the scalability of the system. There has been a lot of research on OPC UA.

Table 1. Comparison between wireless communication technologies.

Technology			Cover Range	Data Rate	Power
WWAN	Cellular Network	4G	**Long-Range** 10 km	**High** 100 Mbps	**High**
		5G	**Long-Range** 1 km	**High** 10 Gbps	**High**
	LPWAN	SigFox	**Long-Range** 10 km (urban), 40 km (rural)	**Low** 100 bps	**Low**
		LoRa	**Long-Range** 5 km (urban), 20 km (rural)	**Low** 50 kbps	**Low**
		NB-IoT	**Long-Range** 1 km (urban), 10 km (rural)	**Low** 200 kpbs	**Low**
WLAN	Wi-Fi		**Short-Range** 50 m	**High** 1 Gbps	**High**
	Passive Wi-Fi		**Short-Range** 30 m	**Low** 11 Mbps	**Low**
WPAN	Bluetooth	Bluetooth Classic	**Short-Range** 100 m	**Low** 3 Mbps	**Moderate**
		Bluetooth LE	**Short-Range** 100 m	**Low** 2 Mbps	**Low**
	LR-WPAN	ZigBee	**Short-Range** 100 m	**Low** 250 kbps	**Low**
		WirelessHART	**Short-Range** 200 m	**Low** 250 kbps	**Low**
		ISA100.11a	**Short-Range** 600 m	**Low** 250 kbps	**Low**
		6LoWPAN	**Short-Range** 100 m	**Low** 250 kbps	**Low**
		Z-wave	**Short-Range** 100 m	**Low** 100 kbps	**Low**
RFID	RFID		**Short-Range** 100 m	**Low** 400 kbps	**Moderate**
	NFC		**Short-Range** 0.04 m	**Low** 400 kbps	**Low**

Liu et al. [91] proposed a cyber-physical machine tools (CPMT) platform based on OPC UA and MTConnect. The authors developed an MTConnect to OPC UA interface to solve the interoperability problem between OPC UA and MTConnect, which converts the MTConnect information model and data into OPC UA counterparts. This platform has enabled standardized, interoperable, and efficient data communication between machine tools and various software applications. Kim and Sung [92] designed an OPC UA wrapper that allows UA clients to access legacy servers with OPC Classic interfaces seamlessly. The OPC UA wrapper consists of a UA server and a classic client that interact with each other through shared memory and semaphore. Martinov et al. [93] presented an OPC UA system to monitor the working process of CAN servo drives. The system consists of servo drives, a motion controller, a CNC kernel, and an OPC UA Server. Servo drives are connected to the motion controller via the CAN bus. The CNC kernel collects the whole data from the motion controller and sends it to the OPC UA server. The OPC UA server will convert this data into a uniform format via the information model so that the OPC UA client can monitor the servo drive remotely. Wang et al. [94] proposed a tool condition monitoring methodology considering tool life prediction based on the Industrial Internet. The OPC UA server collects data from the machine via a private protocol and then converts the data via the information model internally. The converted data are used for tool condition monitoring and life prediction.

A generic OPC UA-based device data acquisition framework can be summarized from the literature review above, as shown in Figure 7. With the implementation of the OPC UA server, sensors under different protocols can communicate with cloud applications. First, the data from the sensor is transferred to the OPC UA server. The information model of the corresponding sensor is created in the OPC UA server. Data are encoded into the standardized message through the information model. Finally, the cloud applications can process these standardized messages in the OPC UA client to enable remote monitoring and maintenance of the equipment.

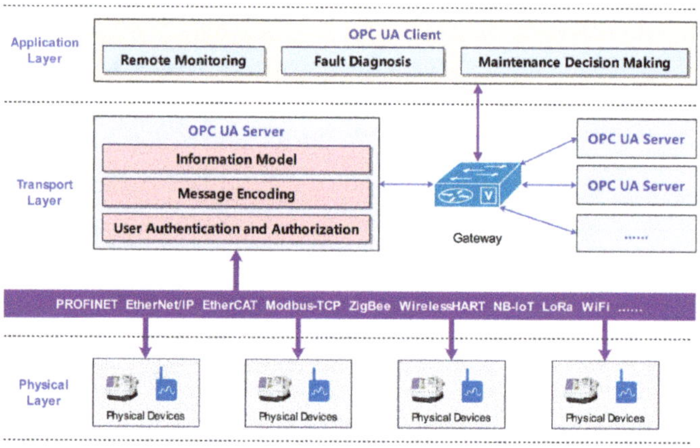

Figure 7. OPC UA-based data acquisition framework.

3.2.2. Fog Computing

Because of the insufficient computing capability of the devices at the edge, it is impossible to perform complex calculations on the data at the edge while guaranteeing the real-time performance of the transmission [95]. However, with the application of artificial intelligence technology in the field of equipment maintenance becoming more advanced, higher demands are placed on the computing capability of the system [96]. Transmitting all the raw data to the cloud for calculation again suffers from the problem of excessive data volume [97]. Therefore, processing the data at the transport layer becomes a solution for these issues.

Fog computing was considered an implementation of edge computing in previous studies [98]. However, fog computing has evolved into a new computing paradigm. It incorporates the concept of edge computing and provides a structured middle layer between the edge and the cloud, bridging the gap between the Internet of Things and cloud computing [99]. The fog node is not necessarily directly connected to the end device; it can be located anywhere between the end device and the cloud [100]. In the Industrial Internet, fog computing and federated learning are usually used in a fusion, as fog computing provides greater computational capacity for artificial intelligence applications.

Liu et al. [101] proposed a wireless signal classification framework based on federated learning. The raw data are processed by frequency reduction and sampling pretreatment, and its intelligent representation is obtained. The intelligent representation is put into the neural network in the node for training. The loss function of each node is sent to the aggregator and aggregated. Then the result is fed back to each node and used for gradient optimization to achieve global aggregation. It can solve the problem of reduced signal classification rates caused by different types of mixed signals in complex industrial environments and protect the privacy of industrial information. Brik et al. [102] proposed a disruption monitoring system based on fog computing and federated learning to monitor production for interruptions. The system collects position and movement data from manufacturing resources (workers, robots, equipment, etc.) via cameras and then transmits the data to the fog nodes for calculation. Each fog node trains a local prediction model and then transfers the model weights to the cloud server only. Federated averaging (FedAvg) algorithm is used in the cloud server to aggregate all local models to generate a global model. Global models are deployed to the fog nodes to predict the location of manufacturing resources. Production is considered to be interrupted once the measured position is found to be inconsistent with the predicted position. This failure information will be transmitted to the cloud, where the production tasks will be rescheduled via a rescheduling algorithm. The tasks after rescheduling are then transmitted to the manufacturing resources for production adjustment.

A fog computing framework fusing federated learning can be summarized from the above literature, as illustrated in Figure 8. The raw data are collected from the devices at the edge, and the collected data can be preprocessed or transmitted directly to the fog nodes. In the fog node, the initial global model is first downloaded and then trained locally. The local model in each fog node will be transferred to the cloud after training. In the cloud, the local models from each fog node will be aggregated to generate a new global model, which can be deployed to the fog node for equipment maintenance.

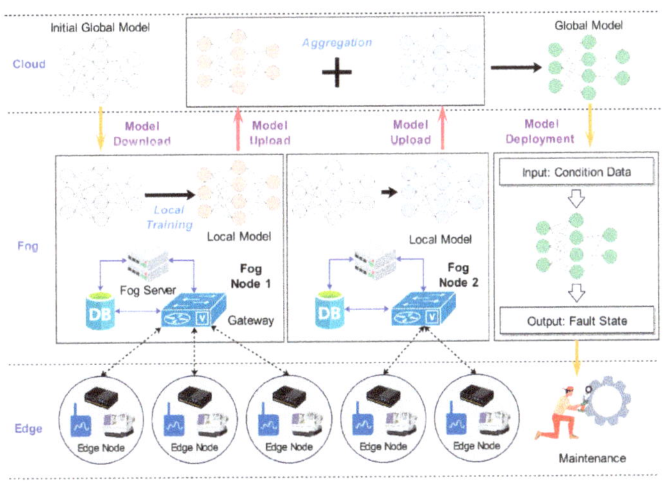

Figure 8. Fog computing framework fusing federated learning.

3.3. *Application Layer*

The application layer runs in the cloud with powerful computing capabilities for the final processing of the transmitted data. The results of the calculations can be presented directly to the engineers to make the correct decisions. The following is an overview of six perspectives: big data, artificial intelligence, digital twins, data analytics, O&M optimization, and sustainability.

3.3.1. Big Data

In the Industrial Internet, massive amounts of data are generated by a large number of sensors and controllers [103]. These data are characterized by their large volume, fast transmission speed, and variety of types and are known as 'big data' [14,15]. Traditional approaches to data storage and computation are unable to process such large volumes of data. Hence a new computing paradigm is required.

Yu et al. [104] presented a global manufacturing Big Data ecosystem for predictive maintenance, which involves the acquisition, storage, processing, and visualization of data. It is applied to detect abnormal patterns in the syngas reciprocating compressor. The system continuously collects signals such as vibration, temperature, pressure, and speed from the turbine syngas compressor through hundreds of sensors, averaging approximately 57 million entries per day. After such a large amount of data has been collected, it is replicated in triplicate and stored randomly on cloud nodes via an optimized Hadoop Distributed File System (HDFS) to avoid data loss. During data analysis, the data are first converted into DataFrame format and stored in the Apache Hive Central Data Warehouse and MapR Binary Database, respectively. The data are then computed using the MapReduce-based distributed PCA algorithm with Apache Spark as the data processing engine to enable the identification of equipment failures. The identification results are presented to the engineers on a visualization screen. Wan et al. [105] proposed a Spark-based parallel ant colony optimization (ACO)-K-means clustering algorithm for fault diagnosis of large amounts of rolling bearing operating condition monitoring data. The collected 119.8 GB of raw bearing vibration data was stored in the HDFS. The data are clustered on the Spark efficient computing platform with the ACO-K-means clustering algorithm to obtain a fault diagnosis model. The results demonstrate that the big data computing framework can improve the efficiency of model computation and fault diagnosis.

In the acquisition of big data, the high concurrency problem must be solved because a large amount of data arrives at the database simultaneously, causing blocking. Park and Chi [106] introduced a high throughput data ingestion system for machine logs in the manufacturing industry. Machine log stream data from a group of milling machines are first sent to a set of pre-assigned distributed buffers called Topic. Apache Kafka manages these Topics so they can be stored in the database orderly. Sahal et al. [107] studied a big-data-based predictive maintenance case in wind energy. Wind farms are geographically distributed, and data from wind turbines need to be aggregated with a standard data storage model. Therefore, RabbitMQ is well suited to solve this problem with its federated queues. RabbitMQ is a distributed queuing management technology based on the Advanced Message Queuing Protocol (AMQP), which can ensure receiving data from sensors in the correct order. Under the AMQP protocol, the publisher's messages are transferred to the exchange. The exchange distributes the received messages to the bound queues according to the routing rules. Finally, the AMQP agent delivers the messages to the consumers who have subscribed to this queue. Consumers can also retrieve these messages by themselves as needed. Liu et al. [108] employed the publish/subscribe communication protocol Message Queuing Telemetry Transport (MQTT) to realize the data exchange between different equipment. After the publisher's message is transmitted to the MQTT Broker, it is routed directly to the subscriber and is not stored in the queue. Therefore, MQTT has low energy consumption and is perfect for small devices.

The collected industrial big data are massive, multi-source, and heterogeneous—containing structured, semi-structured, and unstructured data—thus creating a huge challenge for data storage. Traditional relational databases are excellent for storing structured data. However, with the explosive growth of data volume, Structured Query Language (SQL)-based information query has become unable to meet the demand due to its inherent limitations in terms of scalability and fault tolerance [109]. Hence, NoSQL databases are gradually becoming the solution for storing big data. Martino et al. [110] compared the performance of three popular NoSQL Database Management Systems—namely Cassandra, MongoDB, and InfluxDB—in storing Industrial big data. The results show that InfluxDB has the best performance because the data streams from industry devices can be considered a collection of time series. In order to support Online Analytical Processing (OLAP) of big data rather than just storing data, data warehouse has been proposed. Silva et al. [111] demonstrated a logistics big data warehouse for the automotive industry. The data warehouse stores current and historical logistics data to support real-time monitoring of logistics status and online prediction of on-time delivery using machine learning algorithms. The data warehouse dramatically improves the efficiency of online data analysis. Data lake is proposed to store heterogeneous data from different sources. Munirathinam et al. [112] designed a semiconductor manufacturing data lake with Hadoop. It stores all data from multiple business units, including batch data, process data, and quality data from wafer production. The authors demonstrate a visual analysis of the data with hive tables and tableau. Various organizations of the company—such as the Manufacturing and Quality departments—use it to benchmark across different design IDs and fabs, and set yield/quality ramp and maturity targets for newer product generation. A data lake is a repository of data from disparate sources stored in their original format and can contain everything from relational data to JSON documents to PDFs to audio files. The data can be stored without conversion leading to highly efficient. Its flexible character allows business analysts and data scientists to look for unexpected patterns and insights. Data lakes are becoming the most advanced solution for big data storage.

The above literature shows that big data computing frameworks mainly address the problem of distributed storage and computation of massive amounts of data, as shown in Figure 9. The enormous amount of data collected from the equipment and production lines will be stored. The dominant distributed storage system is HDFS. The data are then processed on the computing cluster with Hadoop MapReduce, Spark, and other computing engines. They employed machine learning, data analysis, and other approaches to monitor equipment and production lines remotely.

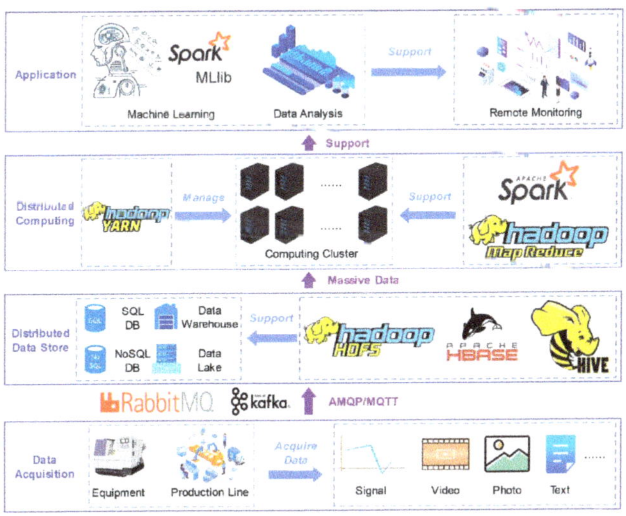

Figure 9. Big data application framework.

3.3.2. Artificial Intelligence

Fault diagnosis and anomaly detection are the main applications of artificial intelligence in equipment maintenance. Fault diagnosis can identify the reason for a fault to occur. Anomaly detection can only determine the occurrence of a fault, but it has advantages in terms of dataset generation. This subsection takes the literature review from these two perspectives.

A. Fault Diagnosis

Pursuing the relationship between monitoring data and machine health states is always a widespread concern in machine health management, and fault diagnosis plays a significant role in solving such issues [113]. Machine learning-based fault diagnosis has been widely used in the monitoring and maintenance of equipment and production lines [114].

Han and Li [115] developed a novel out-of-distribution (OOD) detection-assisted trustworthy machinery fault diagnosis method. At first, a deep ensembled fault diagnosis system is established by integrating multiple deep neural networks. Then, a trustworthiness analysis is performed with an uncertainty-aware depth ensemble to detect OOD samples and give warnings about potentially unreliable diagnoses. Lastly, the deep ensembles' prediction and uncertainty are carefully considered to achieve trustworthy decisions. The proposed method was validated with a wind turbine fault diagnosis case and a gearbox fault diagnosis case. The wind turbine fault dataset consists of one set of normal data and three sets of faults, each with 1000 samples. The gearbox fault dataset consists of one set of normal data and four sets of fault data, each with 1000 samples. The result demonstrates that it exhibits significant advantages in diagnosing OOD samples and obtaining trustworthy fault diagnosis results. Liu et al. [116] proposed a deep feature-enhanced generative adversarial network (GAN) for rolling bearing fault diagnosis. A new generator objective function integrated with a pull-away function was designed to avoid mode collapse phenomena and improve the stability of the GAN. A self-attention mechanism is used in the GNN to enhance the learning of the features of the original vibration signal. In order to guarantee the accuracy and diversity of the generated samples, an automatic data filter was constructed. At last, a convolutional neural network is added as a classifier for fault diagnosis. The method was validated on a rolling bearing vibration signal dataset from an electric locomotive. The dataset contains one set of normal data with a 126,000 sample size and five sets of fault data with a 12,600 sample size. The results demonstrate the better performance of this method in unbalanced sample fault diagnosis. Ferracuti et al. [117] proposed a fault diagnosis algorithm for rotating machinery based on the Wasserstein distance. The authors extracted frequency- and time-based features from the vibration signals and then considered the Wasserstein distance in the learning phase to differentiate the different equipment operating conditions. The statistical distance-based fault diagnosis technique permits obtaining an estimation of fault signature without training a classifier. Therefore, it is very efficient and can be used for embedded hardware. This algorithm can solve the problem of fault diagnosis for rotating machinery at low signal-to-noise ratios and different operating conditions. It can also be applied to system monitoring and prognostics, allowing for predictive maintenance of rotating machinery. Ferracuti et al. [118] studied the problem of defect detection and diagnosis of induction motors based on motor current signature analysis. The researchers estimate the probability density functions of data related to healthy and faulty motors with a Clarke–Concordia transformation and kernel density estimation. Kullback–Leibler divergence is used as an index for the automatic identification of defects because it identifies the dissimilarity between two probability distributions. Fast Gaussian transform improves kernel density estimation. This method has a low computational cost and enables real-time quality control at the end of the production line. The experiments show that the proposed method can detect and diagnose different induction motor faults and defects.

As can be seen from the above reviews, fault diagnosis methods require the training of machine learning models on labeled fault datasets. The trained model can identify the occurrence of a fault and diagnose the type of fault according to the acquired data. However, it is difficult to collect sufficient fault data in a real production scenario, so there are limitations in the application of this method.

B. Anomaly Detection

Equipment failures are rare in real production scenarios, so we can only obtain a tiny amount of equipment failure data. Therefore, it is impossible to acquire enough fault data to train the machine learning models, which brings a huge challenge for the application of machine learning techniques in predictive maintenance [119]. Anomaly detection algorithms are an effective way to solve this problem [120].

Zhao et al. [121] proposed a one-class classification model based on extreme learning machine boundary (ELM-B) to detect bearing failures. The dataset is a NASA-bearing dataset provided by the Center for Intelligent Maintenance Systems (IMS) at the University of Cincinnati. The model is a single-layer feed-forward neural, with an input layer, a hidden layer and an output layer. The researchers calculated the RMS, kurtosis, peak–peak, crest factor, and skewness of the healthy bearing vibration signals in the dataset as inputs to the model. The model is trained to produce 1 at the output. The vibration signals of the bearings are collected by sensors and fed into the trained model. If the output is not equal to 1, it is assumed that a fault has occurred. Tanuska et al. [122] proposed an anomaly detection algorithm for detecting anomalies in conveyor carrier wheel bearings in automotive assembly lines. The researchers collected 16,000 bearing temperature data, of which there were only 18 abnormalities. They designed a multi-layer perceptron (MLP) with 13 neurons in the input layer, 18 neurons in the hidden layer, and 2 neurons in the output layer. The MLP can detect bearing anomalies based on the minimum and average temperature of the bearing. Kähler et al. [123] presented an anomaly detection approach based on a convolutional autoencoder (CAE) to detect surface defects in aircraft landing gear components. The CAE consists of an encoder and a decoder. The encoder comprises an input layer and a convolutional layer for compressing the image. The decoder reconstructs the compressed representation of the input using transposed convolution and convolution layers. The researchers collected 600 defect-free images and 300 defective images of aircraft landing gear surfaces. From this sample, 500 of the defect-free images are randomly selected for training, while the remaining 100 defect-free images and 300 defective images are used for testing.

From the above literature, it can be concluded that anomaly detection approaches can contribute to solving the problem of insufficient fault samples in the industry because the method only requires normal data for the training. However, the shortcoming of anomaly detection is that the exact cause of the fault cannot be identified.

3.3.3. Digital Twin

After the collected data has been processed, the results should be presented to the engineers for monitoring. Traditional monitoring methods include simple charts, pictures, two-dimensional electronic kanban or videos, etc. These methods suffer from poor visibility, low interactivity, and limited scalability, making it difficult for engineers to have a comprehensive understanding of the operating conditions of equipment and production lines. Digital twin technology can solve these problems to a certain degree [124,125].

Fan et al. [126] proposed a generic architecture and implementation method for digital-twin visualization. Data collected in the production line is encapsulated in Automation Markup Language (AML), a digital twin data exchange format, and then cached, managed, and transformed in real-time by Cyber Engine for visualization. Digital mock-up (DMU) enables the physical data of a production line to be constructed as virtual 3D scenario motions and state changes. The digital twin can present complete and visual information about the production line to the engineers, improving their decision-making. Fera et al. [127] proposed a novel digital twin framework to evaluate the production line performance. This framework collects data on the body posture and working hours of workers in production with a wearable sensor. The digital twin of the worker is created by binding human motion data to the digital human model via a special interface. This human motion data will be tied to a digital model of a human to generate a digital twin model of the worker. The management team can evaluate the efficiency of the production line accordingly and optimize the assignment of production tasks. Liu et al. [108] proposed a digital twin-based cyber-physical production system (CPPS). Digital geometry models of 3D printing lines are encapsulated using web technology. An ontology-based information model was designed to bind the data to the geometry model for 3D visualization and remote control of the production line.

As shown in Figure 10, the digital twin is combined with data collection techniques, artificial intelligence, and big data computing to map equipment and production lines in the physical space to the digital space. It allows remote monitoring of equipment and production lines, giving a comprehensive insight into the production scenario so that the correct decisions can be made.

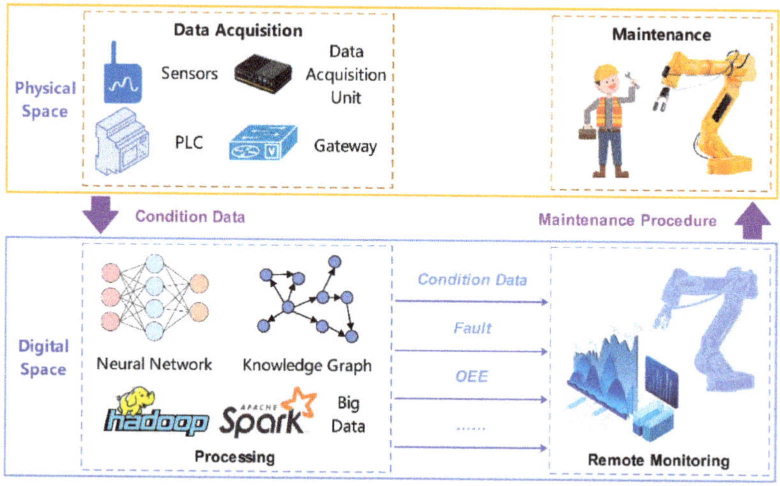

Figure 10. Digital twin.

3.3.4. Data Analytics

Data analytics is the process of collecting, managing, processing, analyzing, and visualizing evolving data [128] which has a wide range of applications in the Industrial Internet.

Zuo et al. [129] proposed an Internet-of-Things (IoT) and cloud-based novel approach for product energy consumption evaluation and analysis (ECEA). Data related to product energy consumption is dynamically collected in real-time. The system analyzes the energy consumption of the products in the pre-production, production, and post-production stages—including transportation energy consumption, processing energy consumption, auxiliary energy consumption, usage energy consumption, etc. A design solution for a bearing bracket in a toy aircraft was optimized with this method and gave a design solution with minimum energy consumption, qualified functional quality, and within-budget cost. Zhong et al. [130,131] proposed a real-time big data analytics framework to monitor intelligent manufacturing shop floors. Researchers tracked workpieces in production in real-time with RFID. RFID-cuboids were introduced to represent logistics information, thus mining trajectory knowledge and associated indexes for evaluating various manufacturing objects such as workers and machines. The steps of knowledge mining include data cleaning, compression, classification, and pattern recognition. This knowledge supports differentiated decision-making, for example, logistics planning, production planning and scheduling, as well as enterprise-oriented strategy.

The above literature shows that the collected data needs to be cleaned and compressed. Data mining, statistical analysis, and other methods are applied to obtain useful knowledge for production monitoring and optimization.

3.3.5. Operations and Maintenance Optimization

Operations and maintenance optimization is a significant application in industrial production. There are several studies that have been conducted to address this issue.

Yang et al. [132] developed a weather-centered opportunistic O&M framework to enable a flexible maintenance resource allocation for wind turbines. This framework quantifies the negative (delays caused by severe wind conditions) and positive (maintenance opportunities) impacts of wind conditions on production and maintenance processes. The advantage of this framework is the consideration of providing additional maintenance opportunities when wind velocities are too small to keep the turbines running. Maintenance downtime and production losses are significantly reduced because these spare times are fully utilized. The authors developed a renewal scenario for turbine components, built a maintenance cost model, and derived the optimal maintenance age for minimizing the maintenance cost of wind turbines with sufficient maintenance resources. An improved performance-based contracting (PBC) model was established to capture the comprehensive effect of both production and maintenance processes. This framework collects wind turbine condition data and wind velocity data for real-time monitoring so that maintenance tasks are optimized with the support of the PBC. A case study shows that this framework is more flexible in resource allocation, significantly reducing maintenance costs and increasing revenue. Hu et al. [133] proposed a joint decision-making strategy for job scheduling and preventive maintenance (PM) planning for a two-machine flow shop with resumable jobs, where both job-dependent operating conditions (OC) and imperfect maintenance (IM) are considered. A hybrid processing time model is built to obtain the optimal sequence when the failure rate of a machine is constant under a fixed OC. The authors presented a joint optimization model for job scheduling and PM planning when the machine failure rate is time-varying at a fixed OC and calculated it with a genetic algorithm. The advantage of this method is that it considers the OC data collected during the production. The results demonstrate that the approach is effective in reducing production completion time and also in reducing the frequency of failures.

From the above review, it can be concluded that the current research for O&M optimization not only considers the maintenance tasks and production tasks of the equipment but also the operating conditions of the equipment. Real-time monitoring of working conditions enables further optimization of production and maintenance task scheduling issues, improving productivity and reducing maintenance costs.

3.3.6. Sustainability

Sustainability is an essential issue in the context of climate change, where industrial production plays an important role. There are several applications for improving sustainability in manufacturing with remote monitoring and maintenance on Industrial Internet.

Rojek et al. [134] proposed a digital twins system for manufacturing and maintenance sustainability. The authors obtained real data from some companies engaged in eco-design, process planning, and process supervision. Several artificial intelligence models were built based on this data and fused into the digital twins. The digital twins monitor production processes with artificial intelligence models for process parameter optimization, production planning, and equipment maintenance to improve manufacturing and maintenance sustainability. Caterino et al. [135] defined a new remanufacturing framework based on cloud computing technology called cloud remanufacturing (CRMfg). The framework translates remanufacturing resources and capabilities into services delivered via the Internet, allowing for the mutually beneficial connection of remanufacturing service providers and customers in different locations. The CRMfg can monitor registered remanufacturers and their equipment, thus enabling the scheduling and real-time tracking of product remanufacturing tasks. It can significantly improve the efficiency of remanufacturing, thus contributing to achieving economic and environmental sustainability. Çınar et al. [136] introduced the application of machine learning-based predictive maintenance in sustainable smart manufacturing. The collected real production data are used to train machine learning models for fault diagnosis, thus enabling predictive maintenance. Predictive maintenance can significantly reduce hidden problems, failures, and accidents in production, resulting in less breakdown maintenance and lower maintenance costs, ultimately achieving sustainable manufacturing.

From the above literature, it can be seen that the sustainability of the manufacturing process is improved after the introduction of modern technologies related to the Industrial Internet. Remote monitoring of equipment and production lines optimizes process parameters and reduces waste caused by suboptimal processes; optimizes production scheduling and reduces economic consumption caused by waiting time; and minimizes the frequency of failures and reduces damage caused by equipment downtime.

4. Industrial Application

Remote monitoring and maintenance technology has been widely implemented in the industrial area. This section demonstrates several real industrial application cases.

Yang et al. [23] demonstrate a case of monitoring and maintenance for a grinding and polishing robot. The robots are used for grinding and polishing prebaked carbon anodes, a key consumable for aluminum electrolysis production. The proposed framework enables data-driven maintenance, intelligent fault diagnosis and prediction, and knowledge-based maintenance and fault diagnosis service for the robots. The authors developed a B/S-architecture industrial internet service platform. The platform includes the following functions: maintenance of equipment according to the manufacturer's predefined maintenance schedule; identification of process parameter anomalies; monitoring of machining accuracy; and answering questions about equipment failures, process parameters, and machining accuracy through knowledge graph Q&A technology.

Li et al. [22] developed an Industrial 4.0 platform for equipment monitoring and maintenance in prebaked carbon anode production. This platform is applied for the monitoring and maintaining the kneader, critical equipment for prebaked carbon anode production. The platform can collect equipment working condition data, production process data, and product quality data, enabling production planning, equipment failure maintenance, process parameter optimization, and product quality control.

Scheuermann et al. [137] proposed an example of an Industrial 4.0 manufacturing process called the Agile Factory. The Agile Factory is implemented in mass customization production scenarios. The Agile Factory assembly line is component-based, combining trackable mobile workstations with fixed workstations. Therefore, the products are traceable during the production process. A customer feedback loop was implemented to allow for mass customization of products by permitting change requests during assembly time.

Bonci et al. [138] demonstrate a case study of the application of fault diagnosis technology in industrial packaging machinery. The collected current data of the equipment is pre-processed with a demodulating technique by means of the analytic envelope. The pre-processed signals are subjected to continuous wavelet transform and discrete wavelet transform for fault diagnosis. A real industrial case demonstrates that the method can detect belt failures in packaging machinery running continuously in non-stationary conditions.

As can be seen from the above cases, the main applications in the industry are equipment maintenance, process optimization, product quality control, production planning, troubleshooting, etc.

5. Challenges

Although a great effort has been dedicated to the monitoring and maintenance of equipment and production lines on the Industrial Internet, there are still many challenges that remain to be addressed. Key challenges stem from the requirements in real-time performance, interoperability, security, and intelligence. These challenges will be discussed in this section, as shown in Figure 11.

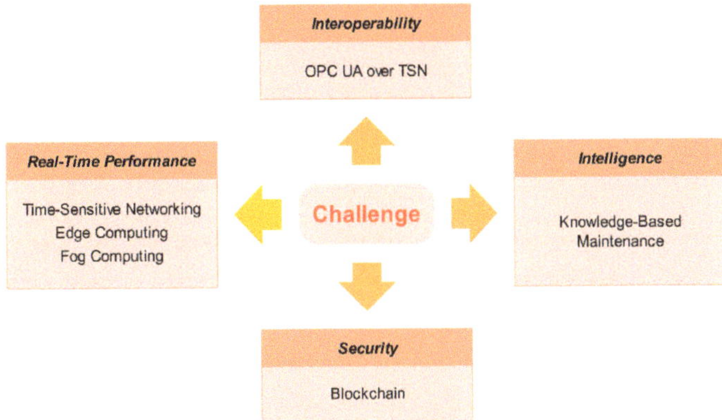

Figure 11. Challenges.

5.1. Real-Time Performance

In order to enable remote monitoring and maintenance of equipment and production lines in the Industrial Internet, sensors, actuators, machines, and other computing devices need to cooperate with each other. Low latency and reliable data transmission must therefore be ensured for the reliability and efficiency of the Industrial Internet. Real-time performance is a significant issue for data transport in the Industrial Internet [139].

The emergence of time-sensitive networks (TSN) holds the promise of solving this problem. Time-sensitive networking (TSN) is a set of standards under development by the time-sensitive networking task group of the IEEE 802.1 working group [140,141]. It achieves time synchronization, limited low latency, and high reliability over standard connection technologies such as Ethernet that meet the requirements of time-sensitive applications in industrial systems [142]. It has the potential to become the standard for the next generation of industrial communication and automation [11,143].

The National Institute of Standards and Technology (NIST) has built a collaborative robotic workcell testbed enabled by Wireless TSN technologies. The study shows that the percentage of idle time experienced by the operator robot is lower when TSN is enabled because the robot can receive commands from the controller more rapidly, which increases the productivity of this collaborative robot in industrial environments [144]. Yang et al. [145] proposed a TSN chain flow abstraction, TC-Flow, that solves the problem of coordinated scheduling of multiple data streams in industrial applications such as control and security applications. Nikhileswar et al. [146] present an industrial control system implemented by 5G and TSN and evaluate it. The results show that TSN can significantly reduce the latency of the network. Pop et al. [147] proposed that using TSN as a deterministic transport for the fog computing network layer in industrial automation can reduce the latency and improve the stability of data transmission in the Industrial Internet.

Overall, TSN combined with edge computing and fog computing is expected to be a way to improve the real-time performance of the Industrial Internet.

5.2. Interoperability

In the Industrial Internet, the interconnection of people, machines, and things is to be realized. The transmission of real-time data from industrial equipment and information from network applications such as operations management are separated in existing factory intranets, with the former generally being routed through Fieldbus or industrial Ethernet and the latter relying on conventional Ethernet. OPC UA has been able to interconnect devices at the application layer but not at the data link layer in the Open Systems Interconnection (OSI) model. TSN enables network interconnection and data interoperability at the data link layer. Therefore, the convergence of OPC UA and TSN promises a unified Industrial Internet [8].

Pfrommer et al. [10] presented an approach that combines a non-real-time OPCUA server with a real-time OPC UA Pub Sub, where both have accessibility to the shared information model without dropping the real-time guarantee to the publisher. The publisher can therefore run within hardware-triggered interrupts to guarantee low latency and less jitter. Li et al. [12] present a two-layer manufacturing system communication architecture with OPC UA and TSN technology in heterogeneous networks. The TSN network is used as the communication backbone for realizing the real-time services of the industrial automation system, which connects the heterogeneous industrial automation subsystems at the field level with the upper-layer entities. OPC UA is used to realize the exchange of information between the heterogeneous subsystems in the field layer and the entities in the upper layers. The results prove that different types of devices can communicate with each other in this system with excellent real-time performance.

From the above, it can be seen that OPC UA over TSN will be a promising way to solve interoperability problems in the Industrial Internet.

5.3. Security

The main features of the Industrial Internet are openness, interconnection and sharing, which pose serious security challenges. An attack on the network can lead to loss, leakage, and tampering of industrial data. In the event of an attack on the control network, there would be enormous financial damage and even a threat to the lives of the public. It seems that blockchain holds the promise of alleviating these problems [148].

Gu et al. [149] implemented a functional safety and information security protection mechanism based on blockchain technology in the CPS system. The equipment must be authenticated Safety Integrity Level (STL) before accessing the CPS. A new equipment block is created with a combination of asymmetric and symmetric key encryption methods, and the STL of the equipment will be stored in the block. The researchers propose a fault threshold mechanism based on smart contract technology to ensure functional safety and information security during equipment communication. Qu et al. [150] combine blockchain and federated learning technologies to propose a blockchain-enabled federated learning (FL-Block) model which enables decentralized privacy protection. FL-Block enables decentralized privacy protection through hybrid identity generation, comprehensive authentication, access control, and off-chain data storage and retrieval.

It can be concluded that blockchain technology is emerging as a prospective solution to security issues in the Industrial Internet due to its decentralization, non-tamperability, traceability, and high cryptographic security.

5.4. Intelligence

Artificial intelligence is already widely used in the maintenance of equipment and production lines. The common approach is data-driven fault diagnosis, which can be used to diagnose and isolate faults in specific devices. However, the implementation of data-driven fault diagnosis requires careful design of physical models, signal patterns, and machine learning algorithms to describe faults [17]. A knowledge-based fault diagnosis approach is suitable for complex or multi-component systems/processes without detailed mathematical models, which is becoming another direction of development [151,152].

Cao et al. [153] proposed a novel Knowledge-based System for Predictive Maintenance in Industry 4.0 (KSPMI). KSPMI blends computational intelligence and symbolic intelligence. Firstly, chronicle mining (a special type of sequential pattern mining approach) is used to extract machine degradation models from industrial data. After that, domain ontologies and Semantic Web Rule Language (SWRL) rule-based reasoning use the extracted chronicle patterns to query and reason on system input data with rich domain and contextual knowledge. The system is able to predict future failures of equipment and the time of occurrence. Wang et al. [154] presented a framework for the intelligent operation and maintenance of traction transformers based on knowledge graphs. The framework integrates multiple sources of heterogeneous data from traction transformers into structured knowledge with a unified knowledge representation model. The researchers constructed a knowledge entity graph, concept graph, fault treatment graph, and fault case graph to achieve multi-source condition data fusion and correlation analysis, multi-dimensional differentiated state evaluation, and intelligent assisted maintenance decision-making.

As can be seen, the combination of data-driven fault diagnosis and knowledge-driven fault diagnosis may lead to further progress in the intelligent maintenance of equipment in the future.

6. Conclusions

In this paper, a detailed overview of the monitoring and maintenance of equipment and production lines on the Industrial Internet is proposed. At first, a brief review of its architecture is presented, and a three-layer reference architecture is summarized, containing the physical layer, the transport layer and the application layer. We then provide a detailed literature review of the key enabling technologies involved in each layer, including data acquisition technologies, edge computing, communication technologies, fog computing, big data, artificial intelligence, digital twins, data analytics, O&M optimization, and sustainability. Next, we demonstrate some industrial application cases. We also discuss the challenges in terms of real-time performance, interoperability, security, and intelligence. Overall, we have reviewed the most advanced research in this field and discussed the direction of future research, which is expected to be a reference for researchers addressing this area.

There are still some limitations in this review because remote monitoring and maintenance for equipment and production lines on the Industrial Internet is a very wide field. In the Industrial Internet, many nodes work simultaneously and are prone to failures. Therefore, the fault tolerance of the system is a key issue in the face of various potential failures. Numerous sensors and devices continuously consume large amounts of energy, so research on an energy-efficient 'green Industrial Internet' is also necessary. There are also some traditional industrial software systems in industrial applications, such as enterprise resource planning (ERP), manufacturing execution systems (MES), etc. How to interoperate with these systems to further improve the digitalization and intelligence of production is a huge challenge. In the future, we will conduct special research on the above issues.

Author Contributions: Conceptualization, P.J. and Q.L.; Investigation, Q.L. and Y.Y.; Resources, Y.Y.; Writing—original draft preparation, Q.L.; Writing—review and editing, P.J.; Visualization, Q.L.; Supervision, P.J.; Project administration, P.J.; Funding acquisition, P.J. All authors have read and agreed to the published version of the manuscript.

Funding: This research was funded by the National Key Research and Development Program of China, grant number 2021YFE0116300; the National Natural Science Foundation of China, grant number 51975464.

Institutional Review Board Statement: Not applicable.

Informed Consent Statement: Not applicable.

Data Availability Statement: Not applicable.

Conflicts of Interest: The authors declare no conflict of interest.

References

1. Chen, W. Intelligent manufacturing production line data monitoring system for industrial internet of things. *Comput. Commun.* **2020**, *151*, 31–41. [CrossRef]
2. Zonta, T.; da Costa, C.A.; da Rosa Righi, R.; de Lima, M.J.; da Trindade, E.S.; Li, G.P. Predictive maintenance in the Industry 4.0: A systematic literature review. *Comput. Ind. Eng.* **2020**, *150*, 106889. [CrossRef]
3. Pech, M.; Vrchota, J.; Bednář, J. Predictive Maintenance and Intelligent Sensors in Smart Factory: Review. *Sensors* **2021**, *21*, 1470. [CrossRef] [PubMed]

4. Golmakani, H.R.; Fattahipour, F. Optimal replacement policy and inspection interval for condition-based maintenance. *Int. J. Prod. Res.* **2011**, *49*, 5153–5167. [CrossRef]
5. Wang, Y.; Deng, C.; Wu, J.; Wang, Y.; Xiong, Y. A corrective maintenance scheme for engineering equipment. *Eng. Fail. Anal.* **2014**, *36*, 269–283. [CrossRef]
6. Stenström, C.; Norrbin, P.; Parida, A.; Kumar, U. Preventive and corrective maintenance—cost comparison and cost–benefit analysis. *Struct. Infrastruct. Eng.* **2016**, *12*, 603–617. [CrossRef]
7. Kim, D.-S.; Tran-Dang, H. Wireless Sensor Networks for Industrial Applications. In *Industrial Sensors and Controls in Communication Networks: From Wired Technologies to Cloud Computing and the Internet of Things*; Kim, D.-S., Tran-Dang, H., Eds.; Springer International Publishing: Cham, Switzerland, 2019; pp. 127–140.
8. Bruckner, D.; Stănică, M.P.; Blair, R.; Schriegel, S.; Kehrer, S.; Seewald, M.; Sauter, T. An Introduction to OPC UA TSN for Industrial Communication Systems. *Proc. IEEE* **2019**, *107*, 1121–1131. [CrossRef]
9. Ratasuk, R.; Vejlgaard, B.; Mangalvedhe, N.; Ghosh, A. NB-IoT system for M2M communication. In Proceedings of the 2016 IEEE Wireless Communications and Networking Conference, Doha, Qatar, 3–6 April 2016; pp. 1–5.
10. Pfrommer, J.; Ebner, A.; Ravikumar, S.; Karunakaran, B. Open Source OPC UA PubSub Over TSN for Realtime Industrial Communication. In Proceedings of the 2018 IEEE 23rd International Conference on Emerging Technologies and Factory Automation (ETFA), Torino, Italy, 4–7 September 2018; pp. 1087–1090.
11. Bello, L.L.; Steiner, W. A Perspective on IEEE Time-Sensitive Networking for Industrial Communication and Automation Systems. *Proc. IEEE* **2019**, *107*, 1094–1120. [CrossRef]
12. Li, Y.; Jiang, J.; Lee, C.; Hong, S.H. Practical Implementation of an OPC UA TSN Communication Architecture for a Manufacturing System. *IEEE Access* **2020**, *8*, 200100–200111. [CrossRef]
13. Lavric, A.; Petrariu, A.I.; Popa, V. SigFox Communication Protocol: The New Era of IoT? In Proceedings of the 2019 International Conference on Sensing and Instrumentation in IoT Era (ISSI), Lisbon, Portugal, 29–30 August 2019; pp. 1–4.
14. Li, T.; Song, W.N.; Zhou, J.H. Research on Artificial Intelligence Industrial Big Data Platform for Industrial Internet Applications. In Proceedings of the 2021 IEEE 4th International Conference on Electronics Technology (ICET), Chengdu, China, 7–10 May 2021; pp. 955–959.
15. Wang, J.; Xu, C.; Zhang, J.; Zhong, R. Big data analytics for intelligent manufacturing systems: A review. *J. Manuf. Syst.* **2022**, *62*, 738–752. [CrossRef]
16. Rakic, S.; Pero, M.; Sianesi, A.; Marjanovic, U. Digital Servitization and Firm Performance: Technology Intensity Approach. *Eng. Econ.* **2022**, *33*, 398–413. [CrossRef]
17. Liu, R.; Yang, B.; Zio, E.; Chen, X. Artificial intelligence for fault diagnosis of rotating machinery: A review. *Mech. Syst. Signal Process.* **2018**, *108*, 33–47. [CrossRef]
18. Deng, S.; Zhao, H.; Fang, W.; Yin, J.; Dustdar, S.; Zomaya, A.Y. Edge Intelligence: The Confluence of Edge Computing and Artificial Intelligence. *IEEE Internet Things J.* **2020**, *7*, 7457–7469. [CrossRef]
19. Errandonea, I.; Beltrán, S.; Arrizabalaga, S. Digital Twin for maintenance: A literature review. *Comput. Ind.* **2020**, *123*, 103316. [CrossRef]
20. Wang, G.; Nixon, M.; Boudreaux, M. Toward Cloud-Assisted Industrial IoT Platform for Large-Scale Continuous Condition Monitoring. *Proc. IEEE* **2019**, *107*, 1193–1205. [CrossRef]
21. Yang, H.; Sun, Z.; Jiang, G.; Zhao, F.; Lu, X.; Mei, X. Cloud-Manufacturing-Based Condition Monitoring Platform with 5G and Standard Information Model. *IEEE Internet Things J.* **2021**, *8*, 6940–6948. [CrossRef]
22. Li, Q.; Yang, Y.; Jiang, P. An Industry 4.0 Platform for Equipment Monitoring and Maintaining in Carbon Anode Production. *IFAC-PapersOnline* **2022**, *55*, 37–41. [CrossRef]
23. Yang, Y.; Yang, M.; Jiang, P. The Design of an Integrated Monitoring and Maintenance Framework for Newly Developed Equipment: Using Industrial Robot as Example. *IFAC-PapersOnline* **2022**, *55*, 42–47. [CrossRef]
24. Short, M.; Twiddle, J. An Industrial Digitalization Platform for Condition Monitoring and Predictive Maintenance of Pumping Equipment. *Sensors* **2019**, *19*, 3781. [CrossRef]
25. Xia, B.; Wang, K.; Xu, A.; Zeng, P.; Yang, N.; Li, B. Intelligent Fault Diagnosis for Bearings of Industrial Robot Joints Under Varying Working Conditions Based on Deep Adversarial Domain Adaptation. *IEEE Trans. Instrum. Meas.* **2022**, *71*, 3508313. [CrossRef]
26. Jamshed, M.A.; Ali, K.; Abbasi, Q.H.; Imran, M.A.; Ur-Rehman, M. Challenges, Applications, and Future of Wireless Sensors in Internet of Things: A Review. *IEEE Sens. J.* **2022**, *22*, 5482–5494. [CrossRef]
27. Li, M.; Zhang, L.; Dong, H.; Wang, Y.; Yan, X.; Hao, Z.; Tan, Q. Wireless Passive Flexible Strain Sensor Based on Aluminium Nitride Film. *IEEE Sens. J.* **2022**, *22*, 3074–3079. [CrossRef]
28. Sancho, J.I.; Almandoz, I.; Barandiaran, M.; Diaz, J.; Mendizabal, J. Scalable Wireless Wearing Monitoring System for Harsh Industrial Environment. *IEEE Trans. Ind. Electron.* **2022**, *69*, 1011–1020. [CrossRef]
29. Ahmed, M.F.; Hasan, M.K.; Chowdhury, M.Z.; Hoan, N.C.; Jang, Y.M. Continuous Status Monitoring of Industrial Valve Using OCC-Enabled Wireless Sensor Network. *IEEE Trans. Instrum. Meas.* **2022**, *71*, 1–10. [CrossRef]
30. Walker, J.M.; Prokop, A.; Lynagh, C.; Vuksanovich, B.; Conner, B.; Rogers, K.; Thiel, J.; MacDonald, E. Real-time process monitoring of core shifts during metal casting with wireless sensing and 3D sand printing. *Addit. Manuf.* **2019**, *27*, 54–60. [CrossRef]

31. Lei, X.; Wu, Y. Research on mechanical vibration monitoring based on wireless sensor network and sparse Bayes. *EURASIP J. Wirel. Commun. Netw.* **2020**, *2020*, 225. [CrossRef]
32. Patil, V.S.; Mane, Y.B.; Deshpande, S. FPGA Based Power Saving Technique for Sensor Node in Wireless Sensor Network (WSN). In *Computational Intelligence in Sensor Networks*; Mishra, B.B., Dehuri, S., Panigrahi, B.K., Nayak, A.K., Mishra, B.S.P., Das, H., Eds.; Springer: Berlin/Heidelberg, Germany, 2019; pp. 385–404.
33. Salameh, H.A.B.; Dhainat, M.F.; Benkhelifa, E. An End-to-End Early Warning System Based on Wireless Sensor Network for Gas Leakage Detection in Industrial Facilities. *IEEE Syst. J.* **2021**, *15*, 5135–5143. [CrossRef]
34. Dande, B.; Chang, C.Y.; Liao, W.H.; Roy, D.S. MSQAC: Maximizing the Surveillance Quality of Area Coverage in Wireless Sensor Networks. *IEEE Sens. J.* **2022**, *22*, 6150–6163. [CrossRef]
35. Guqhaiman, A.A.; Akanbi, O.; Aljaedi, A.; Chow, C.E. A Survey on MAC Protocol Approaches for Underwater Wireless Sensor Networks. *IEEE Sens. J.* **2021**, *21*, 3916–3932. [CrossRef]
36. Ahmad, I.; Hee, L.M.; Abdelrhman, A.M.; Imam, S.A.; Leong, M.S. Scopes, challenges and approaches of energy harvesting for wireless sensor nodes in machine condition monitoring systems: A review. *Measurement* **2021**, *183*, 109856. [CrossRef]
37. Yeh, P.C.; Chien, T.H.; Hung, M.S.; Chen, C.P.; Chung, T.K. Attachable Magnetic-Piezoelectric Energy-Harvester Powered Wireless Temperature Sensor Nodes for Monitoring of High-Power Electrical Facilities. *IEEE Sens. J.* **2021**, *21*, 11140–11154. [CrossRef]
38. Scardelletti, M.C.; Kulkarni, S.; Romanofsky, R.R. Real-Time Wireless Pressure Sensing System for Stall and Loading Measurements in the Rotating Frame of Reference for a Low Speed Compressor. *IEEE Sens. J.* **2022**, *22*, 12637–12644. [CrossRef]
39. Hao, C.; Jiang, C. Robust Wireless Sensor Network Against Strong Electromagnetic Pulse. *IEEE Sens. J.* **2021**, *21*, 5572–5579. [CrossRef]
40. Yu, W.; Liang, F.; He, X.; Hatcher, W.G.; Lu, C.; Lin, J.; Yang, X. A Survey on the Edge Computing for the Internet of Things. *IEEE Access* **2018**, *6*, 6900–6919. [CrossRef]
41. Satyanarayanan, M. The Emergence of Edge Computing. *Computer* **2017**, *50*, 30–39. [CrossRef]
42. Shi, W.; Cao, J.; Zhang, Q.; Li, Y.; Xu, L. Edge Computing: Vision and Challenges. *IEEE Internet Things J.* **2016**, *3*, 637–646. [CrossRef]
43. Zhang, T.; Li, Y.; Philip Chen, C.L. Edge computing and its role in Industrial Internet: Methodologies, applications, and future directions. *Inf. Sci.* **2021**, *557*, 34–65. [CrossRef]
44. Qiu, T.; Chi, J.; Zhou, X.; Ning, Z.; Atiquzzaman, M.; Wu, D.O. Edge Computing in Industrial Internet of Things: Architecture, Advances and Challenges. *IEEE Commun. Surv. Tutor.* **2020**, *22*, 2462–2488. [CrossRef]
45. Zhang, J.; Deng, C.; Zheng, P.; Xu, X.; Ma, Z. Development of an edge computing-based cyber-physical machine tool. *Robot. Comput. -Integr. Manuf.* **2021**, *67*, 102042. [CrossRef]
46. Wen, H.; Zhang, G.; Zhang, H. Design and Application of Instrument Remote Operation and Maintenance and Intelligent Analysis Platform Based on Edge Computing Technology. In Proceedings of the 2022 International Conference on Computer Engineering and Artificial Intelligence (ICCEAI), Shijiazhuang, China, 22–24 July 2022; pp. 612–617.
47. Hafeez, T.; Xu, L.; Mcardle, G. Edge Intelligence for Data Handling and Predictive Maintenance in IIOT. *IEEE Access* **2021**, *9*, 49355–49371. [CrossRef]
48. Wang, X.; Han, Y.; Leung, V.C.M.; Niyato, D.; Yan, X.; Chen, X. Convergence of Edge Computing and Deep Learning: A Comprehensive Survey. *IEEE Commun. Surv. Tutor.* **2020**, *22*, 869–904. [CrossRef]
49. Yin, S.; Bao, J.; Li, J.; Zhang, J. Real-time task processing method based on edge computing for spinning CPS. *Front. Mech. Eng.* **2019**, *14*, 320–331. [CrossRef]
50. Boguslawski, B.; Boujonnier, M.; Bissuel-Beauvais, L.; Saghir, F.; Sharma, R.D. IIoT Edge Analytics: Deploying Machine Learning at the Wellhead to Identify Rod Pump Failure. In Proceedings of the SPE Middle East Artificial Lift Conference and Exhibition, Manama, Bahrain, 28–29 November 2018.
51. Lee, K.C.; Villamera, C.; Daroya, C.A.; Samontanez, P.; Tan, W.M. Improving an IoT-Based Motor Health Predictive Maintenance System Through Edge-Cloud Computing. In Proceedings of the 2021 IEEE International Conference on Internet of Things and Intelligence Systems (IoTaIS), Bandung, Indonesia, 23–24 November 2021; pp. 142–148.
52. Bowden, D.; Marguglio, A.; Morabito, L.; Napione, C.; Panicucci, S.; Nikolakis, N.; Makris, S.; Coppo, G.; Andolina, S.; Macii, A. A Cloud-to-edge Architecture for Predictive Analytics. In Proceedings of the Workshops of the EDBT/ICDT 2019 Joint Conference (EDBT/ICDT 2019), Lisbon, Portugal, 26 March 2019.
53. Thomesse, J.P. Fieldbus Technology in Industrial Automation. *Proc. IEEE* **2005**, *93*, 1073–1101. [CrossRef]
54. Wells, T. *MODBUS: A Standard Computer Bus Structure*; Admiralty Surface Weapons Establishment Portsmouth: Haslemere, UK, 1978.
55. Kiencke, U.; Dais, S.; Litschel, M. Automotive serial controller area network. *SAE Trans.* **1986**, *95*, 823–828.
56. Vincent, P.; Parker, J. Digital Communications And Process Control Transmitters. In Proceedings of the IEE Colloquium on Measurements we Couldn't Make Without a Micro, London, UK, 17 May 1988; pp. 17–20.
57. JOST, R.; Mott, P.; Peters, H. INTERBUS–A COMPUTER SYSTEM FOR RADIO DISPATCHING OF BUSES. *VERKEHR UND TECHNIK* **1987**, *40*, 447–451.
58. Aguiar, M.W.C.d. *Análise Comparativa de Desempenho da Camada Enlace de dados do Profibus e do FIP, Propostas Candidatas ao Padrão" Field-Bus"*; Universidade Federal De Santa Catarina: Florianópolis, Brazil, 1989.
59. Glanzer, D. Foundation fieldbus forging high-level network standard. *InTech* **1996**, *43*, 443448.

60. Bengtsson, M.; Örn, R. *Fieldbus Communication Module: An Adapter for the Control & Communication Link Protocol*; Halmstad University: Halmstad, Sweden, 2000.
61. Felser, M.; Sauter, T. Standardization of industrial Ethernet—the next battlefield? In Proceedings of the IEEE International Workshop on Factory Communication Systems, Vienna, Austria, 22–24 September 2004; pp. 413–420.
62. Brooks, P. Ethernet/IP-industrial protocol. In Proceedings of the ETFA 2001 8th International Conference on Emerging Technologies and Factory Automation, Proceedings (Cat. No.01TH8597). Juan les Pins, France, 15–18 October 2001; Volume 502, pp. 505–514.
63. Swales, A. Open modbus/tcp specification. *Schneider Electric* **1999**, *29*, 3–19.
64. Cena, G.; Seno, L.; Valenzano, A.; Vitturi, S. Performance analysis of Ethernet Powerlink networks for distributed control and automation systems. *Comput. Stand. Interfaces* **2009**, *31*, 566–572. [CrossRef]
65. Jansen, D.; Buttner, H. Real-time Ethernet: The EtherCAT solution. *Comput. Control. Eng.* **2004**, *15*, 16–21. [CrossRef]
66. Feld, J. PROFINET-scalable factory communication for all applications. In Proceedings of the IEEE International Workshop on Factory Communication Systems, Vienna, Austria, 22–24 September 2004; pp. 33–38.
67. Wu, X.; Xie, L. Performance evaluation of industrial Ethernet protocols for networked control application. *Control. Eng. Pract.* **2019**, *84*, 208–217. [CrossRef]
68. Seferagić, A.; Famaey, J.; De Poorter, E.; Hoebeke, J. Survey on Wireless Technology Trade-Offs for the Industrial Internet of Things. *Sensors* **2020**, *20*, 488. [CrossRef] [PubMed]
69. Navarro-Ortiz, J.; Sendra, S.; Ameigeiras, P.; Lopez-Soler, J.M. Integration of LoRaWAN and 4G/5G for the Industrial Internet of Things. *IEEE Commun. Mag.* **2018**, *56*, 60–67. [CrossRef]
70. Zhang, K.; Zhu, Y.; Maharjan, S.; Zhang, Y. Edge intelligence and blockchain empowered 5G beyond for the industrial Internet of Things. *IEEE Netw.* **2019**, *33*, 12–19. [CrossRef]
71. Chaudhari, B.S.; Zennaro, M.; Borkar, S. LPWAN Technologies: Emerging Application Characteristics, Requirements, and Design Considerations. *Future Internet* **2020**, *12*, 46. [CrossRef]
72. Bor, M.; Vidler, J.E.; Roedig, U. LoRa for the Internet of Things. In Proceedings of the International Conference on Embedded Wireless Systems and Networks (EWSN), Graz, Austria, 15–17 February 2016.
73. Perahia, E.; Stacey, R. *Next Generation Wireless LANs: 802.11 n and 802.11 ac*; Cambridge University Press: Cambridge, UK, 2013.
74. Seno, L.; Cena, G.; Scanzio, S.; Valenzano, A.; Zunino, C. Enhancing communication determinism in Wi-Fi networks for soft real-time industrial applications. *IEEE Trans. Ind. Inform.* **2016**, *13*, 866–876. [CrossRef]
75. Kellogg, B.; Talla, V.; Gollakota, S.; Smith, J.R. Passive {Wi-Fi}: Bringing Low Power to {Wi-Fi} Transmissions. In Proceedings of the 13th USENIX Symposium on Networked Systems Design and Implementation (NSDI 16), Santa Clara, CA, USA, 16–18 March 2016; pp. 151–164.
76. Siep, T.M.; Gifford, I.C.; Braley, R.C.; Heile, R.F. Paving the way for personal area network standards: An overview of the IEEE P802.15 Working Group for Wireless Personal Area Networks. *IEEE Pers. Commun.* **2000**, *7*, 37–43. [CrossRef]
77. Bilstrup, U.; Wiberg, P.A. Bluetooth in industrial environment. In Proceedings of the 2000 IEEE International Workshop on Factory Communication Systems (Cat. No.00TH8531), Porto, Portugal, 6–8 September 2000; pp. 239–246.
78. Leonardi, L.; Patti, G.; Bello, L.L. Multi-Hop Real-Time Communications Over Bluetooth Low Energy Industrial Wireless Mesh Networks. *IEEE Access* **2018**, *6*, 26505–26519. [CrossRef]
79. Chen, F.; Wang, N.; German, R.; Dressler, F. Performance Evaluation of IEEE 802.15.4 LR-WPAN for Industrial Applications. In Proceedings of the 2008 Fifth Annual Conference on Wireless on Demand Network Systems and Services, Garmisch-Pertenkirchen, Germany, 23–25 January 2008; pp. 89–96.
80. Farahani, S. *ZigBee Wireless Networks and Transceivers*; Newnes: Oxford, UK, 2011.
81. Song, J.; Han, S.; Mok, A.; Chen, D.; Lucas, M.; Nixon, M.; Pratt, W. WirelessHART: Applying Wireless Technology in Real-Time Industrial Process Control. In Proceedings of the 2008 IEEE Real-Time and Embedded Technology and Applications Symposium, St. Louis, MO, USA, 22–24 April 2008; pp. 377–386.
82. Petersen, S.; Carlsen, S. WirelessHART Versus ISA100.11a: The Format War Hits the Factory Floor. *IEEE Ind. Electron. Mag.* **2011**, *5*, 23–34. [CrossRef]
83. Shelby, Z.; Bormann, C. *6LoWPAN: The Wireless Embedded Internet*; John Wiley & Sons: Hoboken, NJ, USA, 2011.
84. Fouladi, B.; Ghanoun, S. Security evaluation of the Z-Wave wireless protocol. *Black Hat USA* **2013**, *24*, 1–2.
85. Weinstein, R. RFID: A technical overview and its application to the enterprise. *IT Prof.* **2005**, *7*, 27–33. [CrossRef]
86. Ramanathan, R.; Imtiaz, J. NFC in industrial applications for monitoring plant information. In Proceedings of the 2013 Fourth International Conference on Computing, Communications and Networking Technologies (ICCCNT), Tiruchengode, India, 4–6 July 2013; pp. 1–4.
87. Qin, W.; Chen, S.; Peng, M. Recent advances in Industrial Internet: Insights and challenges. *Digit. Commun. Netw.* **2020**, *6*, 1–13. [CrossRef]
88. Li, J.Q.; Yu, F.R.; Deng, G.; Luo, C.; Ming, Z.; Yan, Q. Industrial Internet: A Survey on the Enabling Technologies, Applications, and Challenges. *IEEE Commun. Surv. Tutor.* **2017**, *19*, 1504–1526. [CrossRef]
89. Cavalieri, S. A Proposal to Improve Interoperability in the Industry 4.0 Based on the Open Platform Communications Unified Architecture Standard. *Computers* **2021**, *10*, 70. [CrossRef]
90. Mahnke, W.; Leitner, S.-H.; Damm, M. *OPC Unified Architecture*; Springer Science & Business Media: Berlin, Germany, 2009.

91. Liu, C.; Vengayil, H.; Lu, Y.; Xu, X. A Cyber-Physical Machine Tools Platform using OPC UA and MTConnect. *J. Manuf. Syst.* **2019**, *51*, 61–74. [CrossRef]
92. Kim, W.; Sung, M. Standalone OPC UA Wrapper for Industrial Monitoring and Control Systems. *IEEE Access* **2018**, *6*, 36557–36570. [CrossRef]
93. Martinov, G.; Issa, A.; Martinova, L. Controlling CAN Servo Step Drives and Their Remote Monitoring by Using Protocol OPC UA. In Proceedings of the 2019 International Multi-Conference on Industrial Engineering and Modern Technologies (FarEastCon), Vladivostok, Russia, 1–4 October 2019; pp. 1–5.
94. Wang, Y.; Zheng, L.; Wang, Y. Event-driven tool condition monitoring methodology considering tool life prediction based on industrial internet. *J. Manuf. Syst.* **2021**, *58*, 205–222. [CrossRef]
95. Qi, Q.; Tao, F. A Smart Manufacturing Service System Based on Edge Computing, Fog Computing, and Cloud Computing. *IEEE Access* **2019**, *7*, 86769–86777. [CrossRef]
96. Abdulkareem, K.H.; Mohammed, M.A.; Gunasekaran, S.S.; Al-Mhiqani, M.N.; Mutlag, A.A.; Mostafa, S.A.; Ali, N.S.; Ibrahim, D.A. A Review of Fog Computing and Machine Learning: Concepts, Applications, Challenges, and Open Issues. *IEEE Access* **2019**, *7*, 153123–153140. [CrossRef]
97. Bierzynski, K.; Escobar, A.; Eberl, M. Cloud, fog and edge: Cooperation for the future? In Proceedings of the 2017 Second International Conference on Fog and Mobile Edge Computing (FMEC), Valencia, Spain, 8–11 May 2017; pp. 62–67.
98. Yi, S.; Li, C.; Li, Q. A survey of fog computing: Concepts, applications and issues. In Proceedings of the 2015 workshop on mobile big data, Hangzhou, China, 22 June 2015; pp. 37–42.
99. Donno, M.D.; Tange, K.; Dragoni, N. Foundations and Evolution of Modern Computing Paradigms: Cloud, IoT, Edge, and Fog. *IEEE Access* **2019**, *7*, 150936–150948. [CrossRef]
100. Singh, J.; Singh, P.; Gill, S.S. Fog computing: A taxonomy, systematic review, current trends and research challenges. *J. Parallel Distrib. Comput.* **2021**, *157*, 56–85. [CrossRef]
101. Liu, M.; Yang, K.; Zhao, N.; Chen, Y.; Song, H.; Gong, F. Intelligent Signal Classification in Industrial Distributed Wireless Sensor Networks Based Industrial Internet of Things. *IEEE Trans. Ind. Inform.* **2021**, *17*, 4946–4956. [CrossRef]
102. Brik, B.; Messaadia, M.; Sahnoun, M.h.; Bettayeb, B.; Benatia, M.A. Fog-supported Low-latency Monitoring of System Disruptions in Industry 4.0: A Federated Learning Approach. *ACM Trans. Cyber-Phys. Syst.* **2022**, *6*, 1–23. [CrossRef]
103. Ur Rehman, M.H.; Yaqoob, I.; Salah, K.; Imran, M.; Jayaraman, P.P.; Perera, C. The role of big data analytics in industrial Internet of Things. *Future Gener. Comput. Syst.* **2019**, *99*, 247–259. [CrossRef]
104. Yu, W.; Dillon, T.; Mostafa, F.; Rahayu, W.; Liu, Y. A Global Manufacturing Big Data Ecosystem for Fault Detection in Predictive Maintenance. *IEEE Trans. Ind. Inform.* **2020**, *16*, 183–192. [CrossRef]
105. Wan, L.; Zhang, G.; Li, H.; Li, C. A Novel Bearing Fault Diagnosis Method Using Spark-Based Parallel ACO-K-Means Clustering Algorithm. *IEEE Access* **2021**, *9*, 28753–28768. [CrossRef]
106. Park, J.; Su-young, C. An implementation of a high throughput data ingestion system for machine logs in manufacturing industry. In Proceedings of the 2016 Eighth International Conference on Ubiquitous and Future Networks (ICUFN), Vienna, Austria, 5–8 July 2016; pp. 117–120.
107. Sahal, R.; Breslin, J.G.; Ali, M.I. Big data and stream processing platforms for Industry 4.0 requirements mapping for a predictive maintenance use case. *J. Manuf. Syst.* **2020**, *54*, 138–151. [CrossRef]
108. Liu, C.; Jiang, P.; Jiang, W. Web-based digital twin modeling and remote control of cyber-physical production systems. *Robot. Comput.-Integr. Manuf.* **2020**, *64*, 101956. [CrossRef]
109. Moniruzzaman, A.; Hossain, S.A. Nosql database: New era of databases for big data analytics-classification, characteristics and comparison. *arXiv* **2013**, arXiv:1307.0191.
110. Martino, S.D.; Fiadone, L.; Peron, A.; Riccabone, A.; Vitale, V.N. Industrial Internet of Things: Persistence for Time Series with NoSQL Databases. In Proceedings of the 2019 IEEE 28th International Conference on Enabling Technologies: Infrastructure for Collaborative Enterprises (WETICE), Capri, Italy, 12–14 June 2019; pp. 340–345.
111. Silva, N.; Barros, J.; Santos, M.Y.; Costa, C.; Cortez, P.; Carvalho, M.S.; Gonçalves, J.N.C. Advancing Logistics 4.0 with the Implementation of a Big Data Warehouse: A Demonstration Case for the Automotive Industry. *Electronics* **2021**, *10*, 2221. [CrossRef]
112. Munirathinam, S.; Sun, S.; Rosin, J.; Sirigibathina, H.; Chinthakindi, A. Design and Implementation of Manufacturing Data Lake in Hadoop. In Proceedings of the 2019 IEEE International Conference on Smart Manufacturing, Industrial & Logistics Engineering (SMILE), Hangzhou, China, 20–21 April 2019; pp. 19–23.
113. Gao, Z.; Cecati, C.; Ding, S.X. A Survey of Fault Diagnosis and Fault-Tolerant Techniques—Part I: Fault Diagnosis With Model-Based and Signal-Based Approaches. *IEEE Trans. Ind. Electron.* **2015**, *62*, 3757–3767. [CrossRef]
114. Lei, Y.; Yang, B.; Jiang, X.; Jia, F.; Li, N.; Nandi, A.K. Applications of machine learning to machine fault diagnosis: A review and roadmap. *Mech. Syst. Signal Process.* **2020**, *138*, 106587. [CrossRef]
115. Han, T.; Li, Y.-F. Out-of-distribution detection-assisted trustworthy machinery fault diagnosis approach with uncertainty-aware deep ensembles. *Reliab. Eng. Syst. Saf.* **2022**, *226*, 108648. [CrossRef]
116. Liu, S.; Jiang, H.; Wu, Z.; Li, X. Data synthesis using deep feature enhanced generative adversarial networks for rolling bearing imbalanced fault diagnosis. *Mech. Syst. Signal Process.* **2022**, *163*, 108139. [CrossRef]

117. Ferracuti, F.; Freddi, A.; Monteriù, A.; Romeo, L. Fault Diagnosis of Rotating Machinery Based on Wasserstein Distance and Feature Selection. *IEEE Trans. Autom. Sci. Eng.* **2022**, *19*, 1997–2007. [CrossRef]
118. Ferracuti, F.; Giantomassi, A.; Iarlori, S.; Ippoliti, G.; Longhi, S. Electric motor defects diagnosis based on kernel density estimation and Kullback–Leibler divergence in quality control scenario. *Eng. Appl. Artif. Intell.* **2015**, *44*, 25–32. [CrossRef]
119. Khan, S.S.; Madden, M.G. One-class classification: Taxonomy of study and review of techniques. *Knowl. Eng. Rev.* **2014**, *29*, 345–374. [CrossRef]
120. Janakiraman, V.M.; Nielsen, D. Anomaly detection in aviation data using extreme learning machines. In Proceedings of the 2016 International Joint Conference on Neural Networks (IJCNN), Vancouver, BC, Canada, 24–29 July 2016; pp. 1993–2000.
121. Zhao, P.; Kurihara, M.; Tanaka, J.; Noda, T.; Chikuma, S.; Suzuki, T. Advanced correlation-based anomaly detection method for predictive maintenance. In Proceedings of the 2017 IEEE International Conference on Prognostics and Health Management (ICPHM), Dallas, TX, USA, 19–21 June 2017; pp. 78–83.
122. Tanuska, P.; Spendla, L.; Kebisek, M.; Duris, R.; Stremy, M. Smart Anomaly Detection and Prediction for Assembly Process Maintenance in Compliance with Industry 4.0. *Sensors* **2021**, *21*, 2376. [CrossRef]
123. Kähler, F.; Schmedemann, O.; Schüppstuhl, T. Anomaly detection for industrial surface inspection: Application in maintenance of aircraft components. *Procedia CIRP* **2022**, *107*, 246–251. [CrossRef]
124. Rasheed, A.; San, O.; Kvamsdal, T. Digital Twin: Values, Challenges and Enablers From a Modeling Perspective. *IEEE Access* **2020**, *8*, 21980–22012. [CrossRef]
125. Hauge, J.B.; Zafarzadeh, M.; Jeong, Y.; Li, Y.; Khilji, W.A.; Larsen, C.; Wiktorsson, M. Digital twin testbed and practical applications in production logistics with real-time location data. *Int. J. Ind. Eng. Manag.* **2021**, *12*, 129. [CrossRef]
126. Fan, Y.; Yang, J.; Chen, J.; Hu, P.; Wang, X.; Xu, J.; Zhou, B. A digital-twin visualized architecture for Flexible Manufacturing System. *J. Manuf. Syst.* **2021**, *60*, 176–201. [CrossRef]
127. Fera, M.; Greco, A.; Caterino, M.; Gerbino, S.; Caputo, F.; Macchiaroli, R.; D'Amato, E. Towards Digital Twin Implementation for Assessing Production Line Performance and Balancing. *Sensors* **2020**, *20*, 97. [CrossRef] [PubMed]
128. Marjani, M.; Nasaruddin, F.; Gani, A.; Karim, A.; Hashem, I.A.T.; Siddiqa, A.; Yaqoob, I. Big IoT Data Analytics: Architecture, Opportunities, and Open Research Challenges. *IEEE Access* **2017**, *5*, 5247–5261. [CrossRef]
129. Zuo, Y.; Tao, F.; Nee, A.Y.C. An Internet of things and cloud-based approach for energy consumption evaluation and analysis for a product. *Int. J. Comput. Integr. Manuf.* **2018**, *31*, 337–348. [CrossRef]
130. Zhong, R.Y.; Xu, C.; Chen, C.; Huang, G.Q. Big Data Analytics for Physical Internet-based intelligent manufacturing shop floors. *Int. J. Prod. Res.* **2017**, *55*, 2610–2621. [CrossRef]
131. Zhong, R.Y.; Huang, G.Q.; Lan, S.; Dai, Q.Y.; Chen, X.; Zhang, T. A big data approach for logistics trajectory discovery from RFID-enabled production data. *Int. J. Prod. Econ.* **2015**, *165*, 260–272. [CrossRef]
132. Yang, L.; Li, G.; Zhang, Z.; Ma, X.; Zhao, Y. Operations & Maintenance Optimization of Wind Turbines Integrating Wind and Aging Information. *IEEE Trans. Sustain. Energy* **2021**, *12*, 211–221. [CrossRef]
133. Hu, J.; Jiang, Z.; Liao, H. Joint optimization of job scheduling and maintenance planning for a two-machine flow shop considering job-dependent operating condition. *J. Manuf. Syst.* **2020**, *57*, 231–241. [CrossRef]
134. Rojek, I.; Mikołajewski, D.; Dostatni, E. Digital Twins in Product Lifecycle for Sustainability in Manufacturing and Maintenance. *Appl. Sci.* **2021**, *11*, 31. [CrossRef]
135. Caterino, M.; Fera, M.; Macchiaroli, R.; Pham, D.T. Cloud remanufacturing: Remanufacturing enhanced through cloud technologies. *J. Manuf. Syst.* **2022**, *64*, 133–148. [CrossRef]
136. Çınar, Z.M.; Abdussalam Nuhu, A.; Zeeshan, Q.; Korhan, O.; Asmael, M.; Safaei, B. Machine Learning in Predictive Maintenance towards Sustainable Smart Manufacturing in Industry 4.0. *Sustainability* **2020**, *12*, 8211. [CrossRef]
137. Scheuermann, C.; Verclas, S.; Bruegge, B. Agile Factory—An Example of an Industry 4.0 Manufacturing Process. In Proceedings of the 2015 IEEE 3rd International Conference on Cyber-Physical Systems, Networks, and Applications, Hong Kong, China, 19–21 August 2015; pp. 43–47.
138. Bonci, A.; Longhi, S.; Nabissi, G. Fault Diagnosis in a belt-drive system under non-stationary conditions. An industrial case study. In Proceedings of the 2021 IEEE Workshop on Electrical Machines Design, Control and Diagnosis (WEMDCD), Modena, Italy, 8–9 April 2021; pp. 260–265.
139. Liu, Y.; Kashef, M.; Lee, K.B.; Benmohamed, L.; Candell, R. Wireless Network Design for Emerging IIoT Applications: Reference Framework and Use Cases. *Proc. IEEE* **2019**, *107*, 1166–1192. [CrossRef] [PubMed]
140. Messenger, J.L. Time-Sensitive Networking: An Introduction. *IEEE Commun. Stand. Mag.* **2018**, *2*, 29–33. [CrossRef]
141. Finn, N. Introduction to Time-Sensitive Networking. *IEEE Commun. Stand. Mag.* **2018**, *2*, 22–28. [CrossRef]
142. Seol, Y.; Hyeon, D.; Min, J.; Kim, M.; Paek, J. Timely Survey of Time-Sensitive Networking: Past and Future Directions. *IEEE Access* **2021**, *9*, 142506–142527. [CrossRef]
143. Simon, C.; Maliosz, M.; Mate, M. Design Aspects of Low-Latency Services with Time-Sensitive Networking. *IEEE Commun. Stand. Mag.* **2018**, *2*, 48–54. [CrossRef]
144. Sudhakaran, S.; Montgomery, K.; Kashef, M.; Cavalcanti, D.; Candell, R. Wireless Time Sensitive Networking for Industrial Collaborative Robotic Workcells. In Proceedings of the 2021 17th IEEE International Conference on Factory Communication Systems (WFCS), Linz, Austria, 9–11 June 2021; pp. 91–94.

145. Yang, D.; Gong, K.; Ren, J.; Zhang, W.; Wu, W.; Zhang, H. TC-Flow: Chain Flow Scheduling for Advanced Industrial Applications in Time-Sensitive Networks. *IEEE Netw.* **2022**, *36*, 16–24. [CrossRef]
146. Nikhileswar, K.; Prabhu, K.; Cavalcanti, D.; Regev, A. Time-Sensitive Networking Over 5G for Industrial Control Systems. In Proceedings of the 2022 IEEE 27th International Conference on Emerging Technologies and Factory Automation (ETFA), Stuttgart, Germany, 6–9 September 2022; pp. 1–8.
147. Pop, P.; Raagaard, M.L.; Gutierrez, M.; Steiner, W. Enabling Fog Computing for Industrial Automation Through Time-Sensitive Networking (TSN). *IEEE Commun. Stand. Mag.* **2018**, *2*, 55–61. [CrossRef]
148. Latif, S.; Idrees, Z.; e Huma, Z.; Ahmad, J. Blockchain technology for the industrial Internet of Things: A comprehensive survey on security challenges, architectures, applications, and future research directions. *Trans. Emerg. Telecommun. Technol.* **2021**, *32*, e4337. [CrossRef]
149. Gu, A.; Yin, Z.; Cui, C.; Li, Y. Integrated Functional Safety and Security Diagnosis Mechanism of CPS Based on Blockchain. *IEEE Access* **2020**, *8*, 15241–15255. [CrossRef]
150. Qu, Y.; Gao, L.; Luan, T.H.; Xiang, Y.; Yu, S.; Li, B.; Zheng, G. Decentralized Privacy Using Blockchain-Enabled Federated Learning in Fog Computing. *IEEE Internet Things J.* **2020**, *7*, 5171–5183. [CrossRef]
151. Chi, Y.; Dong, Y.; Wang, Z.J.; Yu, F.R.; Leung, V.C.M. Knowledge-Based Fault Diagnosis in Industrial Internet of Things: A Survey. *IEEE Internet Things J.* **2022**, *9*, 12886–12900. [CrossRef]
152. Sarazin, A.; Bascans, J.; Sciau, J.-B.; Song, J.; Supiot, B.; Montarnal, A.; Lorca, X.; Truptil, S. Expert system dedicated to condition-based maintenance based on a knowledge graph approach: Application to an aeronautic system. *Expert Syst. Appl.* **2021**, *186*, 115767. [CrossRef]
153. Cao, Q.; Zanni-Merk, C.; Samet, A.; Reich, C.; Beuvron, F.d.B.d.; Beckmann, A.; Giannetti, C. KSPMI: A Knowledge-based System for Predictive Maintenance in Industry 4.0. *Robot. Comput. -Integr. Manuf.* **2022**, *74*, 102281. [CrossRef]
154. Wang, J.; Zhang, Z.; Gao, S.; Yu, L.; Zhang, D.; Kou, L.; Nie, H.; Tang, X. Framework and Key Technologies of Intelligent Operation and Maintenance of Traction Transformer Based on Knowledge Graph. In Proceedings of the 5th International Conference on Electrical Engineering and Information Technologies for Rail Transportation (EITRT), Singapore, 21–23 October 2021; pp. 476–485.

machines

Review

Secure Blockchain Middleware for Decentralized IIoT towards Industry 5.0: A Review of Architecture, Enablers, Challenges, and Directions

Jiewu Leng [1,2], Ziying Chen [1], Zhiqiang Huang [1], Xiaofeng Zhu [1], Hongye Su [1], Zisheng Lin [1] and Ding Zhang [1,*]

1 State Key Laboratory of Precision Electronic Manufacturing Technology and Equipment, Guangdong University of Technology, Guangzhou 510006, China
2 Shenzhen Research Institute, City University of Hong Kong, Shenzhen 518057, China
* Correspondence: zhangding@gdut.edu.cn

Abstract: Resilient manufacturing is a vision in the Industry 5.0 blueprint for satisfying sustainable development goals under pandemics or the rising individualized product needs. A resilient manufacturing strategy based on the Industrial Internet of Things (IIoT) networks plays an essential role in facilitating production and supply chain recovery. IIoT contains confidential data and private information, and many security issues arise through vulnerabilities in the infrastructure. The traditional centralized IIoT framework is not only of high cost for system configuration but also vulnerable to cyber-attacks and single-point failure, which is not suitable for achieving the resilient manufacturing vision in Industry 5.0. Recently, researchers are seeking a secure solution of middleware based on blockchain technology integration for decentralized IIoT, which can effectively protect the consistency, integrity, and availability of IIoT data by utilizing the auditing and tamper-proof features of the blockchain. This paper presented a review of secure blockchain middleware for decentralized IIoT towards Industry 5.0. Firstly, the security issues of conventional IIoT solutions and the advantages of blockchain middleware are analyzed. Secondly, an architecture of secure blockchain middleware for decentralized IIoT is proposed. Finally, enabling technologies, challenges, and future directions are reviewed. The innovation of this paper is to study and discuss the distributed blockchain middleware, investigating its ability to eliminate the risk of a single point of failure via a distributed feature in the context of resilient manufacturing in Industry 5.0 and to solve the security issues from traditional centralized IIoT. Also, the four-layer architecture of blockchain middleware presented based on the IIoT application framework is a novel aspect of this review. It is expected that the paper lays a solid foundation for making IIoT blockchain middleware a new venue for Industry 5.0 research.

Keywords: decentralized Industrial Internet of Things; blockchain middleware; data security; Industry 5.0; resilient manufacturing

Citation: Leng, J.; Chen, Z.; Huang, Z.; Zhu, X.; Su, H.; Lin, Z.; Zhang, D. Secure Blockchain Middleware for Decentralized IIoT towards Industry 5.0: A Review of Architecture, Enablers, Challenges, and Directions. *Machines* **2022**, *10*, 858. https://doi.org/10.3390/machines10100858

Academic Editor: Panagiotis Kyratsis

Received: 31 August 2022
Accepted: 22 September 2022
Published: 26 September 2022

1. Introduction

The pandemic of COVID-19 has expedited the reshaping of supply chain management. Resilient Manufacturing (RM) is envisaged in the Industry 5.0 concept as a result of either sustainable development goals under pandemics or the rising individualized product needs [1,2]. It is described as a manufacturing system that can sustain possibly severe disturbances and recover from an undesired condition to the desired state. As well, it is considered to have the ability to mitigate the negative impacts of disruptions, such as networking faults, machinery troubleshooting, and material supply breakdowns, and swiftly recover to normal conditions.

Industry 4.0 is driven by technology. Its development has brought many cutting-edge technologies, but it lacks robustness and high resilience when exposed to unknown factors. Industry 5.0, based on the technology driven by Industry 4.0, is looking for common interests and values of workers from different countries, so as to transform into a value-oriented era. Among them, achieving a high level of resilience is an essential capability recognized by Industry 5.0. In the field of the Internet of Things (IoT) in Industry 4.0, the technology-driven IoT has given rise to computers with powerful computing power and cutting-edge data storage and processing technologies. Hence, most IoT architectures rely on a centralized framework, which makes the IoT vulnerable to unknown disturbances in the event of a single-point failure. In contrast, in the value-driven Industry 5.0, where the ability to quickly resume manufacturing when faced with unknown disruptions is a recognized fundamental feature. As well, the decentralized architecture of IIoT plays an essential role in achieving the system resilience in the transition from Industry 4.0 to Industry 5.0. Similarly, a resilient manufacturing strategy based on the Industrial Internet of Things (IIoT) networks plays an essential role in facilitating production and supply chain recovery. IIoT networks comprise billions of devices that generate enormous amounts of data [3,4], which usually contain sensitive information about humans and machines that could pose security issues. The vast majority of modern IIoT frameworks are built on centralized architectures. Users must have faith in the security of these services to process and store their data. Data processing, security, and privacy services are now provided by existing centralized systems, which call for extremely powerful computers from other parties. Data generated from equipment is collected and sent to a central cloud computing center for intense data gathering and translation. It needs a middleware solution to provide a development and execution environment that supports interoperability, distributed decision making, and the effective integration of heterogeneous human-machine systems and devices, which can also make the digital information transparent, immutable, traceable to query, and auditability (in Figure 1). The absence of secure middleware with processing and storage capabilities results in the high complexity of industrial networks. Therefore, researchers are pursuing middleware methods to availably address the mentioned IIoT security issues including integrity, consistency, and availability. However, the information may be utilized inappropriately or disclosed to certain unauthorized parties. The use of IIoT middleware will likely be severely constrained in the future due to the inadequacy of traditional security measures (e.g., cryptographic techniques). A centralized system may also make it difficult for the IIoT to be highly scalable and robust in the event of a single-point failure, which is not suitable for achieving the resilient manufacturing vision in Industry 5.0.

Decentralized techniques may be a preferable option to increase the security and dependability of current IIoT systems [5]. In addition to preventing single points of failure, the decentralized architecture of IIoT enables the long-term expansion of networks, in which blockchain is an important enabler [6]. Blockchain is a distributed ledger [7] that cannot be tampered with or cryptographically faked because of the chain topology that joins data blocks progressively following the chronological order [8]. Blockchain has many advantages to solve the dependability, interoperability, and security concerns of IIoT [9] by supplying decentralized processes. Industrial processes and entities will be able to register and confirm their products and services via the integration of blockchain with IIoT [5]. Establishing blockchain middleware for decentralized IIoT could provide dependable and secure services for industries to store and trade data, addressing the issues mentioned above (e.g., interoperability, heterogeneity, data security) and thus building trust among the value chain's participants [10]. Blockchain middleware can offer software or interface for many platforms and applications to successfully integrate communication between devices and the blockchain network. Through blockchain middleware, programmers with insufficient knowledge of the blockchain can shield the complex and tedious blockchain underlying principle to implement the interoperability between IIoT data and blockchain network, and the deployment of smart contracts via its API interface. In this way, it can promote the implementation of scalability, reliability, real-time, availability, security, and privacy of decentralized IIoT applications towards Industry 5.0.

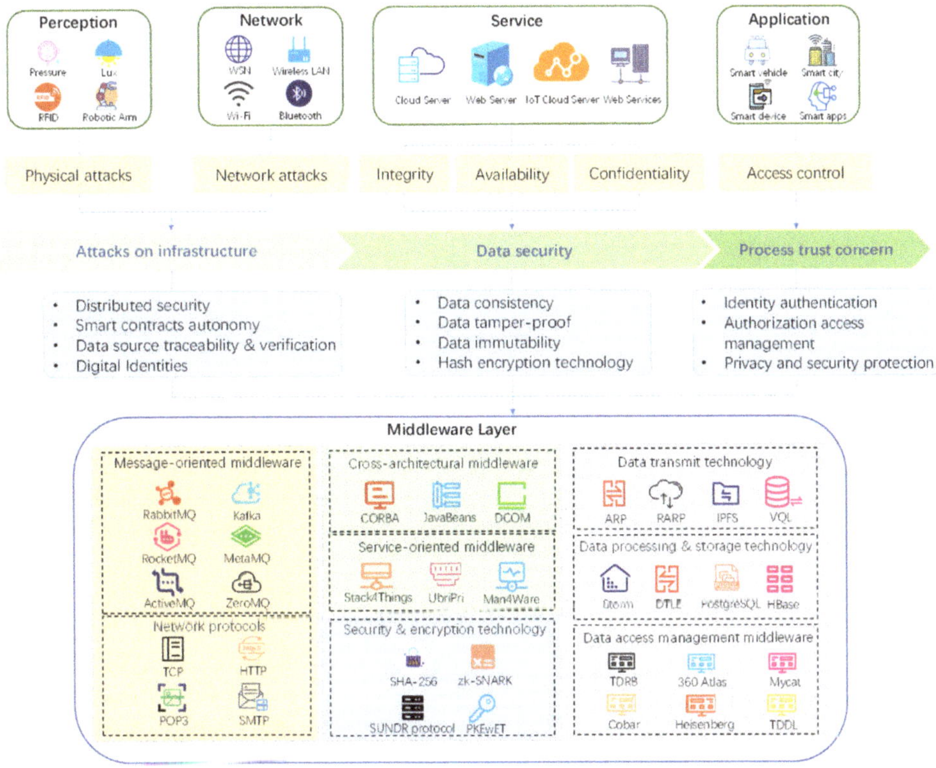

Figure 1. Conventional middleware solution and security issues for IIoT.

Recently, researchers are pursuing a blockchain middleware method to availably address the mentioned IIoT security issues. However, the current exploration of blockchain middleware is still at its initial stage. Muller and Breque [1,2] both specifically discussed the key role of resilience in industry 5.0, but they did not elaborate on the impact of blockchain technology on industry 5.0 in terms of distributed information security from the perspective of resilience. In terms of the IIoT, Xu et al. [3] summarized the application status, enabling technologies, and future directions of the IoT in the industrial field in detail, but the description of the application level of the IoT middleware and blockchain technology is not comprehensive. Wang, Lian, Leng, and Zheng et al. [6–9] summarized the application of blockchain technology in IoT and IIoT in detail. They emphasized that the characteristics of blockchain technology (tamper-proofing, immutable, and decentralized), data structure, and consensus mechanism play a key role in the security issues of IIoT. However, their discussion on the application of IoT middleware is not comprehensive enough, and the key enabling technologies and challenges of IoT middleware are not mentioned in the classification. Latif et al. [5] reviewed the development and application status of blockchain technology in IIoT. The integration of some middleware platforms and blockchain technology were also studied from the perspective of IoT middleware, but the key role of blockchain middleware in Industry 5.0 resilient manufacturing was not analyzed in the paper. Current research findings on secure blockchain middleware are relatively scarce and less systematic. A systematic introduction to blockchain middleware for decentralized IIoT towards Industry 5.0 is absent. Motivated by this observation, this paper tries to review secure blockchain middleware for decentralized IIoT towards Industry

5.0. We searched the Web of Science databases for literature. A three-step process was used to further analyze the articles that were found.

Using appropriate screening criteria, the initial step is to find high-quality articles. To produce high-quality publications, working papers and commentary are not included. Meanwhile, three keywords, namely, "Industrial Internet of Things", "blockchain middleware", and "data security and privacy", were identified for searching publications. This inclusive search yielded 271 publications for further analysis (up to 31 July 2022).

Secondly, to emphasize the architecture and the role of blockchain middleware in IIoT, articles on advanced technologies for IIoT security and blockchain middleware are also included. More specifically, the selection criteria are shown as follows.

1. These studies highlight the security issues, including infrastructure security, data security, and process trust, in IIoT, are selected. These studies highlight the concepts, technologies, architecture, and application of blockchain middleware technology in IIoT is selected.
2. Reviews/frameworks on IIoT blockchain middleware and enabling technologies were evaluated to offer a comprehensive understanding of the trends, functions, technologies, and challenges involved in Industry 5.0.
3. Studies containing concepts and issues related to digital transformation were taken into consideration, including those that did not specifically include blockchain middleware in the title, keywords, or abstract. This makes it possible to identify potential directions for future industrial innovations.

Research that is unrelated to (1) IIoT management or the blockchain domain; (2) studies that were not authored in English; (3) brief papers that are fewer than four pages are eventually excluded.

In the final step, 105 articles that fit the criteria for inclusion were included, and 21 more articles were found after the references were used as a source for literary analysis. Additionally, 21 supplementary references were added to make the review concrete. Therefore, this review consists of 126 articles in total.

Therefore, this review discusses the research status of blockchain middleware applications in decentralized IIoT. Firstly, the security issues of conventional IIoT solutions are analyzed in Section 2. Secondly, the advantages of blockchain middleware and architecture of secure blockchain middleware for decentralized IIoT are proposed in Section 3. Finally, enabling technologies, challenges, and future directions are reviewed in Sections 4 and 5, respectively. Finally, Section 6 presented the concluding remarks obtained from this review.

2. Security Issues in IIoT

2.1. The Characteristics of IIoT

Applications for the Internet of Things can influence many areas of daily life for workers. IIoT devices, networking, and communication technologies vary to fit the goals and requirements of diverse human-machine collaborative applications. Additionally, IIoT refers to the application of specific IIoT technologies and various smart objects (such as smart sensors, smart actuators, and smart manufacturing devices) in an industrial setting for the advancement of goals specific to the industry. New research trends for industrial applications are being introduced by IIoT [11,12]. It incorporates several cutting-edge technologies, including digital twins, big data analytics, robots, artificial intelligence (AI), smart sensors, actuators, and different communications protocols in traditional industrial environments [3,4]. By streamlining the production process, lowering costs, and boosting the productivity of smart businesses, IIoT aims to improve the performance of current industrial operations. It is clear that IIoT has attracted both academic and industrial interest, and as a result, it will significantly influence the design and development of next-generation (Industry 5.0) industrial infrastructures. The following list of IIoT characteristics serves as a summary.

2.1.1. Mass and Jumbled Data

Big data from IIoT sensing has the following qualities: high volume, high velocity, high veracity, and high diversity. Massive volumes of data can be produced in real-time by sensor devices. In this case, the problem with manufacturing data is both inconsistent and unreliable. This difficulty results from the variety of data types used in manufacturing, each of which calls for a specific signal acquisition parameter [13].

Furthermore, results may be predicted using big data processing and machine learning approaches. Due to the fast expansion of IIoT sensing, massive data are produced and stored locally or in cloud-based data repositories. Fundamentally new approaches for large-scale IIoT data management, information processing, and industrial process control are necessary to fully realize the full potential of big data analytics for smart manufacturing. For instance, the IIoT may use a large number of sensors to continually track the state of a machine in real-time and subsequently send data to the cloud. IIoT data includes both real-time data from in-situ monitoring of machines as well as signals gathered from machining tools and units. Easy access to the data from the cloud platform allows for parallel processing on distributed computers, which may be utilized to gather important data and develop prediction-making algorithms. In the end, decision-making is encouraged (e.g., production scheduling) [14].

2.1.2. Distributed Architecture

IIoT is moving towards distributed nature. Its distributed architecture is advantageous to the deployment of edge computing. Data generated in the IIoT is growing exponentially and much faster than that in traditional centralized cloud environments, where data is stored. As well, data storage and transmission issues (such as latency and bandwidth) on cloud devices make transmission speed a core issue. Therefore, a distributed architecture facilitates the deployment of edge computing to effectively solve the problem of inefficient transmission from IIoT to cloud architecture.

Rather than processing centrally in an internally deployed data center or public cloud, the distributed edge architecture brings it closer to the humans and devices in use. Edge computing is critical for manufacturing processes that use large amounts of data and require fast response times while ensuring safety. From IIoT devices to data centers, the cloud features (e.g., data, networking, storage, and computing) are distributed at all levels of the overall edge computing node [15], transferring the storage and calculation to the most efficient location for processing data. Then the key performance of IIoT application, safety, and efficiency requirements can be realized.

2.1.3. Heterogeneity

1. Multiple devices and heterogeneous network communication

Heterogeneity is a key characteristic of the IIoT when acting as manufacturing services [16]. The heterogeneity of the IIoT ranges from machines to humans and networks. When it comes to communication between different devices encapsulated with heterogeneous networks, gateways play an indispensable role. The gateway of the IIoT is mainly used for device access, data collection, and device monitoring. The main function of the gateway is to convert the equipment with two different protocols into the corresponding protocol for two-way data transmission. It is mainly aimed at networking heterogeneous devices that cannot communicate directly. When implementing the IIoT, decisions about how to transfer data are often complex. Currently, the three most common IIoT communication protocols are MQTT, AMQP, and COAP. The standardization of IIoT data connectivity and advance in simulation tools make it easier to make more informed decisions on data connectivity and integration.

Heterogeneous network integration is a promising emerging solution. For network heterogeneity of IIoT in Industry 4.0, many researchers have presented some architectures to integrate diverse networks. However, the presented solutions lack a consistent paradigm for accessing diverse networks and are only partially employed to address the heterogeneity of diverse constrained wireless communication networks [17]. Therefore, unified governance of heterogeneous networks is necessary to enable communication between the various networks. The use of network middleware, which gives clients a single application call interface while hiding the heterogeneity of underlying networks, is one feasible option [17]. Additionally, it is important to standardize the middleware's interface as much as possible so that the device producers can develop adaptive device connections with heterogeneous networks.

2. Interaction of heterogeneous systems and software

With the exponentially-growing network size and heterogeneity in systems, the development of IIoT security faces huge challenges. In such an environment, it is worth considering how to integrate heterogeneous systems to achieve efficient and safe interactive operation. Currently, the middleware solution has become an effective tool for researchers to achieve interoperability between systems. It allows for the on-demand integration of systems and software components. Additionally, it aids in the abstraction of availability and heterogeneity. To add new values for various systems, several middleware platforms were created.

2.2. The Architecture of IIoT

In this section, we first examine the foundation of a fundamental IIoT architecture and how these security concerns might work there before diving into a more in-depth examination of security issues. Several ideas provide different IIoT designs. These architectures are often categorized according to layers. Notably, the IIoT architecture frequently overlaps with these layers.

Sisinni et al. [18] proposed a three-layer design for the IIoT. The layers are sensors, networks, and services. Xu et al. [19] presented another three-layer design for the IIoT. It contains physical, communication, and application layers, unlike [18]. In [20], a four-layer IIoT architecture that consists of layers for perception, network, support, and application is proposed. The support layer can be viewed as a data layer conducting data analytics functions. In [5], different from the support layer of the proposal IIoT architecture in [20], another four-layer IIoT concept that is extensively used was introduced. The four-layer in [5] contains the perception, network, middleware, and application layers. In this section, each of these layers will be briefly discussed. A five-layer reference architecture is approved by the International Telecommunication Union. It includes layers for sensing, gaining access, networking, middleware, and applications.

An IIoT architecture in [5] is introduced. Notably, we choose a four-layer architecture that may be applied to various IIoT systems. It can accommodate the fundamental components of a three-layer IIoT architecture, while also readily expanding this four-layer architecture with additional components to depict a five-layer IIoT architecture with finer granularity. In addition, considering that the discussion focuses on the application of blockchain middleware to IIoT security issues, the enterprise applications and cloud computing services will not be discussed in this paper. That said, as shown in Figure 2, the IIoT security issues we summarized fit into the discussion in a four-layer architecture.

2.2.1. Perception Layer

The perception layer, which includes a variety of sensors for collecting various kinds of human-machine collaborative production context data, is thought to be the lowest physical layer of the IIoT architecture [21]. The perception layer is made up of sensors and actuators that acquire and interpret context data to carry out tasks (e.g., retrieve location and acceleration) [22]. A variety of IIoT applications require the perception layer [23]. To connect the physical and cyber worlds, a variety of end devices can be employed at the perception layer. Near Field Communications (NFC), RFID, wireless sensors and actuators, RFID, and some smart devices are examples of typical end devices. This layer also makes use of many sensing technologies including GPS, WSN, RSN, and others. Multiple sensors built within it collect information and recognize smart items in an industrial setting. With specific computational and energy needs, sensors, actuators, imaging devices, RFID tags, and other technologies are the main technologies of this layer. After being transformed into digital form, the data collected from the environment is sent to the network layer. For instance, an RFID tag consists of a tiny microchip linked to an antenna. In a manufacturing environment, things may be recognized, tracked, and monitored by applying RFID tags to them. Additionally, this layer can be divided into perception nodes and networks [5].

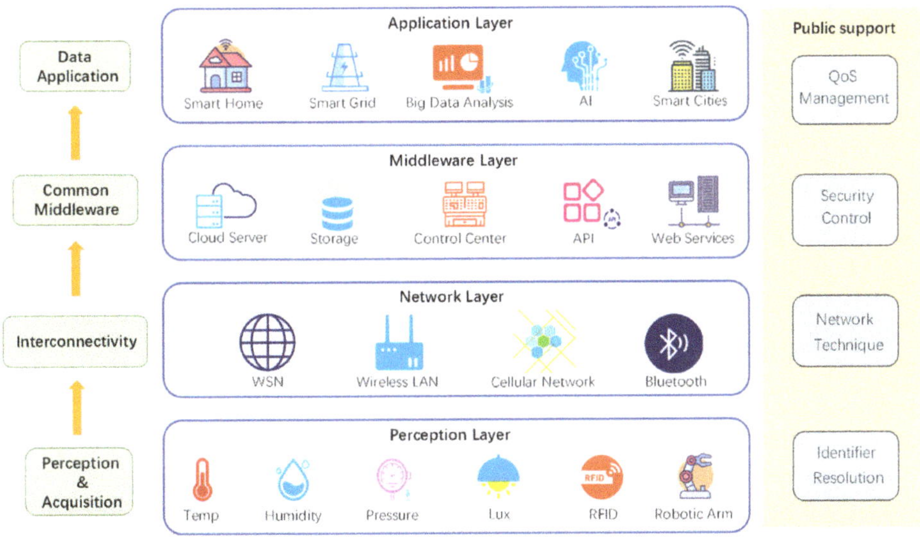

Figure 2. The four-layer architecture of IIoT.

2.2.2. Network Layer

The network layer encapsulates large amounts of protocols (e.g., MQTT, COAP, ZigBee, Ethernet). For the protocols of the IIoT, it can generally be divided into two categories, namely communication protocol (e.g., Bluetooth and ZigBee) and transmission protocol (e.g., High-Speed Ethernet (HSE), Modbus TCP/IP, and ProfiNet), performing secure information sharing [24]. Cloud computing and the Internet are the fundamental components of this layer [25]. Additionally, Internet gateway devices work in this tier by utilizing the most recent communication technologies to deliver network-connected services.

2.2.3. Middleware Layer

This third-level layer, commonly called the support layer, is presented [26]. It offers IIoT systems database and cloud services for the application layer to use further [27]. The middleware layer employs advanced computational techniques to evaluate, process, and store data. It can use cutting-edge technologies such as cloud computing and big data analytics to automatically analyze and compute the information that has been acquired. As described in the previous section. Middleware has become an effective tool for researchers to achieve interoperability between systems [28]. Some middleware models were proposed to provide added value for various industrial systems. The details will be described in later sections.

2.2.4. Application Layer

The termination layer of the IIoT is another name for the application layer. By preserving data integrity, secrecy, and authentication, this layer performs as the bridge between users and applications. This layer accesses the middleware layer's data and offers multiple services to the users. Additionally, it is integrated with commercial organizations to access smart applications [29,30]. Using internet-capable devices such as smartphones, tablets, PCs, wearable technology, and many other smart gadgets, users can access the smart services at this layer. To develop smart applications, the application layer incorporates the IIoT network, such as smart factories, healthcare, and smart grid [31].

2.3. Security Issues in IIoT Architecture

The stable operation of IIoT brings significant improvement in automation level to industry, and at the same time makes it challenging to impose more security issues within it [32]. Several security issues must be solved to offer the client and users a flexible, scalable, and reliable IIoT environment. Attacks on IIoT raise a serious security issue for the industry. These attacks have the potential to seriously harm businesses and occasionally even put lives in danger [33]. One of the most important challenges in an IIoT system, among others, is data security. Usually, the characteristics of an IIoT system make it challenging when the standard heavy-weight security designs cannot be directly used to meet such difficulties (e.g., resource-constrained, heterogeneous, mobile). On the contrary, they need more systems and frameworks that can satisfy the unique needs of an IIoT system. Therefore, a platform for smooth data exchange between enterprises is required to transmit and analyze multiple data sources efficiently, such as in a supply chain network [34]. Consequently, organizational issues, such as cross-enterprise information sharing and cooperation (via data interchange and transparency), are especially demanding in addition to technological considerations. Due to the IIoT-linked network of heterogeneous communication protocols, and various network platforms, a possible vulnerability in one area may cause a more significant effect on the system's entire performance [35].

In general, developing a comprehensive and cohesive system is challenging due to the heterogeneity in the IIoT system. The security issues in IIoT have been compiled and categorized in several surveys. For example, four primary categories of security attacks in the IIoT were Physical Attacks, Network Attacks, Software Attacks, and Data Attacks [31]. For IIoT, [31] has examined relevant studies on IIoT security challenges from the two perspectives of defending/preventing attacks and authentication/authorization. Jayalaxmi et al. [20] presented a security taxonomy for the IIoT system based on six distinct security services: authentication, confidentiality, non-repudiation, availability, integrity, and privacy. Pal and Jadidi [33] summarized the data confidentiality, they also considered the communication, performance, and heterogeneity in users and devices. Further, dynamic infrastructure as one of the IIoT characteristics and cascading services are discussed. Leng et al. [8] considered the combination of IIoT security issues and blockchain technology. They proposed a PDI framework for surveying blockchain security.

In this paper, we considered the comprehensive survey on IIoT attacks proposed by Jayasree Sengupta et al. [31] and introduced cyber-attacks from four layers in IIoT proposed by Shahid Latif et al. [5]. Moreover, as shown in Table 1, based on the proposed attacks categorized by Shahid Latif et al. [5], we further supplement the classification of these security issues (including cyber-attacks, threats, and potential security [33]), and categorize the attributes of them corresponding to the layer they are in. Especially, based on the middleware layer mentioned by Shahid Latif et al. [5], we combined the analysis of security issues in the support layer from Shantanu Pal et al. [33] and in the processing layer from Jayasree Sengupta et al. [31] and supplement the security issues of the middleware layer. At last, we reclassified the security issues into three categories (i.e., attacks on infrastructure, data security, and process trust concern) based on the related works [31,33,36]. The details are described as followed in Figure 3 and Table 1.

Table 1. The overview of IIoT security issues.

Security Type		Attack Name	Description
Attacks on infrastructure	Physical attacks	Tampering [37]	The act of physically modifying a device (e.g., RFID) or communication link. Such a type of attack will lead to the consequence of access to sensitive information and gain access.
		Attack device performance [38,39]	The target of these attacks is mainly to affect or interrupt system operation by affecting device performance, for instance, a heatstroke attack, DoS/DDoS, and replay attacks.
		Side channel attack [37]	Attackers collect the encryption keys by adopting timing, power, and fault attack on devices to encrypt/decrypt confidential data.
		Permanent Denial of Service (PDoS, phlashing) [31]	Attacks are launched by destroying firmware or uploading a corrupted BIOS using malware.
		RF interference/jamming [40]	Attackers create and send noise signals over the Radio Frequency (RF)/WSN signals to initiate DoS attacks on the tags/sensor nodes thereby jamming communications.
	Network attacks	Injection attacks [40]	Include malicious code injection and fake node injection.
		Traffic analysis attacks [37]	Confidential data flowing to and from the devices are sniffed by the attackers.
		DoS/DDoS [31]	Attackers make the network traffic unavailable to the users, or multiple compromised nodes attack a specific target by flooding messages to crash the server/resources.
		Control over communication attacks [41,42]	Attackers target routing protocols and redirect the routing path from the original receiver to an insecure one by misconfiguring routers, gateways, and DNS servers. Include blackhole, wormholes, Sybil, and pharming attacks.
Data security	Integrity	Dynamic Infrastructure [43–45]	The interactions in IIoT are dynamic. It requires better security protections that could securely integrate a real-time IIoT state in cyberspace into the physical space.
		Data (resource) security [43,46,47]	Attackers manipulate sensitive information and perform unauthorized data access, breaching the trust relationships between users.
	Confidentiality	MITM attack [48]	Attackers join the network as legitimate users and gain control over the nodes via the active attack path to sniff information.
		Data breach [31]	The disclosure of personal, sensitive, or confidential information in an unauthorized manner.
		Phishing site [49]	Attackers obtain a user's private information by sending emails that contain a malicious link (e.g., malware or spyware).
		Sniffing attacks [48]	Attackers take control over a device through device tracking or tag tracking and then attack devices that are connected to them.
	Availability	Data Inconsistency [31]	Attack on data integrity results in the inconsistency of data in transit or data stored in a central database.
		Malware [50]	An adversary infects the system via malicious software to achieve tamper with data or steal information, or launch DoS.

Table 1. *Cont.*

Security Type	Attack Name	Description
Process trust concern	RFID Unauthorized Access [37]	An attacker can read, modify or delete data presented on RFID nodes because of the lack of proper authentication mechanisms.
	Unauthorized Access [31]	Access control gives access to authorized users and denies access to unauthorized users. With unauthorized access, malicious users can gain data ownership or access sensitive data.
	Device impersonation [51–53]	Using identity fabrication to disrupt the integrity of a database by data forgery.
	Service interruption [54,55]	The failure of cascading services and misuse of the services by malicious actors lead to the failure of interconnected services.

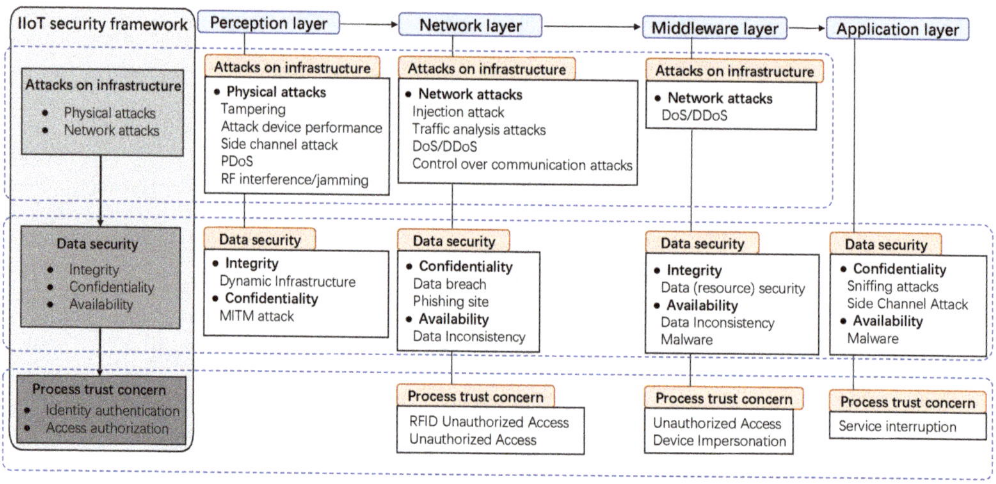

Figure 3. The framework of IIoT security issues.

3. Blockchain Middleware for Decentralized IIoT

This section discusses how blockchain is integrated into a middleware framework for many IIoT security issues. Typically, distributed resources and services from many technologies are used in smart manufacturing applications [56,57]. These resources and services could come from a single huge manufacturing company or a group of linked companies that work to support a targeted value chain. In this way, the distributed ledger services can then be connected to the blockchain-based middleware to create verifiable and immutable transaction logs. These transactions on the blockchain network can also be authenticated. Reliable, traceable records and resources to ensure that these transactions are accurate. The whole process will be supported by the blockchain services, which will also replicate and distribute the finished transaction across the involved entities as well as encrypt and append it to the chain. Finally, blockchain middleware for decentralized IIoT provides the ability to mitigate the negative impacts of disruptions, such as networking faults, machinery troubleshooting, and material supply breakdowns, and swiftly recover to normal conditions.

3.1. Advantages of Blockchain Middleware in IIoT

The resilient manufacturing vision in Industry 5.0 blueprint implies the ability to mitigate the negative impacts of disruptions, such as networking faults, machinery troubleshooting, and material supply breakdowns, and swiftly recover to normal conditions. A resilient manufacturing strategy based on the IIoT networks plays an essential role in facilitating production and supply chain recovery.

The advance of heterogeneous devices/technologies/applications of IIoT brings new challenges to developing applications in the industrial environment [58]. In this case, a middleware solution can integrate heterogeneous computing and communications devices, facilitate interoperability between applications and services, and provide common services for applications while simplifying application development [59,60]. As one of the application branches of the IIoT, the development of the IIoT is closely related to the development of Industry 5.0 [56]. Multiple functional types of devices or applications should be coordinated to achieve a shared goal.

However, there are pressing issues (e.g., interoperability, heterogeneity, and data security) that currently impede its effective development. Additionally, conventional security measures such as cryptographic methods [61] alone are unable to maintain data integrity on this massive scale. Furthermore, it is unrealistic to simply extend costly Internet security methods into the IIoT [62–65]. On the one hand, IIoT services such as centralized cloud storage could inherent insecurity from cyber-attacks, malicious code injection, and tampering that have a great influence on data security issues [31,33]. In addition, the cloud services storage is vulnerable to single-node failure [65–67]. On the other hand, in the context of IIoT, things and resources are heterogeneous and require specific programming [68]. Furthermore, the deployed protocols, together with conversion mechanisms, need to interoperate at different layers of the IIoT network safely [8]. In general, traditional IIoT services cloud cannot ensure data security such as integrity, consistency, and availability. It needs a middleware solution to provide a development and execution environment that supports interoperability, decentralized decision-making, and the effective integration of heterogeneous systems and devices, which can also make digital information transparent, immutable, traceable to query, and auditability. Therefore, researchers are pursuing a middleware method to availably address the mentioned IIoT security issues.

This distributed ledger cannot be tampered with or cryptographically faked because of the chain structure that links data blocks sequentially following the chronological order [8]. Blockchain enables distributed (peer-to-peer) and trusted (software application) transmission and recording of transactions and events. By incorporating blockchain technology into a middleware that combines various manufacturing processes with other value chain elements, it is possible to protect applications and build trust amongst the participants in the value chain [56,69]. This blockchain middleware approach enables addressing the issues mentioned above (e.g., interoperability, heterogeneity, and data security) and promotes the benefits of IIoT for achieving system resilience towards Industry 5.0.

As shown in Table 2, the metrics of blockchain middleware can be categorized in several aspects as follows:

Table 2. Metrics for application of blockchain middleware in IIoT.

Type	Metrics	Instance
Digital identities	Access permissions management	Develop distributed access control policies for the Internet [70]
		Data Authentication and privacy protection [71]
	Identities verification	No need to buy cryptocurrency or protect private keys [72]
		Digital asset management [73]

Table 2. *Cont.*

Type	Metrics	Instance
Distributed security	Privacy preservation	Point-to-point encrypted transmission and digital signature [74]
		Authorization, communication, and subject matching encryption [65]
	Data security	Data tamper-resisting [7,75]
Smart contracts	Autonomous application	Without the requirement for significant paperwork or third-party registration [56]
		Delegation of access permissions [76]
	Trust support	No need to verify whether participating on both sides is trustworthy [71]
Micro-controls	Data transmission	Enhance the data synchronization [77]
	Data storage	Reorganize the data from the database [78]
	Data tracking	Enrich the data query function based on the blockchain data provenance [79,80]

3.1.1. Digital Identities

Blockchain provides a digital analog that can be utilized to identify various entities, including corporations, in addition to machines [81]. Such characteristic makes it possible to verify the identity of individuals and organizations taking part in industrial operations through a public network. Hence, every entity participating in the manufacturing process, including machines can be given a digital identity.

Blockchain middleware can utilize a digital Identity authentication mechanism to realize. In the way of integrating a Networked smart object (NOS) (i.e., a flexible and cross-domain middleware) with the blockchain network, Rizzardi et al. [70] presented a secure and reliable distributed cross-domain access control. Genes-Duran et al. [72] proposed a blockchain middleware, which allows users to purchase services. Tapas et al. [76] proposed a model, which supports distributed resource access authorization and delegating responsibilities through the combination of the Ethereum blockchain network, smart contracts, and Stack4Things. Park et al. [71] proposed a framework. Based on digital identity authentication, it supports automatic off-chain operation verification, data authentication, and privacy protection. Hasan et al. [73] introduced a blockchain-based distributed digital manufacturing asset platform.

3.1.2. Distributed Security

The capability of blockchain to use a distributed approach to preserve the data is one of its fundamental success factors. Integrating IIoT middleware with blockchain networks allows for the protection of individual transactions. For instance, Lian et al. [7] proposed tamper-proof detection middleware. It ensured the integrity and consistency of data and the security of confidential data recorded in the blockchain and relational database. Tapas et al. [76] proposed a model. By integrating the middleware Stack4Things with the Ethereum blockchain network, the model realizes a distributed resource access authorization. Ochoa et al. [74] proposed a blockchain middleware called PriChain, which leverages the Ethereum blockchain to achieve the decentralization of the UbiPri middleware. Lv et al. [65] implemented distributed privacy protection for the publish/subscribe model, this approach effectively avoids a centralized single point of failure. Additionally, it becomes impossible for any of these organizations to subsequently dispute being engaged or in agreement because each transaction is documented with complete consent from all parties involved and relies on confirmed digital identities [75]. The method employed enables greater trust in the accuracy of the recorded transactions as well as improved transaction protection, and reduced exposure risks in the event of security breaches.

3.1.3. Smart Contracts

The use of blockchained smart contracts can effectively empower manufacturing industries and it might potentially boost several industrial sectors in diverse manners. Automation of agreement procedures between businesses and their partners and consumers is one of these improvements in the framework of Industry 4.0 [56]. Eliminating the need for third-party registrations or extensive documentation, would greatly cut administrative expenses and offer a more effective model for initiating, negotiating, and finalizing contracts. Smart contracts can be used to undertake several agreements along the value chain for smart manufacturing. Smart contracts can be completed more quickly and inexpensively while maintaining the legitimacy and validity of standard contracts.

In the application of IIoT, many researchers have proposed blockchain middleware driven by smart contracts. Tapas et al. [76] combined the functions of smart contracts to enable resource access authorization and delegation responsibilities. Park et al. [71] enable smart contracts to make the middleware support automate off-chain operation verification. Ochoa et al. [74] integrated a smart contract with a distributed storage service IPFS, which effectively ensured users' privacy.

3.1.4. Micro-Controls

Allowing for fine-grained modifications is another way that blockchain capabilities might benefit smart manufacturing [56]. The capacity to safely record activities without external verifications and guarantees will enhance the quantity of data and activities that are recorded and enable enterprises to create comprehensive ledgers of their operations. To give quality controls at any degree of detail, they may be simply examined. Additionally, it allows the generation of precise records that are simple to use as audit trails and assessment criteria for a manufacturer's operations [82]. This will make it possible for processes and activities to be continuously recorded. For instance, blockchain enables the constant and precise collection of data on situations involving safety. By doing this, the authenticity of data, authenticity, and creation of an unchangeable record are all guaranteed. Following that, data can be mined to examine any recorded information, including occurrences, consequences, and reactions. As a result of the analysis, these accidents will be better understood, patterns and the causes of issues will be found, and information will eventually be used to improve operations and develop better safety procedures.

In recent years, researchers are also pursuing a middleware method that enhances the data transmission and storage of the ledger data recorded in blockchain to facilitate micro-control of industrial manufacturing. For instance, Wang et al. [77] use middleware as an intermediary to facilitate the synchronization of database data to the blockchain. Peng et al. [78] proposed a blockchain-based middleware layer, which can extract transactions from the blockchain and efficiently reorganize them into the database. In addition, the middleware also provides various query services for users. Zhou et al. [79] proposed a distributed ledger data query platform, it not only supports querying blocks and transactions in a variety of ways but also provides a function to track its running history, which enables users to query blockchain blocks and transactions by shielding underlying principles of the blockchain. Hasan et al. [73] proposed a blockchain-based client middleware, in which the customers can track the source of data generated based on blockchain architecture in manufacturing systems. Liu et al. [80] proposed a blockchain-based middleware to process and store heterogeneous data from multiple sources at different stages of the product lifecycle. The research effectively addressed cross-enterprise access, processing, and analysis of production information problems. In addition, the integrity and confidentiality of product data are also considered. Based on the existing data storage middleware, Lu et al. [83] realized a blockchain-based cloud data acquisition and processing system with a large flow, high concurrency, and high availability.

3.2. Architecture of Secure Blockchain Middleware for Decentralized IIoT

This paper proposed the architecture of the blockchain middleware for decentralized IIoT towards Industry 5.0 as shown in Figure 4. In general, the IIoT middleware integrated with the blockchain network is connected to the application layer. In Figure 4, there is an interaction of four processes. In the process I, it shows the data transmission between the database and the middleware layer. The middleware layer acts as an intermediary between the blockchain network and the database. In the process II, for the connection between the middleware layer and the blockchain network, the middleware layer can utilize the service of blockchain, realizing data consistency, integrity, traceability, and auditing in IIoT. In the process III, data for object identification, tracking, and monitoring in a manufacturing environment can interact at the network layer through transport protocols between different devices. The design, the manufacturing, the manufacturing schedule, and other information of products can be packaged as a transaction recorded in the blockchain layer. In the process IV, for the connection between the application layer and the middleware layer. IIoT applications enable invoking related services of the middleware layer, such as query service, message pub/sub, and authentication and authorization. At the same time, the application layer in the upper can provide technical support for the middleware layer. For example, deep learning can be used to select the storage strategy of the middleware, and big data extraction and analysis technology is used to promote the middleware to make the data reorganized and to be on-chain. Table 3 is the related work on four processes in the architecture of blockchain middleware.

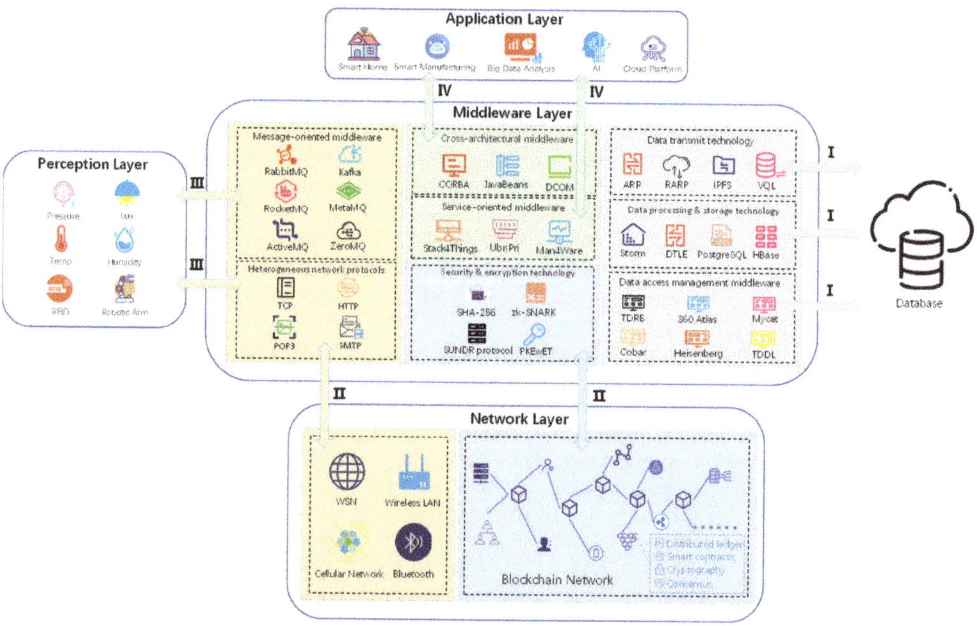

Figure 4. The architecture of the blockchain middleware for decentralized IIoT.

3.3. The Review of Blockchain Middleware

Based on the description of blockchain middleware in the previous section and the contribution of blockchain middleware mentioned in 23 references, we summarize the mentioned blockchain middleware in Table 4. By combining the architecture of blockchain middleware proposed in the last section, we divide them into five categories according to their functional type, including distributed data storage, data synchronism, security and privacy, function integration, and blockchain IIoT cloud platform. The details of the review are described below.

Table 3. Related work on four processes in the architecture of blockchain middleware.

Process	Related work	References
Process I	Create a verifiable query layer to make transactions in the underlying blockchain system efficient to extract and reorganize.	[78]
	Combining distributed storage services with IPFS data transfer technology for data storage security and IIoT performance.	[74]
	Using the four-module model to facilitate the synchronization between database and blockchain system.	[77]
	Using TDRB middleware to achieve the tamper-proof monitoring of data transmission between blockchain and relational database.	[7]
Process II	Combining middleware with the blockchain Hyperledger Fabric to achieve data traceability and queryable.	[79]
	Using blockchain cryptography to protect pub/sub-system from centralized single points of failure.	[65]
	Leveraging blockchain transaction validation technology to achieve efficient and secure heterogeneous networks.	[84]
	Using cloud storage combined with blockchain technology to ensure security and prevent forking attacks.	[85]
Process III	Information (e.g., product design, manufacturing progress, and data for tracking and monitoring) in the manufacturing system is packaged into transaction records via middleware between the perception layer and the blockchain system.	[73]
	The multi-source heterogeneous manufacturing data of the product life cycle is packed on-chain through the perception layer, and the manufacturing process autonomy is completed by leveraging smart contracts.	[80]
Process IV	Using machine learning algorithms to implement on-chain storage strategy selection.	[64]
	Leverage big data collection and storage technology to achieve high throughput and concurrency of the system.	[83]
	Using a combination of service-oriented middleware Stack4Things for distributed resource access authorization and responsibilities delegation.	[76]
	Using Byzantine consensus algorithm to achieve distributed fault tolerance of pub/sub system application.	[66]

Table 4. Review of blockchain middleware in recent years.

Type	Authors	Functional Features	Advantages
Distributed data storage	Danish et al. [64]	Make auditable, traceable, and immutable cloud storage decisions	Data traceability, auditability, accountability, integrity
	Ochoa et al. [74]	Decentralized implementation of UbiPri middleware using the Ethereum blockchain	Data integrity and privacy
	Lu et al. [83]	Integrates with data processing technology and distributed message queue technology to implement data collection and storage of the HBase system based on IIoT big data architecture.	Data availability, integrity, and stability
Data Synchronism	Zhou et al. [79]	Allows users to mask the underlying principles of blockchain to query blocks and transactions	Queryable and traceability
	Peng et al. [78]	Extract transactions stored in the underlying blockchain system and efficiently reorganize them into a database	Provide efficient query services for blockchain data and make query data results authentical
	Wang et al. [77]	As an intermediary to facilitate the synchronization of database data to the blockchain	Improves throughput and speed of transaction synchronization and ensures consistency between database and blockchain
	Lian et al. [7]	Provide efficient tamper-proof detection for relational database	Tamper-proof and ensures the integrity, confidentiality, and consistency of data
Function integration	Zupan et al. [86]	Decentralized pub/sub messaging for a multi-federated, licensed environment	Security, validating, and privacy-preserving messaging
	Tapas et al. [76]	Distributed resource access authorization and delegating responsibilities through the Ethereum blockchain network, smart contracts, and Stack4Things	Make the data trusted and auditing
	Ramachandran et al. [66]	A distributed fault-tolerant pub/sub broker with blockchain-based immutability	Avoiding a single point of failure
	Lv et al. [65]	A distributed publish/subscribe model for privacy protection based on blockchain technology to avoid a centralized single point of failure	Confidentiality, privacy preservation, and resistance to DDOS attacks
	Rizzardi et al. [70]	NOS integrated with blockchain to achieve secure and reliable distributed access control for IIoT resource	Integrity and Confidentiality Resist DOS/DDOS attacks Tamper-proof
Security and privacy	Zou et al. [85]	The lowest trust blockchain is used to ensure the security of cloud storage services	Prevent forking attacks and MITM attacks
	Samaniego et al. [87]	Mining is distributed to edge components and is divided into levels	Eliminates the limitation of low computing power
	Sanwar et al. [84]	Provides a delay-sensitive, time-sensitive transaction authentication technology and security and privacy solutions	Minimizing the delay of the transaction, ensuring security and privacy
	Genes et al. [72]	Users can access to easily create blockchain transactions, securing the management of their identity in IIoT	avoiding user impersonation
	Park et al. [71]	Enable smart contracts to automatically validate off-chain operations while supporting data authentication and privacy protection	Provides authentication and privacy preservation

Table 4. *Cont.*

Type	Authors	Functional Features	Advantages
Blockchain IIoT cloud platform	Hasan et al. [73]	The resource of data generated based on blockchain architecture in manufacturing systems can be traced	Data privacy and security
	Liu et al. [80]	Process multi-source heterogeneous data at different stages of the product life cycle and broadcast the processed data to the blockchain network	Data integrity. Supports cross-enterprise access, processing, and analysis of production information

3.3.1. Distributed Data Storage

To achieve high traffic and high concurrency, Lu et al. [83] proposed a middleware framework, namely Hadoop. The middleware layer integrates with a data processing technology (Storm) and a distributed message queue technology (Kafka) to implement data collection and storage of the HBase system. Concerning distributed data storage, Danish et al. [64] proposed a blockchain-based adaptive middleware for IIoT data storage decision selection. In Danish architecture [64], the storage decision and cryptographic hash of the IIoT data are stored on the blockchain network and allow the IIoT application owners in the application layer to audit the decision and data integrity through the adaptive middleware. At the same time, Machine Learning (ML) is applied to make decisions on the storage methods. However, there was no emphasis on users' security and privacy in [64].

3.3.2. Data Synchronism

The synchronization of information between the database and the blockchain is critical. To solve this problem, Zhou et al. [79] proposed a data analysis middleware framework called Ledgerdata Refiner to extract and synchronize transaction data from the blockchain network directly and then parses data relationships to provide unified interfaces for users in the application layer. This middleware, allows users to mask the underlying principles of blockchain to query blocks and transactions. Based on the research of the data on-chain process, Wang et al. [77] proposed a blockchain middleware model, which achieves the efficiency of the IIoT transaction data synchronization. The four-phase model [77] ensures synchronization consistency when the system fails. Peng et al. [78] proposed a Verifiable Query Layer architecture that can effectively reorganize transactions that are recorded in the underlying blockchain system to give application users a variety of query services.

3.3.3. Security and Privacy

A blockchain-based middleware named Amatista was presented by Samaniego et al. [87] for managing the IIoT in a zero-trust mode. Amatista has no trust in the transactions or the infrastructure. It validates both participant resources and the transactions they generate. As a result, the data flow is not only a data-centric reading but also a resource-centric communication that enables access to controlled resources without the usual requirement for a central authority. Furthermore, the hierarchical mining method is also given a context parameter by Amatista. As well, the mining process in [87] is designed on two different levels. For the research on data storage security, Zou et al. [85] proposed a blockchain middleware system to enhance the security of cloud storage, called ChainFS. But there is no mention of heterogeneous device gateways security and low latency validation. While researching identity authentication and access authorization, Genes et al. [72] proposed the Key Management System (KMS) and combined it with Man4Ware [56] to address the identity management problem. In [72], the middleware layer encapsulates APIs, communication protocols, and key management technologies. The application layer can send requests to the middleware through the APIs, and the middleware connects with the blockchain network to complete the creation of a blockchain transaction. Park et al. [71] proposed a blockchain middleware framework called Ziraffe, which can support authentication of origin for external data and protection of user privacy. The Ziraffe framework consists of four parts: users, Ziraffe in the middleware layer, the blockchain network, and the data resource server. In Ziraffe, programmers can utilize the framework to distribute smart contracts, users can acquire privacy protection according to their credentials through the framework, and the server of the data source supports a middleware-based signature. Users have access to external data sources and are enabled to import values into the blockchain. Finally, when a user downloads data from a data source, the server can confirm the external origin by signing the issue.

3.3.4. Function Integration

Many researchers integrate blockchain technology and create the capabilities to enable advanced services for the application layer of IIoT. Zupan et al. [86] implemented decentralized message publishing/subscription in a multi-federal and permissioned environment. In [86], the application layer communicates to Kafka (a distributed messaging model) of the middleware layer via proxies. The operation performed by publishers and subscribers in the application layer is received by a proxy. Moreover, to validate the operations sent from Kafka, Zupan has modeled the pub/sub semantics using smart contracts in the blockchain network. Lv et al. [65] proposed a blockchain-based privacy-preserving publish/subscribe model. To protect the subscribers' privacy, they used lightweight public key encryption to encrypt topics.

In distributed authorization and business delegation. Tapas et al. [76] proposed a blockchain middleware model that integrates a blockchain-based authorization and delegation mechanism with a middleware, Stack4Things (S4T). In [76], the application layer connects to the middleware layer that contains S4T middleware and its built-in database, and the middleware layer interacts with the blockchain network that encapsulates smart contracts with different functions. Eventually, access to the user is granted or not depending on the result. The user's request is sent to the middleware layer and waits for validation. Then smart contracts on the blockchain network record the resulting data on the chain. As well, the results are returned to the middleware layer and the user's access is confirmed. But in the face of the single point of collapse problem, Tapas does not take it into account. Ramachandran et al. [66] implemented Byzantine fault tolerance on distributed authorization and authentication. On cross-domain problems, integration of the existing Network Smart Objects (NOS) platform with a blockchain network is proposed by Rizzardi et al. [70] to allow decentralized or peer-to-peer operations. In [70], the application layer sends a subscription request to the corresponding NOS unit via the broker. After the processing of resource data and application layer data, the data of different NOSs units are packaged as transactions into a blockchain network through a consensus mechanism. In this way, NOSs can be deployed in an environment where members do not trust each other.

3.3.5. Blockchain IIoT Cloud Platform

Hasan et al. [73] introduced a client middleware of the CMaaS platform that enables the client-side program to submit HTTP requests including model parameter modification, and toolpath regeneration requests, directly to the CMaaS platform. Having considered processing heterogeneous production information across different enterprises, Liu et al. [80] proposed an IIoT blockchain middleware for PLM. The presented framework divides the information flow into three stages, each of which includes different product-related activities such as product design, product production, and warehousing management, among others. Users can conduct a search using the industrial blockchain-based middleware before they need to access the appropriate design files. Then, using the specified application programming interface, users can access additional huge design files. This can boost group decision-making to complete collaborative design activities promptly and accurately. Additionally, it promotes the traceability of particular documentation across several companies.

4. Enablers in Blockchain Middleware

This section focuses on some important enabling technologies for the integration of IIoT middleware with the blockchain network. It is divided into four levels from the scope of blockchain to industrial applications for discussion. According to the architecture of the blockchain middleware mentioned in the previous sections, we put forward the level enabling technical framework of the blockchain middleware, as shown in Figure 5.

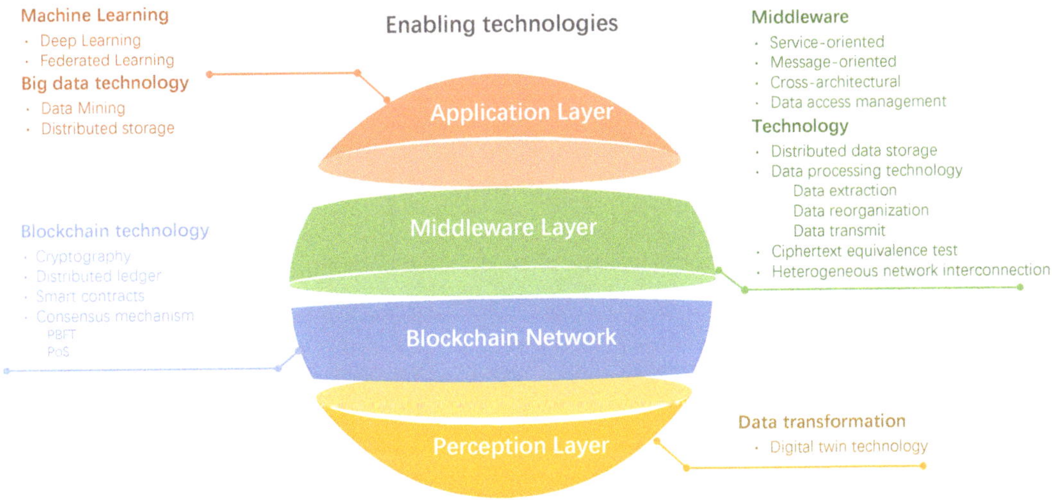

Figure 5. Key enabling technologies in blockchain middleware for IIoT.

4.1. Enablers for the Application Layer

4.1.1. Distributed Machine Learning

The management and processing of increasing data are generating new challenges in the industrial environment [88]. ML as the core of artificial intelligence (AI) technology, can protect data security and data privacy in edge services of IIoT through the integration with blockchain smart contracts [89]. The blockchain network is directly connected to edge nodes as well as IIoT middleware and communicates through the smart contract, which acts as a linkage between ML and blockchain network data interaction. Machine learning algorithms classify off-chain data resources and then store them in the blockchain network through blockchain middleware. In this way, it makes the middleware decision efficiently by classifying raw data as important through a machine learning algorithm [64]. In addition, the efficiency of data processing is improved and the overhead is reduced.

In IIoT, the constant growth of smart devices leads to privacy leakage and insufficient model accuracy of edge services. Another important factor that can have an impact on latency and reliability in industries is quick and efficient computing. To compute and evaluate the massive amounts of data acquired from the different network platforms the IIoT needs strong tools, as mentioned in [90]. Deep Learning (DL), however, is a recent study area in the field of ML. It is included in ML to go towards the original objective (i.e., AI). At the moment, DL-Enabled cloud/edge computing provides intelligent computing infrastructure for IIoT platforms while performing quick and efficient calculations. The utilization of edge-based computing infrastructure is acceptable due to the low power and limited storage of the devices [90].

However, due to insecure access to insufficient or low-quality data, the DL algorithm cannot fully meet the needs of IIoT. Thereby, Google proposed a decentralized DL model named Federated Learning (FL) for secure computing on end-user devices in wireless networks. FL in IIoT networks achieves a collaborative artificial intelligence training model based on the distributed devices' massive data without moving them into the host server. FL can coordinate distributed computational devices to train collaboratively on a shared prediction/classification task. For instance, Hiessl et al. [91] proposed an FL model for updating knowledge. Similarly, the case study in [92] uses FL to detect anomalies in IIoT. Lu et al. [93] presented a secure data-sharing architecture, which can be combined with FL. This distributed blockchain secure data sharing architecture can model the data sharing into an ML problem. It enables the integration of FL into the permissioned blockchain during the sharing process, which solves the security and privacy issues (e.g., data leakage) in wireless networks and achieves a balance of accuracy, efficiency, and security of data management.

4.1.2. Secure Big Data Analytics

With the advent of 5G, the IIoT has developed rapidly [94]. The massive data generated in smart industrial manufacturing needs to be mined and selected to enable producers to check the process quality of products and workpiece or equipment defects earlier. However, traditional data mining methods are inefficient and the validity of the data is insufficient. From the perspective of ML, Mathias et al. [95] proposed a data mining method to analyze limited samples of electrical signals. They developed an Open Platform Communication (OPC UA)-based simulation IIoT application to monitor the data mining mechanism for welding processes. Wang et al. [94] presented an online support vector machine-based data cleaning approach in the data collection from mobile edge nodes, to maintain information reliability/integrity while reducing the networking bandwidth and energy consumption of industrial sensing data acquisition. In [83], streaming big data processing technology (such as Storm) and big data storage technology (such as HBase) are used to make up the distributed processing blockchain-based middleware.

Based on the existing middleware, integrating with big data processing and storage technology can simplify the processing of off-chain data and reduce the cost of data system maintenance.

4.2. *Enablers for the Middleware Layer*

Many of the blockchain middleware mentioned in previous sections are integrated with the blockchain network by the IIoT middleware (e.g., Service-oriented middleware, Message-oriented middleware, and Cross-domain middleware), which acts as the underlying technical support of the blockchain middleware. In previous sections, for example, Kafka, a distributed message queue middleware, is integrated with a blockchain network with smart contracts to allow users to audit and validate the consumed data in the pub/sub-system, achieving distributed security. Network Smart Objects (NOS), a kind of cross-domain middleware, integrates with blockchain networks to achieve point-to-point operations across domains for applications in an environment where members do not trust each other [70]. In [56], based on Man4Ware, a blockchain-enabled service-oriented middleware is designed for protecting intelligent manufacturing applications and building trust between the parties involved in the value chain.

In addition to the integration of IIoT middleware, some underlying technologies can also support the integration of blockchain networks with IIoT middleware. They are as shown in Table 5. For instance, some data storage and transmission technology promote middleware to extract and process blockchain network data [74]. Some encryption technology [7,65,71] ensures application data security and privacy and can also reduce system maintenance costs. As well, there are also APIs open for middleware and blockchain network interaction. All these underlying technologies can act as key technologies to support efficient operation and functional integration of blockchain middleware.

Table 5. Enabling technologies for the middleware layer.

Type		Techniques	Functional Features of Integration
Middleware	Service-oriented	Stack4Things [76]	Focus on authentication, authorization, and delegation mechanisms
		Man4Ware [72]	Distributed ledgers created and maintained through the Man4Ware service can be used as a trusted, traceable record and source to verify the correctness of transactions
		UbiPri [74]	User privacy centralized management middleware
	Message-oriented	Kafka [83,86]	Efficiently process data streams in real-time and store them persistently in distributed replication clusters while maintaining high throughput
Security and encryption		SHA-256 [7]	Protect the private data in blockchain middleware and facilitate the retrieval of the data from the blockchain network
		zk-SNARK [71]	A zero-knowledge proof that can perform computations after a validator with weak computations outsources the computation to an unreliable validator and feedback results with evidence that results are correct in off-chain computations.
		PKEwET (Public Key Encryption with Equality Test) primitive [65]	A ciphertext equivalence test to determine whether two ciphertexts encrypted by different public keys are equal without decrypting the ciphertext, effectively reducing user storage costs
		SUNDR protocol [85]	A remote file system to ensure fork consistency to the client and prevent forking attacks
Data processing, storage, and management		Networked smart object (NOS) [70]	Enabled to manage the data provided by heterogeneous sources in a distributed manner and evaluate, utilizing proper algorithms
		Strom [96]	An open-source distributed, scalable, and fault-tolerant real-time computing system to simplify parallel real-time data processing
		HBase [97]	A distributed database has good compatibility with distributed storage, aggregated computing, and random access to massive semi-structured or unstructured data in real-time.
		PostgreSQL [79]	A database can parse out information and reorganize it as a third-party database to provide multiple query functions
Data transmit technology		IPFS [74]	A files system for distributed storage and P2P shared files to implement other middleware modules
		TiDB [77]	Convert database data to key-value pairs for easy storage in the blockchain
Others		SDN-Gateway [84]	Act as a linkage between LLN and blockchain, provide networking control operations, and execute different actions against vulnerabilities and cyberattacks.
		Trinity APIs [66]	Through the APIs, data can be sent to the blockchain to initiate the consensus and block creation process to complete the interaction with the middleware and blockchain network

4.3. Enablers for the Blockchain Network

4.3.1. Consensus Mechanism

Generally, blockchain is a distributed ledger system, and its key issue is consistency. The consensus mechanism is widely used in the distributed system, which allows all the nodes in the blockchain with an accounting problem to agree on an accounting. Unlike the consensus mechanism form of the traditional blockchain (e.g., Bitcoin), in the IIoT blockchain network, blocks are validated by decentralized nodes [98]. IIoT infrastructure that is transformed into virtual digital assets by digital-twin technology can be recorded on the blockchain network in the form of the cryptographic hash value. All manufacturing processed events are registered by the machine tools onto the blockchain as transactions [99]. In this way, users can take the initiative in the blockchain through the token they hold. This is different from the traditional consensus algorithm, like PoW. By contrast, it can reduce the power cost of industrial computing equipment to a certain extent.

Furthermore, some IIoT middleware can also integrate certain consensus algorithms like PBFT to achieve distributed fault tolerance, and realize distributed message distribution broker, avoiding a single point of collapse. Ramachandran et al. [66] implemented Byzantine fault tolerance on distributed authorization and authentication to avoid central points of failure. For the integration of the PBFT consensus algorithm, the distributed middleware can expose a set of APIs to interact with a blockchain following the interface architecture. As well, they can utilize the consensus mechanism of a blockchain network to realize distributed fault-tolerances.

4.3.2. Smart Contracts

Smart contract in the blockchain is trackable, secure, and unalterable. It can be built based on a distributed ledger to achieve authentication and access control without the third party and be scripts stored in the blockchain [100]. The off-chain resource can interact with blockchain network data through smart contracts [101]. Unlike the Bitcoin finance system, smart contracts in IIoT deal with transactions of virtual digital assets converted by industrial infrastructure entities. Once manufacturers deploy a smart contract, they are permitted to store the hash of the latest industrial entity updating on the blockchain. Then they can retrieve the industrial entity, and request interesting production information. In addition, the blockchain network where cryptographic token for pay-as-you work paves the way for a community of manufacturing and product services among IIoT devices. Furthermore, in the onchain-off-chain interaction process, the authentication of the data feed is necessary. For instance, Town Crier was one of the first works to look at authenticated linking smart contracts to external off-chain HTTPS-enabled data resources [102].

4.3.3. Cryptography and Distributed Ledger

As the core technology of blockchain, cryptography and distributed ledger support the data entry and storage of blockchain. The underlying data architecture is determined by blockchain cryptography. The packaged data blocks are processed into a chain structure using cryptography's hash functions. Because the hash algorithm is unidirectional and tamper-resistant, the data can not be tampered with and can be traced only in the blockchain network. In addition, accounts in the blockchain network will be encrypted by asymmetric encryption, thus ensuring the security of data. Distributed ledger builds the framework of blockchain. It is essentially a distributed database. When a piece of data is generated, it will be stored in this database after everyone processes it, so distributed ledger plays the role of data storage in the blockchain.

Blockchain middleware can utilize cryptography and distributed ledger to achieve traceability of historical data, auditing, and data security (in Figure 6). Commonly, in the applications of the manufacturing sector, the distributed ledger includes a series of transactions capturing manufacturing event data associated with machine tools. As well, the set of all transactions can represent a chain of manufacturing events for a product. The blockchain middleware utilizes encapsulated data extraction and data processing techniques to interact with transaction data from a distributed ledger database. At the same time, blocks of data packaged using cryptography are recorded into the blockchain network as key-value pairs, increasing the capacity of data storage and ensuring the security of IIoT middleware.

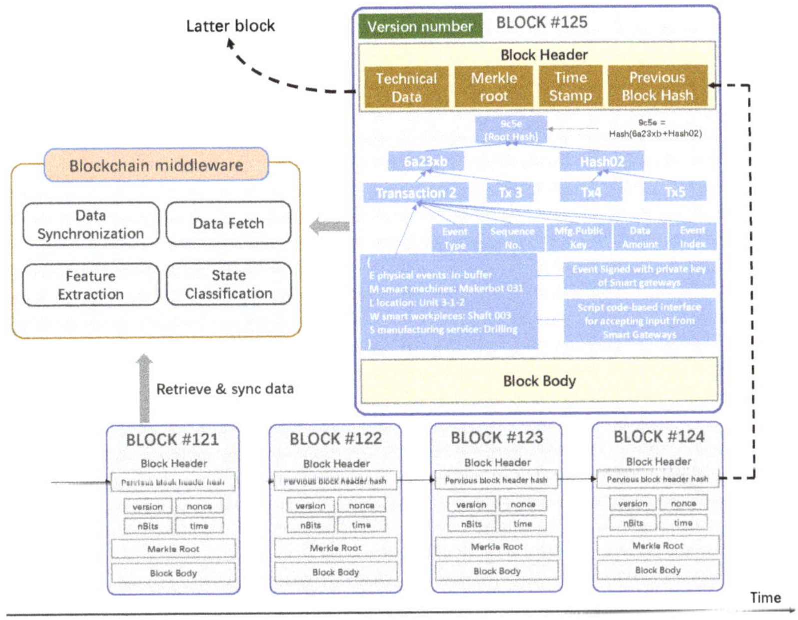

Figure 6. Illustration of Blockchain data structure storage.

4.4. Digital Transformation for the Perception Layer

In IIoT, as for the record and synchronization of manufacturing data, the original raw data are usually not stored directly in the blockchain, in which a digital twin system is built to realize the transformation of industrial infrastructure from physical/digital data into blockchained assets [103–105].

On the simplification of manufacturing information into data-tag of limited spaces, Leng et al. adopted an abbreviation schema to accommodate this limitation [106]. Through a mapping algorithm, the digital twin identification of the data tag on each IIoT entity is linked to the corresponding blockchain transaction on the blockchain network, and it is the anchor for obtaining the entity's lifecycle activities [107]. In the interaction (synchronization) process between the cyber and physical space of the IIoT, it illustrates a lifecycle digital twin model of individualized products [108]. The data tag is a mapping to its associated digital twin [109]. By collecting the transactions of workpieces/products' related events, the digital twin of products is therefore securitized for lifecycle tracking. Thus, users can interact with a digital twin model on their demands via the middleware layer. When the users invoke the blockchain middleware interface to send off manufacturing data, transactions reveal a continuous trail of the products being fabricated.

5. Research Directions

Based on the discussion of enabling technologies in the previous section, we offer some suggestions for research directions according to the four-layer architecture (in Figure 7). It is expected that it lays a foundation for making IIoT blockchain middleware a new venue for Industry 5.0 research innovation.

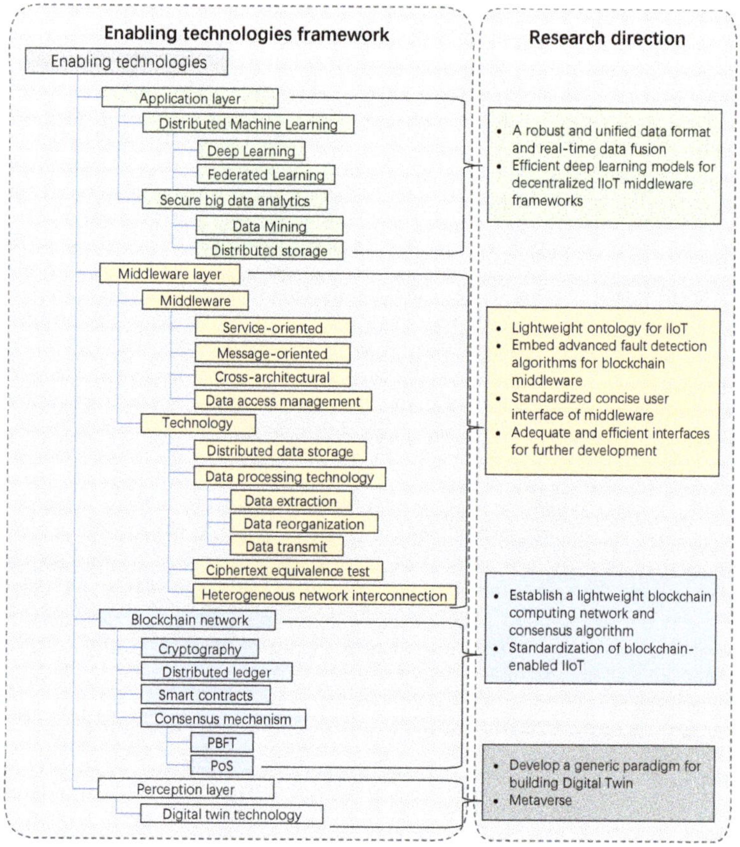

Figure 7. Research direction framework of blockchain middleware for IIoT.

5.1. Directions for the Application Layer

5.1.1. A robust and Unified Data Format and Real-Time Data Fusion

Information is distributed among multiple devices and is difficult to aggregate due to geographical or mapping barriers. A robust and cost-effective data format, as well as an integration method, is desirable [110] for managing data to achieve resilient manufacturing for Industry 5.0. The integration of Information Communication Technologies (ICT) into the industrial Internet of Things framework is also a favorable development direction for real-time data fusion [111], such as data preparation, crawl scheduling, multi-level indexing, and human and machine query.

5.1.2. Efficient Deep Learning Models for Decentralized IIoT Middleware Frameworks

Concerning the overall complexity of management issues and the finite storage and computation capabilities of the IIoT devices, existing implementations of distributed machine learning based on edge design uses additional optimization information to obtain higher productivity, self-organization capabilities, lower running times, and energy consumption [112,113]. For instance, federated learning (FL, also termed federated machine learning) has been proposed to coordinate distributed computational devices to train collaboratively on a shared prediction/classification task. However, the implementation of the FL in the smart industry is still not simple, and future research directions need to identify a better way to manage distributed computing resources to meet the needs of complex IIoT environments in Industry 5.0.

5.2. Directions for the Middleware Layer

5.2.1. Lightweight Ontology for IIoT

In integrating a blockchain-based data sharing framework to preserve security and privacy [114], IIoT interoperability affects its scalability and performance [115]. When a large number of machines interact in industrial networks [116], provisioning interoperability in IIoT from semantic conflicts, global heterogeneity, and new/unknown devices is of significance for achieving resilience due to lacking standard architecture, in which ontology methods are useful [117]. Existing ontology-based semantic approaches are cumbersome for satisfying resource flexibility needs. Therefore, in the processing of distributed middleware sensory data, designing a lightweight ontology for decentralized IIoT so that semantic interoperability will save the processing and annotation time, which is in urgent need for achieving resilient manufacturing towards Industry 5.0.

5.2.2. Embed Advanced Fault Detection Algorithms for Blockchain Middleware

As increasingly heterogeneous devices are involved in the decentralized IIoT, the possibilities of failures and faults increase [118]. Blockchain middleware for decentralized IIoT is supposed to be robust in not only detecting and withstanding failures but also detecting faults promptly. Advanced self-configured fault detection algorithms should be embedded into the blockchain middleware for efficiently and securely coordinating different devices. Accuracy and timeliness in detecting faults will accelerate the industrial process in achieving system resilience towards Industry 5.0.

5.2.3. Standardized Concise User Interface of Middleware

The user interface of the blockchain middleware is supposed to be as concise as possible so that engineers with different implementation fields are aware of the IIoT application he/she is using [118]. In the distributed environment, engineers will be adopting or operating the industrial blockchain with limited knowledge of networking and communications. Seamless integration of the decentralized IIoT with a user-friendly interface for blockchain middleware will help its acceptance, facilitating users to do the work without the complex underlying principles.

5.2.4. Adequate and Efficient Interfaces for Further Development

Developing efficient interfaces can not only enrich the functions of middleware but also facilitate the development and deployment of middleware and reduce the coupling between codes. For example, in ledger data analysis middleware, providing a sufficient interface for further function development can promote searching blocks or transactions efficiently. Design interfaces in databases or decentralized file sharing systems can solve the low-quality data issues in smart manufacturing [119]. Design adequate API interfaces to support more complicated SQL queries and integrate more diverse databases as well as blockchain systems.

5.3. Directions for the Blockchain Network

5.3.1. Establish a Lightweight Blockchain Computing Network and Consensus Algorithm

Introducing blockchain into the decentralized IIoT suffers from two obvious disadvantages. First, time latency in industrial controls is at a microsecond level [120], while the blockchain network usually cannot satisfy microsecond resolution demands in IIoT. Second, controllers in current IIoT devices are generally of low performance and limited storage spaces. Collecting massive industrial data into blockchain nodes hosting the IIoT devices may result in the collapse of the entire network, and thereby is impossible under this circumstance. Therefore, integrating blockchain with middleware should be lightweight to solve the storage and performance issues.

Existing consensus methods usually rely on costly computing tasks and puzzles to make the participants liberally add new blocks to a blockchain [121]. However, in industrial applications, the need for crash fault tolerance is much higher than that for

Byzantine fault tolerance. Innovations in storage structure and consensus mechanism for industrial usage should be performed to improve the throughput [122,123]. Establishing an effective lightweight consensus algorithm in IIoT systems for industrial applications is a future direction for improving the efficiency of the IIoT blockchain network under disruptions or disturbances.

5.3.2. Standardization of Blockchain-Enabled IIoT

Standardizing blockchain middleware for decentralized IIoT, as well as synchronizing with existing standards, is still in its early stages. Without clear regulations, coordination between different IIoT blockchain systems is challenging [124]. Blockchain middleware standards are supposed to provide guidelines for either developers or engineering clients. Furthermore, blockchained smart contracts are supposed to be legally enforceable for eliminating conflicts between participants. Although the integration of IIoT middleware with blockchain networks has inherent security features, there are still some exploitable loopholes in smart contracts in blockchain networks.

5.4. Directions for the Perception Layer

5.4.1. Develop a Generic Paradigm for Building Digital Twin

A unified data model or a generic Digital Twin architecture is in great need for digital asset/information conversion, concerning a lack of consensus on how to build a Digital Twin system for heterogeneous systems in a distributed IIoT network [125]. Therefore, designing a new paradigm for establishing a Digital Twin is supposed to be a generic paradigm of implementing a basic Digital Twin system more compatible with IIoT blockchain middleware.

5.4.2. Metaverse

According to Dr. Yu Yuan's definition, the Metaverse may refer to a kind of experience in which the outside world is perceived by the users (human or non-human) as being a universe that is built upon digital technologies as a different universe ("Virtual Reality"), a digital extension of our current universe ("Augmented Reality" or "Mixed Reality"), or a digital counterpart of our current universe ("Digital Twin"). Named after the universe, a metaverse shall be persistent and should be massive, comprehensive, immersive, and self-consistent. Described as "meta", a metaverse should be ultra-realistic, accessible, pervasive, and may be decentralized. In a narrow sense, metaverse may be simply defined as Persistent Virtual Reality (PVR). In a broad sense, the metaverse is the advanced stage and long-term vision of Digital Transformation.

Metaverse may be a promising research direction for the development of blockchain middleware [126]. The biggest difference between Industry 4.0 and the era of steam engines, electrification, and information technology is that the concept of Industry 4.0 places more emphasis on intelligence and the use of digital technology to minimize human involvement in the entire production process (i.e., the link between automated equipment and IT systems, which originally required human involvement, can be completed automatically without the need for human participation). In fact, under the concept of the digital twin. The human is only responsible for building the virtual world and defining the way of data collection, management and optimization, and then a continuous learning, optimization, and intelligent interaction will be formed between the physical space and the virtual space, and the role of the human is to supervise this association to keep it undefined and normal. Therefore, the Metaverse may be a promising research direction for the development of blockchain middleware for enhancing system resilience.

6. Concluding Remarks

This paper presented a review of secure blockchain middleware for decentralized IIoT towards Industry 5.0. The security issues of conventional IIoT solutions and the advantages of blockchain middleware are analyzed. Key enabling technologies in blockchain middleware are categorized respectively based on the corresponding layers, namely, the perception layer, the blockchain network, the middleware layer, and the application layer. The application of digital transformation in blockchain middleware is discussed. Future research directions for blockchain middleware in IIoT are outlined. The purpose of this paper is to research and analyze the abilities of distributed blockchain middleware to minimize the threat of a single point of failure in the context of resilient manufacturing in Industry 5.0. Compared with the traditional centralized IIoT, the distributed characteristics of secure blockchain middleware are more resistant to various disturbances (e.g., security issues) or other unknown factors in the industrial manufacturing environment, and can play an essential role in Industry 5.0 resilient manufacturing. It is expected that the paper lays a solid foundation for making IIoT blockchain middleware a new venue for Industry 5.0 research.

Author Contributions: Conceptualization, J.L. and D.Z.; methodology, J.L.; formal analysis, Z.C.; investigation, Z.C. and Z.H.; writing—original draft preparation, J.L., Z.C., Z.H., X.Z., H.S. and Z.L.; writing—review and editing, J.L. and D.Z.; supervision, J.L.; funding acquisition, J.L. All authors have read and agreed to the published version of the manuscript.

Funding: This work was supported by the National Natural Science Foundation of China under Grant No. 52075107, U20A6004 and 72271067; Natural Science Fund of Guangdong Province under Grant No. 2022B1515020006; the State Administration of Science, Technology and Industry for National Defense, PRC under Grant No. JCKY2020209B005; and the Shenzhen Special Fund for the Development of Strategic Emerging Industries under Grant No. JCYJ20170818100156260.

Data Availability Statement: The data that support the findings of this study are available from the corresponding author upon reasonable request.

Conflicts of Interest: The authors declare no conflict of interest.

References

1. Müller, J. Enabling Technologies for Industry 5.0. 2020. Available online: https://www.4bt.us/wp-content/uploads/2021/04/INDUSTRY-5.0.pdf (accessed on 19 July 2022).
2. Breque, M.; De Nul, L.; Petridis, A. Industry 5.0: Towards a Sustainable, Human-Centric and Resilient European Industry. 2021. Available online: https://msu.euramet.org/current_calls/documents/EC_Industry5.0.pdf (accessed on 19 July 2022).
3. Da Xu, L.; He, W.; Li, S. Internet of Things in Industries: A Survey. *IEEE Trans. Ind. Inform.* **2014**, *10*, 2233–2243. [CrossRef]
4. Li, S.; Zhao, S.; Yang, P.; Andriotis, P.; Xu, L.; Sun, Q. Distributed Consensus Algorithm for Events Detection in Cyber-Physical Systems. *IEEE Internet Things J.* **2019**, *6*, 2299–2308. [CrossRef]
5. Latif, S.; Idrees, Z.; e Huma, Z.; Ahmad, J. Blockchain Technology for the Industrial Internet of Things: A Comprehensive Survey on Security Challenges, Architectures, Applications, and Future Research Directions. *Trans. Emerg. Telecommun. Technol.* **2021**, *32*, e4337. [CrossRef]
6. Wang, X.; Zha, X.; Ni, W.; Liu, R.P.; Guo, Y.J.; Niu, X.; Zheng, K. Survey on Blockchain for Internet of Things. *Comput. Commun.* **2019**, *136*, 10–29. [CrossRef]
7. Lian, J.; Wang, S.; Xie, Y. TDRB: An Efficient Tamper-Proof Detection Middleware for Relational Database Based on Blockchain Technology. *IEEE Access* **2021**, *9*, 66707–66722. [CrossRef]
8. Leng, J.; Zhou, M.; Zhao, J.L.; Huang, Y.; Bian, Y. Blockchain Security: A Survey of Techniques and Research Directions. *IEEE Trans. Serv. Comput.* **2022**, *15*, 2490–2510. [CrossRef]
9. Zheng, Z.; Xie, S.; Dai, H.N.; Chen, X.; Wang, H. Blockchain Challenges and Opportunities: A Survey. *Int. J. Web Grid Serv.* **2018**, *14*, 352–375. [CrossRef]
10. Sun, Q.; Wang, N.; Li, S.; Zhou, H. Local Spatial Obesity Analysis and Estimation Using Online Social Network Sensors. *J. Biomed. Inform.* **2018**, *83*, 54–62. [CrossRef] [PubMed]
11. Yli-Ojanperä, M.; Sierla, S.; Papakonstantinou, N.; Vyatkin, V. Adapting an Agile Manufacturing Concept to the Reference Architecture Model Industry 4.0: A Survey and Case Study. *J. Ind. Inf. Integr.* **2018**, *15*, 147–160. [CrossRef]
12. Zhou, B.; Maines, C.; Tang, S.; Shi, Q.; Yang, P.; Yang, Q.; Qi, J. A 3-D Security Modeling Platform for Social IoT Environments. *IEEE Trans. Comput. Soc. Syst.* **2018**, *5*, 1174–1188. [CrossRef]
13. Bukkapatnam, S.T.; Lakhtakia, A.; Kumara, S.R. Analysis of Sensor Signals Shows Turning on a Lathe Exhibits Low-Dimensional Chaos. *Phys. Rev. E* **1995**, *52*, 2375–2387. [CrossRef] [PubMed]
14. Yang, H.; Kumara, S.; Bukkapatnam, S.T.; Tsung, F. The internet of Things for Smart Manufacturing: A Review. *IISE Trans.* **2019**, *51*, 1190–1216. [CrossRef]

15. Leng, J.; Chen, Z.; Sha, W.; Ye, S.; Liu, Q.; Chen, X. Cloud-Edge Orchestration-Based Bi-Level Autonomous Process Control for Mass Individualization of Rapid Printed Circuit Boards Prototyping Services. *J. Manuf. Syst.* **2022**, *63*, 143–161. [CrossRef]
16. Issarny, V.; Georgantas, N.; Hachem, S.; Zarras, A.; Vassiliadist, P.; Autili, M.; Gerosa, M.A.; Hamida, A.B. Service-Oriented Middleware for the Future Internet: State of the Art and Research Directions. *J. Internet Serv. Appl.* **2011**, *2*, 23–45. [CrossRef]
17. Zeng, J.; Yang, L.T.; Lin, M.; Ning, H.; Ma, J. A Survey: Cyber-Physical-Social Systems and Their System-Level Design Methodology. *Future Gener. Comput. Syst.* **2020**, *105*, 1028–1042. [CrossRef]
18. Sisinni, E.; Saifullah, A.; Han, S.; Jennehag, U.; Gidlund, M. Industrial Internet of Things: Challenges, Opportunities, and Directions. *IEEE Trans. Ind. Inform.* **2018**, *14*, 4724–4734. [CrossRef]
19. Xu, H.; Yu, W.; Griffith, D.; Golmie, N. A Survey on Industrial Internet of Things: A Cyber-Physical Systems Perspective. *IEEE Access* **2018**, *6*, 78238–78259. [CrossRef]
20. Jayalaxmi, P.; Saha, R.; Kumar, G.; Kumar, N.; Kim, T.-H. A Taxonomy of Security Issues in Industrial Internet-of-Things: Scoping Review for Existing Solutions, Future Implications, and Research Challenges. *IEEE Access* **2021**, *9*, 25344–25359. [CrossRef]
21. Jose, A.C.; Malekian, R. Improving Smart Home Security: Integrating Logical Sensing into Smart Home. *IEEE Sens. J.* **2017**, *17*, 4269–4286. [CrossRef]
22. Al-Fuqaha, A.; Guizani, M.; Mohammadi, M.; Aledhari, M.; Ayyash, M. Internet of Things: A Survey on Enabling Technologies, Protocols, and Applications. *IEEE Commun. Surv. Tutor.* **2015**, *17*, 2347–2376. [CrossRef]
23. Perera, C.; Zaslavsky, A.; Christen, P.; Georgakopoulos, D. Context Aware Computing for The Internet of Things: A Survey. *IEEE Commun. Surv. Tutor.* **2013**, *16*, 414–454. [CrossRef]
24. Silva, B.N.; Khan, M.; Han, K. Internet of Things: A Comprehensive Review of Enabling Technologies, Architecture, and Challenges. *IETE Technol. Rev.* **2017**, *35*, 205–220. [CrossRef]
25. Alaba, F.A.; Othman, M.; Hashem, I.A.T.; Alotaibi, F. Internet of Things security: A survey. *J. Netw. Comput. Appl.* **2017**, *88*, 10–28. [CrossRef]
26. Latif, S.; Zou, Z.; Idrees, Z.; Ahmad, J. A Novel Attack Detection Scheme for the Industrial Internet of Things Using a Lightweight Random Neural Network. *IEEE Access* **2020**, *8*, 89337–89350. [CrossRef]
27. Muhammad, F.; Anjum, W.; Mazhar, K.S. A Critical Analysis on the Security Concerns of Internet of Things (IoT). *Int. J. Comput. Appl.* **2015**, *111*, 1–6.
28. Mohamed, N.; Al-Jaroodi, J. A Survey on Service-Oriented Middleware for Wireless Sensor Networks. *Serv. Oriented Comput. Appl.* **2011**, *5*, 71–85. [CrossRef]
29. Mohiuddin, I.; Almogren, A. Workload Aware VM Consolidation Method in Edge/Cloud Computing for Iot Applications. *J. Parallel Distrib. Comput.* **2019**, *123*, 204–214. [CrossRef]
30. Alam, M.G.R.; Hassan, M.M.; Uddin, M.Z.; Almogren, A.; Fortino, G. Autonomic Computation Offloading in Mobile Edge for Iot Applications. *Future Gener. Comput. Syst.* **2019**, *90*, 149–157. [CrossRef]
31. Sengupta, J.; Ruj, S.; Das Bit, S. A Comprehensive Survey on Attacks, Security Issues and Blockchain Solutions for IoT and IIoT. *J. Netw. Comput. Appl.* **2019**, *149*, 102481. [CrossRef]
32. Attri, T.; Bhushan, B. Enabling Technologies, Attacks, and Machine Learning-Based Countermeasures for IoT and IIoT. In *Integration of WSNs into Internet of Things*; CRC Press: Boca Raton, FL, USA, 2021; pp. 249–272. [CrossRef]
33. Pal, S.; Jadidi, Z. Analysis of Security Issues and Countermeasures for the Industrial Internet of Things. *Appl. Sci.* **2021**, *11*, 9393. [CrossRef]
34. Ghadge, A.; Kara, M.E.; Moradlou, H.; Goswami, M. The Impact of Industry 4.0 Implementation on Supply Chains. *J. Manuf. Technol. Manag.* **2020**, *31*, 669–686. [CrossRef]
35. Müller, J.M.; Veile, J.W.; Voigt, K.-I. Prerequisites and Incentives for Digital Information Sharing in Industry 4.0—An International Comparison Across Data Types. *Comput. Ind. Eng.* **2020**, *148*, 106733. [CrossRef]
36. Mahmoud, R.; Yousuf, T.; Aloul, F.; Zualkernan, I. Internet of Things (Iot) Security: Current Status, Challenges and Prospective Measures. In Proceedings of the 2015 10th International Conference for Internet Technology and Secured Transactions (ICITST), London, UK, 14–16 December 2015; pp. 336–341.
37. Andrea, I.; Chrysostomou, C.; Hadjichristofi, G. Internet of Things: Security vulnerabilities and challenges. In Proceedings of the 2015 IEEE Symposium on Computers and Communication (ISCC), Larnaca, Cyprus, 6–9 July 2015; pp. 180–187.
38. Yang, J.C.; Fang, B.X. Security Model and Key Technologies for the Internet of Things. *J. China Univ. Posts Telecommun.* **2011**, *18*, 109–112. [CrossRef]
39. Xiao, B.; Chen, W.; He, Y.; Sha, E.M. An Active Detecting Method Against SYN Flooding Attack. In Proceedings of the 11th International Conference on Parallel and Distributed Systems (ICPADS'05), Fukuoka, Japan, 20–22 July 2005; Volume 1, pp. 709–715.
40. Ahemd, M.M.; Shah, M.A.; Wahid, A. IoT Security: A Layered Approach for Attacks & Defenses. In Proceedings of the 2017 International Conference on Communication Technologies (ComTech), Rawalpindi, Pakistan, 19–21 April 2017; pp. 104–110.
41. Airehrour, D.; Gutierrez, J.; Ray, S.K. Secure Routing for Internet of Things: A Survey. *J. Netw. Comput. Appl.* **2016**, *66*, 198–213. [CrossRef]
42. Pongle, P.; Chavan, G. A Survey: Attacks on RPL and 6LoWPAN in IoT. In Proceedings of the 2015 International Conference on Pervasive Computing (ICPC), Pune, India, 8–10 January 2015; pp. 1–6.

43. Elkhodr, M.; Shahrestani, S.; Cheung, H. The Internet of Things: Vision & Challenges. In Proceedings of the IEEE 2013 Tencon-Spring, Sydney, NSW, Australia, 17–19 April 2013; pp. 218–222.
44. Palattella, M.R.; Dohler, M.; Grieco, A.; Rizzo, G.; Torsner, J.; Engel, T.; Ladid, L. Internet of Things in the 5G Era: Enablers, Architecture, And Business Models. *IEEE J. Sel. Areas Commun.* **2016**, *34*, 510–527. [CrossRef]
45. Guinard, D.; Trifa, V.; Karnouskos, S.; Spiess, P.; Savio, D. Interacting with the SOA-Based Internet of Things: Discovery, Query, Selection, and On-Demand Provisioning of Web Services. *IEEE Trans. Serv. Comput.* **2010**, *3*, 223–235. [CrossRef]
46. O'Leary, D.E. 'Big Data', The 'Internet of Things' and the 'Internet of Signs. *Intell. Syst. Account. Financ. Manag.* **2013**, *20*, 53–65. [CrossRef]
47. Yan, Z.; Zhang, P.; Vasilakos, A.V. A Survey on Trust Management for Internet of Things. *J. Netw. Comput. Appl.* **2014**, *42*, 120–134. [CrossRef]
48. Li, S.; Xu, L.D.; Zhao, S. The Internet of Things: A Survey. *Inf. Syst. Front.* **2015**, *17*, 243–259. [CrossRef]
49. Gupta, B.B.; Tewari, A.; Jain, A.K.; Agrawal, D.P. Fighting Against Phishing Attacks: State of the Art and Future Challenges. *Neural Comput. Appl.* **2016**, *28*, 3629–3654. [CrossRef]
50. Varga, P.; Plosz, S.; Soos, G.; Hegedus, C. Security Threats and Issues in Automation IoT. In Proceedings of the 2017 IEEE 13th International Workshop on Factory Communication Systems (WFCS), Trondheim, Norway, 31 May–2 June 2017; pp. 1–6.
51. Babar, S.; Mahalle, P.; Stango, A.; Prasad, N.; Prasad, R. Proposed Security Model and Threat Taxonomy for the Internet of Things (IoT). In *International Conference on Network Security and Applications*; Springer: Berlin/Heidelberg, Germany, 2010; pp. 420–429.
52. Pal, S.; Hitchens, M.; Varadharajan, V. Modeling Identity for the Internet of Things: Survey, Classification and Trends. In Proceedings of the 2018 12th International Conference on Sensing Technology (ICST), Limerick, Ireland, 4–6 December 2018; pp. 45–51.
53. Sarma, A.C.; Girão, J. Identities in the Future Internet of Things. *Wirel. Pers. Commun.* **2009**, *49*, 353–363. [CrossRef]
54. Lhaksmana, K.M.; Murakami, Y.; Ishida, T. Analysis of Large-Scale Service Network Tolerance to Cascading Failure. *IEEE Internet Things J.* **2016**, *3*, 1159–1170. [CrossRef]
55. Lhaksmana, K.M.; Murakami, Y.; Ishida, T. Cascading Failure Tolerance in Large Scale Service Networks. In Proceedings of the 2015 IEEE International Conference on Services Computing, New York, NY, USA, 27 June–2 July 2015; pp. 1–8.
56. Mohamed, N.; Al-Jaroodi, J. Applying Blockchain in Industry 4.0 Applications. In Proceedings of the 2019 IEEE 9th Annual Computing and Communication Workshop and Conference (CCWC), Las Vegas, NV, USA, 7–9 January 2019; pp. 0852–0858.
57. Leng, J.; Ruan, G.; Jiang, P.; Xu, K.; Liu, Q.; Zhou, X.; Liu, C. Blockchain-Empowered Sustainable Manufacturing and Product Lifecycle Management in Industry 4.0: A Survey. *Renew. Sustain. Energy Rev.* **2020**, *132*, 110112. [CrossRef]
58. Razzaque, M.A.; Milojevic-Jevric, M.; Palade, A.; Clarke, S. Middleware for Internet of Things: A Survey. *IEEE Internet Things J.* **2015**, *3*, 70–95. [CrossRef]
59. Ngu, A.H.; Gutierrez, M.; Metsis, V.; Nepal, S.; Sheng, Q.Z. IoT Middleware: A Survey on Issues and Enabling Technologies. *IEEE Internet Things J.* **2017**, *4*, 1–20. [CrossRef]
60. Fremantle, P.; Scott, P. A Survey of Secure Middleware for the Internet of Things. *PeerJ Comput. Sci.* **2017**, *3*, e114. [CrossRef]
61. Hassan, A. Lightweight Cryptography for the Internet of Things. In Proceedings of the Future Technologies Conference (FTC) 2020, Vancouver, BC, Canada, 5–6 November 2020; Arai, K., Kapoor, S., Bhatia, R., Eds.; Advances in Intelligent Systems and Computing. Springer: Cham, Switzerland, 2020. [CrossRef]
62. Lyu, X.; Ni, W.; Tian, H.; Liu, R.P.; Wang, X.; Giannakis, G.B.; Paulraj, A. Optimal Schedule of Mobile Edge Computing for Internet of Things Using Partial Information. *IEEE J. Sel. Areas Commun.* **2017**, *35*, 2606–2615. [CrossRef]
63. Atlam, H.F.; Alenezi, A.; Alharthi, A.; Walters, R.J.; Wills, G.B. Integration of cloud Computing with Internet of Things: Challenges and Open Issues. In Proceedings of the 2017 IEEE International Conference on Internet of Things (Ithings) and IEEE Green Computing and Communications (Greencom) and IEEE Cyber, Physical and Social Computing (Cpscom) and IEEE Smart Data (SmartData), Exeter, UK, 21–23 June 2017; pp. 670–675.
64. Danish, S.M.; Assoc, C.M. A Blockchain-Based Adaptive Middleware for Large Scale Internet of Things Data Storage Selection. In Proceedings of the Middleware'19: 20th International Middleware Conference Doctoral Symposium, Davis, CA, USA, 9–13 December 2019; pp. 17–19.
65. Lv, P.; Wang, L.; Zhu, H.; Deng, W.; Gu, L. An IOT-Oriented Privacy-Preserving Publish/Subscribe Model Over Blockchains. *IEEE Access* **2019**, *7*, 41309–41314. [CrossRef]
66. Ramachandran, G.S.; Wright, K.L.; Zheng, L.; Navaney, P.; Naveed, M.; Krishnamachari, B.; Dhaliwal, J. Trinity: A Byzantine Fault-Tolerant Distributed Publish-Subscribe System with Immutable Blockchain-Based Persistence. In Proceedings of the 2019 IEEE International Conference on Blockchain and Cryptocurrency (ICBC), Seoul, Korea, 14–17 May 2019; pp. 227–235.
67. Kshetri, N. Can Blockchain Strengthen the Internet of Things? *IT Prof.* **2017**, *19*, 68–72. [CrossRef]
68. Al-Jaroodi, J.; Mohamed, N. Service-Oriented Middleware: A Survey. *J. Netw. Comput. Appl.* **2012**, *35*, 211–220. [CrossRef]
69. Leng, J.; Yan, D.; Liu, Q.; Xu, K.; Zhao, J.L.; Shi, R.; Wei, L.; Zhang, D.; Chen, X. ManuChain: Combining Permissioned Blockchain with a Holistic Optimization Model as Bi-Level Intelligence for Smart Manufacturing. *IEEE Trans. Syst. Man, Cybern. Syst.* **2019**, *50*, 182–192. [CrossRef]
70. Rizzardi, A.; Sicari, S.; Miorandi, D.; Coen-Porisini, A. Securing the Access Control Policies to the Internet of Things Resources Through Permissioned Blockchain. *Concurr. Comput. Pract. Exp.* **2022**, *34*, e6934. [CrossRef]

71. Park, J.; Kim, H.; Kim, G.; Ryou, J. Smart Contract Data Feed Framework for Privacy-Preserving Oracle System on Blockchain. *Computers* **2020**, *10*, 7. [CrossRef]
72. Duran, R.G.; Yarleque-Ruesta, D.; Belles-Munoz, M.; Jimenez-Viguer, A.; Munoz-Tapia, J.L. An Architecture for Easy Onboarding and Key Life-Cycle Management in Blockchain Applications. *IEEE Access* **2020**, *8*, 115005–115016. [CrossRef]
73. Hasan, M.; Starly, B. Decentralized Cloud Manufacturing-As-A-Service (Cmaas) Platform Architecture with Configurable Digital Assets. *J. Manuf. Syst.* **2020**, *56*, 157–174. [CrossRef]
74. Ochôa, I.S.; Silva, L.A.; de Mello, G.; da Silva, B.A.; de Paz, J.F.; González, G.V.; Garcia, N.M.; Leithardt, V.R.Q. PRICHAIN: A Partially Decentralized Implementation of UbiPri Middleware Using Blockchain. *Sensors* **2019**, *19*, 4483. [CrossRef]
75. Underwood, S. Blockchain Beyond Bitcoin. *Commun. ACM* **2016**, *59*, 15–17. [CrossRef]
76. Tapas, N.; Merlino, G.; Longo, F. Blockchain-based IoT-cloud authorization and delegation. In Proceedings of the 2018 IEEE International Conference on Smart Computing (SMARTCOMP), Taormina, Italy, 18–20 June 2018; pp. 411–416.
77. Wang, N.; Wang, B.; Liu, T.; Li, W.; Yang, S. A Middleware Approach to Synchronize Transaction Data to Blockchain. In Proceedings of the 2020 29th International Conference on Computer Communications and Networks (ICCCN), Honolulu, HI, USA, 3–6 August 2020; pp. 1–8.
78. Peng, Z.; Wu, H.; Xiao, B.; Guo, S. VQL: Providing Query Efficiency and Data Authenticity in Blockchain Systems. In Proceedings of the 2019 IEEE 35th International Conference on Data Engineering Workshops (ICDEW), Macao, China, 8–12 April 2019; pp. 1–6.
79. Zhou, E.; Sun, H.; Pi, B.; Sun, J.; Yamashita, K.; Nomura, Y. Ledgerdata Refiner: A Powerful Ledger Data Query Platform for Hyperledger Fabric. In Proceedings of the 2019 Sixth International Conference on Internet of Things: Systems, Management and Security (IOTSMS), Granada, Spain, 22–25 October 2019; pp. 433–440.
80. Liu, X.; Wang, W.; Guo, H.; Barenji, A.V.; Li, Z.; Huang, G.Q. Industrial Blockchain Based Framework for Product Lifecycle Management in Industry 4.0. *Robot. Comput. Manuf.* **2019**, *63*, 101897. [CrossRef]
81. Maxim, J. Onename Launches Blockchain Identity Product Passcard. *Bitcoin Mag.* **2015**. Available online: https://bitcoinmagazine.com/business/onename-launches-blockchain-identity-product-passcard-1431548450 (accessed on 19 July 2022).
82. Takahashi, R. *How Can Creative Industries Benefit from Blockchain?* McKinsey & Company: Hong Kong, China, 2017; Available online: http://bloomen.io/portfolio-item/mckinsey-can-creative-industries-benefit-blockchain/ (accessed on 19 July 2022).
83. Lu, Y.; Fu, Q.; Xi, X.; Chen, Z. Cloud Data Acquisition and Processing Model Based on Blockchain. *J. Intell. Fuzzy Syst.* **2020**, *39*, 5027–5036. [CrossRef]
84. Zupan, N.; Zhang, K.W.; Jacobsen, H.A. Demo: Hyperpubsub: A Decentralized, Permissioned, Publish/Subscribe Service Using Blockchains. In Proceedings of the Middleware '17: 18th International Middleware Conference, Las Vegas, NV, USA, 11–15 December 2017; pp. 15–16.
85. Tang, Y.; Zou, Q.; Chen, J.; Li, K.; Kamhoua, C.A.; Kwiat, K.; Njilla, L. ChainFS: Blockchain-Secured Cloud Storage. In Proceedings of the 2018 IEEE 11th International Conference on Cloud Computing (CLOUD), San Francisco, CA, USA, 2–7 July 2018; pp. 987–990.
86. Samaniego, M.; Deters, R. Zero-Trust Hierarchical Management in IoT. In Proceedings of the 2018 IEEE International Congress on Internet of Things (ICIOT), Bhimtal, India, 23–24 February 2018; pp. 88–95.
87. Hosen, A.S.M.S.; Singh, S.; Sharma, P.K.; Ghosh, U.; Wang, J.; Ra, I.-H.; Cho, G.H. Blockchain-Based Transaction Validation Protocol for a Secure Distributed IoT Network. *IEEE Access* **2020**, *8*, 117266–117277. [CrossRef]
88. Walas, F.; Redchuk, A. IIoT/IoT and Artificial Intelligence/Machine Learning as a Process Optimization Driver under Industry 4.0 Model. *J. Comput. Sci. Technol.* **2021**, *21*, e15. [CrossRef]
89. Tian, Y.; Li, T.; Xiong, J.; Bhuiyan, M.Z.A.; Ma, J.; Peng, C. A Blockchain-Based Machine Learning Framework for Edge Services in IIoT. *IEEE Trans. Ind. Inform.* **2021**, *18*, 1918–1929. [CrossRef]
90. Khalil, R.A.; Saeed, N.; Masood, M.; Fard, Y.M.; Alouini, M.-S.; Al-Naffouri, T.Y. Deep Learning in the Industrial Internet of Things: Potentials, Challenges, and Emerging Applications. *IEEE Internet Things J.* **2021**, *8*, 11016–11040. [CrossRef]
91. Hiessl, T.; Schall, D.; Kemnitz, J.; Schulte, S. Industrial Federated Learning—Requirements and System Design. In Proceedings of the PAAMS 2020: Highlights in Practical Applications of Agents, Multi-Agent Systems, and Trust-worthiness, the PAAMS Collection, L'Aquila, Italy, 7–9 October 2020; Communications in Computer and Information Science. Springer: Cham, Switzerland, 2020; Volume 1233. [CrossRef]
92. Liu, Y.; Garg, S.; Nie, J.; Zhang, Y.; Xiong, Z.; Kang, J.; Hossain, M.S. Deep Anomaly Detection for Time-Series Data in Industrial Iot: A Communication-Efficient on-Device Federated Learning Approach. *IEEE Internet Things J.* **2020**, *8*, 6348–6358. [CrossRef]
93. Lu, Y.; Huang, X.; Dai, Y.; Maharjan, S.; Zhang, Y. Blockchain and Federated Learning for Privacy-Preserved Data Sharing in Industrial IoT. *IEEE Trans. Ind. Inform.* **2019**, *16*, 4177–4186. [CrossRef]
94. Wang, T.; Ke, H.; Zheng, X.; Wang, K.; Sangaiah, A.K.; Liu, A. Big Data Cleaning Based on Mobile Edge Computing in Industrial Sensor-Cloud. *IEEE Trans. Ind. Inform.* **2019**, *16*, 1321–1329. [CrossRef]
95. Mathias, S.G.; Schmied, S.; Grossmann, D. Monitoring of Discrete Electrical Signals from Welding Processes Using Data Mining and Iiot Approaches. In Proceedings of the 2020 IEEE 32nd International Conference on Tools with Artificial Intelligence (ICTAI), Baltimore, MD, USA, 9–11 November 2020; pp. 911–916.
96. Rahmstorf, S. Rising hazard of storm-surge flooding. *Proc. Natl. Acad. Sci. USA* **2017**, *114*, 11806–11808. [CrossRef]

97. Yang, F.; Milosevic, D.; Cao, J. Optimising Column Family for OLAP Queries in HBase. *Int. J. Big Data Intell.* **2017**, *4*, 23–35. [CrossRef]
98. Leng, J.; Jiang, P.; Xu, K.; Liu, Q.; Zhao, J.L.; Bian, Y.; Shi, R. Makerchain: A blockchain with chemical signature for self-organizing process in social manufacturing. *J. Clean. Prod.* **2019**, *234*, 767–778. [CrossRef]
99. *ERC-721*; Non-Fungible Token Standard. Ethereum Foundation: Zuger, Switzerland, 2018.
100. Christidis, K.; Devetsikiotis, M. Blockchains and Smart Contracts for the Internet of Things. *IEEE Access* **2016**, *4*, 2292–2303. [CrossRef]
101. Leng, J.; Sha, W.; Lin, Z.; Jing, J.; Liu, Q.; Chen, X. Blockchained Smart Contract Pyramid-Driven Multi-Agent Autonomous Process Control for Resilient Individualised Manufacturing Towards Industry 5.0. *Int. J. Prod. Res.* **2022**, 1–20. [CrossRef]
102. Zhang, F.; Cecchetti, E.; Croman, K.; Juels, A.; Shi, E. Town Crier: An Authenticated Data Feed for Smart Contracts. In Proceedings of the CCS'16: 2016 ACM SIGSAC Conference on Computer and Communications Security, Vienna, Austria, 24–28 October 2016; pp. 270–282.
103. Liu, Q.; Zhang, H.; Leng, J.; Chen, X. Digital Twin-Driven Rapid Individualised Designing of Automated Flow-Shop Manufacturing System. *Int. J. Prod. Res.* **2018**, *57*, 3903–3919. [CrossRef]
104. Leng, J.; Chen, Z.; Sha, W.; Lin, Z.; Lin, J.; Liu, Q. Digital Twins-Based Flexible Operating of Open Architecture Production Line for Individualized Manufacturing. *Adv. Eng. Inform.* **2022**, *53*, 101676. [CrossRef]
105. Leng, J.; Zhou, M.; Xiao, Y.; Zhang, H.; Liu, Q.; Shen, W.; Su, Q.; Li, L. Digital Twins-Based Remote Semi-Physical Commissioning of Flow-Type Smart Manufacturing Systems. *J. Clean. Prod.* **2021**, *306*, 127278. [CrossRef]
106. Leng, J.; Jiang, P.; Liu, C.; Wang, C. Contextual Self-Organizing of Mass Individualization Process Under Social Manufacturing Paradigm: A Cyber–Physical–Social System Approach. *Enterp. Inf. Syst.* **2018**, *14*, 1124–1149. [CrossRef]
107. Zhao, R.; Yan, D.; Liu, Q.; Leng, J.; Wan, J.; Chen, X.; Zhang, X. Digital Twin-Driven Cyber-Physical System for Autonomously Controlling of Micro Punching System. *IEEE Access* **2019**, *7*, 9459–9469. [CrossRef]
108. Zhang, Y.; Ren, S.; Liu, Y.; Sakao, T.; Huisingh, D. A Framework for Big Data Driven Product Lifecycle Management. *J. Clean. Prod.* **2017**, *159*, 229–240. [CrossRef]
109. Leng, J.; Ye, S.; Zhou, M.; Zhao, J.L.; Liu, Q.; Guo, W.; Cao, W.; Fu, L. Blockchain-Secured Smart Manufacturing in Industry 4.0: A Survey. *IEEE Trans. Syst. Man, Cybern. Syst.* **2020**, *51*, 237–252. [CrossRef]
110. Rehman, M.H.U.; Ahmed, E.; Yaqoob, I.; Hashem, I.A.T.; Imran, M.; Ahmad, S. Big Data Analytics in Industrial IoT Using a Concentric Computing Model. *IEEE Commun. Mag.* **2018**, *56*, 37–43. [CrossRef]
111. Younan, M.; Houssein, E.H.; Elhoseny, M.; Ali, A.A. Challenges and Recommended Technologies for the Industrial Internet of Things: A Comprehensive Review. *Measurement* **2020**, *151*, 107198. [CrossRef]
112. Angelopoulos, A.; Michailidis, E.T.; Nomikos, N.; Trakadas, P.; Hatziefremidis, A.; Voliotis, S.; Zahariadis, T. Tackling Faults in the Industry 4.0 Era—A Survey of Machine-Learning Solutions and Key Aspects. *Sensors* **2019**, *20*, 109. [CrossRef] [PubMed]
113. Leng, J.; Ruan, G.; Song, Y.; Liu, Q.; Fu, Y.; Ding, K.; Chen, X. A Loosely Coupled Deep Reinforcement Learning Approach for Order Acceptance Decision of Mass-Individualized Printed Circuit Board Manufacturing in Industry 4.0. *J. Clean. Prod.* **2020**, *280*, 124405. [CrossRef]
114. Gao, Y.; Chen, Y.; Hu, X.; Lin, H.; Liu, Y.; Nie, L. Blockchain Based IIoT Data Sharing Framework for SDN-Enabled Pervasive Edge Computing. *IEEE Trans. Ind. Inform.* **2020**, *17*, 5041–5049. [CrossRef]
115. Derhamy, H.; Eliasson, J.; Delsing, J. IoT Interoperability—On-Demand and Low Latency Transparent Multiprotocol Translator. *IEEE Internet Things J.* **2017**, *4*, 1754–1763. [CrossRef]
116. Shahzad, Y.; Javed, H.; Farman, H.; Ahmad, J.; Jan, B.; Zubair, M. Internet of Energy: Opportunities, applications, architectures and challenges in smart industries. *Comput. Electr. Eng.* **2020**, *86*, 106739. [CrossRef]
117. Rahman, H.; Hussain, M.I. A Comprehensive Survey on Semantic Interoperability for Internet of Things: State-of-The-Art and Research Challenges. *Trans. Emerg. Telecommun. Technol.* **2020**, *31*, e3902. [CrossRef]
118. Aazam, M.; Zeadally, S.; Harras, K.A. Deploying Fog Computing in Industrial Internet of Things and Industry 4.0. *IEEE Trans. Ind. Inform.* **2018**, *14*, 4674–4682. [CrossRef]
119. Uddin, M.A.; Stranieri, A.; Gondal, I.; Balasubramanian, V. A Survey on the Adoption of Blockchain in Iot: Challenges and Solutions. *Blockchain Res. Appl.* **2021**, *2*, 100006. [CrossRef]
120. Huo, R.; Zeng, S.; Wang, Z.; Shang, J.; Chen, W.; Huang, T.; Wang, S.; Yu, F.R.; Liu, Y. A Comprehensive Survey on Blockchain in Industrial Internet of Things: Motivations, Research Progresses, and Future Challenges. *IEEE Commun. Surv. Tutor.* **2022**, *24*, 88–122. [CrossRef]
121. Guan, Z.; Lu, X.; Wang, N.; Wu, J.; Du, X.; Guizani, M. Towards Secure and Efficient Energy Trading in Iiot-Enabled Energy Internet: A Blockchain Approach. *Future Gener. Comput. Syst.* **2020**, *110*, 686–695. [CrossRef]
122. Wang, E.K.; Liang, Z.; Chen, C.M.; Kumari, S.; Khan, M.K. PoRX: A Reputation Incentive Scheme for Blockchain Consensus of IIoT. *Future Gener. Comput. Syst.* **2020**, *102*, 140–151. [CrossRef]
123. Khan, M.Z.; Alhazmi, O.H.; Javed, M.A.; Ghandorh, H.; Aloufi, K.S. Reliable Internet of Things: Challenges and Future Trends. *Electronics* **2021**, *10*, 2377. [CrossRef]
124. Singh, S.; Sharma, P.K.; Yoon, B.; Shojafar, M.; Cho, G.H.; Ra, I.-H. Convergence of Blockchain and Artificial Intelligence in Iot Network for the Sustainable Smart City. *Sustain. Cities Soc.* **2020**, *63*, 102364. [CrossRef]

125. Fuller, A.; Fan, Z.; Day, C.; Barlow, C. Digital Twin: Enabling Technologies, Challenges and Open Research. *IEEE Access* **2020**, *8*, 108952–108971. [CrossRef]
126. Lim, W.Y.B.; Xiong, Z.; Niyato, D.; Cao, X.; Miao, C.; Sun, S.; Yang, Q. Realizing the Metaverse with Edge Intelligence: A Match Made in Heaven. *IEEE Wirel. Commun.* **2022**, 1–9. [CrossRef]

machines MDPI

Article

Research on the High Precision Synchronous Control Method of the Fieldbus Control System

Lingyu Chen, Jieji Zheng, Dapeng Fan and Ning Chen *

College of Intelligence Science and Technology, National University of Defense Technology, Changsha 410073, China
* Correspondence: chenning007xm@126.com; Tel.: +86-138-0748-0956

Abstract: The synchronization control performance of the Fieldbus control system (FCS) is an important guarantee for the completion of multi-axis collaborative machining tasks, and its synchronization control accuracy is one of the decisive factors for the machining quality. To improve the synchronization control accuracy of FCS, this paper first makes a comprehensive analysis of the factors affecting synchronization in FCS. Secondly, by analyzing the communication model of linear Ethernet, a distributed clock compensation method based on timestamps is proposed to solve the asynchronous problem of communication data transmission in the linear ethernet bus topology. Then, based on the CANopen application layer protocol, the FCS communication and device control task collaboration method is proposed to ensure the synchronous control of multiple devices by FCS. Finally, an experimental platform is built for functional verification and performance testing of the proposed synchronization method. The results show that the proposed synchronization method can achieve a communication synchronization accuracy of 50 ns and a device control synchronization accuracy of 150 ns.

Keywords: fieldbus control system; synchronization control; distributed clock; ethernet fieldbus; CANopen

Citation: Chen, L.; Zheng, J.; Fan, D.; Chen, N. Research on the High Precision Synchronous Control Method of the Fieldbus Control System. *Machines* **2023**, *11*, 98. https://doi.org/10.3390/machines11010098

Academic Editors: Huosheng Hu, Pingyu Jiang, Gang Xiong, Timo R. Nyberg, Zhen Shen, Maolin Yang and Guangyu Xiong

Received: 24 November 2022
Revised: 4 January 2023
Accepted: 10 January 2023
Published: 11 January 2023

1. Introduction

With the development of Industry 4.0, intelligent, Fieldbus-based, and open industrial automation control systems have become the development direction in industrial manufacturing [1]. The FCS adopts industrial ethernet Fieldbus and standardized communication protocols, which can make the central controller (master), node controllers (slave), HMI, and other devices from a control network through a single Fieldbus. It realizes various functions such as industrial device control, data monitoring and acquisition, and open interconnection and is widely used in CNC processing, industrial robotics, and automatic production lines [2–4]. However, synchronization control of multiple devices in a Fieldbus network without a uniform time reference becomes an important challenge for the development of FCS [5]. In high-end equipment manufacturing industries such as numerical control processing with nano-level interpolation requirements, processing tasks need to be completed by multi-axis collaboration. The asynchronous control time of each axis will cause the actual machining path to deviate from the set path, which is related to the success or failure of the machining task [6,7]. Therefore, the synchronization accuracy of device control is the core index that determines the system's performance. Traditional industrial control systems mostly use centralized multi-axis motion control cards with simultaneous axis control and signal output in the same processor, thus ensuring the synchronization accuracy of control [8,9]. This system architecture is not in line with the development directions of standardization, openness, and interconnection of industrial control systems.

Establishing a uniform time reference is an effective means to improve the synchronization control accuracy of the system [10]. Time synchronization methods can be broadly classified into three types according to the principle of action: high-precision clock sources, satellite timing services, and network time synchronization [11]. The high-precision clock source approach is to integrate high-precision crystal oscillation circuits in each slave station to provide additional high-precision time information [12]. However, the system will generate unacceptable cumulative errors over long periods of operation. The satellite time service method uses the high-precision clock provided by GPS satellites as the unified time reference for the system [13]. Although the time accuracy meets the industrial control requirements, the cost is too high to be suitable for industrial production applications. The network time synchronization method, represented by the precision network time synchronization method, represented by the precision time protocol (PTP), achieves time synchronization by continuously correcting and compensating the time of each slave station in the network [14]. PTP is based on a real-time ethernet Fieldbus and requires no additional hardware, making it an effective means of achieving high-precision synchronous control of FCS. More and more researchers have adopted the PTP method to improve the synchronous control performance of industrial control systems [15–17]. Lam D. et al. proposed a method to eliminate the frequency drift factor of the master clock crystal oscillator based on the PTP method, which improved the time synchronization accuracy of industrial ethernet by 30% [18]. Chen C. et al. designed the Modbus protocol and PTP protocol together to solve the time synchronization problem of industrial wireless sensor actuators [19]. Seo Y. et al. conducted an adaptive estimation of network noise and clock drift interference in the PTP-based system, which improved the robustness of the system synchronization clock [20]. This method of recording physical layer hardware transceiver time can obtain 100 ns level synchronization accuracy and has been widely used in major real-time Ethernet buses, such as PROFINET IRT [21], POWERLINK [22], and EtherCAT [23]. In addition, some scholars have further improved the time synchronization accuracy by using clock dynamic compensation. Buhr S. et al. proposed a timestamp measurement method combining the PI controller and PHY chip clock phase relationship, which can achieve sub-nanosecond synchronization accuracy [24]. Gong F. et al. used the Kalman filtering method to model the clock offset variation characteristics, which can improve synchronization accuracy in a temperature-variation environment [25]. Qing L. et al. proposed a dynamic delay-corrected clock synchronization algorithm for the transmission delay asymmetry between the system clock and the local clock of each slave station to improve the clock synchronization accuracy [26]. The above time synchronization methods can establish a uniform time reference and achieve high time synchronization accuracy. However, the synchronized control of industrial equipment by FCS is also influenced by its system architecture, especially the timing sequence between system communication and control behavior. For this paper, based on analyzing the factors affecting the synchronization of FCS device control, a time reference based on distributed clocks is established, and a collaborative method of communication scheduling and device control is designed to achieve high-precision synchronization control of devices.

This paper is organized as follows. In Section 2, the overall architecture and working principle of FCS are analyzed, and the influencing factors of unsynchronized control of system equipment are obtained. In Section 3, the mechanism of real-time Ethernet bus communication transmission delay is analyzed, and a distributed clock synchronization method based on timestamp compensation is proposed. In Section 4, the communication scheduling and device control timing sequence in FCS is designed based on the CANopen application layer protocol and forms the FCS device synchronization control method. In Section 5, a typical FCS experimental platform is built to verify the distributed clock synchronization accuracy and device synchronization control accuracy. Finally, Section 6 concludes this paper and points out follow-up work. The proposed FCS device synchronization control method can achieve 50 ns communication transmission synchronization accuracy and 150 ns device synchronization control accuracy. This method can improve the multi-axis collaborative processing accuracy of the FCS system and also provide a reference for the design of the synchronization control function of a similar industrial automation control system.

2. Analysis of Synchronization Factors in FCS

This section analyzes the overall system architecture and communication and data transmission principles of FCS. Then, by analyzing the timing sequence of the system control task execution, the influencing factors that lead to the desynchronization of device control by FCS are summarized, which provides a theoretical basis for the synchronization method proposed subsequently.

2.1. Basics of FCS

Typical FCS adopts an architecture that separates system logic control from device control. The master executes the user program logic and sends the control instructions to each slave through the real-time ethernet bus, and the slaves control their connected devices according to the control instructions. Figure 1 illustrates the principle of FCS architecture. The integrated development environment (IDE) is a comprehensive software platform for users to configure system hardware, edit applications, compile code, download, debug, and perform human-machine interaction. The master receives the logic code downloaded from the IDE, compiles it as a program organization unit (POU), and runs it periodically in a real-time thread of the operating system. The results of the POU operation are sent to the slaves via the real-time Ethernet bus. The slave receives the command from the master, runs the device control algorithm to control its peripherals and uploads the sensor data and its status parameters to the master. In addition, for application scenarios with network expansion requirements, the OPC-UA module of the master station can be interconnected with the cloud through the OPC-UA gateway. It is worth noting that since the master integrates all slave databases and the real-time ethernet bus adopts standardized communication protocols, the system function implementation does not depend on the hardware, reflecting the standardized characteristics of FCS. This paper focuses on the synchronization problem of FCS for device control.

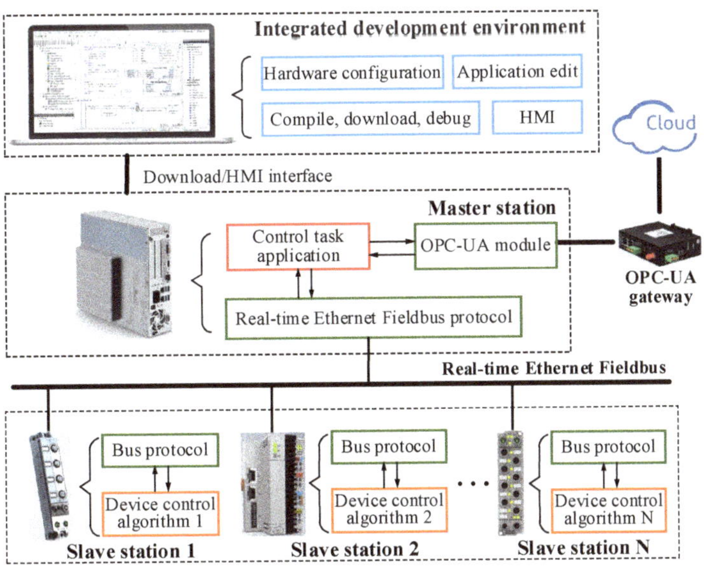

Figure 1. Schematic diagram of the FCS architecture.

2.2. Communication and Control Principles of FCS

The control of various devices in the system by the FCS is essentially the process of periodic data interaction between master and slave in a real-time ethernet field bus. Ethernet buses can be divided into linear, tree, star, and other forms according to the topology. Due to the simple structure and high real-time data transmission characteristics of linear topology, this paper uses a linear ethernet architecture for synchronous analysis.

The communication structure and data transmission principle of FCS using linear topology are shown in Figure 2. The FCS communication structure can be divided into a physical layer, a data link layer, and an application layer. The functions and data transmission principles of each layer are as follows.

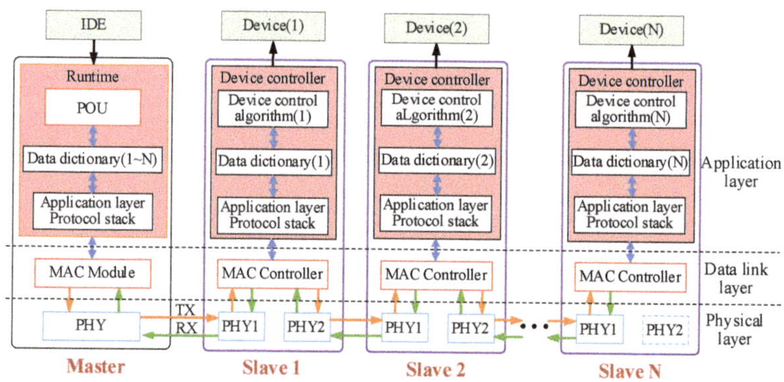

Figure 2. Schematic diagram of the communication structure of FCS.

- Physical layer: Provide physical media for ethernet data transmission between master and slave, mainly including a PHY circuit, Category 5 cable (CAT-5), registered jack 45 (RJ45), etc. To ensure the independence of each slave in the linear topology, the physical layer of the slave includes two independent network ports.
- Data link layer: primarily consists of a MAC controller that integrates a real-time data transmission protocol. The real-time data transmission protocol processes physical layer data and transmits it directly to the application layer, bypassing the network layer and transport layer.
- Application layer: It mainly consists of the master runtime and the slave device controller. Runtime consists mainly of POU, data dictionary, and an application layer protocol stack. The POU runs the user application periodically and stores the operation results in the master data dictionary. The application layer protocol stack updates each slave data dictionary by sending ethernet frames according to the set communication period, where the master data dictionary is essentially the sum of the data dictionaries of all slaves. The slave device controller updates its own data dictionary by exchanging communication data with the MAC controller and executes device control algorithms to control its connected devices according to the data dictionary parameters.

According to the system communication architecture and data transmission principle, the typical timing sequence of system control task execution is shown in Figure 3. When the master enters cycle T, it sends one ethernet frame to obtain the status of each slave and sensor data as the input condition for POU. Then, the application layer stack sends the results of the POU to each slave. Each slave reads the data from the ethernet frame and outputs the control signal to the device after processing the protocol and running the control algorithm. The above process is repeated for each cycle until the system stops.

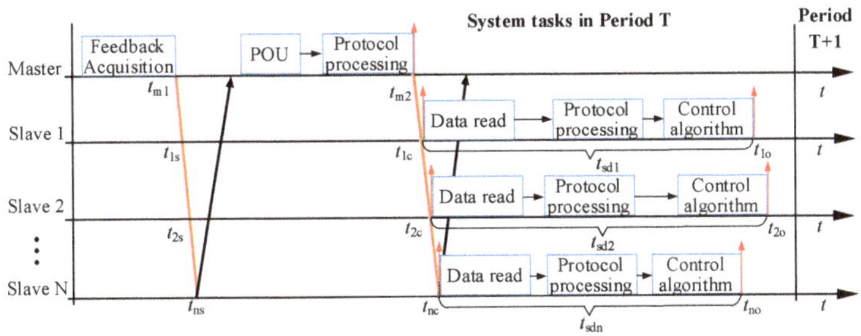

Figure 3. Schematic diagram of a typical FCS task timing sequence.

2.3. Synchronization Error Analysis

Synchronous control mainly refers to the time consistency of the motion state of the devices in the industry controlled by FCS under the same command, such as the simultaneous action of switches and synchronous motion of motors. However, since the response time of different hardware to the command varies, this paper only considers the time synchronization of the control signal output from the slave. According to the task timing sequence analysis in Figure 3, the main factors affecting system synchronization control are as follows:

- Factor 1: Communication transmission delay. The master sends data frames containing the control commands for this cycle to the slaves at t_{m2}. Since the communication link length of each slave is different, the time t_{1c}~t_{nc} for receiving ethernet frames from slaves 1~N is different. This causes the start time of the slave control task to be out of synchronization.
- Factor 2: Slave task processing time differences. After receiving the master data frame, the slave station needs to perform data reading, protocol processing, and control algorithm operation tasks. However, different functional types of slaves may receive different lengths of data and experience differences in protocol processing times and inconsistent levels of control algorithm complexity. This leads to different task processing times (t_{sd1}~t_{sdn}) for slaves 1~N as well, which affects the synchronization of the slave control signal output.
- Factor 3: Timeliness differences in slave feedback data. Due to the different sensor sampling times of each slave, the feedback data of each slave read by the master at t_{m1} via Ethernet frame may not be consistent in time. This may lead to some deviation in the POU operation results, causing the control logic to be out of synchronization.

By analyzing the above synchronization influencing factors, the main reason leading to Factor 1 is that the communication data transmission is not synchronized, while Factor 2 and Factor 3 are mainly caused by the unreasonable coordination of system communication and control tasks in time. Therefore, to eliminate the asynchronous factor of Factor 1~3, this paper focuses on the synchronization of communication transmission and the coordination of communication and device control.

3. Fieldbus Communication Synchronization Method Based on Distributed Clock

3.1. Analysis of Communication Transmission Delay

Figure 4 shows the communication delay principle between two adjacent slaves A and B in a linear network consisting of one master and N slaves. During the forward transmission, the Ethernet frame is received by the PHY0_rx port of slave A, then processed by the MAC controller and sent out from the PHY1_tx port. The PHY0_rx port of slave B receives the data sent from the PHY1_tx port of slave A and performs the same processing as slave A. When the ethernet frame is transmitted to the slave N, it starts to be transmitted in the reverse direction and finally returns to the master.

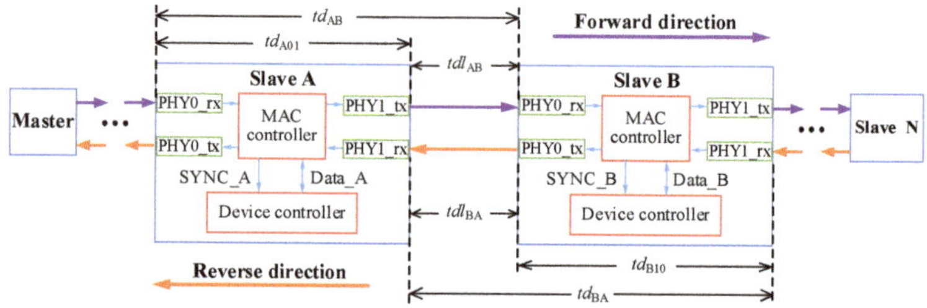

Figure 4. Schematic diagram of the communication transmission delay in a linear network.

During forward transmission, after the Ethernet frame arrives at slave A, the MAC controller needs to perform data reading, feedback data uploading, and retransmission operations. The time consumed in this process is called station processing delay, which is represented by td_{A01}. The time consumed for the ethernet frame to travel between the PHY_tx port of slave A and the PHY0_rx port of slave B is called the cable transmission delay and is denoted by tdl_{AB}. Therefore, the overall delay time td_{AB} for two adjacent slaves A and B to receive Ethernet frames sent by the master can be expressed as follows.

$$td_{AB} = td_{A01} + tdl_{AB} \tag{1}$$

Similarly, during reverse transmission, the transmission delay td_{BA} of slave stations A and B can be expressed as follows:

$$td_{BA} = td_{B10} + tdl_{BA} \tag{2}$$

The above analysis shows that communication frame transmission in a linear network is mainly affected by cable transmission delay and station processing delay, resulting in different times for each slave station to receive the same communication frame. The phenomenon is objective and unavoidable. However, the communication data ultimately serves the device controller. A SYNC output signal with adjustable output time is designed in the MAC controller. The synchronization of SYNC signal generation is ensured by accurately measuring the cable transmission delay and station processing delay of each slave station and compensating for the time of the SYNC output signal. The device controller of each slave station reads communication data from the MAC controller according to the SYNC signal, which ensures synchronized communication.

3.2. *Method of Communication Delay Measurement*

Define the first slave connected to the master as the reference slave. The communication delay of each other slave is based on the reference slave. To analyze the time consumption relationship of the communication transmission process more intuitively, a linear network communication transmission model is established, as shown in Figure 5.

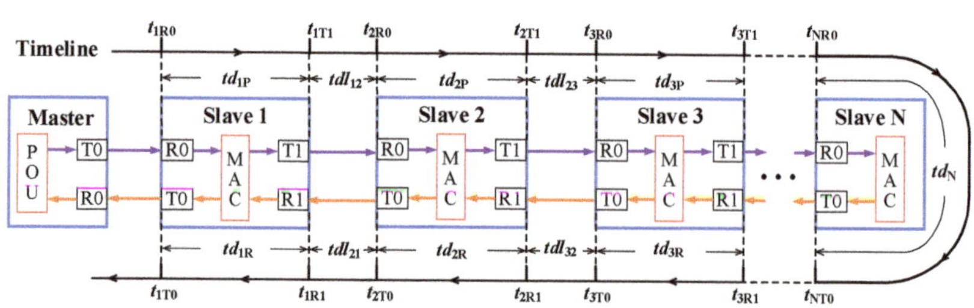

Figure 5. Communication transmission delay model for the linear network.

The model describes the time relationship of the communication frame transmitted in the Fieldbus network consisting of 1 master and N slaves. The definition of each parameter in Figure 5 is shown in Table 1.

Table 1. Parameter definition table for communication transmission delay model.

NO.	Name	Definition
1	t_{NR0}	The time the data frame arrives at port R0 of slave N.
2	t_{NT1}	The time the data frame leaves the T1 port of slave N.
3	t_{NR1}	The time the data frame arrives at port R1 of slave N.
4	t_{NT0}	The time the data frame leaves the T0 port of slave N.
5	td_{NP}	Station processing delay of slave N during forward transmission.
6	td_{NR}	Station processing delay of slave N during reverse transmission.
7	td_{N}	Station processing delay of the last slave.
8	tdl_{AB}	Transmission delay of communication cables between slaves A and B.

According to the model analysis, the cable transmission delay and station processing delay are calculated as follows.

1. Cable transmission delay calculation

The connection distance between slave stations in industrial applications is generally short, and the transmission asymmetry time of CAT-5 is about 0.1 ns/m, which can be considered symmetrical for cable transmission delay. Therefore, for the adjacent slaves i and j, the following relationship is satisfied.

$$tdl_{ij} = tdl_{ji} \tag{3}$$

Considering the different time bases of each slave, let the time deviation between slave 1 and slave 2 be $t_{offset12}$, and then the cable transmission delay between slave 1 and slave 2 can be represented by Equation (4).

$$\begin{cases} tdl_{12} = (t_{2R0} + t_{offset12}) - t_{1T1} \\ tdl_{21} = t_{1R1} - (t_{2T0} + t_{offset12}) \end{cases} \tag{4}$$

Substituting Equation (3) into Equation (4), the cable transmission delay between slave 1 and slave 2 can be obtained as:

$$tdl_{12} = tdl_{21} = [(t_{1R1} - t_{1T1}) - (t_{2T0} - t_{2R0})]/2 \tag{5}$$

The time deviation $t_{offset12}$ in the result of Equation (5) is counteracted. Without loss of generality, the cable transmission delay from slave i to the reference slave is the sum of the cable transmission delays between all neighboring slave nodes in its forward link. The calculation method is shown in Equation (6).

$$tdl_{1i} = \begin{cases} \sum\limits_{j=1,k=2}^{j=i-1,k=i} \frac{[(t_{jR1} - t_{jT1}) - (t_{kT0} - t_{kR0})]}{2}, i \geq 2 \\ 0 \qquad\qquad , i = 1 \end{cases} \tag{6}$$

2. Station processing delay calculation

The station processing delay reflects the time that the communication frame stays inside the slave station. It is the total time from the moment when the read signal of the PHY chip at the receiving port is valid to the moment when the write signal of the PHY chip at the sending port is valid. Since all station processing occurs within the slave station, it can be recorded directly using local time. According to the model, the station processing delay of slave station i can be calculated as follows:

$$\begin{cases} td_{iP} = t_{iT1} - t_{iR0} \\ td_{iR} = t_{iT0} - t_{iR1} \end{cases} \quad (7)$$

When communication frames are transmitted in the reverse direction, the slave stations do not exchange data. Therefore, only td_{iP} is considered in the station delay compensation. All station processing delays td_{1i} from station i to the reference slave can be expressed by Equation (8).

$$td_{1i} = \sum_{j=1}^{j=i-1} (t_{jT1} - t_{jR0}) \quad (8)$$

According to the calculation results of Equations (6) and (8), it is easy to obtain the communication delay between slave station i and the reference slave station. The master can write the cable transmission delay and station processing delay of each slave into each MAC controller as important parameters for slave clock synchronization.

3.3. Distributed Clock Synchronization

In the Fieldbus network, the local clock of each slave station is affected by the power-on time, environment temperature, crystal precision, and other conditions showing dynamic change characteristics. The distributed clock makes the system time consistent by correcting the time of each slave to match the moment of the reference slave. Figure 6 shows the principle of distributed clock compensation.

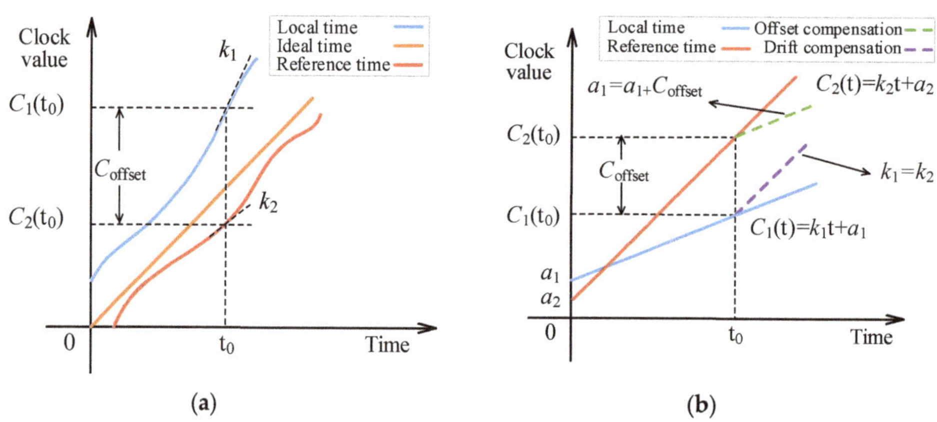

Figure 6. Schematic Diagram of Distributed Clock Compensation: (**a**) actual curve of local time and reference time slave; (**b**) curve of local time and reference time after compensation.

In Figure 6a, the local and reference clocks exhibit dynamic changes with time. The offset of the local and reference clocks at time t_0 is C_{offset}. In addition, there may be a difference between the change rate of the local clock k_1 and the change rate of the reference clock k_2 at t_0. The difference in the change rate of the clock will cause the clock curve to drift and may cause C_{offset} to increase over time.

Although the clock curve is not an ideal straight line, it can be approximated as a straight line over a short period, as shown in Figure 6b. Calculate the difference between the offset value and the drift value of the local clock and the reference clock, and compensate the local clock to synchronize the local clock with the reference clock.

1. Clock offset compensation

The basic idea of clock offset compensation is to add the offset $C_{offset}(t)$ of the local clock and reference clock at time t to the local clock as the compensation value. The expression of the local time C_{local} of the slave station is as Equation (9).

$$C_{local}(t) = kt + a \tag{9}$$

where k is the initial value of the local time change rate, and a is the initial value of the local time.

The ethernet frame containing the clock compensation message sent by the master arrives at the reference slave at t_{ref} time. The reference slave local time, $C_{ref}(t_{ref})$, is written into the data frame as timestamp information. When the frame arrives at slave i at time t_i, the time deviation $C_{offset}(t_i)$ between slave i and the reference slave can be expressed as Equation (10).

$$C_{offset}(t_i) = C_{ref}(t_i) - C_{local}(t_i) \tag{10}$$

The local time $C_{ref}(t_i)$ of the reference slave at time t_i cannot be directly obtained from slave station i. However, $C_{ref}(t_i)$ can be calculated using timestamp information, $C_{ref}(t_{ref})$, and Equations (6) and (8), as follows:

$$C_{ref}(t_i) = C_{ref}(t_{ref}) + tdl_{1i} + td_{1i} \tag{11}$$

where tdl_{1i} and td_{1i} are the cable transmission delay and station processing delay between slave i and the reference slave, respectively.

According to Equations (9)–(11), the time deviation $C_{offset}(t_i)$ of slave i from the reference slave at time t_i can be obtained as follows:

$$C_{offset}(t_i) = C_{ref}(t_{ref}) + tdl_{1i} + td_{1i} - kt_i - a \tag{12}$$

The value of $C_{offset}(t_i)$ was added to the initial value of local time and used as the new initial synchronization time a_{new}. In this way, the new local time can be formed as follows:

$$C_{local}(t) = kt + a + C_{offset}(t_i) = kt + a_{new} \tag{13}$$

2. Clock drift compensation

The clock drift phenomenon is mainly caused by the difference between the slave local clock change rate and the reference slave change rate. Periodically calculating the reference slave time change rate and using it as the slope of the local clock curve is an effective method for drift compensation.

Let t_k and t_{k+1} be the times at which the two clock compensation messages arrive at slave i, respectively. The local time change rate k_i of the slave during this period can be expressed as follows:

$$k_i = [C_{local}(t_{k+1}) - C_{local}(t_k)]/(t_{k+1} - t_k) \tag{14}$$

According to Equation (11), the reference slave time change rate k_{ref} can also be obtained as Equation (15).

$$k_{ref} = [C_{ref}(t_{k+1}) - C_{ref}(t_k)]/(t_{k+1} - t_k) \tag{15}$$

The change rate of local time can be corrected to k_{new} according to the relationship of the ratio of k_{ref} and k_i. The local time after drift compensation and offset compensation can be expressed as Equation (16).

$$C_{local}(t) = k\frac{[C_{ref}(t_{k+1}) - C_{ref}(t_k)]}{[C_{local}(t_{k+1}) - C_{local}(t_k)]}t + a_{new} = k_{new}t + a_{new} \tag{16}$$

In summary, the use of time drift and offset compensation methods to establish a system-distributed clock can make each slave and the reference slave time maintain a high degree of synchronization and solve the bus communication transmission delay problem.

4. Collaborative Method of System Communication and Device Control

In an architecture where logic control and device control are separated, ensuring the temporal rationality of master communication tasks and slave control tasks is the key to synchronous control of FCS. This section proposes a system communication and device control collaboration method based on the CANopen application layer protocol. Through the design of the system communication data transmission method, communication scheduling behavior, and device control timing sequence, the synchronization of device control by FCS is ensured.

4.1. CANopen Data Transmission Method over Ethernet Links

The transmission mode of the CANopen protocol is based on the CAN bus, and each communication frame can only carry data information for one slave station. In this way, the master station needs to send multiple communication frames at a time to complete the command transmission within a communication cycle. There is no doubt that this will cause the time spent acquiring communication data from each slave station to be out of sync.

To ensure synchronous transmission of communication data, each slave's data can be packaged into different messages and integrated into ethernet frames. However, this method will generate more invalid data in ethernet frames, such as message headers, interrupt responses, etc. Therefore, this section uses the data mapping method to design the ethernet data frame format to improve the data transmission efficiency of the communication frame and to achieve more slave data transmission in the limited Ethernet data space. The mapping principle is shown in Figure 7.

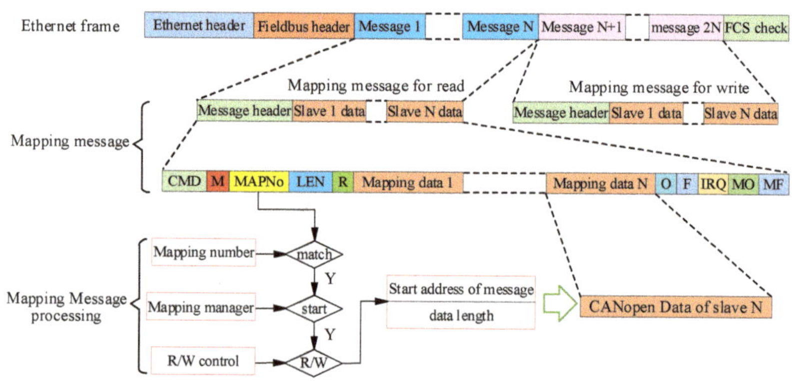

Figure 7. Schematic diagram of the Ethernet frame data structure.

In Figure 7, the messages sent by the master to each slave are combined into one read message, and the upload messages of each slave are combined into one write message. The slave reads and writes the mapped message using the configuration information, such as the mapping number, starting address, and data length. This data mapping transmission method improves the data transmission efficiency of ethernet frames and allows for a larger loading space for CANopen protocol data.

In the mapping message, the data of each slave is organized according to the standard CANopen data frame format, which consists of 11 bytes of space for Cobid, data length, and data information, respectively. The format of the CANopen frame transmission in the mapping message is shown in Figure 8.

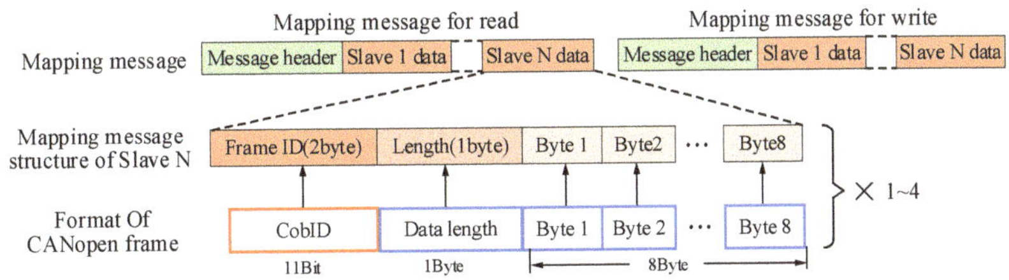

Figure 8. Schematic diagram of CANopen frames integrated with the mapping message.

CANopen specifies a variety of communication objects, including NMT, SDO, PDO, etc. The number of CANopen frames required for NMT, SDO, and PDO transmission varies, and the master needs to configure different lengths of read/write message mapping space according to the type of frames to be sent. NMT and SDO only need one CANopen frame space for a single transmission. However, PDO can be configured into 1–4 groups as required, including T_{PDO} and R_{PDO}. When composing ethernet frames, the master needs to reserve data space of the appropriate size according to the number of PDOs of each slave. In this way, the CANopen application layer data can be transmitted over the ethernet Fieldbus.

4.2. System Communication Scheduling Method

To ensure real-time communication, the answer to the current command in the linear network can only be transmitted back through the next ethernet frame. This delayed response mechanism is not conducive to the timely upload of slave data, which may cause the output logic of the POU to be out of sync. To make the POU and the device control algorithm time reasonable, this section designs the communication frame scheduling methods for system initialization, pre-operation, and operation status according to CANopen.

1. Initialization state scheduling

In this stage, after the slave completes the initialization of the protocol stack, a bootup frame containing the slave node information will be generated and written to the MAC controller. The master sends a query frame to read the bootup information of each slave in the bus network. If the node information conforms to the hardware configuration file, it will enter the pre-operation state. If it does not conform, it indicates that the bus connection or slave station is abnormal.

2. Pre-operation state scheduling

In this state, the master performs status queries, parameter configuration, PDO configuration, and other operations on the slave station through SDO. Figure 9 shows the communication timing sequence of the slave device controller, MAC controller, and POU in the pre-operation state.

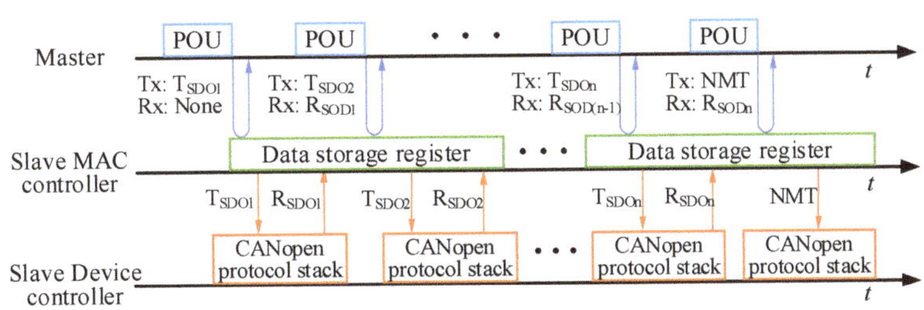

Figure 9. Schematic diagram of system communication timing sequence in the pre-operation state.

After entering the pre-operation state, the master sends the T_{SDO1} command to each slave. At this time, the read-back information may still be bootup information, which can be ignored. After the device controller of the slave reads the T_{SDO1} data in the MAC controller, it is processed by the CANopen protocol stack, and the SDO response R_{SDO1} is written to the MAC controller. When the master sends the T_{SDO2} command, it can read back the R_{SDO1} information in the MAC controller. At the end of the pre-operation state, the master station sends an NMT command to read back R_{SDOn} information and switch the slave station to the operation state.

3. Operation state scheduling

The communication data on operation status is time-sensitive and closely related to the POU control logic. In this state, the master periodically sends a synchronization frame (syncf) to read the R_{PDO} information of the slave and sends instructions after the POU operation to each slave through T_{PDO} frames. The key to system synchronization control is to ensure that the POU input parameters are the latest data of the slave in this cycle and that the POU output instructions can be executed by the slave in this cycle.

Figure 10 shows the system communication timing sequence in the operating state. The communication scheduling task in a period can be divided into three stages, and the tasks of each stage are the same. Taking communication period 1 as an example, the communication scheduling process of master and slave at each stage is described in detail below.

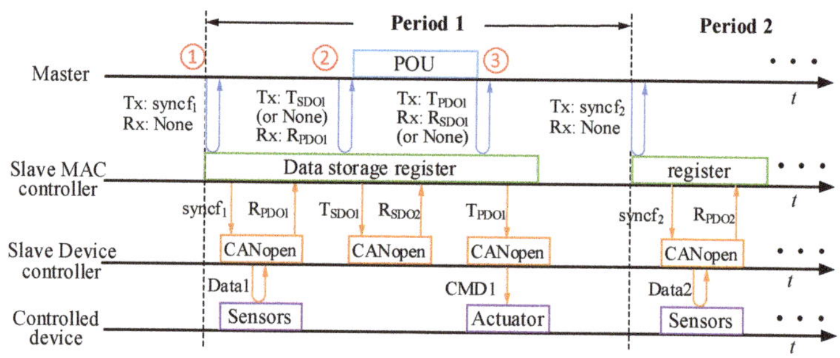

Figure 10. Schematic diagram of system communication timing sequence in operation.

- Stage 1: At the beginning of communication period 1, the master sends $syncf_1$ to each slave station, and the read-back information is ignored. After the device controller of the slave reads $syncf_1$, it updates the sensor data and writes it to the MAC controller through the R_{PDO1} frame of CANopen, waiting for the next frame to be read back.
- Stage 2: The master sends a query frame to read back the R_{PDO1} information in the MAC controller of each slave. If an SDO frame is requested to be sent during the period, the query frame is replaced by a T_{SDO1} data frame. After reading back R_{PDO1}, the master takes the R_{PDO1} data as a POU input parameter and runs POU.
- Stage 3: The master sends the POU operation result through the T_{PDO1} frame. The slave runs the device control algorithm according to the instructions of T_{PDO1} and controls the actuator.

The above communication scheduling mechanism can ensure the logic and independence of sensor data acquisition, POU operation, and the device control algorithm in time.

4.3. Slave Control Timing Sequence Design

The CANopen-based communication scheduling method provides a good timing sequence reference for the implementation of the slave control algorithm. Combined with the distributed clock synchronization signal SYNC, the execution time of each slave task can be unified. The timing sequence relationship between system communication and slave station control tasks under operation status is shown in Figure 11.

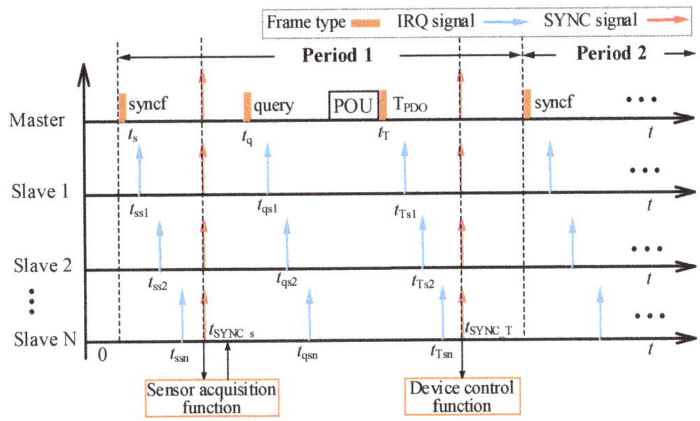

Figure 11. Slave control task timing sequence diagram in operation state.

In Figure 11, the MAC controller of each slave station will generate an IRQ signal after receiving the communication frame, informing the device controller that the communication data can be read. Due to the communication transmission delay, the interrupt signal IRQ time generated by each slave MAC controller after receiving the communication frame is inconsistent. However, the distributed clock can be used to generate a SYNC signal simultaneously after the communication frame is received by all slaves. The SYNC signal of the synchronization frame is used as the trigger for the sensor acquisition function of each slave, and the SYNC signal of the T_{PDO} frame is used as the trigger for the device control algorithm.

It is worth noting that when setting the SYNC generation time, it must be ensured that the last slave in the Fieldbus has finished receiving the data frames. Otherwise, it will cause some slaves to be out of synchronization with the control commands of other slaves. According to Equations (6)–(8), the SYNC signal generation time t_{SYNC} of any frame should satisfy the following condition:

$$t_{SYNC} > t_{send} + tdl_{1n} + td_{1n} + (t_{nT0} - t_{nR0}) \tag{17}$$

where n is the last slave node number, and t_{send} is the time when the communication frame is received by the reference slave.

Due to the different complexity of the control algorithms of the slave, the execution time is quite different. To have the same time for each slave to output control signals, the control signal generated by the control algorithm in this cycle is output at the same time as the SYNC signal in the next cycle, which can ensure output synchronization. Figure 12 shows the timing sequence principle of slave communication and control task execution.

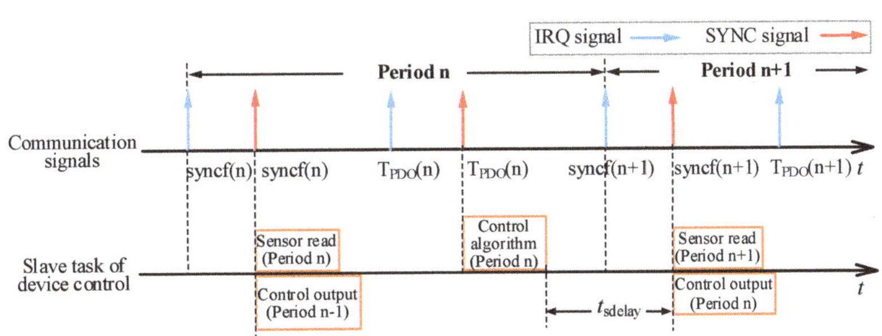

Figure 12. Slave control task timing sequence diagram in operation state.

During period n, the slave station executes the sensor reading task in this period according to the SYNC signal of syncf(n) and simultaneously outputs the control signal of the previous period. When the SYNC signal of the T_{PDO}(n) frame is valid, the slave executes the control algorithm of this cycle and latches the output signal. The latched control signal will be output when the SYNC signal of the syncf(n + 1) frame of the n + 1 period is valid. The synchronous output method after the latch will generate the output delay of t_{sdelay}. However, the communication period of a real-time Ethernet bus-based control system is generally in the order of milliseconds, and t_{sdelay} does not affect the control performance of the system.

In summary, the logical synchronization of the system and the synchronization of the device control can be ensured by the collaborative design of the system communication scheduling and device control tasks.

5. Experiment

In this section, an experimental platform is constructed to verify the high-precision synchronization control method of the device proposed in this paper. Figure 13 shows the self-developed FCS experimental platform, which consists of a power supply, a master station, four slave stations, an HMI display, and a PC with IDE installed. The master adopts Hi3559a as a CPU chip that integrates the Linux Ubuntu16.04 operating system and forms a linear network with 4 slaves via ethernet. The slave uses the FPGA of HME-P1P6060N0TF784C as the MAC controller and the Soc chip HME-M7A12N0F484I7 with the integrated cortex-m3 core as the device controller. The clock synchronization protocol proposed in this paper is designed in Verilog, and integrated in the HME-P1P60600N0TF784C chip. It forms the Ethernet physical layer with the general-purpose 100 M PHY chip, network transformer, and M12 industrial connector. The Codesys software is installed on the PC as the system IDE. The HMI display provides the human-machine interaction interface. The power supply provides 24 V DC power to the master station and four slave stations. The device interface of each slave is led from the interface board to facilitate oscilloscope observation of the distributed clock synchronization signal SYNC as well as the device control signal.

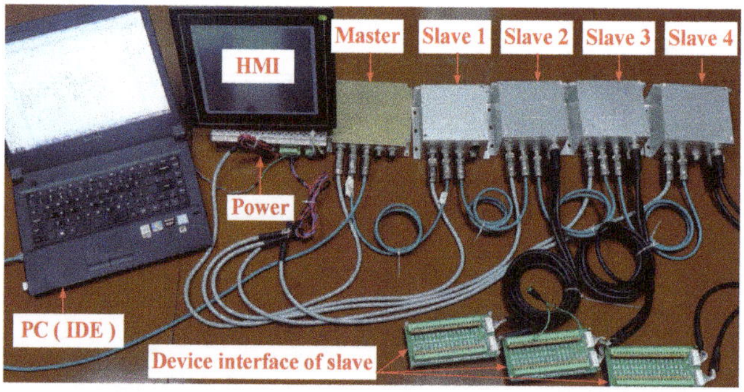

Figure 13. FCS experimental platform diagram.

A multi-axis robot application scenario is simulated using the experimental platform shown in Figure 13. The master controls each slave to generate the orthogonal pulse signal required by the robot joint driver at the same time to control each joint axis. Set the system communication cycle to 1 ms, and design the POU logic in the IDE to make the slave generate 20 pulses with a frequency of 50 KHz in each communication cycle.

The SYNC signal is directly generated by the distributed clock, and the time error of the SYNC signal of each slave directly reflects the synchronization accuracy of the distributed clock. After the system is powered on and enters the operation state, the SYNC signals of slave 1~4 are shown in Figure 14.

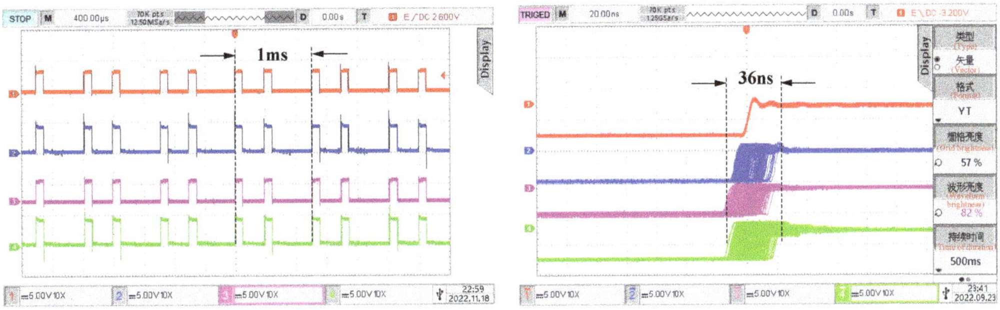

(**a**) (**b**)

Figure 14. System distributed clock synchronization test result diagram: (**a**) SYNC signal waveform of each slave under a 400 μs time scale; (**b**) SYNC signal waveform of each slave under a 20 ns time scale.

Figure 14a shows the oscilloscope waveform of the SYNC signal of each slave on a 400 μs time scale. It can be found that the SYNC signal generates two edge jumps within 1 ms, representing the synchronization frame and T_{PDO} frame, respectively. SYNC signals show consistency and periodicity in time. Figure 14b shows the rise edge detail of the SYNC signal under a 20 ns time scale. The rise time of the SYNC signal of each slave has deviation and jitter, and the jitter time error is about 36 ns. This time jitter is caused by uncertainty factors such as timestamp measurement accuracy, data transmission delay jitter, PHY chip data reception, transmission jitter, etc.

Time synchronization of the device control signal output of each slave is the key to multi axis coordinated control. Use the oscilloscope to observe the pulse output ports of each slave station to test the synchronization performance of the equipment. The test results are shown in Figure 15.

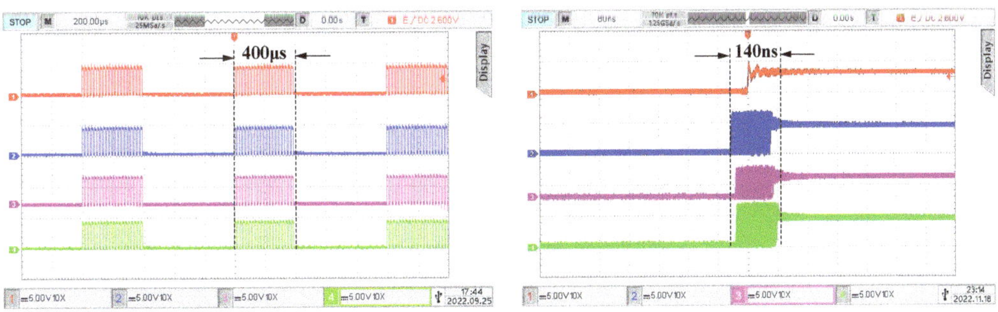

(**a**) (**b**)

Figure 15. System device control synchronization test result diagram: (**a**) pulse output waveform of each slave under a 200 μs time scale; (**b**) pulse output waveform of each slave under a 80 ns time scale.

When the time scale of the oscilloscope is 200 us, slave 1~4 generates pulse signals with a total time width of 400 μs in a communication cycle. The pulse output time of each slave is the same, and the number and frequency of pulses are the same as the logic settings of the master station, as shown in Figure 15a. When the time scale of the oscilloscope is amplified to 80 ns, it can be observed that there is deviation and jitter in the jumping time of each slave station pulse signal, and the overall jitter time range is about 140 ns, as shown in Figure 15b. According to the slave control timing sequence in Section 4, the time error of the output pulse of each slave shall be theoretically consistent with the SYNC signal error. However, because the pulse output involves MCU software modules and related hardware circuits, the synchronization error is larger than the theoretical value.

To verify the independence of system synchronization accuracy and running time, the experimental platform in Figure 13 was tested for seven consecutive days. The synchronization error of the SYNC signal and pulse output signal is tested several times a day, and its maximum value is recorded. The experimental results are shown in Table 2.

Table 2. Synchronization performance test table during continuous system operation.

Test Time	Synchronization Error of SYNC	Synchronization Error of Pulse
Day 1	36 ns	140 ns
Day 2	37 ns	143 ns
Day 3	37 ns	140 ns
Day 4	35 ns	137 ns
Day 5	40 ns	146 ns
Day 6	37 ns	141 ns
Day 7	36 ns	140 ns

From the test data in Table 2, it can be seen that the system communication transmission synchronization error and device control synchronization error remain stable during the continuous operation of the system. The fluctuation range of communication transmission synchronization error is 4 ns, and the fluctuation range of device control synchronization error is 8 ns.

In summary, the synchronization control method proposed in this paper can ensure that the synchronization accuracy of FCS communication transmission is less than 50 ns and the synchronization accuracy of device control is less than 150 ns.

6. Conclusions

This paper focuses on the research of the device synchronization control method of a Fieldbus control system. Through the analysis of the composition principle of typical FCS and the communication control process of FCS based on a real-time ethernet field bus, it is concluded that the main factors affecting the synchronization control of the system are the communication transmission delay and the desynchrony of communication and control logic. To eliminate the impact of communication transmission delay on synchronization control, this paper analyzes the communication transmission mechanism in the linear ethernet topology and uses timestamps to measure the cable transmission delay and static processing delay. The distributed synchronization clock of the system is established according to the delay measurement results, which provides a unified time benchmark for each slave of the system. In addition, aiming at the problem that the communication tasks and control logic of the system are not synchronized, this paper uses the CANopen application layer protocol to design the system communication data transmission, communication task scheduling, and device control timing, forming a collaborative method of system communication and device control. Finally, a typical FCS experimental platform is built to verify the synchronization control method proposed in this paper. The experimental results show that the synchronization method proposed in this paper can make the communication transmission synchronization accuracy less than 50 ns and the equipment control synchronization accuracy less than 150 ns. The synchronization method proposed in this paper meets the requirements of industrial applications for the synchronization accuracy of the control system and provides a reference for FCS synchronization-related research. In the future, we will further study the improvement of synchronization accuracy.

Author Contributions: Conceptualization, L.C. and D.F.; methodology, L.C. and D.F.; software, L.C.; validation, L.C., J.Z. and N.C.; formal analysis, D.F.; investigation, L.C.; resources, D.F. and N.C.; data curation, J.Z.; writing—original draft preparation, L.C.; writing—review and editing, L.C.; visualization, N.C.; supervision, D.F.; project administration, D.F.; funding acquisition, N.C. All authors have read and agreed to the published version of the manuscript.

Funding: This research was funded by the National Natural Science Foundation of China (Grant No. U19A2072).

Data Availability Statement: Not applicable.

Conflicts of Interest: The authors declare no conflict of interest.

References

1. Zeng, P.; Wang, Z.; Jia, Z.; Kong, L.; Li, D.; Jin, X. Time-slotted software-defined Industrial Ethernet for real-time Quality of Service in Industry 4.0. *Futur. Gener. Comput. Syst.* **2019**, *99*, 1–10. [CrossRef]
2. Erwinski, K.; Paprocki, M.; Grzesiak, L.M.; Karwowski, K.; Wawrzak, A. Application of Ethernet Powerlink for Communication in a Linux RTAI Open CNC system. *IEEE Trans. Ind. Electron.* **2012**, *60*, 628–636. [CrossRef]
3. Wang, S.; Ouyang, J.; Li, D.; Liu, C. An Integrated Industrial Ethernet Solution for the Implementation of Smart Factory. *IEEE Access* **2017**, *5*, 25455–25462. [CrossRef]
4. Danielis, P.; Skodzik, J.; Altmann, V.; Schweissguth, E.; Golatowski, F.; Timmermann, D.; Schacht, J. Survey on real-time communication via ethernet in industrial automation environments. In Proceedings of the 2014 IEEE Emerging Technology and Factory Automation, Barcelona, Spain, 16–19 September 2014.
5. Liang, G.; Li, W. Some Thoughts and Practice on Performance Improvement in Distributed Control System Based on Fieldbus and Ethernet. *Meas. Control.* **2016**, *49*, 109–118. [CrossRef]
6. Jia, Z.-Y.; Ma, J.-W.; Song, D.-N.; Wang, F.-J.; Liu, W. A review of contouring-error reduction method in multi-axis CNC machining. *Int. J. Mach. Tools Manuf.* **2018**, *125*, 34–54. [CrossRef]
7. Li, Y.-H.; Zheng, Q.; Yang, L. Multi Hydraulic Motors Synchronized Control Based on Field Bus with FlexRay Protocol. *Adv. Sci. Lett.* **2012**, *9*, 603–608. [CrossRef]
8. Zhong, G.; Shao, Z.; Deng, H.; Ren, J. Precise Position Synchronous Control for Multi-Axis Servo Systems. *IEEE Trans. Ind. Electron.* **2017**, *64*, 3707–3717. [CrossRef]
9. Shi, T.; Liu, H.; Geng, Q.; Xia, C. Improved relative coupling control structure for multi-motor speed synchronous driving system. *IET Electr. Power Appl.* **2016**, *10*, 451–457. [CrossRef]
10. Liang, G. Control and communication co-design: Analysis and practice on performance improvement in distributed meas-urement and control system based on fieldbus and Ethernet. *ISA Trans.* **2015**, *54*, 169–192. [CrossRef]
11. Kurata, N. Development and application of an autonomous time synchronization sensor device using a chip scale atomic clock. *Sens. Transducers* **2018**, *219*, 17–25.
12. Chen, R.; Liu, Y.; Li, X.; Fan, D.; Yang, Y. High-precision time synchronization based on common performance clock source. In Proceedings of the 2019 14th IEEE International Conference on Electronic Measurement & Instruments (ICEMI), Changsha, China, 1–3 November 2019.
13. Guo, H.; Crossley, P. Design of a Time Synchronization System Based on GPS and IEEE 1588 for Transmission Substations. *IEEE Trans. Power Deliv.* **2016**, *32*, 2091–2100. [CrossRef]
14. Bello, L.L.; Steiner, W. A Perspective on IEEE Time-Sensitive Networking for Industrial Communication and Automation Systems. *Proc. IEEE* **2019**, *107*, 1094–1120. [CrossRef]
15. Pedretti, D.; Bellato, M.; Isocrate, R.; Bergnoli, A.; Brugnera, R.; Corti, D.; Corso, F.D.; Galet, G.; Garfagnini, A.; Giaz, A.; et al. Nanoseconds Timing System Based on IEEE 1588 FPGA Implementation. *IEEE Trans. Nucl. Sci.* **2019**, *66*, 1151–1158. [CrossRef]
16. Idrees, Z.; Granados, J.; Sun, Y.; Latif, S.; Gong, L.; Zou, Z.; Zheng, L. IEEE 1588 for Clock Synchronization in Industrial IoT and Related Applications: A Review on Contributing Technologies, Protocols and Enhancement Methodologies. *IEEE Access* **2020**, *8*, 155660–155678. [CrossRef]
17. Popescu, D.A.; Moore, A.W. Measuring Network Conditions in Data Centers Using the Precision Time Protocol. *IEEE Trans. Netw. Serv. Manag.* **2021**, *18*, 3753–3770. [CrossRef]
18. Lam, D.K.; Yamaguchi, K.; Nagao, Y.; Kurosaki, M.; Ochi, H. An improved precision time protocol for industrial WLAN communication systems. In Proceedings of the 2016 IEEE International Conference on Industrial Technology (ICIT), Taipei, Taiwan, 14–17 March 2016.
19. Chen, C.-H.; Lin, M.-Y.; Tew, W.-P. Wireless fieldbus networking with precision time synchronization for a low-power WSAN. *Microprocess. Microsyst.* **2022**, *90*, 104509. [CrossRef]
20. Seo, Y.; Son, K.; An, G.; Nam, K.; Chang, T.; Kang, S. Improved Time-Synchronization Algorithm Based on Direct Compen-sation of Disturbance Effects. *Sensors* **2019**, *19*, 3499. [CrossRef]
21. Gutierrez-Rivas, J.L.; Torres-Gonzalez, F.; Ros, E.; Diaz, J. Enhancing White Rabbit Synchronization Stability and Scalability Using P2P Transparent and Hybrid Clocks. *IEEE Trans. Ind. Inform.* **2021**, *17*, 7316–7324. [CrossRef]

22. Romanov, A.; Slepynina, E. Real-time Ethernet POWERLINK communication for ROS. Part II. Hardware and software. In Proceedings of the 2020 Ural Smart Energy Conference (USEC), Ekaterinburg, Russia, 13–15 November 2020.
23. Paprocki, M.; Erwiński, K. Synchronization of Electrical Drives via EtherCAT Fieldbus Communication Modules. *Energies* **2022**, *15*, 604. [CrossRef]
24. Buhr, S.; Kreisig, M.; Protze, F.; Ellinger, F. Subnanosecond Time Synchronization Using a 100Base-TX Ethernet Transceiver and an Optimized PI-Clock Servo. *IEEE Trans. Instrum. Meas.* **2020**, *70*, 1–8. [CrossRef]
25. Gong, F.; Sichitiu, M.L. Temperature compensated Kalman distributed clock synchronization. *Ad Hoc Netw.* **2017**, *62*, 88–100. [CrossRef]
26. Qing, L.; Xinyang, R. IEEE 1588 and a dynamic delay correction clock synchronization algorithm. In Proceedings of the 2019 IEEE 5th International Conference on Computer and Communications (ICCC), Chengdu, China, 6–9 December 2019.

machines

Article

A Novel Method for LCD Module Alignment and Particle Detection in Anisotropic Conductive Film Bonding

Tengyang Li [1,2], Feng Zhang [1,3,*], Huabin Yang [1,3], Huiyuan Luo [1] and Zhengtao Zhang [1,3]

1 CAS Engineering Laboratory for Intelligent Industrial Vision, Institute of Automation, Chinese Academy of Sciences, Beijing 100190, China
2 The School of Artificial Intelligence, University of Chinese Academy of Sciences, Beijing 100049, China
3 CASI Vision Technology Co., Ltd., Luoyang 471000, China
* Correspondence: feng.zhang@ia.ac.cn

Abstract: In this paper, we propose a misalignment correct method and a particle detection algorithm to improve the accuracy in the quality inspection of the LCD module after the anisotropic conductive film (ACF) bonding. We use only one camera to acquire images of multiple positions in order to establish the transformation from the image space to the world coordinate. Our method can accurately determine the center of rotation of the carrier table and calculate the deviation of position and angle of the tested module. Compared to traditional ways that rely on multiple cameras to align the large-sized product, our method has the advantages of simple structure, low cost, and fast calibration process. The particle detection is performed after positioning all bumps of the bonded module. The gray morphology-based algorithm is developed to detect the extreme point of every particle and refine the particle result through blob analysis. This method reduces the over-checking rate and performs better on the detection precision for dense particles. We verify the effectiveness of our proposed methods in our experiments. The alignment error can be less than 0.05 mm, and the accuracy of the particle detection is 93% while the recall rate is 92.4%.

Keywords: conductive particles; LCD module; ACF bonding; automated optical inspection; visual-alignment

Citation: Li, T.; Zhang, F.; Yang, H.; Luo, H.; Zhang, Z. A Novel Method for LCD Module Alignment and Particle Detection in Anisotropic Conductive Film Bonding. *Machines* **2023**, *11*, 49. https://doi.org/10.3390/machines11010049

Academic Editors: Pingyu Jiang, Gang Xiong, Timo R. Nyberg, Zhen Shen, Maolin Yang and Guangyu Xiong

Received: 30 November 2022
Revised: 27 December 2022
Accepted: 28 December 2022
Published: 1 January 2023

1. Introduction

Liquid crystal display (LCD) panels have been widely used in smartphones, car monitors, and other industries. To fit the proliferated applications with different sizes and shapes, LCD modules are expected to be more integrated, thinner, and of higher resolution. Anisotropic conductive film (ACF) bonding technology, as the key process in the production of LCD modules, is extremely important in order to achieve a higher signal density and smaller overall package, which can enable the LCD modules more light-weighted and miniaturized. To reduce the size of the module, circuit components such as integrated circuits (ICs) and flexible printed circuits (FPCs) are expected to be connected to display panels at a higher level of integration.

ACF is an important material that serves as a connector to achieve such a level of integration. The authors of [1–3] examined the mechanical reliability of the ACF and considered it as a low-cost and reliable material that can be used in semiconductors packaging. As shown in Figure 1, with the ACF, multiple components can be integrated into the display panel to form a compact LCD module. Such a material can achieve both a mechanical and electrical link between the substrates of peripheral circuits to that of the LCD.

Figure 2a shows a detailed bonding process, in which the ACF is first attached to the panel substrates, and the substrates of ICs or FPCs can be further glued to it to realize a reliable connection between components. This process is called ACF bonding. As Figure 2b shows, the ACF mainly composes of conductive particles and polymer resin. The conductive particles provide the ability of conductivity, while the resin works as an adhesive to hold the LCD panel tightly with other circuits. ACF allows the two components, namely, the LCD panel and ICs or FPCs in our application, to create a reliable bond and enable electrical interconnection. Similar conducting polymer and metal nanoparticle ink have also been developed in [4,5] to fabricate materials with higher conductivity, thus improving the electrical performance of printed electronics. Mainly used processes include bonding integrated circuits (ICs) or flexible printed circuits (FPCs) onto the glass substrates, which are called chip-on-glass (COG) and flex-on-glass (FOG), respectively.

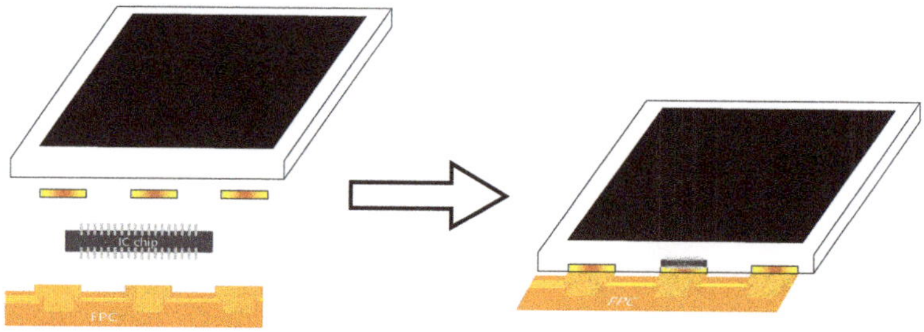

Figure 1. ACF can reduce the size of the LCD module and achieve a higher level of integration.

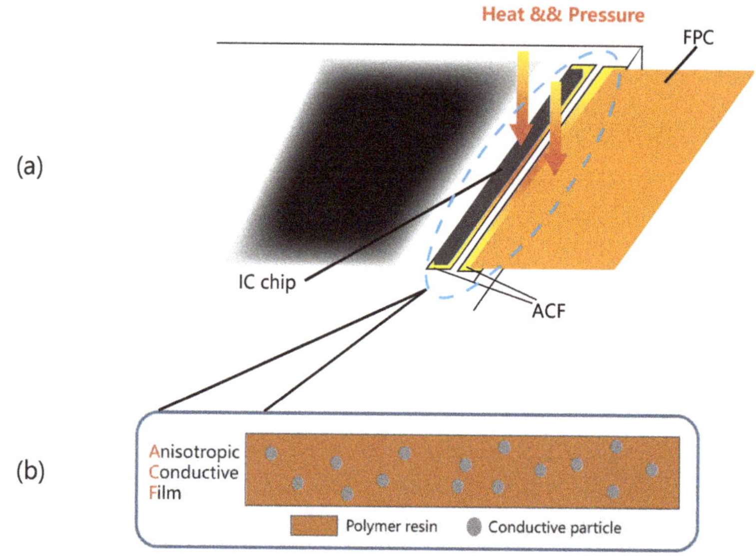

Figure 2. (**a**) Demonstrates the diagram of the bonding process; (**b**) is zoomed in to show the composition of the ACF.

The conductivity of the ACF is explained in Figure 3a, where the ACF is sandwiched between the two substrates. After a period of heat and pressure, the indium tin oxide (ITO) bumps of the substrates are bonded closer, thus trapping some conductive particles. The interconnection is established by these trapped particles, and the power and signal can be transmitted from external components into the display module. After the ACF bonding process, the appearance of the bump areas taken by the camera is shown in Figure 3b. Conductive particles on the bump areas can be viewed from those taken images. As conductivity is a significant indicator of the quality of LCD modules, extensive studies have been conducted to investigate the factors that contribute to poor conductivity. The traditional method directly tests the resistance of the LCD module after the bonding process, which is time-consuming and may inflict damage to test modules. According to studies [6–11], the number and distribution of conductive particles are essential factors to the conductivity. Therefore, it has become a popular method for LCD manufacturing to verify the conductivity by calculating the number of conductive particles. However, the manual inspection method depends on sophisticated instruments to perform the detection, and due to the low efficiency of labor work, random sampling is usually used in the industry. Such a sampling method is less accurate, and with the mass production of LCD panels, inspection methods that rely on human labor cannot satisfy the growing manufacturing scale. Meanwhile, due to the small size of conductive particles, whose diameter is normally 3–5 μm, it is difficult to obtain high-resolution information about particles using human eyes or traditional instruments.

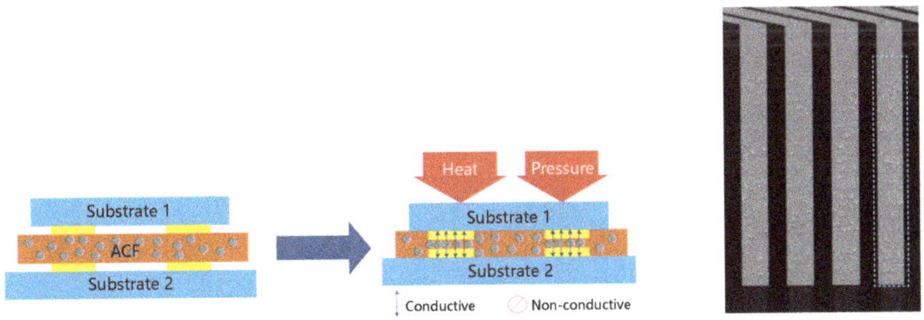

(a) (b)

Figure 3. (**a**) Explains the conductivity principle of the ACF; (**b**) shows part of the bump areas after the bonding process.

To improve the detection efficiency and accuracy without potential contact damage, the automated optical inspection (AOI), equipped with novel optical imaging methods and machine vision algorithms, has been developed by studies [12–20]. The AOI primarily uses the image sensor to acquire high-quality digital images of products and then draws upon a machine vision algorithm to identify the target objects in those images based on the features of the target. Many studies attempt to align the object module to the fiducial position to ensure a consistent imaging condition thus helping improve the detection result. For a large module, multiple cameras-based methods [21–24] are preferred because of higher precision. However, such methods usually repeatedly calculated the offset to reduce the alignment error, for example, an iterative algorithm proposed in [21]. Moreover, studies including [22,24] required multiple shots and movements for alignment, and they assumed a fixed center of rotation of the platform, which cannot be applied to other applications when the module sizes change. The utilization of a specialized marker and alignment layers were employed by the author of [25] in order to achieve precise positioning of the microlens within the array. In our proposed misalignment correction method, one single camera is used to take images of the feature mark on the both ends of the LCD module. This method can enhance the accuracy of alignment while reducing the alignment time by determining the center of rotation and calculating the displacement of the tested module in one stage.

It remains challenging to accurately and quickly detect the number of particles on the small bumps relative to a large entire image of the LCD module. The author of [13] combined differential interference contrast (DIC) prism with the CCD camera and achieved effective and high-contrast imaging of particles. The particles in the image represented spheres composing a bright part and a dark part. The author of [12] improved the Prewitt mask to calculate the image gradient and extract the extreme points using Otsu thresholding. Another study [26] also took the gray distribution of the particles as the feature to separate the original image into a light and a dark part. They then selected the more informative part by comparing the image entropy of the two parts. After clustering gray values in the selected image part, the center regions of particles were determined by the first two values in clusters. However, the aforementioned methods perform poorly in detecting particles when imaging illumination is weak or the particles are heavily overlapped. The gradient-based method is sensitive to the gradient change which makes it easy to recognize more particles than are actually there, especially along edge areas of the image. The clustering-based method has an unstable result in varying illumination conditions and leads to a low detection rate compared to the total number of particles to be checked. In this paper, we also propose a novel particle detection algorithm to improve the detection accuracy of the ACF task. This algorithm emphasizes the region of each particle based on the gray dilation morphology and then restores all particles by finding the extreme points of particles. The main contributions of this article are listed below.

(1) A novel misalignment correction method is proposed to determine the transition and rotation of the carrier table to ensure a consistent imaging area, which can be achieved by taking twelve images in one iteration. The requirement for assembly accuracy of the alignment module is reduced through the proposed method.
(2) A robust and fast detection algorithm is presented for checking the number of conductive particles.
(3) A complete AOI system is constructed to meet the demand for in-line process inspection, which can perform miscellaneous tasks including alignment calibration and correction and particle detection.

This paper is organized as follows. In Section 2, the system design is presented, followed by the misalignment correction method and particle detection algorithm. Details of the experiment results are provided in Section 3. Finally, a conclusion and limitation of the current work are given in Section 4.

2. Materials and Methods

2.1. Automatic Inspection System

The architecture and major modules of the inspection system are shown in Figure 4. Table 1 lists the parameters and manufacturers of the main modules of the automatic inspection system. The loading and unloading modules are placed at both ends to connect with an existing assembly line. Specifically, the load unit receives LCD modules from the upstream line after they complete the bonding process, while non-defective products can be conveyed to the next stage through unload unit. As shown in Figure 4, the system mainly includes five core modules. The conveyor module contains a robotic gripper equipped with a suction cup to softly attach the detected material from upstream and then lay it on the carrier table. The pre-alignment camera is then triggered to acquire the image of the LCD module at both ends of the corner, where we search the mark pattern as the feature point for calibration and alignment calculation afterward. As shown in Figure 5, the mark pattern is a special shape usually printed on the circuit board of the module and can be used as a feature point because of its distinctiveness. The carrier table can be controlled to translate and rotate according to the result of misalignment correction. After the alignment process, the CCD camera is initiated to scan the LCD module and acquire the high-resolution image of the entire bump region where the particle detection algorithm is used to check the number and distribution of conductive particles. The CCD line scan camera is adopted because of the heavily disproportionate aspect ratio of the bonding area of the LCD module, for example, can be 3200 × 60,000. Finally, the good product is conveyed to the unloading unit while the not-good product is filtered to the NG output basket.

As we can see, the imaging quality of conductive particles is important to the detection result. In the industrial setting, however, the line scan camera may struggle to maintain a consistent distance from the module being tested due to the constraints of the practical mechanical structure. Fluctuations in distance can cause the camera to lose focus, resulting in a blurry image of particles. In order to maintain a proper depth of field, typically within a range of 5–8 μm for our instrument, a laser displacement sensor is implemented to enable non-contact distance measurement. This sensor is able to help the line scan camera adaptively adjust the distance relative to the target region and ensure the quality of the image.

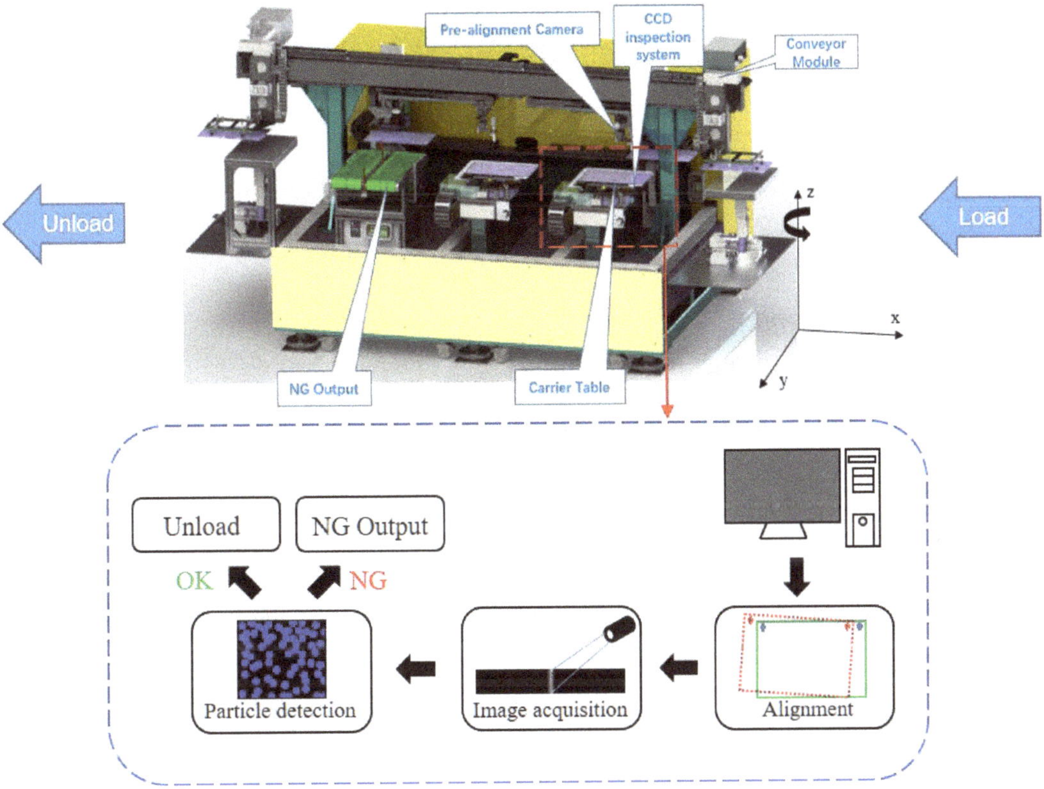

Figure 4. An overview of the automatic inspection system which mainly includes conveyor modules, pre-alignment modules, CCD inspection modules, carrier modules, and NG output modules. Tested LCD modules are first aligned by the misalignment correction algorithm, then the bump images are acquired by the CCD line scan camera, and finally, the conductive particles are detected.

Table 1. Parameters and manufacturers of major modules in the inspection system.

Category	Model Number	Manufacturer	Properties
X-axis linear actuator	GTH5 series	TOYO	Stroke length: 500 mm
Y-axis linear actuator	ETH14 series	TOYO	Stroke length: 500 mm
Θ-axis direct drive rotary motor	LD series	ZCOE	Angular accuracy: ±30″ Angular repeatability: ±2.5″
Pre-alignment camera	MV-CA013 series	HIKROBOT	Resolution: 1280 × 1024 Max. frame rate: 90 fps
TDI line scan camera	VT-3K7C series	Vieworks	Resolution: 3200 Line rate: 100 khz
Laser displacement sensor	ZS-HLDC/LDC Series	Omron	Max. resolution: 0.25 μm/px Max. response speed: 110 μs

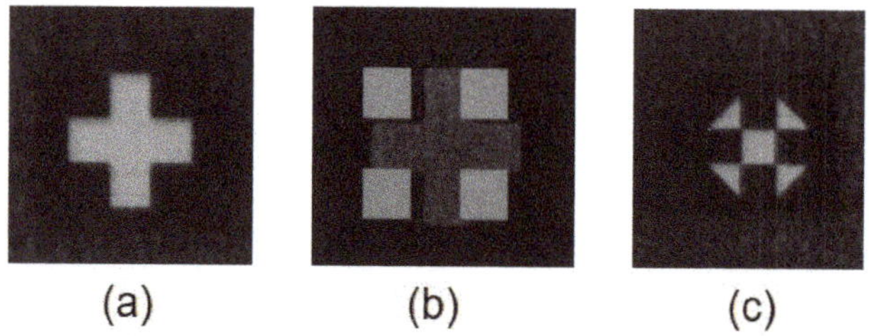

Figure 5. Different shapes of marks: (**a**) represents a cross mark; (**b**) represents a dark cross sandwiched between four bright squares; (**c**) represents a square surrounded by four triangles. The mark shape can be used according to practical manufacturer setting. We can extract the center of the marker as the key point for calibration and alignment.

2.2. Misalignment Correction

The pre-alignment camera installed on the carrier table is responsible for this process. The major procedure of the alignment correction is presented in Figure 6. A reference image is selected to prepare a matching template as a shape model. It normally chooses a distinguished shape in the region of interest (ROI) of the image as the feature mark. Figure 5 shows some different types of marks in the COG and FOG bonding process. Shape-based pattern search is a technique that detects reference marks automatically. We use this way to generate a shape model by specifying the range of rotations of the mark, the contrast of the local gray value, and the minimum size of the object. In most cases, we choose the center of the mark as the feature point after the matching process.

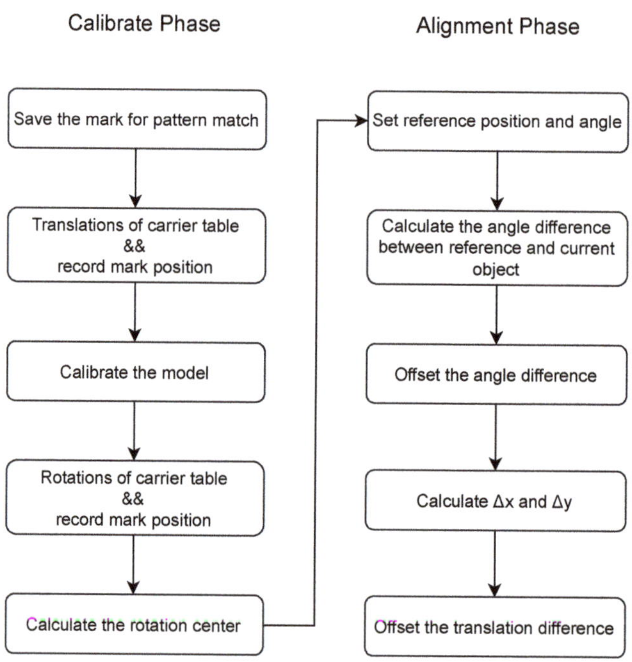

Figure 6. This flowchart indicates the steps to simultaneously calibrate the camera and calculate the center of rotation of the machine.

A 12-point calibration method is introduced to auto-calibrate the camera system and calculate the rotation center of the carrier table simultaneously, from which nine points are used to calculate a homogeneous transformation matrix based on Zhang's method [27], and three points for finding the center of rotation. Such twelve points are derived from different images that are acquired in nine positions and three different angles. Figure 7 shows a sample of twelve images, in which the feature points of the mark are searched based on the described pattern search technique. The center of the mark region is chosen as the feature point. After recording the coordinates of feature points, the corresponding positions of these feature points in the world coordinate were then calculated through the homogeneous transformation matrix.

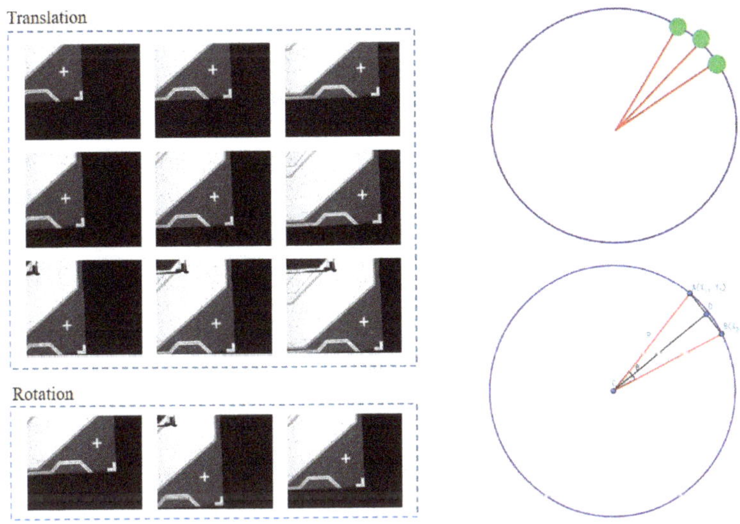

(a) (b)

Figure 7. (**a**) Uses the 12 point misalignment correction method to determine the center of rotation and to calculate the offset of translation. These twelve sample images are taken by a pre-alignment camera installed above the carrier table; (**b**) calculates the physical rotation center on the occasion of a large radius and a small angular interval.

The homogeneous transformation matrix stands for a correspondence between different coordinate systems which can be shown in Equation (1). The image point and the transformed point in the real world are denoted as (x_i, y_i), (X_i, Y_i) in homogeneous coordinates.

$$\begin{pmatrix} X_i \\ Y_i \\ 1 \end{pmatrix} = HMat2d \cdot \begin{pmatrix} x_i \\ y_i \\ 1 \end{pmatrix} \quad (1)$$

A simplified transformation matrix that only consists of translation and rotation is represented as the Equation (2). T_x and T_y are the translation in the X- and Y- axis directions and ra–rd denote the rotation parameters.

$$HMat2d = \begin{pmatrix} ra & rb & T_x \\ rc & rd & T_y \\ 0 & 0 & 1 \end{pmatrix} \quad (2)$$

For a test subject, nine translation images and three rotation images are captured by controlling the motion of the carrier table. In each position, the coordinate of the feature point in the image space is measured by searching reference marks in view of an image while the corresponding location in the real world is recorded. The rotation angle of the carrier table is also included during this procedure. Nine of these pairs of points are used to approximate the transformation matrix which is expected to minimize the distance error between those pairs of points following the Equation (3).

$$min\left(\sum_i \left\| \begin{pmatrix} X_i \\ Y_i \\ 1 \end{pmatrix} - HMat2d \cdot \begin{pmatrix} x_i \\ y_i \\ 1 \end{pmatrix} \right\|^2\right) \tag{3}$$

Once the calibration process is finished, one point in image space can be transformed into that in the real world, thus enabling the calculation of the physical rotation center.

The rest three points are then offered to find the center of rotation. As a rotating angle can have an effect on the position of points in both X- and Y-axis directions, it is necessary to correct the angle deviation before translating the tested object to align with the reference one. Due to a different design of the mechanic structure, the rotation center of the carrier table can vary a lot, which leads to different impacts on the deviation of points. The main idea of calculating the rotation center of the carrier table, known as fitting a circle, is to obtain multiple points by rotating the carrier table and calculating the point of the circle which is supposed to cover those points in the real world. This method is based on an assumption that a pure rotational motion should form a track of a circle for a point on the plane. Given that we can acquire the physical rotation angle of the carrier table, the center of the circle can be calculated by each two of the three points as well as their included angle.

As demonstrated in Figure 7b, $A(X_1, Y_1)$ and $B(X_2, Y_2)$ are two points transformed from image space to the real-world coordinates. The midpoint $D(X_0, Y_0)$ of the line AB through (X_1, Y_1) and (X_2, Y_2) is $(\frac{X_1+X_2}{2}, \frac{Y_1+Y_2}{2})$, from which we can form the parametric equation of perpendicular bisector through C and D.

$$\begin{aligned} X &= X_0 + t\Delta X \\ Y &= Y_0 + t\Delta Y \end{aligned} \tag{4}$$

Because line CD is perpendicular to line AB, the slope of CD is reciprocal to that of AB, which can be expressed as

$$\frac{\Delta Y}{\Delta X} = -\frac{1}{\frac{Y_1 - Y_2}{X_1 - X_2}} = -\frac{X_1 - X_2}{Y_1 - Y_2} \tag{5}$$

Therefore, Equation (4) can be transformed into

$$\begin{aligned} X &= \frac{X_1 + X_2}{2} + t(Y_1 - Y_2) \\ Y &= \frac{Y_1 + Y_2}{2} + t(X_1 - X_2) \end{aligned} \tag{6}$$

Additionally, we can obtain the following equation from the lengths of AD and CD and their included $\angle ACD$

$$\begin{aligned} AD &= \left(X_1 - \frac{X_1 + X_2}{2}\right)^2 + \left(Y_1 - \frac{Y_1 + Y_2}{2}\right)^2 \\ &= \frac{(X_1 - X_2)^2 + (Y_1 - Y_2)^2}{4} \\ CD &= (t(Y_1 - Y_2)^2) + (-t(X_1 - X_2)^2) \\ &= t^2\left[(X_1 - X_2)^2 + (Y_1 - Y_2)^2\right] \\ \angle ACD &= \frac{\theta}{2} = \arctan\frac{AD}{CD} = \arctan\frac{1}{2t} \end{aligned} \tag{7}$$

Therefore, we can derive the relationship between t and the included angle θ as

$$t = \frac{\cos\frac{\theta}{2}}{2\sin\frac{\theta}{2}} = \frac{\sin\theta}{2(1+\cos\theta)} \tag{8}$$

Therefore, we have

$$\begin{aligned} C_x &= \frac{X_1 + X_2}{2} + \frac{sin\theta(Y_1 - Y_2)}{2(1 - cos\theta)} \\ C_y &= \frac{Y_1 + Y_2}{2} + \frac{-sin\theta(X_1 - X_2)}{2(1 - cos\theta)} \\ R &= \sqrt{(X_1 - C_x)^2 + (Y_1 - C_y)^2} \end{aligned} \tag{9}$$

where (C_x, C_y) represents the center of the fitting circle and R specifies the radius. Three centers are given through the formula above and averaged to improve the accuracy of the center of the fitting circle.

According to the flowchart in Figure 6, in the alignment phase, we first measure the angle of the reference image whereby we can obtain the deviation of the angle between the test image and the reference one. The camera can take images on the left and right side of the LCD module, respectively. Through searching for the feature marks printed on both sides, two feature points can be drawn to form a line that can act as a baseline. To compare the different angles of two lines, points have to be transformed to a uniform world coordinate. The transformed positions are then normalized by rotation center (C_x, C_y) to take the center as the origin point. By calculating θ_t and θ_c for reference angle and current angle, the difference of angle can be represented as $\Delta\theta = \theta_t - \theta_c$. Secondly, the offset of translation is determined. Based on the design that the center of rotation is the origin of the coordinate, the radius and angle of a point in polar coordinate can be expressed as

$$r = \sqrt{X_c^2 + Y_c^2}, \quad \theta = \arctan\frac{Y_c}{X_c} \tag{10}$$

Then, the rotated points (X_r, Y_r) can be obtained by

$$\begin{aligned} X_r &= rcos(\theta + \Delta\theta) \\ Y_r &= rsin(\theta + \Delta\theta) \end{aligned} \tag{11}$$

Either point of left or right side feature point can be used to calculate the offset of translation. We use (X_r, Y_r) and (X_t, Y_t) to denote rotated point by $\Delta\theta$ and the reference point, respectively. Finally, the translation offset can be obtained by

$$\begin{aligned} \Delta X &= X_r - X_t \\ \Delta Y &= Y_r - Y_t \end{aligned} \tag{12}$$

After determining the rotation center and the translation offset, the tested module can be aligned to the reference position as Figure 8 shows.

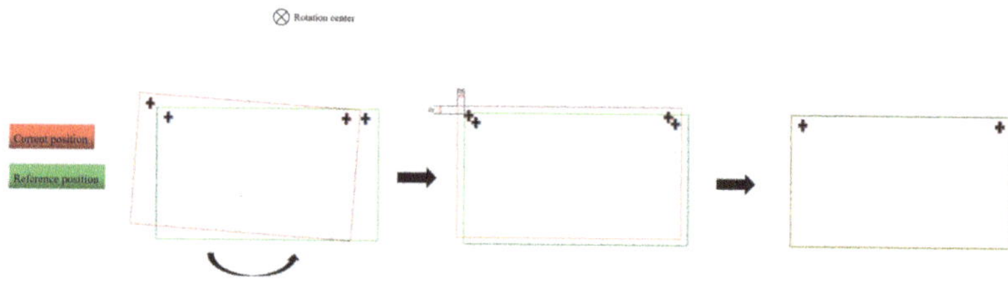

Figure 8. Expected result after the misalignment correction process.

2.3. *Bump Positioning*

After aligning the test object with the reference one, the positional deviation can be reduced thus contributing to accurately positioning bump areas containing the particles to be detected. The aligned LCD module is scanned by a line scan camera installed under the carrier table. Figure 9 shows a scanned image in the COG process. The cross-shaped pattern on the top right of the image serves as the positioning mark, while the red and blue regions in the image stand for ITO bumps. Two major types of bumps are annotated in Figure 9, which are block-like regions for type I and bar-like regions for type II, respectively.

As a whole imaging region consists of numerous bumps and other irrelevant pixels that make up a large portion of the image, it is difficult and time-consuming to extract valid information. Bump positioning intends to segment essential regions from the whole image and allows the detection algorithm to operate on every smaller region, which can reduce the size of detection areas and improve efficiency. Considering that the bump positions of a type of component are relatively fixed, we can construct a standard file with position information of all bump areas in it. Such information can be stored in the form of values relative to the position of the mark, which can be quickly determined by the shape-based match method. The feature mark is first searched in the test image, then the bump areas can be annotated with information in the standard file. The standard file can also be finetuned to adapt to the bump distribution in different LCD modules.

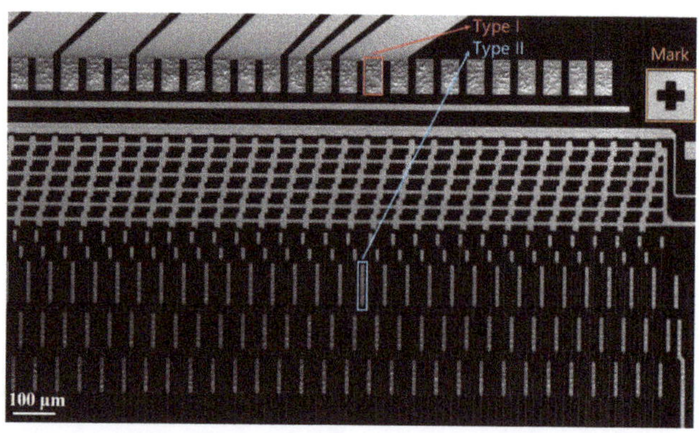

Figure 9. The local image of LCD module containing bump and mark areas.

2.4. *Particle Detection*

The number of particles captured is calculated in every segmented area of the bump. ACF conductive particles represent microspheres and deform with different levels of press strength. During the bonding process, these particles are trapped between the bump areas, causing those areas to show uneven pits. Figure 10a extracts a sample of well-deformed conductive particles, while (b) shows a local image of particles taken under the inappropriate pressure condition.

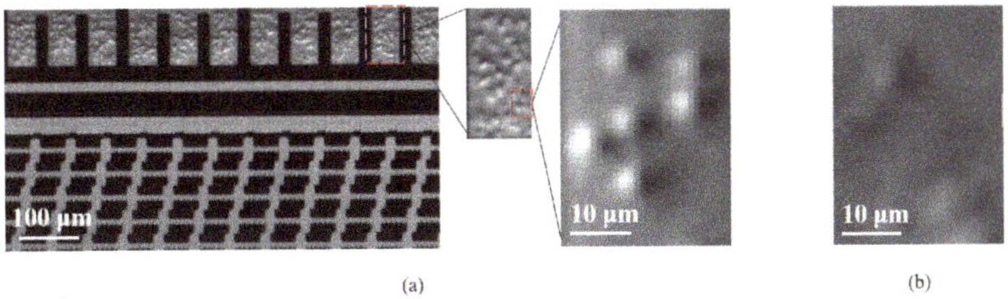

Figure 10. (**a**) Well deformed conductive particles; (**b**) over-deformed particles.

We can observe that the properly pressed particles appear as distinct light and dark parts in the image, while over-deformed particles have a weak contrast compared with neighbor areas. Based on such an observation, the particle areas are intended to be recognized by locating the extreme point of gray value of every particle. However, many false extreme points occur possibly caused by sharp changes in the gray-scale values of the edge of the bump or the intersection area between particles.

To reduce the effect of edge and intersection area on the detection accuracy, we first smooth the original image Figure 11a by the gaussian kernel, as shown in Figure 11b. We denote the original image as $I(x,y)$ and the smoothed image as $I_g(x,y)$. During the smoothing process, the original image is filtered by the 2D gaussian kernel $G(x,y)$ which can be formulated as the Equation (13)

$$G(x,y) = \frac{1}{2\pi\sigma^2} \exp^{-\frac{x^2+y^2}{2\sigma^2}} \tag{13}$$

where x is the distance from the origin in the horizontal axis, y is the distance from the origin in the vertical axis. The smoothed image can be denoted as

$$I_g(x,y) = I(x,y) * G(x,y). \tag{14}$$

Subsequently, we perform a gray value dilation on the smoothed image with a structuring element b in order to highlight the region of each particle. An example of output image after the gray value dilation is shown in Figure 11c. The gray-scale value of the dilated image is defined with the maximum value of the overlapped region between the structuring element and the target image. Such processing can be demonstrated as

$$[I \oplus b](x,y) = \max_{(s,t)\in b} I(x-s, y-t) \tag{15}$$

where s and t denote the distance from the origin in both directions. According to the size of structuring element b, the s and t should be in $[-size, size]$. We can regard the structuring element as another filtering kernel because it serves to further reduce the variation of local gray values. As a result, the gray values of a particle region are equivalent to the maximum value of this region, which is usually the gray value of the center of the particle. Following the process in Figure 11d, the dilated image is compared with the smoothed image to create a differential image where the extreme points of every particle are expected to be emphasized. To achieve this, we normalize the mean gray value of the smoothed image to 128 following the Equation (16)

$$I_g(x,y) = I_g(x,y) \times \frac{128}{Mean} \tag{16}$$

then the differential image $I_d(x,y)$ is created by the Equation (17).

$$I_d(x,y) = I_g(x,y) - [I \oplus b](x,y) + 128 \tag{17}$$

In the differential image, the central parts of particles are brighter than neighbor pixels indicating the existence of particles.

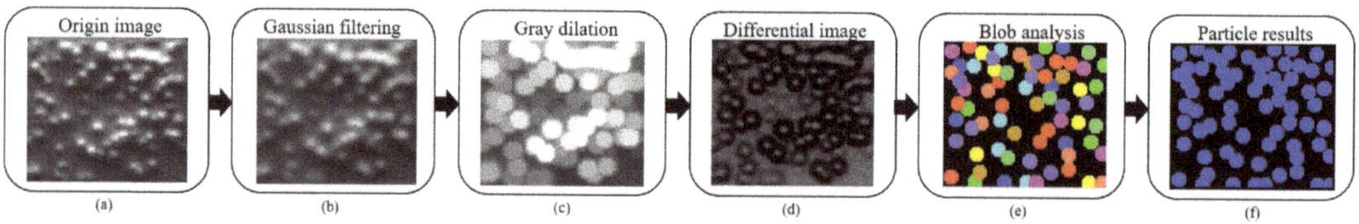

Figure 11. This flowchart shows the procedure of particle detection. The original image (**a**) is smoothed to image (**b**), (**b**) is dilated by a structuring element to produce (**c**), (**d**) highlights the particle centers, (**e**) remove noisy objects by features of the particle, and the final result is presented in (**f**). The processing and visualization of images are supported by OpenCV.

To refine the detection result, we also adopt the blob analysis to exclude noisy objects. During the blob analysis, detected objects are checked case by case, such as a sample shown in Figure 11e. Some gray features of particles can be used to enhance the reliability of the detection result. For example, the gray strength of the particle is assumed to exceed a certain threshold while the result with a lower strength should be removed. In addition, since the standard deviation of the brightness in the particle region should be within a certain range, we can extract the particle objects based on this criterion. After excluding noisy objects, the detection results in Figure 11f can be visualized by the detected centers of particles and a specified particle radius.

The number and density of particles in a given region can be determined from the results depicted in Figure 11f. Normal particles tend to be numerous and have a relatively uniform and dense distribution, while over-deformed particles can be identified by a small number of particles and a sparse distribution in the image analysis. In practical production, the threshold for distinguishing between normal and over-deformed particles should be determined based on the specific criteria of the plant.

3. Results and Discussion

3.1. Experiment Setup

As depicted in Figure 4, the majority of the system consists of mechanical components and sensor sets. The mechanical components primarily consist of robotic grippers responsible for loading and unloading LCD modules, a carrier platform equipped with translation and rotation motors, and a large horizontal track providing X-axis movement for pre-alignment cameras. The sensor sets, including the pre-alignment camera, line scan camera, and laser displacement sensor, are essential tools in the detection system. A detailed configuration can be found in Table 1. In terms of the parameters for the detection algorithm, a Gaussian kernel with a size of 3×3 and a σ of 0.6 was chosen, and the structuring element employed a circle with a radius of 8 pixels. In the blob analysis, the strength threshold of particles is set to 8, and the range of brightness within a particle region is 20–220.

3.2. Calibration Result

In our experiment, we first calibrated the homogeneous transformation matrix between the image space and the world coordinate and calculated the center of rotation of the carrier table. In Figure 12, we show the relationship between the captured images in terms of physical position, where subfigures 1–9 are captured by translating the carrier by 1 mm, while subfigures 10–12 are captured by rotating the carrier by 0.25°.

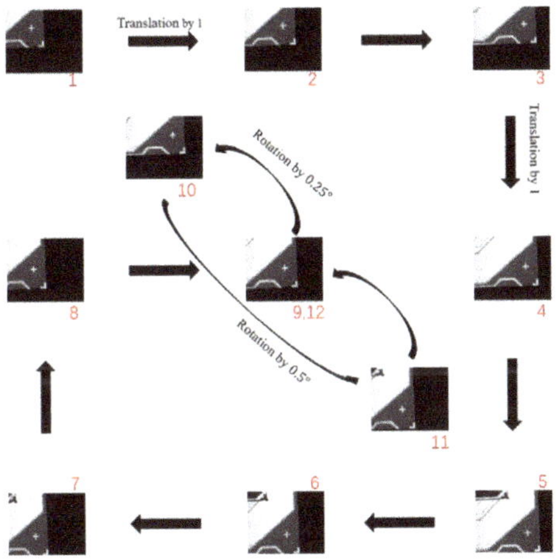

Figure 12. The relationship between the captured images in terms of physical position.

Table 2 shows the pixel location and the physical location of twelve images. The calibration result is shown in the last column. The location and rotation in the world coordinate were determined by the relative distance to the origin of the carrier table. We derived the homogeneous transformation matrix *HMat2d* through images 1–9 while calculating the center of rotation through images 10–12. As the center of rotation of the carrier platform was far from the location of captured images, three centers were calculated by two images and the included angle following the Equation (9) and then averaged to the final center.

Table 2. The pixel location and the physical location of twelve images and the calibration result.

Image Index	X in Pixel (px)	Y in Pixel (px)	X in Carrier (mm)	Y in Carrier (mm)	θ in Carrier (°)	Result (mm)
1	451.501	703.999	−26.4627	325.3494	0	
2	657.752	695.463	−27.4627	325.3494	0	
3	862.519	688.233	−28.4627	325.3494	0	
4	855.523	481.747	−28.4627	324.3494	0	
5	846.937	278.084	−28.4627	323.3494	0	HMat2d [1]
6	642.295	285.407	−27.4627	323.3494	0	
7	436.527	292.669	−26.4627	323.3494	0	
8	444.942	497.018	−26.4627	324.3494	0	
9	649.725	490.896	−27.4627	324.3494	0	
10	780.087	301.71	−27.4627	324.3494	−0.25	C_x = 160.812
11	518.451	678.46	−27.4627	324.3494	0.25	C_y = 182.984
12	649.441	490.792	−27.4627	324.3494	0	R = 235.439

[1] $HMat2d = \begin{pmatrix} -0.00486334 & 0.000181669 & -24.3918 \\ 0.000180196 & 0.00486517 & 321.846 \\ 0 & 0 & 1 \end{pmatrix}$.

3.3. Alignment Result

We selected some test samples to check the precision of our proposed method. Samples were acquired by controlling the machine to move different offsets in Y-axis or around the rotating axis while related information is recorded. This motion was conducted to simulate the occasion when the robotic gripper fails to make sure that the LCD module keeps the same position as the standard module.

In our experiments, the line scan camera was installed at the X-axis. We moved the line scan camera and took images of the left panel and right panel containing the feature marks. The mark locations in both panels were recorded and compared with those in the reference panel. Table 3 presents the positions of the left and right markers in various settings, with the positions being unified to a coordinate system with the center of rotation of the carrier table serving as the origin. For example, Y0.3 moves the carrier platform in the Y-axis direction by 0.3 mm, and Y0.6-R0.3 moves 0.6 mm alongside the Y-axis and rotates 0.3° around the center of rotation of the carrier table.

Table 3. The positions of the left and the right mark (unit:mm).

Setting	Left Mark	Right Mark
Ref.	$(-188.223, 141.469)$	$(-538.615, 141.469)$
Y0.3	$(-187.458, 143.135)$	$(-537.734, 141.295)$
Y0.6	$(-188.265, 142.068)$	$(-538.535, 142.064)$
R0.1	$(-187.995, 141.828)$	$(-538.264, 141.21)$
R0.3	$(-187.455, 142.542)$	$(-537.729, 140.702)$
Y0.6-R0.3	$(-187.458, 143.135)$	$(-537.734, 141.295)$

Table 4 demonstrates the alignment precision as below. The rightmost column shows the difference between the alignment result and previously recorded information, which represents $\Delta X, \Delta Y$, and $\Delta\theta$, respectively. Such results show that our method achieves high precision in restoring the offset of translation and rotation of tested modules.

Table 4. The result of misalignment correction in different settings (unit: $\Delta X(mm), \Delta Y(mm), \Delta\theta(°)$).

Setting	Left Panel	Right Panel	Result
Ref.		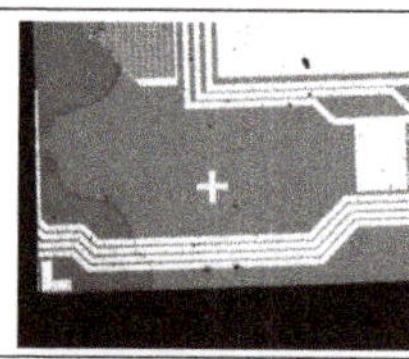	$\Delta X = 0$ $\Delta Y = 0$ $\Delta\theta = 0$
Y0.3	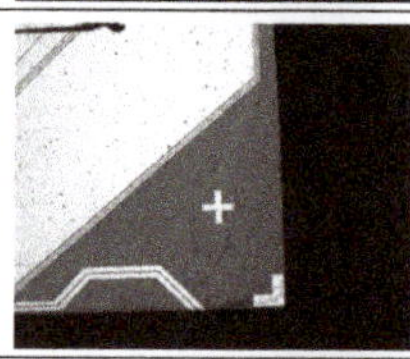	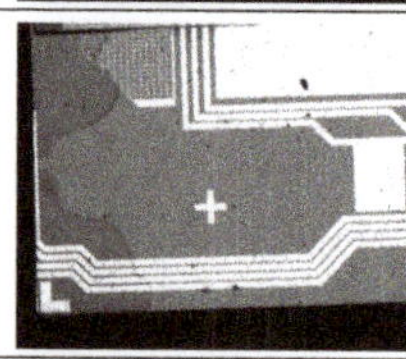	$\Delta X = -0.046$ $\Delta Y = 0.294$ $\Delta\theta = -0.001$
Y0.6		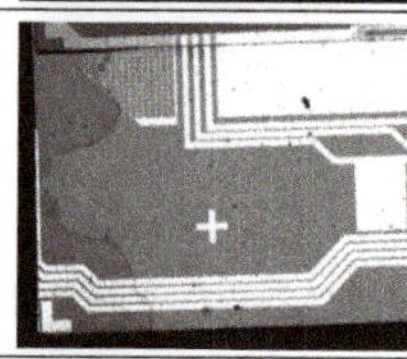	$\Delta X = -0.043$ $\Delta Y = 0.596$ $\Delta\theta = -0.001$
R0.1	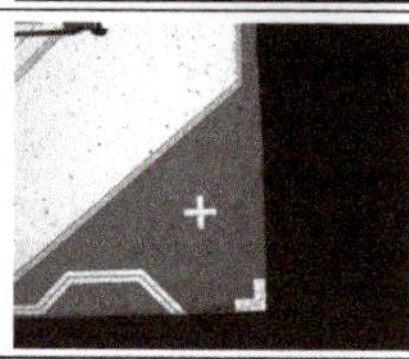	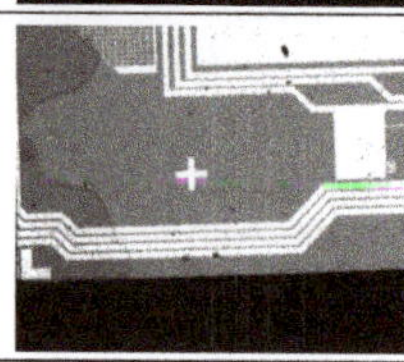	$\Delta X = -0.042$ $\Delta Y = -0.002$ $\Delta\theta = -0.1$

Table 4. *Cont.*

Setting	Left Panel	Right Panel	Result
R0.3	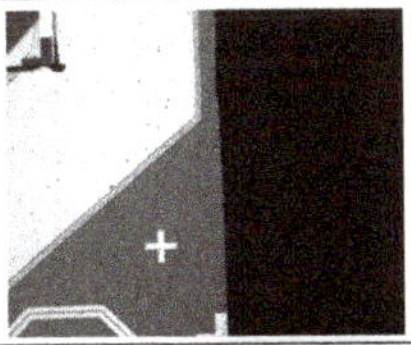	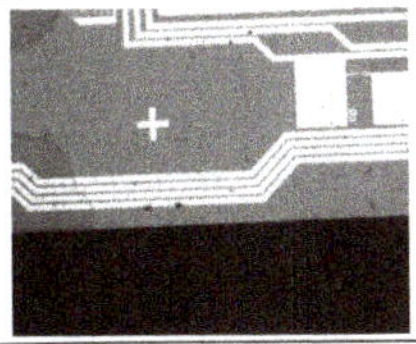	$\Delta X = -0.035$ $\Delta Y = -0.004$ $\Delta\theta = -0.301$
Y0.6-R0.3	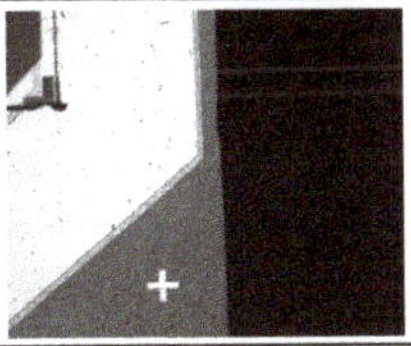	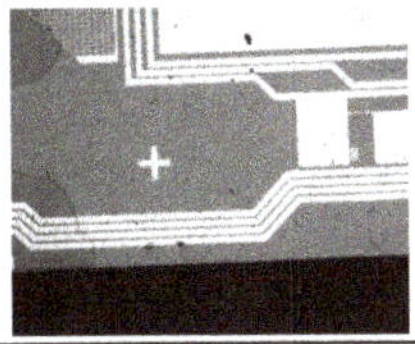	$\Delta X = -0.035$ $\Delta Y = 0.588$ $\Delta\theta = -0.301$

Although the setting of the X-axis was not verified separately, the offset of the X-axis was also calculated simultaneously during the verification of Y and θ. Considering the similarity of the X-axis and Y-axis, the experimental results can also illustrate the alignment accuracy of the X-axis.

Furthermore, we collected all data in three groups listed in Table 5 and analyzed how different offsets can affect the accuracy of our proposed method. We can observe from this table that our alignment method can keep a stable result in pure translation or rotation.

Table 5. The impact of different offset configurations on the accuracy of the alignment algorithm (unit: $X(mm), Y(mm), \theta(°)$).

Configuration	X	Y	θ	ΔX	ΔY	$\Delta\theta$	diff X	diff Y	diff θ
Y0.3 [1]	0	0.3	0	0.046	0.294	0.002	0.046	0.006	0.002
Y0.6	0	0.6	0	0.044	0.596	0.001	0.044	0.004	0.001
Y0.9	0	0.9	0	0.045	0.897	0.001	0.045	0.003	0.001
R0.1	0	0	0.1	0.043	0.002	0.101	0.043	0.002	0.001
R0.2	0	0	0.2	0.04	0.002	0.201	0.04	0.002	0.001
R0.3	0	0	0.3	0.036	0.004	0.301	0.036	0.004	0.001
Y0.3-R0.2	0	0.3	0.2	0.038	0.298	0.201	0.038	0.002	0.001
Y0.6-R0.2	0	0.6	0.2	0.042	0.591	0.2	0.042	0.009	0
Y0.6-R0.3	0	0.6	0.3	0.035	0.588	0.301	0.035	0.012	0.001

[1] Different colors are used to denote the relevance of the results. For instance, the red text "Y0.3" signifies an offset of 0.3 mm along the Y-axis, with a corresponding difference in the result of "0.006".

3.4. Detection Results

We evaluated the effectiveness of the proposed method in bump areas acquired from both COG and FOG samples. Figure 13 shows the results of particle detection in COG and FOG, respectively. Bump areas were segmented according to the saved file about the relative position of bumps to the feature mark in different types of samples such as COG and FOG. Two major types of bumps were categorized the bar-shape and the block-shape. The bar-shape bump is long and narrow, while the block-shape bump is short and wide that contains relatively dense particles. The target area of the block-shaped bump is smaller in size and the visibility of particles is weaker, thus making it more difficult to accurately inspect the number of particles.

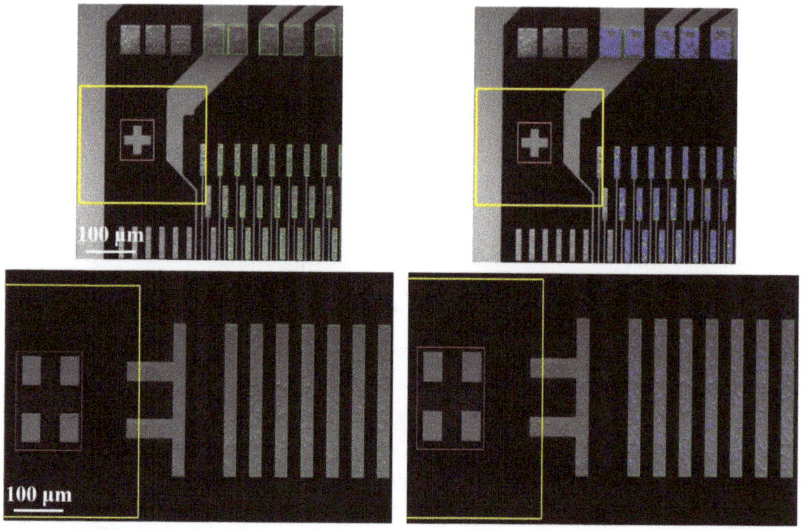

Figure 13. The first row is a part of the bump area of the COG sample (**left**) and the corresponding result (**right**), while the second row displays the input and output of the FOG test in the same order.

Figure 14 looks closer at bar-like bump areas. Bumps with an insufficient number of particles would be sent to NG output based on the manufacturer's requirement.

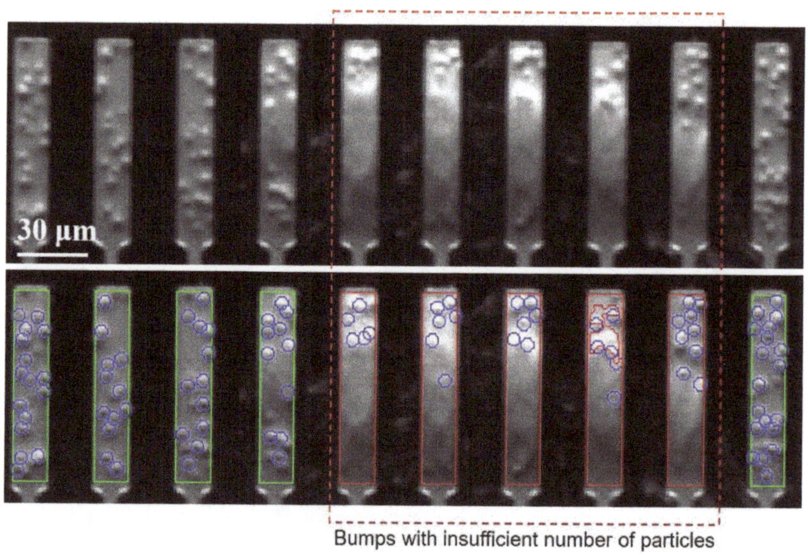

Figure 14. A closer view of the result of particle counts in bar-shape bump areas. The first row is the raw image while the second row is the detection result.

We also compared our method with Lin's [12] and Chen's [26] method and verified the effectiveness of our method as shown in Figure 15. Lin's method was sensitive to slight changes in gray value, thus causing the algorithm to detect more particles in regions where particles overlap densely. Such a method treated the NG module as the OK module and may produce many false positive samples, which is inappropriate in the practical industry. Contrarily, Chen's method recognized the number of overlapped particles more accurately.

However, the performance of Chen's method depended on the initialization of the cluster center and thus may lose a number of particles on some occasions.

The detection result in 433 particle samples of bar-shape bumps and 629 samples of block-shape bumps are shown in Table 6. Lin's method demonstrates a slightly improved recall rate for bar bumps due to its tendency to generate more particle candidates than are actually present, thereby covering a larger portion of the region on samples with smaller dimensions. Conversely, this method exhibits a significant decline in recall performance on samples with larger dimensions, such as the block-shape bumps. Combined with the low precision, Lin's method shows a high over-checking rate which will allow a number of false products to be passed and affect negatively quality control. Chen's precision rate is good but the number of detected particles is lower than that in other methods when the gray-scale value of the image is weak. In contrast, our method balances the precision and recall rate of the detection result and performs better than the aforementioned methods.

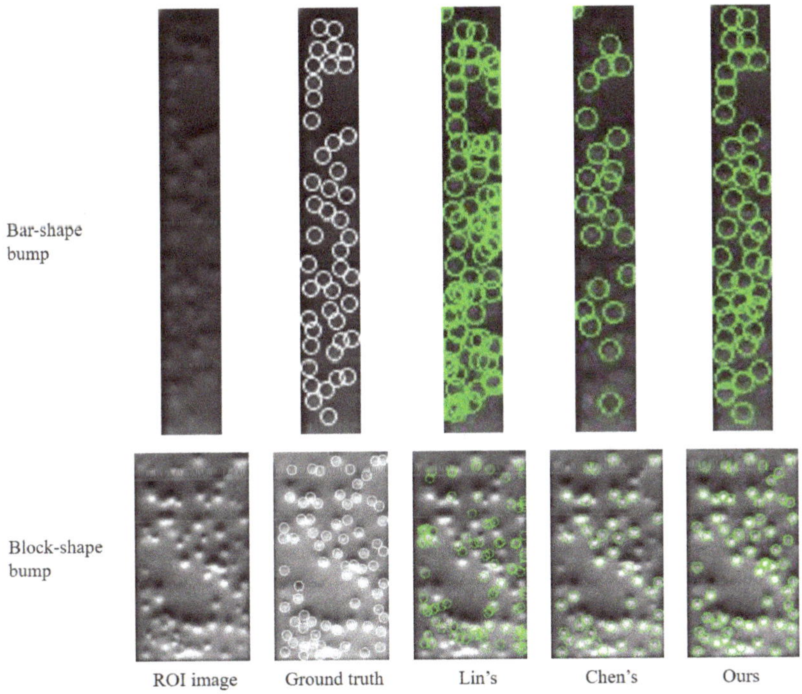

Figure 15. Examples of different detection methods on bar-shape bumps and block-shape bumps.

Table 6. The comparison of detection results of different methods.

Bump Type	Method	Detected Number	Right Result	Precision	Recall	Calculation Time (ms)
Bar-shape	Lin's	884	402	45.5%	92.8%	13.3
	Chen's	256	218	85.2%	50.3%	43.2
	Ours	430	400	93%	92.4%	1.2
Block-shape	Lin's	870	464	53.3%	73.8%	15.15
	Chen's	382	353	92.4%	56.1%	57.7
	Ours	564	537	95.2%	85.4%	5.5

4. Conclusions

In this paper, we establish an automatic inspection system to serve multiple bonding processes such as COG and FOG in LCD module manufacturing. A 12-point calibration method is developed to adaptively control the position and rotation of the carrier table and to alleviate the high requirement of assembly accuracy. This method can complete the alignment by taking images once, which calculates a homogeneous transformation matrix with nine pairs of points and determines the center of rotation with extra three pairs of points. In our experiments, the alignment error can be less than 0.05 mm.

This study also proposes an automatic particle detection method based on gray morphology which achieves a fast and robust inspection of the number of conductive particles trapped in bump areas of the anisotropic conductive film. Based on the observation that the central part of conductive particles is brighter than the neighbor region in grayscale under our DIC imaging model, we apply the gray dilation method to the whole image and subtract dilated image from the original input, from which the center regions of particles are stressed out. Our experiments have examined the effectiveness of this method, in which the precision rate is 93%, while the recall rate is 92.4%. As our proposed detection method depends on finding the difference between the central and peripheral regions of particles, the overlap of multiple particles may intervene in this difference, thus creating an obstacle to accurate segmentation. In the future, further research will be conducted to improve the representation of particles and segment those overlapped particles more accurately.

Author Contributions: Conceptualization, T.L.; methodology, F.Z.; software, T.L. and F.Z.; validation, T.L.; formal analysis, F.Z.; resources, Z.Z.; writing—original draft preparation, T.L.; writing—review and editing, F.Z., H.Y. and H.L.; visualization, T.L.; supervision, F.Z. and Z.Z.; funding acquisition, Z.Z. All authors have read and agreed to the published version of the manuscript.

Funding: This work is supported by the National Key Research and Development Program of China (2022YFB3303800), the National Natural Science Foundation of China (U21A20482, U1909218), the Science Technology Project funding from Binzhou Weiqiao Guoke Advanced Technology Research Institute(E2D21709, E2D21710, E2D21711).

Data Availability Statement: Not applicable.

Conflicts of Interest: The authors declare no conflict of interest.

Abbreviations

The following abbreviations are used in this manuscript:

LCD	Liquid Crystal Display
ACF	Anisotropic Conductive Film
IC	Integrated Circuit
FPC	Flexible Printed Circuit
ITO	Indium Tin Oxide
COG	Chip-on-Glass
FOG	Flex-on-Glass
DIC	Differential Interference Contrast
ROI	Region of Interest

References

1. Gao, L.L.; Chen, X.; Zhang, S.B.; Gao, H. Mechanical properties of anisotropic conductive film with strain rate and temperature. *Mater. Sci. Eng.-Struct. Mater. Prop. Microstruct. Process.* **2009**, *513–514*, 216–221. [CrossRef]
2. Paik, K.W.; Yim, M.J. New anisotropic conductive adhesives for low cost and reliable flip chip on organic substrates applications. In Proceedings of the International Symposium on Electronic Materials and Packaging (EMAP2000), Hong Kong, China, 30 November–2 December 2000; pp. 282–288.
3. Paik, K.W.; Yim, M.J.; Kwon, W.; Smta, S. Significant reliablity enhancement using new anisotropic conductive adhesives for flip chip on organic substrates applications. In Proceedings of the 6th Annual Pan Pacific Microelectronics Symposium, Kauai, HI, USA, 13–16 February 2001; pp. 217–223.

4. Mou, Y.; Zhang, Y.; Cheng, H.; Peng, Y.; Chen, M. Fabrication of highly conductive and flexible printed electronics by low temperature sintering reactive silver ink. *Appl. Surf. Sci.* **2018**, *459*, 249–256. [CrossRef]
5. Mou, Y.; Cheng, H.; Wang, H.; Sun, Q.; Liu, J.; Peng, Y.; Chen, M. Facile preparation of stable reactive silver ink for highly conductive and flexible electrodes. *Appl. Surf. Sci.* **2019**, *475*, 75–82. [CrossRef]
6. Shiozawa, N.; Isaka, K.; Ohta, T. Electric properties of connections by anisotropic conductive film. *J. Electron. Manuf.* **1995**, *5*, 33–37. [CrossRef]
7. Yim, M.J.; Hwang, J.; Paik, K.W. Anisotropic conductive films (ACFs) for ultra-fine pitch Chip-On-Glass (COG) applications. *Int. J. Adhes. Adhes.* **2007**, *27*, 77–84. [CrossRef]
8. Chin, M.; Hu, S.J. A multiple particle model for the prediction of electrical contact resistance in anisotropic conductive adhesive assemblies. *IEEE Trans. Components Packag. Technol.* **2007**, *30*, 745–753. [CrossRef]
9. Shi, F.G.; Abdullah, M.; Chungpaiboonpatana, S.; Okuyama, K.; Davidson, C.; Adams, J.M. Electrical conduction of anisotropic conductive adhesives: Effect of size distribution of conducting filler particles. *Mater. Sci. Semicond. Process.* **1999**, *2*, 263–269. [CrossRef]
10. Yen, Y.W.; Lee, C.Y. ACF particle distribution in COG process. *Microelectron. Reliab.* **2011**, *51*, 676–684. [CrossRef]
11. Xie, B.; Shi, X.Q.; Ding, H. Investigation of mechanical and electrical characteristics for cracked conductive particle in anisotropic conductive adhesive (ACA) assembly. *IEEE Trans. Components Packag. Technol.* **2008**, *31*, 361–369.
12. Lin, C.S.; Huang, K.H.; Lin, T.C.; Shei, H.J.; Tien, C.L. An automatic inspection method for the fracture conditions of anisotropic conductive film in the TFT-LCD assembly process. *Int. J. Optomechatronics* **2011**, *5*, 286–298. [CrossRef]
13. Ni, G.; Liu, L.; Du, X.; Zhang, J.; Liu, J.; Liu, Y. Accurate AOI inspection of resistance in LCD Anisotropic Conductive Film bonding using differential interference contrast. *Optik* **2017**, *130*, 786–796. [CrossRef]
14. Ni, G.; Liu, L.; Zhang, J.; Liu, J.; Liu, Y. Analysis of trapped conductive microspheres in LCD FOG anisotropic conductive film bonding. In Proceedings of the 2nd IEEE Advanced Information Technology, Electronic and Automation Control Conference (IAEAC), Chongqing, China, 25–26 March 2017; pp. 1414–1420.
15. Ni, G.; Liu, L.; Zhang, J.; Liu, J.; Wang, X.; Du, X.; Yue, H.; Liu, Y. Automatic optical inspection of bump offsets in flex on glass bonding using differential interference contrast imaging. *Int. J. Precis. Eng. Manuf.* **2020**, *21*, 177–187. [CrossRef]
16. Sheng, X.; Jia, L.; Xiong, Z.; Wang, Z.; Ding, H. ACF-COG interconnection conductivity inspection system using conductive area. *Microelectron. Reliab.* **2013**, *53*, 622–628. [CrossRef]
17. Yim, M.J.; Paik, K.W. Design and understanding of anisotropic conductive films (ACF's) for LCD packaging. *IEEE Trans. Components Packag. Manuf. Technol. Part A* **1998**, *21*, 226–234.
18. Ni, G.; Du, X.; Liu, L.; Zhang, J.; Liu, J.; Liu, Y. Automated optical inspection of liquid crystal display anisotropic conductive film bonding. *Opt. Eng.* **2016**, *55*, 103109. [CrossRef]
19. Ni, G.; Liu, L.; Zhang, J.; Liu, J.; Liu, Y. High-resolution imaging optomechatronics for precise liquid crystal display module bonding automated optical inspection. *J. Electron. Imaging* **2018**, *27*, 013013. [CrossRef]
20. Lin, C.S.; Huang, K.H.; Lay, Y.L.; Wu, K.C.; Wu, Y.C.; Lin, J.M. An improved pattern match method with flexible mask for automatic inspection in the LCD manufacturing process. *Expert Syst. Appl.* **2009**, *36*, 3234–3239. [CrossRef]
21. Kim, H.T.; Yang, H.J.; Baek, S.Y. Iterative algorithm for automatic alignment by object transformation. *Microelectron. Reliab.* **2007**, *47*, 972–985. [CrossRef]
22. Lin, C.T.; Yu, Y.C. Visual servoing for mask alignment in photolithography positioning system. In Proceedings of the IEEE International Conference on Mechatronics (ICM), Taipei, Taiwan, 10–12 July 2005; pp. 762–767.
23. Sun, W.; Zhang, Z.; Zhang, W. A coaxial alignment method for large flange parts assembly using multiple local images. *IEEE Access* **2021**, *9*, 16716–16727. [CrossRef]
24. Kim, J. New wafer alignment process using multiple vision method for industrial manufacturing. *Electronics* **2018**, *7*, 39. [CrossRef]
25. Algorri, J.F.; Bennis, N.; Urruchi, V.; Morawiak, P.; Sánchez-Pena, J.M.; Jaroszewicz, L.R. Tunable liquid crystal multifocal microlens array. *Sci. Rep.* **2017**, *7*, 17318. [CrossRef] [PubMed]
26. Chen, Y.; Xiao, K.; Guo, Z.; He, J.; Liu, C.; Chen, S. Detection of conducting particles bonding in the circuit of liquid crystal display. *Chin. J. Liq. Cryst. Displays* **2017**, *32*, 553–559. [CrossRef]
27. Zhang, Z.Y. A flexible new technique for camera calibration. *IEEE Trans. Pattern Anal. Mach. Intell.* **2000**, *22*, 1330–1334. [CrossRef]

machines

MDPI

Article

Mask-Guided Generation Method for Industrial Defect Images with Non-uniform Structures

Jing Wei [1,2], Zhengtao Zhang [1,2,3], Fei Shen [1,2,3] and Chengkan Lv [1,2,3,*]

1 Institute of Automation, Chinese Academy of Sciences, Beijing 100190, China
2 The School of Artificial Intelligence, University of Chinese Academy of Sciences, Beijing 100049, China
3 CASI Vision Technology Co., Ltd., Luoyang 471000, China
* Correspondence: chengkan.lv@ia.ac.cn

Abstract: Defect generation is a crucial method for solving data problems in industrial defect detection. However, the current defect generation methods suffer from the problems of background information loss, insufficient consideration of complex defects, and lack of accurate annotations, which limits their application in defect segmentation tasks. To tackle these problems, we proposed a mask-guided background-preserving defect generation method, MDGAN (mask-guided defect generation adversarial networks). First, to preserve the normal background and provide accurate annotations for the generated defect samples, we proposed a background replacement module (BRM), to add real background information to the generator and guide the generator to only focus on the generation of defect content in specified regions. Second, to guarantee the quality of the generated complex texture defects, we proposed a double discrimination module (DDM), to assist the discriminator in measuring the realism of the input image and distinguishing whether or not the defects were distributed at specified locations. The experimental results on metal, fabric, and plastic products showed that MDGAN could generate diversified and high-quality defect samples, demonstrating an improvement in detection over the traditional augmented samples. In addition, MDGAN can transfer defects between datasets with similar defect contents, thus achieving zero-shot defect detection.

Keywords: industrial manufacturing; deep learning; data augmentation; defect generation; defect detection

Citation: Wei, J.; Zhang, Z.; Shen, F.; Lv, C. Mask-Guided Generation Method for Industrial Defect Images with Non-uniform Structures. *Machines* **2022**, *10*, 1239. https://doi.org/10.3390/machines10121239

Academic Editor: Dimitrios Manolakos

Received: 6 November 2022
Accepted: 15 December 2022
Published: 18 December 2022

1. Introduction

Methods based on machine learning and deep learning have remarkably improved industrial defect detection performance [1–3]. However, practical industrial scenarios pose challenges to the current detection methods, such as data problems. Acquiring large datasets for manufacturing applications remains a challenging proposition, due to the time and costs involved [4]. The small number of defect samples and data imbalances can lead to overfitting during the training of supervised deep-learning methods and poor performance in testing [5].

Data augmentation is a very powerful method for building useful deep-learning models and for reducing validation errors [6]. Image data augmentation mainly includes traditional and learning-based methods. Traditional methods can increase the number of samples, but cannot create new defect samples. In contrast, learning-based methods such as GAN (Generative Adversarial Nets) [7], AAE (Adversarial AutoEncoders) [8], and VAE (Variational Auto-Encoder) [9] can model the distribution of a real dataset and synthesize new samples that are different from the original dataset, which increases both the number and the diversity of the dataset. Based on cutting-edge work in image synthesis [10–13], industrial image generation can be carried out, to achieve the augmentation of few-sample datasets. Currently, diverse learning-based augmentation methods are emerging, to alleviate data problems in industrial defect detection [14–20].

However, there are still some challenges that need to be addressed in the current learning-based defect image augmentation methods.

Insufficient retention of realistic background textures. Textures provide important and unique information for intelligent visual detection and identification systems [5]. In industrial defect detection, any slight change to real textures can disturb the detection results. Researchers usually perform defect image generation based on non-defective samples, due to their easy availability in industrial manufacturing. This requires generation methods that preserve the real normal backgrounds to the maximum extent possible. Many works [15–17,19] have used CycleGAN [21] to generate a defective image for an input normal image, where normal backgrounds may be excessively falsified, since they do not constrain the treatment of the normal background.

The independent controls of the normal backgrounds, defect shapes, and defect textures are rarely considered. If independent control and operation of the three are achieved, then they can be arbitrarily combined to obtain an infinite number of defective images from a normal image. However, current effective methods such as SDGAN [15], Defect-GAN [16], and SIGAN [17] control the three as a whole and can only obtain one defect image from a normal image based on well-trained models, whose randomness and diversity are insufficient. Moreover, the pixel-level annotations of the generated defect images can be acquired if we separately process the normal backgrounds and defect regions. Defect-GAN generates a spatial distribution map, to indicate what is modified in the source image compared with the generated image, but it does not decouple the backgrounds and defect regions and cannot obtain accurate binary annotations from the spatial distribution map.

Lack of exploration of the generation of non-uniform complex structure defects with binary annotations. There are multiple semantic regions in a non-uniform structure image, where texture features are different and the corresponding defects are distinct. As shown in Figure 1, there are two types of textures in a zipper image: fabric and zipper teeth, whose defect contents are significantly disparate. In terms of such non-uniform structure images, networks must generate conforming defects at the specified locations, to obtain realistic synthetic results. In addition, obtaining binary annotations for the generated complex texture defect images is also a challenging task. Shuanlong N. et al. [18] used random seeds to construct input masks, while this is limited to simple stripe defects in some uniform textures. Du-Ming T. et al. [19] adopted two CycleGANs to preserve normal backgrounds and threshold segmentation to obtain binary annotations. However, this work also only synthesized defects with uniform textures, and the overall networks contained four generators and four discriminators, which were too complicated.

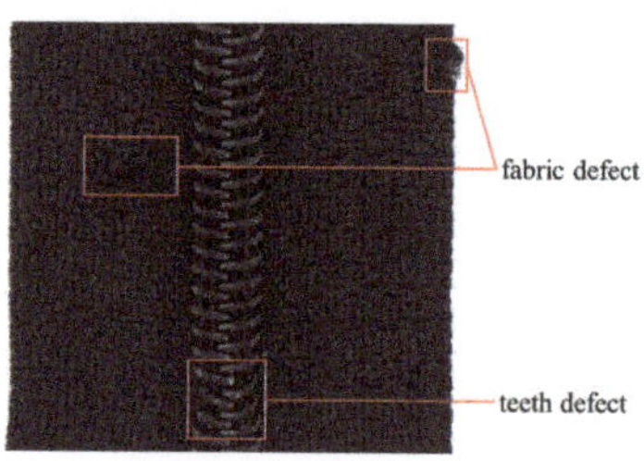

Figure 1. An example of a defect sample with non-uniform structures. The top two boxes indicate a type of fabric defect. The bottom box indicates a type of teeth defect.

To tackle these challenges, we proposed a MDGAN (mask-guided defect generation adversarial network) based on CGAN [22]. The MDGAN can generate realistic defects in regions specified by the input binary mask. First, we introduce a BRM (background replacement module) to extract normal backgrounds using a binary mask to replace the contents at the corresponding positions in the feature maps. The BRM achieves the preservation of the normal backgrounds and facilitates the separate control of the normal backgrounds and the shape of defects. In addition, the generated defect textures can be controlled by training MDGAN separately for different types of defects. Second, we proposed a DDM (double discrimination module), to extract the defect features from the whole feature map with the guidance of binary masks and measure the authenticity of the whole and the local based on one discriminator. In addition, we constructed a pseudo-normal background for each defect image, to provide paired training inputs. This preprocess ensures MDGAN generates defects according to normal features in the same regions, thus enabling generation of defects with non-uniform structures. Finally, the outputs of MDGAN and the input binary masks were combined to construct our pixel-level annotated synthetic datasets.

In summary, the main contributions of this work are as follows.

(1) We constructed corresponding pseudo-normal backgrounds for defective images, which solves the problem of the lack of paired training inputs in industrial defect generation and avoids the dependence on CycleGAN.

(2) We proposed a MDGAN, to achieve independent control of the normal backgrounds, defect shapes, and defect textures of images. The addition of BRM achieves the preservation of normal backgrounds and enables the acquisition of binary annotations. Our DDM focuses on the defect region and the whole image simultaneously, ensuring the quality of the generated results.

(3) Since BRM can achieve total preservation of the normal background in the generated defect images, our MDGAN can also achieve defect transfer between datasets with similar defect contents.

The subsequent sections of this article are organized as follows: Section 2 proposes the MDGAN. Section 3 introduces the related datasets used in this work. Section 4 details the generation, ablation, comparison, and segmentation experiments. Finally, we summarize our work in Section 5.

2. Methods

This section introduces the construction of the paired training inputs, the architectures, and the training process of MDGAN.

2.1. Pseudo Normal Background

Image-to-image (I2I) translation [23] is the most effective method to convert images in the source domain into the target domain, where paired source-target images are needed in training. Defect synthesis based on normal images is an I2I task, where the source domain consists of normal industrial images and the defect images constitute the target domain. However, it is almost impossible to obtain exactly corresponding normal–defective pairs in industrial manufacturing. Many works relied on CycleGAN to avoid this problem. Unfortunately, CycleGAN lacks randomness and cannot retain the background, as shown in Section 1. Therefore, we construct pseudo-normal backgrounds for the defect images. First, we select a similar normal image N for the defect image D, whose binary mask is M. N contains areas where the defect contents in D appears. Then calibrate N to obtain N^c by affine transformation matrix T,

$$N^c_{i,j} = T * N_{i,j} \tag{1}$$

$$T = \begin{bmatrix} cos\theta & -sin\theta & T_x \\ sin\theta & cos\theta & T_y \\ 0 & 0 & 1 \end{bmatrix} \tag{2}$$

where $N^c_{i,j}$, $N_{i,j}$ are points of homogeneous form in N^c and N, respectively, θ is the rotation angle in the anticlockwise direction, and T_x and T_y are translation parameters. Affine transformation can ensure that the area used for filling in the normal image is aligned with the defect area. Finally, use N^c to replace the defect regions in D,

$$B = (D \odot M_-) \oplus (N^c \odot M) \tag{3}$$

where M_- means (1–M), and ⊙, ⊕ means element-level multiplication and addition, respectively. Then B is used as the pseudo-normal background in the source domain to train the MDGAN.

2.2. MDGAN

As shown in Figure 2, MDGAN consists of a generator and a discriminator. A BRM is proposed to modulate the background in the generator using binary masks, as a way to avoid the modification of the background and control the generated defect shapes. A DDM is proposed, to divide the feature map into a whole feature branch and a defect feature branch, which can guide the discriminator to focus on both the whole and the local regions. Only a discriminator is needed to facilitate the generator to output high-quality defect images. Overall, MDGAN is able to generate images with defects appearing in regions specified by binary masks and preserve the normal backgrounds, combining them with input masks to obtain defect samples with pixel-level annotations.

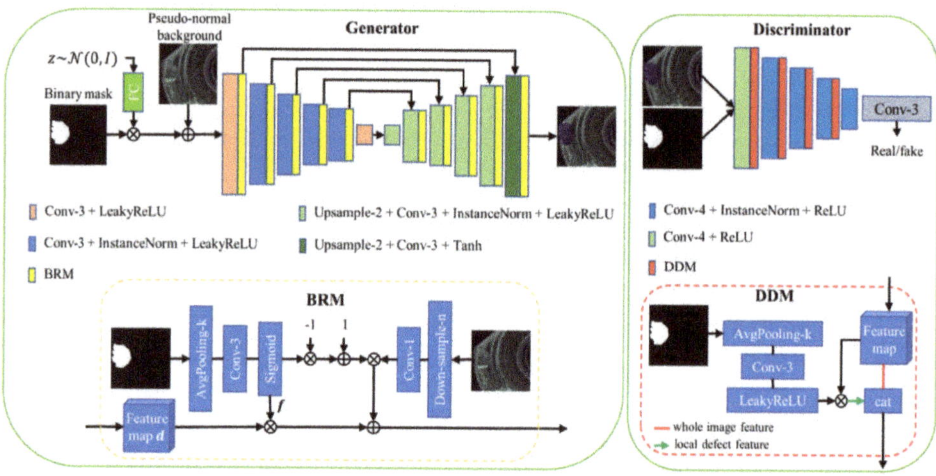

Figure 2. Architectures of MDGAN. *Conv-3* means the convolution kernel size is 3 × 3, *Upsample-2* means the feature map is upsampled two times. *AvgPooling-k* means the average pooling kernel size is k × k, which depends on the size of ***d***. *cat* means the channel-level connection of inputs.

2.2.1. Architectures

As shown in Figure 2, the generator is a UNet-like [24] architecture, whose inputs are Gaussian noise, pseudo-normal background, and a 0-1 binary mask. The size of the output is same as the input images. In the output image, the defective contents are at the "1" positions specified by the mask, and the original normal backgrounds are at the "0" positions. The discriminator adopts patchGAN [23] to process the input defect image and binary mask, and its output indicates the probability that the input defect image annotated by the mask is real.

Background Replacement Module (BRM). As shown in Figure 2, BRM employs a binary mask to fuse the real background to the specific positions of the feature map. First, the input binary mask $\boldsymbol{M}$ is average pooling down-sampled, convolved, and activated by Sigmoid, to obtain a weight map f with values of [0, 1]. f has the same size as the feature map $\boldsymbol{d}$ of the current layer. The input normal background $\boldsymbol{B}$ is downsampled and convolved with 1 × 1 kernels, to obtain a background feature map $\boldsymbol{B}'$ with the same size as f. $\boldsymbol{B}'$ is modulated into the generator to obtain the input of the next layer and the skip-connected $\boldsymbol{F}$,

$$\boldsymbol{F} = (\boldsymbol{B}' \odot f_-) \oplus (\boldsymbol{d} \odot f), \tag{4}$$

Double Discrimination Module (DDM). As shown in Figure 2, DDM divides the current feature map of the discriminator into whole image features and local defect features for processing. First, $\boldsymbol{M}$ is processed to obtain a weight map with the same size as the feature map, and then the weight map is used to extract the local defect information. Finally, the extracted contents are connected with the original feature map and input into the next layer. Thus, DDM assists the discriminator in processing both the local and global information and ensures the realism of the generated defects. Therefore, we need two discriminators to achieve the above tasks without DDM, which slows down the training and makes the architectures more complex.

2.2.2. Training Objectives

First, a binary mask $\boldsymbol{M}$; a real defect image $\boldsymbol{D}$, which is the ground truth of the generator; and a constructed pseudo-normal background $\boldsymbol{B}$ are given, and we sample random noise $\boldsymbol{z}$ from the Gaussian distribution $\mathcal{N}(0,\ I)$ as code from latent space. Next, $\boldsymbol{z}$ is mapped and reshaped to obtain $\boldsymbol{x_z}$ with the same size as $\boldsymbol{B}$. The input of the first convolution layer of the generator is

$$\boldsymbol{x} = \boldsymbol{x_z}\ \odot\ \boldsymbol{M} \oplus \boldsymbol{B}, \tag{5}$$

$\boldsymbol{x}$, $\boldsymbol{M}$, and $\boldsymbol{B}$ are processed by the generator to obtain the output $\boldsymbol{D^r} = G(\boldsymbol{M},\ z,\ \boldsymbol{B})$ and the input feature map $\boldsymbol{d_l}^r$ of the last BRM.

Defect region losses. First, we calculate the defect reconstruction loss of $\boldsymbol{D^r}$ and $\boldsymbol{D}$,

$$\mathcal{L}_{r1} = 5 \parallel (\boldsymbol{D}^r - \boldsymbol{D})\ \odot\ M \parallel_1, \tag{6}$$

where $\parallel \parallel_1$ is the L1 loss. In order to improve the diversity of the generated defects, we sample another $z' \sim \mathcal{N}(0, I)$ to obtain the new outputs $\boldsymbol{D}' = G(\boldsymbol{M},\ \boldsymbol{z}', \boldsymbol{B})$ and $\boldsymbol{d_l}'$, and calculate the diversity loss of the defect regions,

$$\mathcal{L}_{div} = -\parallel (\boldsymbol{D}^r - \boldsymbol{D}')\ \odot\ \boldsymbol{M} \parallel_1. \tag{7}$$

Essentially, $\boldsymbol{D}^r$ and $\boldsymbol{D}'$ are synthetic defect images with the same background and annotations, but different defect contents.

Normal background losses. According to the above description, we calculate the reconstruction loss of the normal backgrounds,

$$\mathcal{L}_{r2} = \parallel (\boldsymbol{D}^r - \boldsymbol{D}) \odot \boldsymbol{M}_- \parallel_1 + \parallel (\boldsymbol{D}' - \boldsymbol{D}) \odot \boldsymbol{M}_- \parallel_1. \tag{8}$$

In addition, when the normal background of $\boldsymbol{d_l}$ (input of the last BRM) is as similar as possible to $\boldsymbol{D}$, the last BRM can only modulate the details of $\boldsymbol{D}$ into $\boldsymbol{d_l}$. On the contrary, MDGAN will depend on the last BRM to substantially replace the normal background of $\boldsymbol{d_l}$, which may lead to a large incoherence between the normal and the defect regions. Therefore, in order to avoid a dependence on the last BRM and obtain a more coherent transition between the defect contents and the normal background of the final output, we calculate $\mathcal{L}_{r3}$ to constrain the backgrounds of $\boldsymbol{d_l}$ to be as similar as possible to $\boldsymbol{D}$,

$$\mathcal{L}_{r3} = \parallel (\boldsymbol{d_l}^r - \boldsymbol{D})\ \odot\ \boldsymbol{M}_-\ \parallel_1 + \parallel (\boldsymbol{d_l}' - \boldsymbol{D})\ \odot\ \boldsymbol{M}_- \parallel_1, \tag{9}$$

Whole image losses. To ensure the coherence of the defect and background regions after replacement, the gradient loss [25] at the boundary between the two regions is calculated,

$$\mathcal{L}_{grad} = \parallel F(\nabla \boldsymbol{M})_{!0\to1}\ \odot\ (\nabla \boldsymbol{D} - \nabla \boldsymbol{D}^r) \parallel_1 + \parallel F(\nabla \boldsymbol{M})_{!0\to1}\ \odot\ (\nabla \boldsymbol{D} - \nabla \boldsymbol{D}') \parallel_1, \tag{10}$$

where ∇ is the gradient and $F(X)_{!0\to1}$ sets all the non-zero elements in X to 1.

In order to stabilize the training process, we add the gradient penalty of WGAN-GP [26,27] to the discriminator. We construct two hybrid samples based on the real and the generated defect samples,

$$\boldsymbol{D}_{\alpha}^{r} = \alpha_1 \boldsymbol{D} + (1 - \alpha_1)\boldsymbol{D}^{r}, \tag{11}$$

$$\boldsymbol{D}_{\alpha}^{\prime} = \alpha_2 \boldsymbol{D} + (1 - \alpha_2)\, \boldsymbol{D}^{\prime}, \tag{12}$$

where α_1 and α_2 are random numbers [0, 1] and the discriminator needs to process the two hybrid samples, which only back-propagate gradients to the discriminator *Dis*. The gradient penalty loss is

$$\mathcal{L}_{gp} = \parallel Dis(\boldsymbol{D}_{\alpha}^{r}, \boldsymbol{M}) \parallel_2 + \parallel Dis(\boldsymbol{D}_{\alpha}^{\prime}, \boldsymbol{M}) \parallel_2. \tag{13}$$

Then we calculate the adversarial loss, to constrain the reality of the images,

$$\mathcal{L}_{adv} = 2 \parallel Dis(\boldsymbol{D}, \boldsymbol{M}) \parallel_2 + \parallel Dis(\boldsymbol{D}^{r}, \boldsymbol{M}) - 1 \parallel_2 + \parallel Dis(\boldsymbol{D}^{\prime}, \boldsymbol{M}) - 1 \parallel_2. \tag{14}$$

Full objective. Ultimately, we obtain the final training objective

$$G, Dis = \min_{G} \max_{Dis} \lambda_r \sum_{i=1}^{3} \mathcal{L}_{ri} + \lambda_d \mathcal{L}_{div} + \lambda_g \mathcal{L}_{grad} + \mathcal{L}_{adv} + \lambda_{gp} \mathcal{L}_{gp} \tag{15}$$

where λ_r, λ_d, λ_g, and λ_{gp} control the contribution of each loss to the whole. Based on the trained MDGAN, inputting the normal background and the binary mask to the generator, the output is the synthesized defect image whose pixel-level annotation is the binary mask.

3. Datasets

3.1. Training Sets

MVTec-AD [28] is a commonly used public dataset in industrial vision tasks. MVTec-test contains multiple classes of defect images with pixel-level annotations, and MVTec-train contains many normal images. Therefore, we employed MVTec-AD to construct the training and testing sets of MDGAN.

In this work, experiments were conducted on four items, including the grid, zipper, capsule, and metal nut in MVTec-AD. Figure 3 shows the training images cropped from the original large images. It can be seen that there are multiple complex texture regions in these items. In addition, the phone band images from the actual production line were used in this work. We classified the phone band defects into dirty, roll, and scratch, to facilitate the network in reducing unnecessary blending. In addition, the defect samples for subsequent segmentation testing needed to be pre-preserved. The numbers of the relevant datasets are shown in Table 1. The training sets for generation and segmentation were taken from the same original images. *Zipper-combined* means there are multiple classes of defects in one image, which were only used in segmentation testing. As shown in Table 1, we replaced some long defect names with initials, to simplify the description.

3.2. Testing Sets

We employed two methods to obtain binary masks to construct the rich MDGAN test sets. First, MVTec-AD provided many binary masks of defect samples, which characterized the shapes of industrial defects and could be cropped as inputs for MDGAN. Second, to generate more defects with different shapes, we acquired a large number of binary masks based on Perlin noise [29]. Figure 4 shows the process of Perlin-noise-based mask generation. The normal images and binary masks were cropped to obtain image pairs. Based on the above two methods, we could obtain an arbitrary number of test sets.

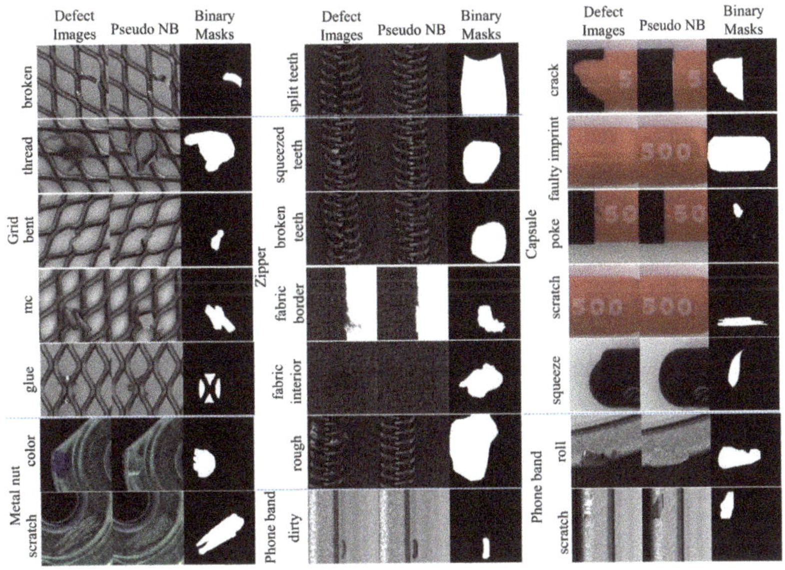

Figure 3. Non-uniform structure images in training sets. *Defect Images* (256^2) are cropped from the original real images. *Pseudo NB* (256^2) are the cropped constructed pseudo-normal backgrounds. *Binary Masks* (256^2) are the cropped binary annotation images. In particular, the size of training images in *Capsule* and *phone band* were 320^2 and 128^2, respectively.

Table 1. The number of datasets (original/cropped images). *Training* means the training sets of MDGAN. *EL/AUG* means the number of augmented samples from MDGAN (*EL*) and traditional methods (*AUG*). *RAW* and *Seg-test* are the real training and testing sets of segmentation, respectively.

Item Defect \ Dataset		Training	EL/AUG	RAW	Seg-Test
Metal nut	scratch	18/1282	900	18/848	5/23
	color	17/1108	800	17/919	5/19
	bent			19/656	6/23
	flip			19/931	4/19
Grid	bent	10/820	500	10/507	2/26
	broken	10/775	500	10/500	2/33
	glue	9/542	700	9/323	2/29
	mc	9/580	800	9/200	2/21
	thread	8/677	500	8/493	3/20
Zipper	broken teeth (bt)	14/771	800	14/597	5/23
	fabric border (fb)	12/817	900	12/521	5/44
	fabric interior (fi)	11/741	1000	11/465	5/38
	rough	12/1105	600	12/814	5/14
	split teeth (spt)	13/867	800	13/620	5/37
	squeezed teeth (sqt)	12/658	1000	12/479	4/22
	combined				16/87
Capsule	crack	23/643	700	23/459	5//12
	faculty imprint (fm)	22/672	900	22/309	4/7
	poke	21/518	800	21/358	4/13
	scratch	23/806	600	23/509	4/19
	squeeze	20/861	500	20/650	4/25
Phone band	scratch	24/496	550	24/258	5/11
	roll	52/1031	550	52/924	12/27
	dirty	49/497	550	49/344	12/24

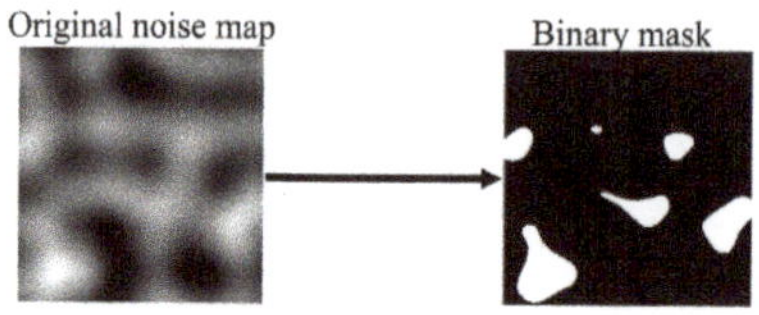

Figure 4. Example of generating a binary mask from Perlin noise. Set the values greater than 0.4 in the *Original noise map* (256^2) to 1 and other values to 0 to obtain a *Binary mask* (256^2).

4. Experiments

MDGAN was used to synthesize defect images based on the above datasets. Except for the metal nut and capsule, which involved three-channel color images, the images consisted of single-channel grayscale images. Section 4.2 shows the generated quality, diversity, and annotation accuracy of MDGAN. The effectiveness of BRM and DDM are demonstrated in Section 4.3. We also compare MDGAN with the most commonly used CycleGAN in Section 4.5, to certify the effectiveness of our methods. Then we assess the advantages of our synthetic samples over traditional augmented results. Finally, we explore the possibility of defect transfer using MDGAN in Section 4.6.

4.1. Implementations

To construct the pseudo-normal backgrounds, we first obtain the main area where defects may occur in defect image D and select normal image N by thresholding, and draw the minimum external rectangular boxes of these areas. Then we conduct affine transformation for D and N, so that the length and width of the two boxes are parallel and the center points are overlapped. Finally, the defect contents in D are filled by N, and the filled image is reversely transformed to the attitude of the original D. These two affine transformations can be simplified as Equations (1)–(3). We performed the above processes based on the OpenCV library, where functions such as cv2.getRotationMatrix2D() and cv2.warpAffine() were used.

MDGAN was trained for each type of defect under each item. We augmented the training image pairs by rotation, flipping, and random cropping. The binary mask was normalized to $[-1, 1]$ when inputting to the MDGAN and $[0, 1]$ when calculating losses. The number of output channels of the 12 convolution layers of the generator from input to output were 64, 128, 256, 256, 512, 512, 512, 256, 256, 128, and 64, respectively; those of the discriminator were 32, 128, 256, 256, 512, and 1, respectively. The dimension of latent space was 8, and the hyperparameters of Equation (15) were set as $\lambda_r = 10$, $\lambda_d = 15$, $\lambda_g = 10$, and $\lambda_{gp} = 10$. The Adam optimizer was used with $\beta_1 = 0.5$, $\beta_2 = 0.999$ to train MDGAN, with a batch size = 20 and learning rate = 0.0004. We trained the MDGAN for 500 iterations on one NVIDIA GeForce RTX 3090 GPU of a sever with Intel(R) Xeon(R) Gold 622306R CPU @ 2.90 GHz. During training, backgrounds could be retained without the last BRM in some datasets, which could simplify the architecture and obtain a better transition between normal backgrounds and defect regions. Therefore, we eliminated the last BRM in the defects of the grid.

4.2. Synthetic Results

Some synthetic defect samples are shown in Figures 5 and 6. As shown in Figure 5, MDGAN achieved defect image generation for all categories. Good synthesis results were obtained for complex and weak defects, such as the fine-grained texture defects (zipper-fb, zipper-fi, grid-thread, etc.), color defects (metal nut-color), metallic defects (grid-mc), and weak defects (phone band). As seen from the *Defect contents* in Figure 5, the generated defect contents were realistic and accurately distributed in the locations marked by the mask, achieving an accurate correspondence between the synthetic defect images and the input binary masks. In addition, MDGAN preserved the background structures in the *Grid* reasonably well, despite canceling the last BRM. In summary, with the help of BRM and DDM, MDGAN was able to generate realistic defect images, while preserving the original real backgrounds outside the annotations and achieved a natural transition between generated defects and real background, so that the generated defect images were accurately labeled by input binary masks.

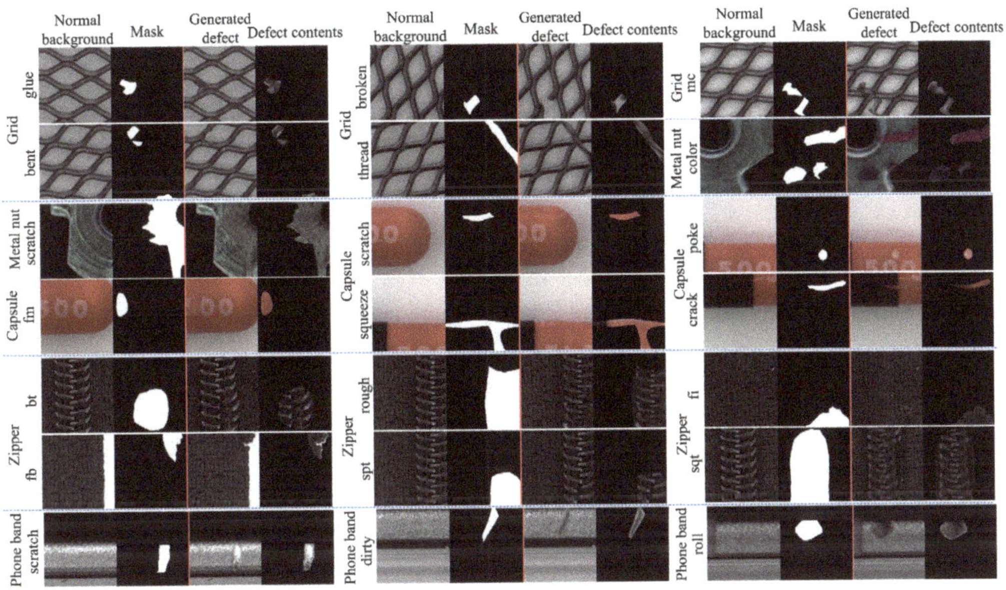

Figure 5. Synthesis results for each item. *Normal background* indicates the real background image, *Mask* is the corresponding binary mask, and both were input into the generator to obtain the synthetic defect image *Generated defect*. *Defect contents* were segmented by the *Mask* from the *Generated defect*.

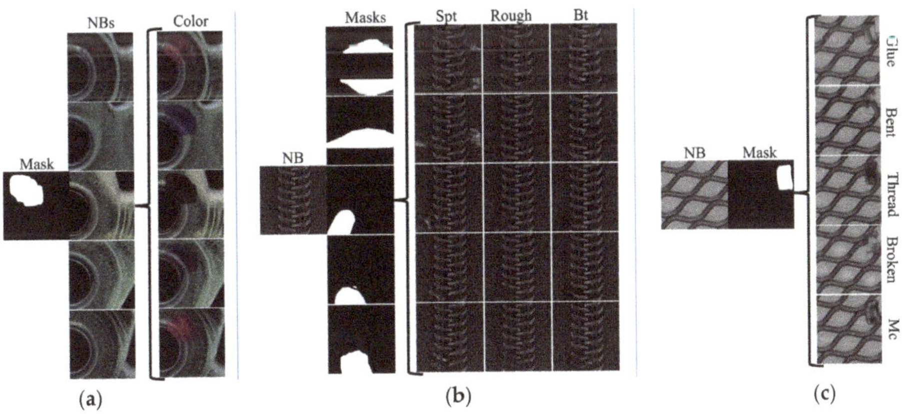

Figure 6. Examples of one-to-many generation. *NB* and *mask* denote the normal background and the binary mask input into MDGAN, respectively. (**a**) Background variation. *Color* denotes the generated metal nut color defects. (**b**) Defect shape and texture variation. *Spt, Rough*, and *Bt* denote the three types of zipper defects generated from the same background. (**c**) Defect texture variation.

Controllability of the backgrounds, defect shapes, and defect textures. We separately processed these three aspects, to verify that MDGAN could generate various defect images for a normal background and obtain accurate annotations. Figure 6 shows the generated results when normal backgrounds, defect shapes, and defect textures were changed, respectively. First, the defect images in Figure 6a shared the same binary mask. This shows that BRM added different backgrounds to the generator, to obtain multiple defect samples, whose annotations were the same, and the defects varied with the background. Second, Figure 6b shows the generated defect images of the zipper, with the same background *NB* and five different masks. Defect images in the same row shared the same mask on the left side. It can be seen that the BRM could replace different background regions with different masks, which assisted MDGAN in generating multiple defects with different shapes and contents on the same normal image. Third, Figure 6c shows multiple categories of generated defects in the grid, where the defect images shared the same normal background (*NB*) and annotations (*Mask*), but had different defective textures. As seen from Figure 6c and each row in Figure 6b, MDGAN could produce multiple types of defect in the same specified region of the same normal image when the training sets were different.

In summary, based on BRM and DDM, MDGAN achieved independent control of the background, defect shape, and defect texture, and was able to generate a huge number of diverse and high-quality defect samples for a normal image. In reality, various defects may appear in a normal image, and our experimental results fit actual situations. Moreover, since MDGAN accurately controls defect shapes and preserves backgrounds using the given binary mask, the output defect regions are precisely labeled by the mask. Thus, our synthetic samples can be used to train segmentation models, which is beneficial for detecting defects.

4.3. Ablation Experiments

To verify the effect of BRM on the background retention and DDM on generated quality, we separately removed the two modules and performed ablation experiments on the above datasets, where the remaining settings were exactly the same as in the formal experiments.

Without BRM. Figure 7 shows the comparison results of the with/without (wo) BRM for the same pairs of test images. As can be seen, the results generated without BRM have drastically and unreasonably modified backgrounds, while MDGAN retained the details of the backgrounds well. In particular, substantial modifications to the original backgrounds led to the loss of real structures in the *phone band*. The generated defect contents did not appear accurately in the annotated positions without BRM, resulting in inaccurate binary annotations.

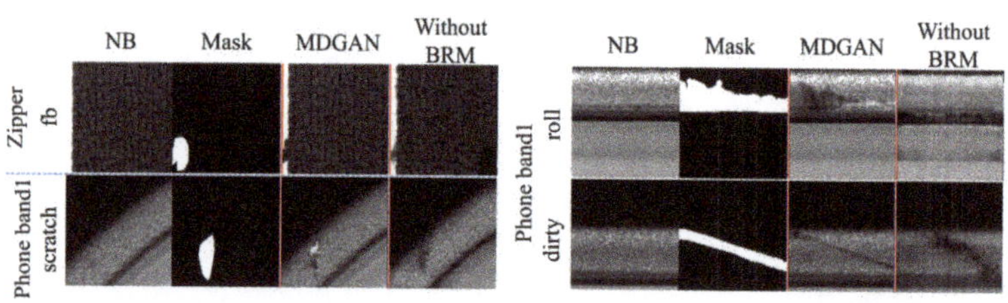

Figure 7. Comparison of with and without BRM. *Background* and *Mask* are input normal backgrounds and binary masks, *MDGAN* is the synthetic result of MDGAN. *Without BRM* is the synthetic result after canceling BRM.

In order to quantify the background retention ability of BRM, we employed the structure similarity index measure (SSIM) to evaluate the background similarity. The SSIM value is between [0, 1], and the larger the value, the more similar the two images. Using the same test set, and based on the with/wo BRM models, to synthesize 1000 defective samples, the SSIM between the normal background areas of the output and the input was calculated. The mean SSIM (mSSIM) is shown in Table 2. It can be seen that the backgrounds generated by MDGAN were highly similar to the real backgrounds. Both the qualitative and quantitative results showed that BRM can modulate the real backgrounds in the feature map of the generator, which restricted the location of the generated defects, avoided the loss of backgrounds in the training, and finally facilitated MDGAN in generating defect samples with realistic backgrounds and accurate binary annotations.

Table 2. Quantitative results of the ablation experiments. Bold fonts indicate better results.

Dataset		mSSIM MDGAN	mSSIM wo BRM	Dataset		mSSIM MDGAN	mSSIM wo BRM
Capsule	crack	**0.998**	0.915	Zipper	fb	**0.996**	0.736
	fm	**0.998**	0.888		fi	**0.990**	0.793
	poke	**0.998**	0.921		rough	**0.890**	0.832
	scratch	**0.998**	0.937		bt	**0.932**	0.849
	squeeze	**0.998**	0.927		sqt	**0.996**	0.800
Grid	bent	**0.975**	0.917		spt	**0.899**	0.813
	broken	**0.977**	0.927	Metal nut	color	**0.997**	0.716
	glue	**0.964**	0.935		scratch	**0.997**	0.725
	mc	**0.977**	0.913	Phone band	dirty	**0.996**	0.734
	thread	**0.972**	0.902		roll	**0.997**	0.707
					scratch	**0.996**	0.665

Without DDM. All DDMs were removed in the discriminator. The mask and image were channel-level connected and input into the discriminator. To match the formal experiments, the number of output channels of the first three convolutional layers was doubled. As shown in Figure 8, the discriminator's binding on the defect quality decreased after removing the DDM. As shown in Figure 8a, the generated defects without DDM contained unrealistic stripes and the zipper teeth were not smooth enough, being totally different from the real images. In addition, as shown in Figure 8b, there were unreasonable black contents in the generated *capsule-crack* without DDM, while MDGAN smoothly stripped the black blocks and generated clear red defects on them. On *capsule-fm*, without DDM covered the annotated region with only fuzzy uniform white blocks, while MDGAN generated realistic and detailed scratches. That is, the networks could not inherit and generate the real defects without DDM. Overall, the addition of DDM assisted the discriminator in focusing on both on the whole image and the local defects, to improve judgment and enable MDGAN to accurately capture the stripes and grayscale distribution of defects and generate more realistic and higher quality images.

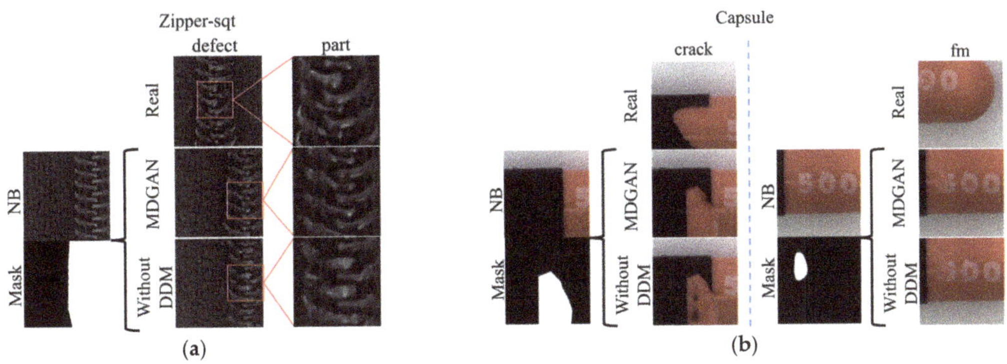

Figure 8. Comparison of with and without DDM. *Real* is the real defect image. *MDGAN* and *Without DDM* are the generated defect samples with and without DDM, respectively. (**a**) One type of effect without DDM. *part* is the enlarged defect in the red box of the *defect*. (**b**) Another effect.

4.4. Comparison with CycleGAN

CycleGAN-based methods are leading the way in defect synthesis. To verify the advantages of MDGAN over other methods, CycleGANs were trained on the above datasets.

To help CycleGAN retain the normal background, we added a L1 loss between the input and output to the original losses [17]. Some of results generated by CycleGAN and MDGAN based on the same backgrounds are shown in Figure 9. Despite the addition of the L1 loss, CycleGAN still modified the normal backgrounds, and it could convert the input normal images into defect images for the zipper-fi, zipper-sqt, zipper-spt, and grid-broken. Moreover, there were stretches and artifacts in the generated results of the capsule and phone band. This indicated that CycleGAN could only convert the source image into the most similar target image seen during the training for few-sample datasets, resulting in either a failed generation or loss of structures in the source image. In contrary, MDGAN generated realistic defect images for each category of input normal backgrounds and retained the original normal textures.

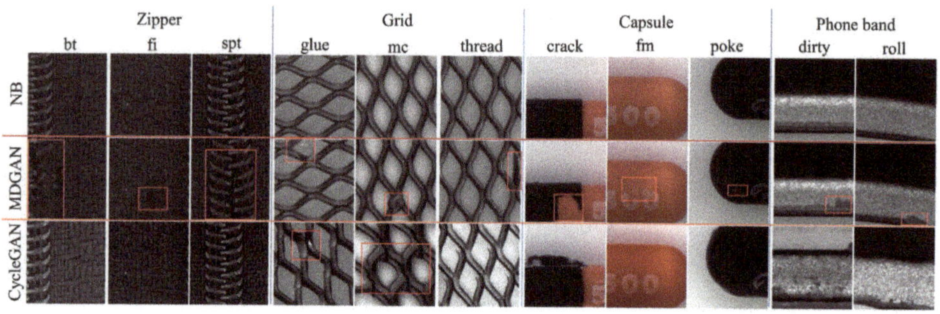

Figure 9. Generated results of *MDGAN* and *CycleGAN* with L1 loss based on the same normal backgrounds *NB*. *MDGAN* was generated from the *NB* and a binary mask by MDGAN, and *CycleGAN* was generated from the *NB* by CycleGAN. Defects are indicated by red boxes.

To quantify the generation quality, 1000 samples were generated on the same testing set based on CycleGAN and MDGAN, and the FID [30] between the generated and the real defect datasets was calculated. The lower the FID, the closer the generated and real features. As shown in Table 3, MDGAN obtained smaller FID than CycleGAN for most defect types. Nevertheless, CycleGAN had a lower FID for the grid-glue, grid-bent, and zipper-fb. In grid-glue, as shown in Figure 9, CycleGAN simply memorized training sets and translated all test images into the most similar training images, resulting in a low FID. In the other two items, MDGAN generated defects in the given annotated regions, and when the mask shapes used in testing differed significantly from the real, the generated features were a little far away from the real features, resulting in a high FID.

Table 3. FID between generated results and real images. Bold fonts indicate better results.

Dataset	Model	MDGAN	CycleGAN + L1
Capsule	squeeze	**81.72**	119.2
	crack	**51.99**	111.45
	fm	**39.99**	94.96
	scratch	**45.02**	99.39
	poke	**60.86**	97.01
Grid	bent	78.84	**63.90**
	broken	**75.98**	112.13
	glue	114.23	**94.76**
	mc	**99.39**	102.92
	thread	**79.74**	89.63
Zipper	bt	**45.69**	50.67
	rough	**48.81**	56.47
	fb	88.08	**86.93**
	fi	**54.96**	59.15
	sqt	**29.13**	136.25
	spt	**38.4**	42.93
Metal nut	color	**69.93**	95.52
	scratch	**66.90**	81.85
Phone band	dirty	**115.31**	129.83
	scratch	**112.54**	169.62
	roll	**107.88**	147.69

Overall, MDGAN constructed pseudo-normal images, to efficiently acquire pairs of inputs without relying on CycleGAN and obtained a higher quality than CycleGAN for most items. Moreover, MDGAN imports random noise to construct latent representations for real defects; in contrast with CycleGAN, which improves both the randomness of the generation and the diversity of results, and is thus more suitable for few-sample synthesis. Furthermore, due to the background preservation with BRM and the constraints on quality from DDM, MDGAN successfully generated defect images with pixel-level annotations, while preserving the real normal backgrounds. Meanwhile, CycleGAN generated samples with only image-level annotations, which could not be used for defect segmentation experiments.

4.5. Detection Performance

To verify the advantages of MDGAN for detection over traditional augmentation, synthetic samples from MDGAN were added to the real segmentation training set RAW, to construct EL; and the traditional brightness adjustment, rotation, and noise injection were adopted to obtain the training set AUG. The number of datasets is shown in Table 1. Two types of segmentation network, UNet and sResNet, were trained using the above three training sets. sResNet is a UNet-like segmentation model, where the skip-connections are removed and the convolution layers are replaced by the Res-blocks in ResNet [31]. UNet consists of six downsampling and six upsampling layers whose number of output channels are 32, 64, 128, 256, 512, 256, 512, 256, 128, 64, 32, and 1. sResNet consists of four downsampling layers, four res-blocks, and four upsampling layers, whose number of output channels are 32, 64, 128, 256, 256, 256, 256, 256, 128, 64, 32, and 1. The output of the two networks is a single-channel map with the same size as the input. We used cross entropy loss to train the two models. The Adam optimizer was used with $\beta_1 = 0.5$, $\beta_2 = 0.999$, batch size = 50, and learning rate = 0.0005 on an NVIDIA GeForce RTX 3090 GPU. The mIoU (mean Intersection over Union) and F1 coefficient were calculated on the same test set at 500th iterations.

The mean testing results of each item are shown in Table 4, where higher values indicate a better detection performance. It can be seen that EL greatly improved the detection performance in contrast to AUG and RAW. The mean results were substantially improved in EL. Among the mean results of the various items, EL outperformed RAW and AUG, with an mIoU improvement up to **7.3%** (UNet-capsule) and F1 up to **6.2%** (sResNet-capsule). On the contrary, AUG reduced the overall detection results (zipper). Figure 10 shows a qualitative comparison of the segmentation results; the results of EL had less over-kill and escape versus RAW and AUG and were closer to the ground truths.

Table 4. Comparison of the mean test results for the three training sets. Bold fonts indicate the optimal IoU and F1 among the three results. R and A denote RAW and AUG, respectively.

Network	**UNet**						**sResNet**					
Index	**mIoU**			**F1**			**mIoU**			**F1**		
Dataset	R	A	EL	R	A	EL	R	A	EL	R	A	EL
Metal nut	0.742	0.748	**0.768**	0.846	0.848	**0.86**	0.733	0.738	**0.751**	0.844	0.84	**0.847**
Grid	0.619	0.603	**0.652**	0.755	0.714	**0.782**	0.636	0.652	**0.657**	0.769	0.783	**0.786**
Zipper	0.732	0.719	**0.775**	0.970	0.959	**0.975**	0.745	0.723	**0.764**	0.977	0.959	**0.979**
Capsule	0.548	0.571	**0.621**	0.692	0.713	**0.748**	0.472	0.483	**0.561**	0.618	0.627	**0.689**
Phone band	0.656	0.646	**0.666**	0.979	0.985	**0.987**	0.65	0.653	**0.681**	0.984	0.981	**0.985**

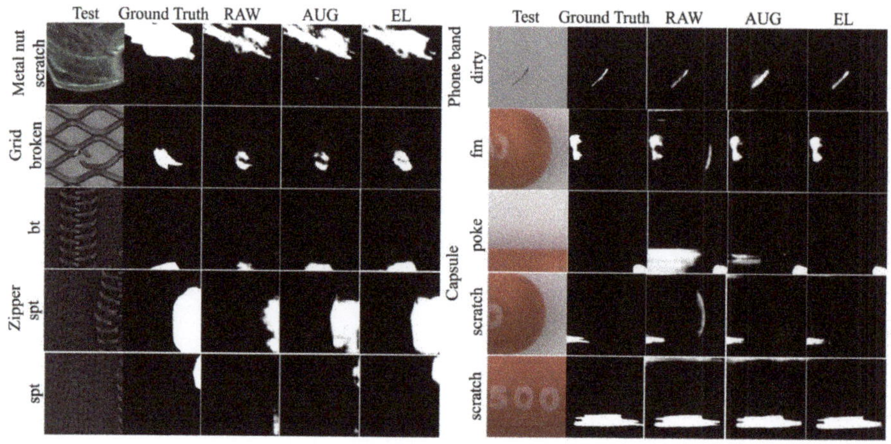

Figure 10. Comparison of UNet-based segmentation results. *Test* and *Ground Truth* denote the test images and the corresponding ground truth (256^2). *RAW*, *AUG*, and *EL* denote the test results obtained by the three training sets (256^2).

Overall, the inclusion of synthetic samples alleviates the data problems of class imbalance, lack of diversity, and few samples, and improves the detection performance. Compared with the traditional augmented samples, there were various backgrounds, defect contents, and shapes of binary annotations in our synthesized datasets, which were very different from the original training sets and could assist the networks in seeing and remembering richer defect information during training. When testing, the EL-trained models learned more knowledge and were more conducive to detecting unseen samples. This demonstrated that the synthetic samples from MDGAN could be used to train the supervised segmentation networks and that our work has practical application value.

4.6. Defect Transfer

It can be seen from the above that MDGAN achieved defect image generation with complete retention of backgrounds. Therefore, we could try to employ the MDGAN-trained source dataset to perform defect synthesis on a target dataset and used only the synthetic target defect images to train models to detect real defects in the target dataset, thus achieving defect transfer and zero-shot detection of the target dataset.

Figure 11 shows the related images and procedure of defect transfer, where the first and second rows in the *Source* are from phone bands on the reverse side (phone band1) and curved glass, respectively, and those in *Target* are from phone band on the front side (phone band2) and phone cover glass, respectively. It can be seen that the *Source* and *Target* have different structures, but their defect contents were similar. In consequence, we trained MDGAN based on phone band1 and curved glass and tested it on phone band2 and phone cover glass. The generated results are shown in the middle parts of Figure 11. It can be seen that, despite the differences in the backgrounds on the two sides of the defect transfer, MDGAN still generated defects in the annotated areas (*Defect contents*) and obtained defect images (*GD*) similar to the real ones. As shown in *Res*, MDGAN only modified the annotated regions and fully retained the targeted normal backgrounds after transfer.

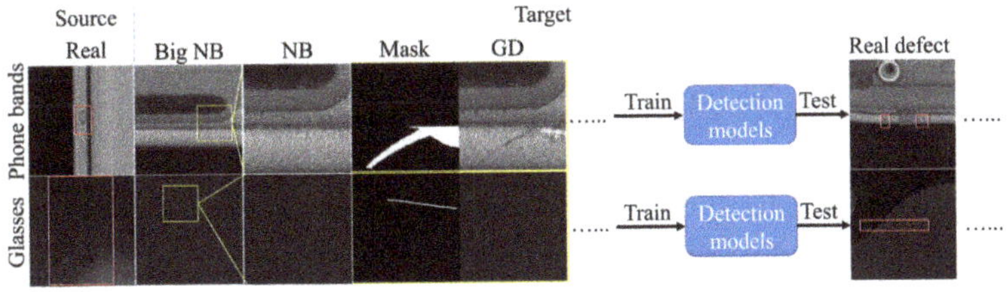

Figure 11. The first row shows the two types of *Phone bands*, and the second is the two types of *glass*. Red boxes indicate real defects. *Source* is the defect image used for training MDGAN. *Target* means the images to be transferred. *NB* is the normal background cut from *Big NB*. *Mask* is the input binary mask. *GD* is the generated target defect image. *Real defect* is defect image of the target datasets to be tested.

In order to verify the effect of the transferred defect samples, we only used the transferred defect samples to train the segmentation networks and adopted the real images as the test set. The number of datasets and the test results are shown in Table 5. It can be seen that good results were obtained on the real test set based on segmentation models trained only using transferred defect samples. The AUC (area under curve) was up to **0.971** (UNet) for phone band1 detection and **0.991** (sResNet) for phone cover glass detection.

Table 5. The numbers of transferred training sets and real test sets, and the test results of segmentation.

Datasets	Train	Test	Models	mIoU	F1	AUC
Phone band2	2508	186	UNet sResNet	0.464 0.425	0.622 0.579	0.971 0.969
Phone cover glass	1302	115	UNet sResNet	0.458 0.439	0.636 0.619	0.987 0.991

Based on the generated samples from MDGAN trained using another dataset, we achieved defect detection on two real industrial surface defect datasets, showing that MDGAN could achieve defect transfer between datasets with similar defects but different backgrounds. Hence, we could train MDGAN based on the existing source defect datasets and then synthesize a large number of defect samples on the new target normal backgrounds when the type of products changed. As long as the defect textures are similar, the resource consumption for recollecting and labeling datasets can be greatly reduced by MDGAN, which is highly valuable for intelligent manufacturing.

5. Conclusions

This paper proposes MDGAN, to tackle the problems of falsifying backgrounds, lack of pixel-level annotations, and less attention being given to non-uniform complex structures in the current defect synthesis methods. Guided by binary masks, MDGAN modulates the real background into the generator using BRM and employs DDM to achieve discrimination of both local and global information. Due to the background-preserving effect of BRM and the quality constraint of DDM, MDGAN solves the problem of falsifying backgrounds and enriches the diversity of datasets. Finally, defect samples with accurate pixel-level annotations on multiple datasets with complex textures were synthesized using MDGAN. In addition, the qualitative and quantitive results showed that MDGAN obtained a better quality than the commonly used CycleGAN. The segmentation results demonstrated that the synthetic samples from MDGAN greatly improved the detection performance, with an improvement of IoU up to **7.3%** and F1 up to **6.2%**. Furthermore, based on the excellent background retention capability of MDGAN, we successfully synthesized target defect images using MDGAN trained on a source dataset and achieved the defect detection of real target samples, based only on the synthesized samples.

There are some aspects that need to be improved in our work. Since the binary masks are directly given by the test sets, the feature and the diversity of the generated defect shapes can be limited. In follow-up work, we will explore the synthesis method of producing both defect images and annotations using networks, as a way to enrich the defect shapes.

Author Contributions: Conceptualization, Z.Z. and F.S.; Formal analysis, J.W. and C.L.; Investigation, J.W. and C.L.; Methodology, J.W., Z.Z. and F.S.; Resources, Z.Z.; Software, Z.Z.; Writing—original draft, J.W.; Writing—review and editing, J.W., Z.Z., F.S. and C.L. All authors have read and agreed to the published version of the manuscript.

Funding: This work was supported by the Youth Innovation Promotion Association, CAS (2020139).

Data Availability Statement: Not applicable.

Conflicts of Interest: The authors declare no conflict of interest.

References

1. Tabernik, D.; Šela, S.; Skvarč, J.; Skočaj, D. Segmentation-based deep-learning approach for surface-defect detection. *J. Intell. Manuf.* **2019**, *31*, 759–776. [CrossRef]
2. Ren, R.; Hung, T.; Tan, K.C. A Generic Deep-Learning-Based Approach for Automated Surface Inspection. *IEEE Trans. Cybern.* **2017**, *48*, 929–940. [CrossRef] [PubMed]
3. Chen, Y.; Ding, Y.; Zhao, F.; Zhang, E.; Wu, Z.; Shao, L. Surface Defect Detection Methods for Industrial Products: A Review. *Appl. Sci.* **2021**, *11*, 7657. [CrossRef]
4. Saksham, J.; Gautam, S.; Paruthi, A.; Soni, U.; Kumar, G. Synthetic data augmentation for surface defect detection and classification using deep learning. *J. Intell. Manuf.* **2022**, *33*, 1007–1020.
5. Czimmermann, T.; Ciuti, G.; Milazzo, M.; Chiurazzi, M.; Roccella, S.; Oddo, C.M.; Dario, P. Visual-Based Defect Detection and Classification Approaches for Industrial Applications—A SURVEY. *Sensors* **2020**, *20*, 1459. [CrossRef] [PubMed]
6. Shorten, C.; Khoshgoftaar, T.M. A survey on Image Data Augmentation for Deep Learning. *J. Big Data* **2019**, *6*, 60. [CrossRef]
7. Goodfellow, I.; Pouget-Abadie, J.; Mirza, M.; Xu, B.; Warde-Farley, D. Generative adversarial nets. *Adv. Neural Inf. Process. Syst.* **2014**, *15*, 2672–2680.
8. Makhzani, A.; Shlens, J.; Jaitly, N.; Goodfellow, I.; Frey, B. Adversarial autoencoders. *arXiv* **2015**, arXiv:1511.05644.
9. Kingma, D.P.; Welling, M. Auto-encoding variational bayes. *arXiv* **2013**, arXiv:1312.6114.
10. Karras, T.; Laine, S.; Aila, T. A Style-Based Generator Architecture for Generative Adversarial Networks. In Proceedings of the 2019 IEEE/CVF Conference on Computer Vision and Pattern Recognition (CVPR), Long Beach, CA, USA, 15–20 June 2019; pp. 4396–4405. [CrossRef]
11. Karras, T.; Laine, S.; Aittala, M.; Hellsten, J.; Lehtinen, J.; Aila, T. Analyzing and improving the image quality of stylegan. In Proceedings of the 2020 IEEE/CVF Conference on Computer Vision and Pattern Recognition (CVPR), Seattle, WA, USA, 16–18 June 2020; pp. 8110–8119. [CrossRef]
12. Choi, Y.; Choi, M.; Kim, M.; Ha, J.W.; Kim, S.; Choo, J. Stargan: Unified generative adversarial networks for multi-domain image-to-image translation. In Proceedings of the IEEE Conference on Computer Vision and Pattern Recognition, Salt Lake City, UT, USA, 18–23 June 2018; pp. 8789–8797.
13. Choi, Y.; Uh, Y.; Yoo, J.; Ha, J.W. Stargan v2: Diversified image synthesis for multiple domains. In Proceedings of the 2020 IEEE/CVF Conference on Computer Vision and Pattern Recognition (CVPR), Seattle, WA, USA, 13–19 June 2020; pp. 8188–8197.
14. Liu, J.; Wang, C.; Su, H.; Du, B.; Tao, D. Multistage GAN for Fabric Defect Detection. *IEEE Trans. Image Process.* **2019**, *29*, 3388–3400. [CrossRef] [PubMed]
15. Niu, S.; Li, B.; Wang, X.; Lin, H. Defect Image Sample Generation with GAN for Improving Defect Recognition. *IEEE Trans. Autom. Sci. Eng.* **2020**, *17*, 1611–1622. [CrossRef]
16. Zhang, G.; Cui, K.; Hung, T.-Y.; Lu, S. Defect-GAN: High-Fidelity Defect Synthesis for Automated Defect Inspection. In Proceedings of the 2021 IEEE Winter Conference on Applications of Computer Vision (WACV), Waikoloa, HI, USA, 3–8 January 2021; pp. 2523–2533. [CrossRef]
17. Su, B.; Zhou, Z.; Chen, H.; Cao, X. SIGAN: A Novel Image Generation Method for Solar Cell Defect Segmentation and Augmentation. *arXiv* **2021**, arXiv:2104.04953.
18. Niu, S.; Li, B.; Wang, X.; Peng, Y. Region- and Strength-Controllable GAN for Defect Generation and Segmentation in Industrial Images. *IEEE Trans. Ind. Inf.* **2021**, *18*, 4531–4541. [CrossRef]
19. Tsai, D.-M.; Fan, S.-K.S.; Chou, Y.-H. Auto-Annotated Deep Segmentation for Surface Defect Detection. *IEEE Trans. Instrum. Meas.* **2021**, *70*, 1–10. [CrossRef]
20. Jalayer, M.; Jalayer, R.; Kaboli, A.; Orsenigo, C.; Vercellis, C. Automatic Visual Inspection of Rare Defects: A Framework based on GP-WGAN and Enhanced Faster R-CNN. In Proceedings of the 2021 IEEE International Conference on Industry 4.0, Artificial Intelligence, and Communications Technology (IAICT), Bandung, Indonesia, 27–28 July 2021; pp. 221–227. [CrossRef]

21. Zhu, J.-Y.; Park, T.; Isola, P.; Efros, A.A. Unpaired image-to-image translation using cycle-consistent adversarial networks. In Proceedings of the IEEE International Conference on Computer Vision, Venice, Italy, 22–29 October 2017; pp. 2223–2232.
22. Mirza, M.; Osindero, S. Conditional generative adversarial nets. *arXiv* **2014**, arXiv:1411.178.
23. Isola, P.; Zhu, J.-Y.; Zhou, T.; Efros, A.A. Image-to-Image Translation with Conditional Adversarial Networks. In Proceedings of the 2017 IEEE Conference on Computer Vision and Pattern Recognition (CVPR), Honolulu, HI, USA, 21–26 July 2017; pp. 5967–5976. [CrossRef]
24. Ronneberger, O.; Fischer, P.; Brox, T. U-Net: Convolutional Networks for Biomedical Image Segmentation. *Int. Conf. Med. Image Comput. Comput. Assist.* **2015**, *9351*. [CrossRef]
25. Zhang, L.; Wen, T.; Shi, J. Deep Image Blending. In Proceedings of the 2020 IEEE Winter Conference on Applications of Computer Vision (WACV), Snowmass, CO, USA, 1–5 March 2020; pp. 231–240. [CrossRef]
26. Arjovsky, M. Soumith Chintala, and Léon Bottou, Wasserstein Generative Adversarial Networks. In Proceedings of the 34th International Conference on Machine Learning, Sydney, NSW, Australia, 6–11 August 2017. [CrossRef]
27. Gulrajani, I.; Ahmed, F.; Arjovsky, M.; Dumoulin, V.; Courville, A.C. Improved Training of Wasserstein GANs. *Adv. Neural Inf. Process. Syst.* **2017**, *30*. [CrossRef]
28. Bergmann, P.; Fauser, M.; Sattlegger, D.; Steger, C. MVTec AD—A Comprehensive Real-World Dataset for Unsupervised Anomaly Detection. In Proceedings of the 2019 IEEE/CVF Conference on Computer Vision and Pattern Recognition (CVPR), Long Beach, CA, USA, 15–20 June 2019; pp. 9584–9592. [CrossRef]
29. Perlin, K. Improving noise. *ACM Trans. Graph.* **2002**, *21*, 681–682. [CrossRef]
30. Heusel, M.; Ramsauer, H.; Unterthiner, T.; Nessler, B.; Hochreiter, S. Gans trained by a two time-scale update rule converge to a local nash equilibrium. *Adv. Neural Inf. Process. Syst. (NeurIPS)* **2017**, *30*, 6629–6640. [CrossRef]
31. He, K.; Zhang, X.; Ren, S.; Sun, J. Deep residual learning for image recognition. In Proceedings of the IEEE Computer Society Conference on Computer Vision and Pattern Recognition (CVPR), Las Vegas, NV, USA, 27–30 June 2016; pp. 770–778. [CrossRef]

machines

MDPI

Article

Dimensional Error Prediction of Grinding Process Based on Bagging–GA–ELM with Robust Analysis

Lei Yang [1,*], Yibo Jiang [2], Hua Liu [3] and Xianna Yang [4]

1 ZJU-Hangzhou Global Scientific and Technological Innovation Center, Hangzhou 311200, China
2 Jiaxing Research Institute, Zhejiang University, Jiaxing 314051, China
3 Wanxiang Qianchao Co., Ltd., Hangzhou 311215, China
4 China Industrial Control Systems Cyber Emergency Response Team, Beijing 100040, China
* Correspondence: lei.yang@zju.edu.cn

Abstract: Grinding, which determines the final dimension of parts, is an important process in manufacturing companies. In praxis, in order to avoid quality problems on the customer's side, an online dimension check is normally used after the grinding process to ensure the product dimensions; however, it is always hysteretic and needs extra space and machine investment. To deal with the issue, dimensional error prediction of the grinding process is highly needed, and does not require extra space or machinery, as well as having better real-time performance. In this paper, a dimensional error prediction algorithm using principal component analysis (PCA), extreme learning machine (ELM), genetic algorithm (GA), and ensemble strategy (bagging algorithm) is designed. Specifically, PCA is used as a pre-treatment method to extract the main relevant components, then a bagging–GA–ELM model is constructed to predict the final product dimensional error after the grinding process, in which extreme learning machine (ELM) is utilized as a basic framework because of its fast calculation speed. GA, with its excellent global optimization capability, is implemented to search optimal input weights and thresholds of ELM, enabling ELM to obtain a better prediction performance. In addition, considering the complex environment of the industrial field, the bagging algorithm is employed to enhance the anti-noise ability of the proposed algorithm. Finally, the proposed algorithm is verified by a case from a bearing company.

Keywords: dimensional error prediction; grinding process; bagging–GA–ELM; robust analysis

Citation: Yang, L.; Jiang, Y.; Liu, H.; Yang, X. Dimensional Error Prediction of Grinding Process Based on Bagging–GA–ELM with Robust Analysis. *Machines* **2023**, *11*, 32. https://doi.org/10.3390/machines11010032

Academic Editor: Kai Cheng

Received: 27 November 2022
Revised: 24 December 2022
Accepted: 26 December 2022
Published: 27 December 2022

1. Introduction

Grinding is a vital manufacturing process, which has a significant impact on a final part's surface quality and cost [1]. With the increasing competition in manufacturing industries, the pursuit of higher quality and lower costs puts the focus on the grinding process. [2–4]. In fact, the grinding process is complex and the final quality and cost are highly related to the process parameters; therefore, many researchers study the influence of the process parameters on quality and cost. Then, artificial intelligence (AI) techniques are used to optimize process parameters to improve product quality and process efficiency [5,6]. In the literature [7], an improved differential evolution algorithm (DE) based on double populations is employed to optimize the grinding operation, and, finally, the surface roughness is effectively reduced. Mondal et al. [8] use particle swarm optimization (PSO) to find the optimal parameters for minimizing the surface roughness of a crane-hook-pin in the grinding process. Furthermore, in order to obtain faster convergence speed and higher performance of models, the idea of a multi-algorithm fusion-based technique has gained a lot of attention in recent years [9–12]. Huang et al. [13] utilize the Taguchi method and grey relational analysis (GRA) to obtain the optimal parameters in the cylindrical grinding process, which successfully solves the problem of minimizing surface roughness and maximizing material removal rate.

On the other side, in practice, a quality check is usually used to avoid quality problems on the customer's side. Currently, the majority of quality checking operates offline, which increases the production throughput time and furthers quality risks. In addition, it also makes the process difficult to trace. Therefore, to reduce grinding cost and increase process control ability, many researchers focus on online quality checks and online quality predictions based on data collected in real-time [14–16].

However, an online quality check requires enterprises to provide extra space and investment. Moreover, the online quality check is always hysteretic; when a quality problem is detected, some further products which are already produced and in the transmission facility (conveyor, Truss manipulator), are already affected. Compared to online quality checks, online quality prediction does not need further space or equipment, and it also has the advantage of instantaneity. Once the quality problem is predicted based on real-time collected data and a prediction algorithm, the process can be stopped to avoid further loss. Therefore, online quality prediction has become a hot topic. Chen et al. [17] propose a novel acoustic signal-based detection method by combining a random forest algorithm (RF) and multiple linear regression model (MLR); at the end, the online prediction problem of tool wear in the grinding process is successfully solved. Li et al. [18] adopt a multi-scale attention convolutional neural network (CNN) to predict the final machining quality of the workpiece. It finally shows the high accuracy of online prediction of the material removal state in experiments.

Recently, researchers also tend to use neural networks with a more complex structure to obtain better prediction accuracy [19,20]. Guo et al. [21] present a novel online prediction system of surface roughness in the grinding process by combining the long short-term memory (LSTM) network with a hybrid feature selection approach; ultimately, it shows good prediction performance. Yin et al. [22] propose a reliable prediction model based on a genetic algorithm (GA) and deep neural network (DNN), and prove its feasibility in the prediction of the roughness and subsurface material damage in the real-time grinding process. However, it also should be noticed that, although the complex neuron network increases the accuracy of prediction, it also makes a significant reduction in calculation efficiency, which has difficulty meeting the requirements of high calculation efficiency in online quality prediction.

Extreme learning machine (ELM) [23] has received more attention for high computational speed and global approximation capability [24]. However, during the prediction process of ELM, some network parameters are determined randomly, which causes the fluctuation of prediction performance and also will lead to an accuracy reduction of product dimensional error prediction. Thus, the fusion of ELM and a heuristic algorithm is needed in order to obtain better prediction performance [25–27]. Liu et al. [28] propose a multi-kernels fault diagnostic model based on ELM and PSO; the suggested fault diagnostic model is tested by well-known rolling bearing data set and achieves successful identification of bearing faults. Sun et al. [29] obtain a prediction model of the extrusion grinding process by combining the ELM model with GA (GA–ELM), based on the comparison results of experiments, the GA–ELM model keeps the actual data prediction error within $\pm 4\%$. However, compared to GA, although the reason for the faster convergence speed of PSO is that the best particles have a significant influence on other particles, it also causes the problem of poor population diversity, which limits the PSO to obtain the global optimum. Therefore, GA has a better ability to acheive a globally optimal solution [30–33].

Moreover, as is well known, in a real industrial environment, the random interference and noise from various processes as well as environments have a non-negligible impact on the online prediction results. For ensuring the reliability of prediction results, it is necessary to improve the robustness of the prediction approach. Currently, the ensemble learning algorithms represented by bagging and boosting have been widely verified because of their ability to improve the accuracy and stability of prediction [34], Gao et al. [35] propose a novel material removal model for belt grinding with the integration of a boosting algorithm, which shows an effective improvement in predicting the material removal despite the complicated grinding environment. Bustillo et al. [36] verify the effectiveness of the bagging algorithm in the case of tool-life prediction in the turning process. The experiments show that the technique of bagging significantly improves the robustness of the prediction model. Furthermore, it should be noted that the bagging algorithm has the characteristics of simple operation, strong interpretability, and excellent performance. In general, the bagging algorithm has more obvious benefits than the boosting algorithm in preventing the overfitting problem and improving the noise immunity of the model [37,38]. Thus, the bagging algorithm is employed to improve the robustness of the proposed algorithm.

As mentioned above, considering the complex environment and processing property, to predict the grinding dimensional error, it is necessary to introduce an algorithm with excellent global search ability, fast calculation speed, high prediction accuracy, and good robustness. Therefore, a dimensional error prediction algorithm which combines the advantages of ELM, GA, and bagging algorithm, is proposed.

2. Materials and Methods

In this section, the theoretical background of the proposed algorithm is introduced first, which consists of PCA, GA, ELM, and bagging algorithm. Then, the proposed algorithm is explained in detail.

2.1. Principal Component Analysis

In order to improve the accuracy and robustness of the proposed algorithm, the feature parameters which are related to the final product dimension should be regarded as inputs of the proposed algorithm. Considering that excessive feature parameters will lead to a calculation efficiency reduction as well as the overlap of information, it is necessary to further extract the main relevant components from feature parameters to simplify the inputs of the proposed algorithm.

Principal component analysis (PCA) [39] is a technique used for reducing the dimension of data, which is widely employed in the analysis of the machining process for removing redundant variables and extracting the main relevant components. The calculation process of PCA is shown in Figure 1 and defined as follows:

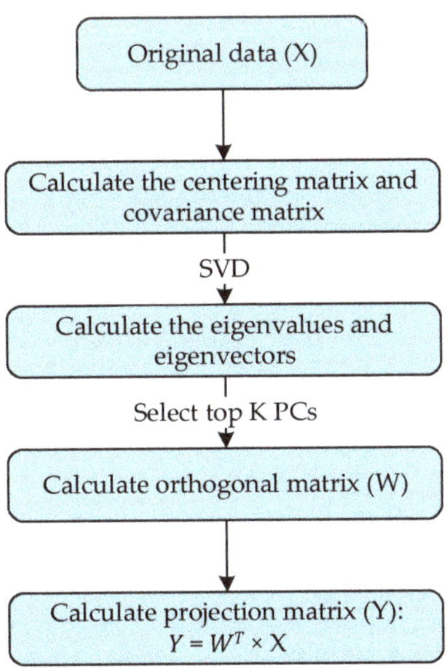

Figure 1. The flowchart of PCA.

Step 1. Suppose original data, named X, consists of N samples with D dimensional features; the centering matrix of original data is given by Equation (1) and the covariance matrix can be further calculated by Equation (2).

Step 2. Via singular value decomposition (SVD), the principal components (PCs) consist of eigenvalues, and eigenvectors are computed from the eigen-decomposition operation of the covariance matrix, which is shown in Equation (3).

Step 3. K (lower than D) eigenvectors are usually picked out to form the orthogonal matrix (W). For this aim, the conventional methodology is used to sort the eigenvalues from maximum to minimum and to select the top K PCs. Finally, the projection matrix Y can be calculated as $Y = W^T \times X$.

$$X_c = \sum_{i=1}^{N} (x_i - \overline{X}) / N \quad (1)$$

$$C = X_c \times X_c{}^T \quad (2)$$

$$C = V \times E \times V^T \quad (3)$$

where $\overline{X}$ and X_c are the mean vector and centering matrix of original data, respectively, C denotes the covariance matrix, V and E represent the eigenvalue matrix and eigenvector matrix, respectively.

2.2. *Extreme Learning Machine*

Due to the characteristics of simple structure, fast calculation speed, and excellent ability in the field of prediction, an extreme learning machine (ELM) [23] is taken as the basic framework of the proposed algorithm.

ELM is a new type of single hidden layer feed-forward neural network (SLFN). The neural network structure of ELM has only three network layers namely the input layer, the hidden layer, and the output layers. The schematic diagram of ELM is shown in Figure 2, and the ELM model can be established as follows:

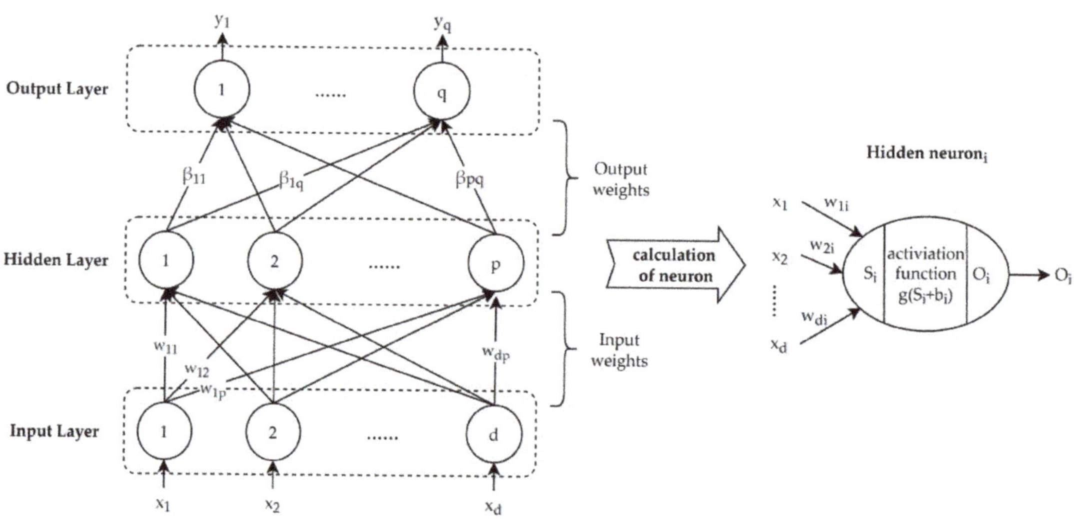

Figure 2. The schematic diagram of ELM.

Firstly, the network parameters of the ELM model should be determined. According to Figure 2, the network parameters mainly consist of the input weights (W), output weights (β) and the threshold of the hidden neuron (b), as shown in Equation (4). The determination of network parameters is introduced by two steps:

Step 1. The input weights and the threshold of the hidden neuron are given randomly;

Step 2. The output weights are obtained by solving the least squares solution of the ELM, which is shown in Equations (5)–(7).

Secondly, based on the determined network parameters, the prediction results can be obtained by Equation (8).

$$W = \begin{bmatrix} w_{11} & \cdots & w_{1p} \\ \vdots & \ddots & \vdots \\ w_{d1} & \cdots & w_{dp} \end{bmatrix},\ \beta = \begin{bmatrix} \beta_{11} & \cdots & \beta_{1q} \\ \vdots & \ddots & \vdots \\ \beta_{p1} & \cdots & \beta_{pq} \end{bmatrix},\ b = \begin{bmatrix} b_1 & \cdots & b_p \end{bmatrix}^T \quad (4)$$

$$X = \begin{bmatrix} x_1 & \cdots & x_d \end{bmatrix}^T,\ Y = \begin{bmatrix} y_1 & \cdots & y_q \end{bmatrix}^T \quad (5)$$

$$O = g\left(X^T W + b^T\right) \quad (6)$$

$$\beta = \left(O^T O\right)^{-1} O^T Y^T \quad (7)$$

$$Y_{pre} = g\left(X^T W + b^T\right)\beta \quad (8)$$

where w_{ij} denotes the connection weight of i-th neuron of the input layer and the j-th neuron of the hidden layer, β_{jm} denotes the connection weight of j-th neuron of the hidden layer and the m-th neuron of the output layer, b_k denotes the threshold of k-th hidden neuron, X and Y are the input and output data of ELM model, W, b, β, $g(\cdot)$, and O represent input weight matrix, threshold matrix, output weight matrix, activation function and output matrix of hidden layer, respectively.

2.3. Genetic Algorithm

Because some network parameters (input weights and threshold) of ELM are determined randomly, this leads to the fluctuation of prediction performance. In order to achieve stable prediction performance and high prediction accuracy, it is necessary to combine the ELM model with another algorithm with excellent global optimization ability.

Based on the selection operation, crossover operation, and mutation operation, the genetic algorithm (GA) [40] can keep the population diversity effectively to avoid the local optimal solution, so that it has a global optimization capability [27,41]. Thus, GA is employed to optimize the ELM by globally searching the optimal value of input weights as well as a threshold in this research, and the flow of GA is shown in Figure 3 and defined as follows:

Step 1. The parameters of GA are determined at first, which consists of the chromosome number in the population, the maximum generation, crossover rate, and mutation rate.

Step 2. Then, the variables of non-linear problems are encoded as the chromosome for obtaining the primary population.

Step 3. Subsequently, the fitness value of each chromosome will be calculated as an evaluation method. By simulating the survival of the fittest in the biological evolution process, the chromosomes with higher fitness value have more possibility to be selected into the next generation and the others will be dropped. The formula is shown in Equation (9).

Step 4. In addition, so as to increase the diversity of the population, crossover operation and mutation operation are necessary. In addition, it is essential to increase the diversity of the population, crossover operation, and mutation operation. The former will randomly select several pairs of chromosomes for crossover based on the crossover rate, which is shown in Equation (10). Furthermore, as shown in Equations (11) and (12), the latter will make the genes of each chromosome mutate by the mutation rate.

Step 5. Finally, repeat step 3 and step 4 until meeting the termination condition. The chromosome with the highest fitness value will be regarded as the best solution to non-linear problems and decoded as the value of related parameters.

$$P_i = f_i / \sum_{j=1}^{M} f_j \quad (9)$$

$$Z_1^{t+1} = \varphi \times Z_1^t + (1-\varphi) \times Z_2^t,\ Z_2^{t+1} = Z_1^t + \varphi \times (Z_1^t + Z_2^t) \tag{10}$$

$$\Delta(t,z) = z \times \theta \times (1 - t/T)^b \tag{11}$$

$$z^{t+1} = \begin{cases} z^t + \Delta(t,(1-z^t)), & \theta > 0.5 \\ z^t - \Delta(t,(z^t+1)), & \theta \le 0.5 \end{cases} \tag{12}$$

where f_i and P_i denote the fitness value and selection probability of i-th chromosome, respectively, Z_1^t and Z_2^t represent the pair of chromosomes in t-th generation, φ and θ are random values between 0 and 1, T is the value of the maximum generation, z^t denotes a gene in t-th generation.

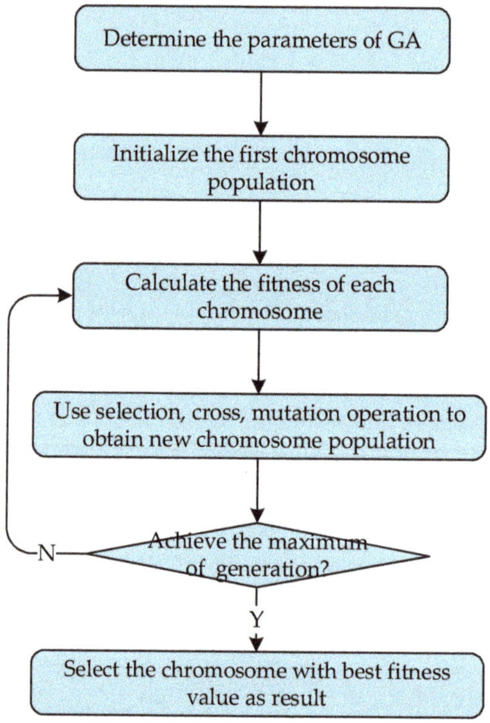

Figure 3. The flowchart of GA.

2.4. Bagging Algorithm

Since the various environmental noise is inevitable and it will greatly influence the final prediction result, therefore, the anti-interference ability of the proposed algorithm needs to be focused. The bagging algorithm [42] is a widely used ensemble strategy. By constructing several base learners and combining the outputs of base learners to obtain a comprehensive evaluation, the bagging algorithm has a powerful ability to solve overfitting problems with the characteristics of high variance. Thus, the bagging algorithm will be utilized as the further optimization algorithm to prevent overfitting problems as well as improve the generalization and robustness of the prediction process.

The mechanism of the bagging algorithm is shown in Figure 4 and illustrated as follows:

Step 1. The bootstrap sampling method is used to obtain several training subsets from the training set, which adopts the strategy of random sampling with replacement.

Step 2. Then, each base learner will be trained by the corresponding training subsets.

Step 3. The multiple prediction results will be generated by base learners and combined to form an aggregated output.

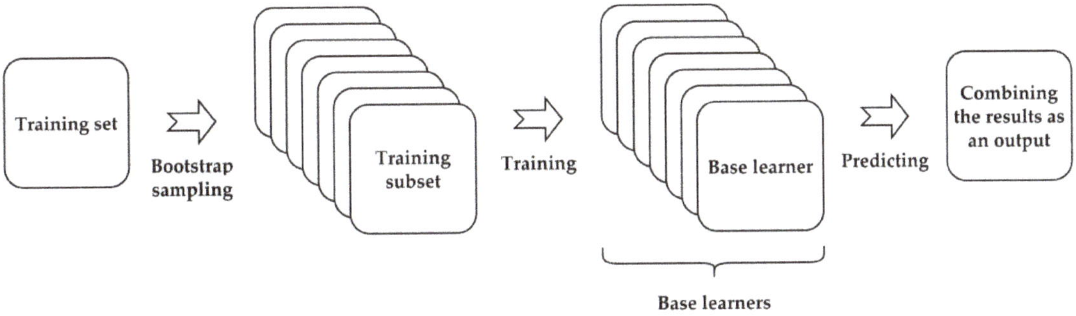

Figure 4. The mechanism of bagging algorithm.

2.5. Proposed Dimensional Error Prediction Algorithm

In order to solve the problem of dimension prediction in the grinding process, a dimensional error prediction approach with the advantages of fast calculation speed, high accuracy, and strong robustness is proposed.

The main idea of the proposed algorithm can be divided into two parts: the raw data is pre-treated at first to remove the redundant information as well as noise. Next, the bagging–GA–ELM algorithm will rapidly learn the complex mapping between parameters and dimensional error to obtain excellent prediction performance, in which the ELM model is taken as a basic framework at first, then it is optimized by the GA to obtain the optimal network parameters. Finally, the bagging algorithm is utilized as further optimization to improve the robustness of ELM model.

The block diagram of proposed algorithm is shown in Figure 5 and described as follows:

Firstly, for obtaining the bagging–GA–ELM model with higher accuracy and robustness, the raw data should be pretreated, which contains three aspects:

1. In the grinding process, some process parameters have a significant impact on the final product dimension, and it also has an obvious influence on the final prediction performance of bagging–GA–ELM [43,44]. Therefore, relevant parameters which are chosen by processing property, previous experiments, and relevant theories, are selected as feature parameters, such as the grinding power of the fine grinding process, the grinding power of the spark out process, etc.
2. Then, normalization is used to eliminate the difference of dimension among different parameters, and PCA is used to extract the main-relevant components from feature parameters. Both of them can improve the calculation speed of the bagging–GA–ELM model. Especially, the negative noise can be avoided by using PCA.
3. At last, the training set is divided into different training subsets by the strategy of random sampling with replacement, which aims to provide the foundation for the subsequent ensemble learning optimization.

Secondly, the bagging–GA–ELM is proposed to predict dimensional error as well as to avoid the scrap with a dimension problem flowing to the customer. The proposed bagging–GA–ELM model consists of the following three parts:

1. ELM is used as a basic framework to predict dimensional error because of its fast calculation speed and excellent ability in the field of industrial prediction. Although ELM just uses the single hidden layer feed-forward network structure, it can also achieve good prediction performance with high accuracy by increasing the width of the network (the number of hidden layer neurons) [45,46]. In addition, compared with other complex neural networks, especially deep neural networks, the design of a single hidden layer greatly reduces the size of network parameters, which makes ELM obtain extremely high calculation efficiency. Hence, the ELM model is selected

in this research to learn the nonlinear relationship between selected features and the dimensional error;

2. The optimization of ELM with GA: the traditional ELM model is characterized by short training time, high execution efficiency, and strong generalization ability. However, because some network parameters (input weights and threshold) of ELM are determined randomly, this results in the fluctuation problem of prediction performance. Therefore, the GA which has global optimization capability, is combined with ELM to search the optimal network parameters, and the optimization detail is delineated as follows:

Step 1. The encoding process of chromosomes: each network parameter (input weight and threshold) will be encoded as a gene and aligned in a row, namely chromosomes. Then, based on the random determination of network parameters, different values of parameters generate different chromosomes until meeting the size of the population.

Step 2. Evaluation of chromosomes: via loading network parameters carried by the chromosome into the GA–ELM part, the prediction accuracy of the training set is used to calculate the fitness value of the chromosome. The formulas are shown in Equation (13).

Step 3. The decoding process of chromosomes: based on the optimization of the GA part, the chromosome with the best fitness value will be decoded into network parameters according to the reverse operation of the encoding process. Then, in accordance with the training set, the output weights will be further decided by solving the least squares solution of the ELM part. Finally, the prediction function of the GA–ELM part is shown in Equation (14).

$$f_i = exp\left(-\sqrt{\sum_{k=1}^{N}(y_k - \hat{y}_k)/N}\right) \tag{13}$$

$$Y_{pre} = \beta^T g\left(W^T X + b\right) \tag{14}$$

where f_i denotes the fitness value of i-th chromosome, y_k and $\hat{y}_k$ represent the observed value and predicted value of k-th sample, N denotes the sample number, W, b, β and $g(\cdot)$ represent input weight matrix, threshold matrix, output weight matrix and activation function of ELM part, respectively, X and Y_{pre} are matrices of parameters and prediction values of dimensional error, respectively.

3. The improvement of prediction robustness by bagging algorithm: due to the unavoidable environmental noise in the real grinding process, which will greatly influence the final prediction result, it is important to ensure the robustness of the proposed algorithm. Since the bagging algorithm has the characteristics of simple operation, strong interpretability, and excellent performance, and it also has significant advantages in preventing the overfitting problem and improving the anti-noise ability of the model, thus, the bagging algorithm is employed to improve the robustness of the proposed algorithm. Based on the bootstrap sampling method, the bagging–GA–ELM model is formed by the base learners trained from different training sets, and the output of bagging–GA–ELM is obtained by averaging the outputs of base learners.

Finally, the bagging–GA–ELM will be obtained and used in the prediction of dimensional error with high accuracy and strong robustness.

3. Results

For proving the feasibility and effectiveness of the proposed algorithm, a case of a heavy cross-axis centerless grinding process in a bearing company in Hangzhou, Zhejiang Province, China, is used to verify the proposed algorithm. The raw data was acquired from a high precision centerless grinder (M1083A), meanwhile, the grinding power was obtained directly by a process monitoring tool which contained the power sensor (IMC, IGTech, Shenzhen, China) installed in the grinding machine.

In this section, the background of experiments is introduced at first, subsequently, the parameters optimization of the proposed algorithm is discussed. Eventually, the dimensional error prediction of the grinding process is explained in detail.

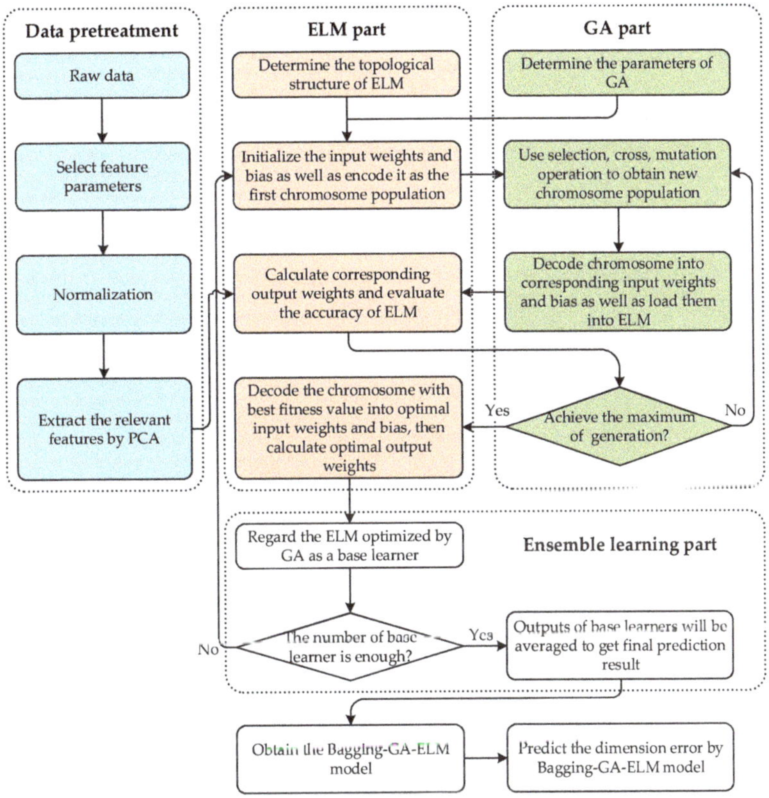

Figure 5. The block diagram of the proposed dimensional error prediction algorithm.

3.1. Background of Experiments

3.1.1. Feature Selection and Sample division

The chosen case is from a centerless grinding process which produces a heavy cross axis; the product drawing is shown in Figure 6. In addition, the grinding machine is shown in Figure 7.

It can be observed from Figure 6 that the dimension tolerance of the products is $[-0.025, 0]$. In addition, according to Figure 6, two diagonal axes are ground at the same time in the grinding process, so that two-dimensional errors (dimensional error of diameter I and dimensional error of diameter II) should be predicted together.

Furthermore, from previous experiments and arguments, the grinding force has a strong influence on the elastic deformation of the product during grinding, hence it is the key influencing factor for dimensional errors. Since grinding force is difficult to be detected and grinding power has a high correlation with grinding force, thus the grinding power is regarded as the main process feature to be studied. Moreover, considering that the fine grinding process and the spark out process have different effects on the final dimension of the product, therefore, after the experimental verification, the grinding power of the above process and its derived variables are selected as important input variables of the proposed algorithm. In addition, in view of the inconsistency of the workpiece, in this case, the 1–20th points of grinding power and other relevant derived variables (peak value of grinding power, etc.) are also selected as input variables.

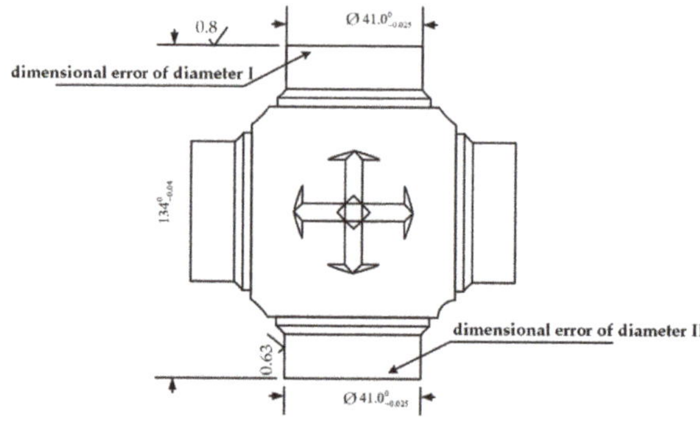

Figure 6. Drawing of heavy cross axis.

Figure 7. The picture of the grinding machine and product.

Finally, the input variables of the proposed algorithm are shown as follows:

1. The grinding power for the fine grinding process and the variables derived from it (average value, area of the power curve, slope, and variance), as recorded in Table 1.
2. The grinding power of spark out process and its derived variables (average value, area of power curve, slope, variance, peak value, mode value, median value, skewness, kurtosis, minimum value, end value, the ratio between the end power of spark out process and average power of fine grinding process as well as the ratio between end power of spark out process and peak of grinding power), as shown in Table 2.
3. Other relevant derived variables of grinding power (peak value, the sum value of 1–20th points, the peak of first twenty points, the variance of first twenty points, the sum value of 40–128th points), as recorded in Table 3.

Moreover, the raw datasets are randomly separated into 60% for the training set, 20% for the validation set, and 20% testing set, in which the training set and validation set are used in the training process of the bagging–GA–ELM model, the training set is used to determine the output weights of ELM model and the validation set is used to evaluate the fitness value of each chromosome in the optimization process of GA. Then a testing set will be used to evaluate the performance of the bagging–GA–ELM model.

Table 1. The grinding power of fine grinding process and its derived variables.

Feature Parameter	Max	Min	Mean
average value	6091.2	4151.9	4694.5
area of power curve	97,482.5	69,439.5	75,425.7
slope	1.5	−85.1	−17.7
variance	540.1	22.2	121.1

Table 2. The grinding power of spark out process and its derived variables.

Feature Parameter	Max	Min	Mean
average value	3747.2	1912.7	2817.4
area of power curve	108,518.5	55,368.5	81,432.0
slope	−55.7	−120.0	−84.8
variance	1188.8	509.9	853.8
peak value	5709.0	3006.0	4434.9
mode value	5295.0	1123.0	2701.9
median value	3616.0	1877.5	2590.7
skewness	1.0	−0.1	0.5
kurtosis	2.9	1.3	2.0
minimum value	2232.0	1015.0	1746.1
end value	2297.0	1061.7	1793.4
the ratio between the end power of spark out process and average power of fine grinding process	0.46	0.24	0.38
the ratio between end power of spark out process and peak of grinding power	0.45	0.21	0.36

Table 3. Other relevant derived variables of grinding power.

Feature Parameter	Max	Min	Mean
peak value	6696.0	4608.0	4975.1
sum of first twenty points	54,432.0	20,082.0	32,668.5
peak of first twenty points	3534.0	1054.0	1909.1
variance of first twenty points	716.6	7.9	192.2
sum of values from 40th to 128th points	457,541.0	280,200.0	348,018.6

3.1.2. Experiment Setup and Evaluation Indicator

The analysis of raw data acquired from the grinding process and all prediction experiments of dimensional error run in the Python 3.8 environment on a 3.20 GHz PC with processor i5-11320H and 16 GB RAM. Since the effect of randomness on the experiment results, all experiments in this research are run repeatedly three times and then averaged to obtain the final result.

In addition, the true positive ratio (TPR), false positive ratio (FPR), and corrected mean absolute percentage error (CMAPE) are used as evaluation indicators of prediction performance, the formulas of which are formed as Equations (15)–(17). Among evaluation indicators, TPR, as well as FPR, are used to evaluate the ability to identify quality problems. In addition, it is worth noticing that the traditional MAPE uses the observed value as the denominator. However, in this case, partially observed values of dimensional error are close to 0 mm, which will make the value of MAPE trend to infinity and make the performance of the prediction model unable to be evaluated accurately. Thus, CMAPE is designed in this case as the corrected value of MAPE, in which the value of tolerance (0.025 mm) will be used as the denominator.

According to Equation (17), if the predicted value deviated significantly from the observed value, the value of CMAPE will become large. On the contrary, when the difference between the predicted value and observed value is small, the value of CMAPE will be close to 0.

$$TPR = 100\% \times TP/(TP + FN) \tag{15}$$

$$FPR = 100\% \times FP/(FP + TN) \tag{16}$$

$$CMAPE = 100\% \times \sum_{i=1}^{n} |y_{pre,i} - y_{obs,i}|/0.025 \tag{17}$$

where TP denotes the number of positive samples predicted as positive, FN denotes the number of positive samples predicted as negative, FP denotes the number of negative samples predicted as positive, TN denotes the number of negative samples predicted as negative, n denotes sample size, $y_{pre,i}$ and $y_{obs,i}$ represent the predicted value and the observed value of ith sample, respectively.

3.2. Parameters Optimization of Proposed Algorithm

It is considered that the different parameters of the proposed algorithm have an essential influence on the prediction performance of the dimensional error. Thus, the determination of main parameters (the main-relevant component number of PCA, the neurons number of hidden layer, and the number of the base learner) are discussed in detail in this section.

3.2.1. The Main Relevant Component Number of PCA

Adopting PCA in the pretreatment of data can extract main relevant components from feature parameters and remove negative noise, but it needs to be noted that excessive main relevant components can deteriorate the elimination effect of noise, while too few main relevant components lead to the loss of effective information. Therefore, the main relevant component number of PCA is analyzed and the performance ranking of different main relevant component numbers from 1 to 22 is shown in detail in Figure 8.

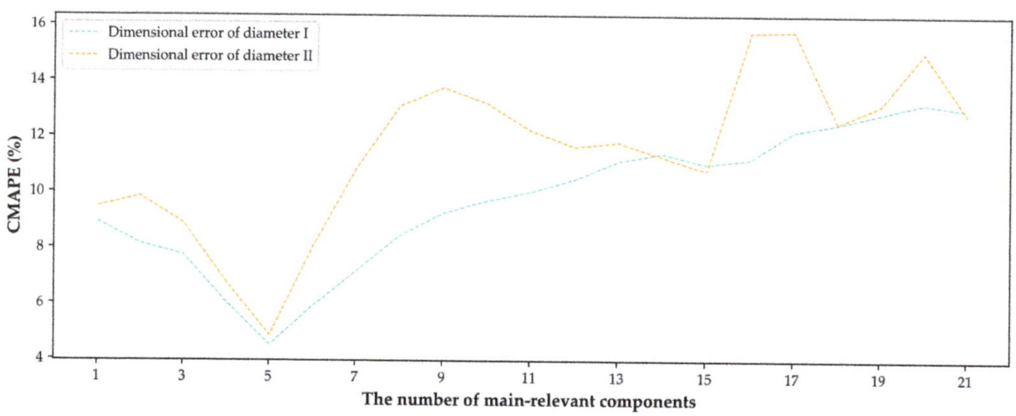

Figure 8. The performance of the main relevant component number on the testing set.

According to Figure 8, as the main relevant component number increases from 1 to 5, it is clear that the CMAPE on the testing set decreases significantly. Then, the optimal CMAPE is obtained when the main relevant component number is 5. With further increase of main relevant component number, the dimensional error of both diameter I and diameter II start to increase gradually, which means that the residual noise results in a reduction of generalization ability. Hence, the top five main relevant components are regarded as the final features extracted from data by PCA.

3.2.2. The Neurons Number of Hidden Layer and Parameters of GA

As a variant of ANN, the ability of ELM to deal with non-linear prediction problems mainly depends on the interconnection of neurons as well as the non-linear transformation of the hidden layer. Thus, the number of neurons in the hidden layer has high relation with the prediction accuracy of the proposed algorithm.

Obviously, the complex association between process parameters and dimensional error is difficult to be learned effectively by the prediction model with few hidden neurons. However, because of the limited information of raw data, too many hidden neurons will make the network parameters (input weights and thresholds) insufficiently trained, which usually results in overfitting problems.

In order to obtain the optimal number of hidden neurons, the prediction performance of the proposed algorithm with different number of hidden neurons are compared by experiments. The result of experiments is shown in Figure 9.

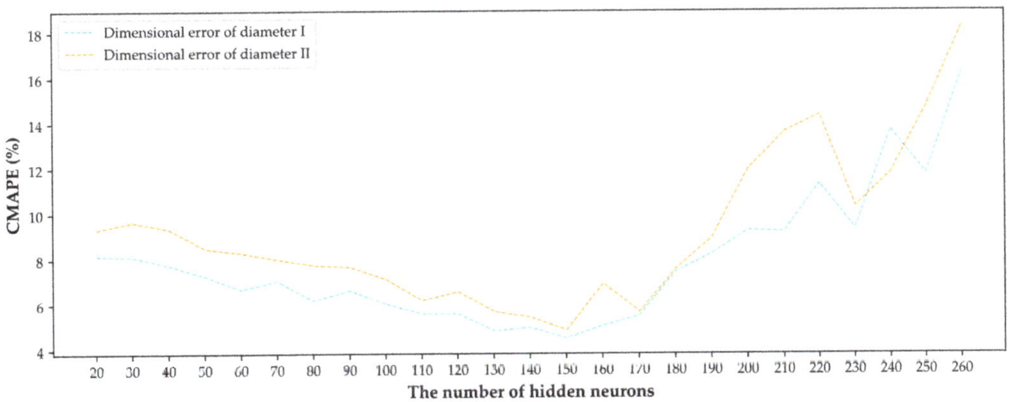

Figure 9. The performance of hidden neuron number on the testing set.

As shown in Figure 9, with the increase of neurons from 20 to 150, the CMAPE of testing set will decline at first. On the contrary, when the number of hidden neurons is more than 150, the increasing neurons start to make the dimensional error fluctuate and increase, which indicates that the prediction model has a symptom of overfitting. As a result, the optimum of neuron number in the hidden layer is 150.

In addition, the genetic algorithm (GA) which has global optimization capability is combined with ELM to search the optimal network parameters, in which the population size and max generation of GA will be discussed in Figure 10.

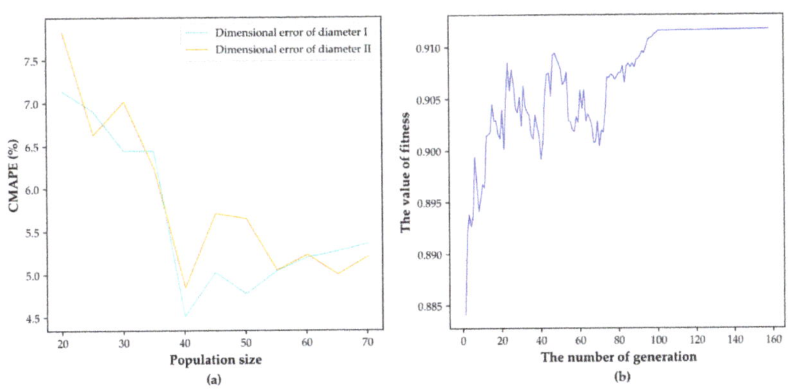

Figure 10. The effects of parameters of GA. (**a**) The effects of population size, (**b**) The effects of generation number.

According to Figure 10a, when the value of population size increases from 20 to 40, the CMAPE on the testing set decreases significantly; it is owing to that the increase in population size improves the diversity of chromosomes and the optimization effects of GA.

However, with a further increase in population size, the curve tendency of CMAPE starts to slow down obviously. Thus, the optimal population size is set as 40. Further, in the view of Figure 10b, when the number of generations increases from 0 to 100, the fitness value increases significantly. However, when the number of generations is larger than 100, the further increase of generation number makes it difficult to obtain an effective improvement of fitness value. Thus, the optimal max generation is set as 100.

Furthermore, the genetic algorithm has been widely used to optimize parameters in the industrial field. Many experiments show that the crossover rate is usually larger than 0.9 to ensure the diversity of the population and the mutation rate is advised to be set lower than 0.05 [47]. Thus, the crossover rate and mutation rate in the evolution of chromosomes are set as 0.95 and 0.05, respectively.

Then, the comparisons between the ELM model and the ELM model optimized with GA (GA-ELM) are shown in Table 4. Compared to the CMAPE of the ELM model, it can be observed that the CMAPE of the GA-ELM model achieves a significant reduction (the reduction rate of CMAPE in diameter I and diameter II are 41.73% and 54.58%, respectively), which shows the significant contribution of GA in accuracy improvement of dimensional error prediction.

Table 4. The comparison between ELM model and GA–ELM model.

Model	CMAPE of Dimensional Error on Testing Set (%)	
	Diameter I	Diameter II
ELM	7.74	10.68
GA–ELM	4.51	4.85

3.2.3. The Number of Base Learners

In the proposed algorithm, the bagging algorithm is used to avoid the impact of noise in the real-time collected data, in which the GA–ELM model is used as the base learner. Then, various base learners are trained depending on different samples. After that, the outputs of these base learners are averaged to reduce the variance of the prediction results.

Since excessive base learners not only increase the time-consuming nature of the calculation process and the waste of computing resources but also have few contributions to the improvement of prediction performance, the optimal number of base learners should be determined carefully.

In this subsection, the influence of different base learner numbers on the prediction performance is studied, as shown in Figure 11.

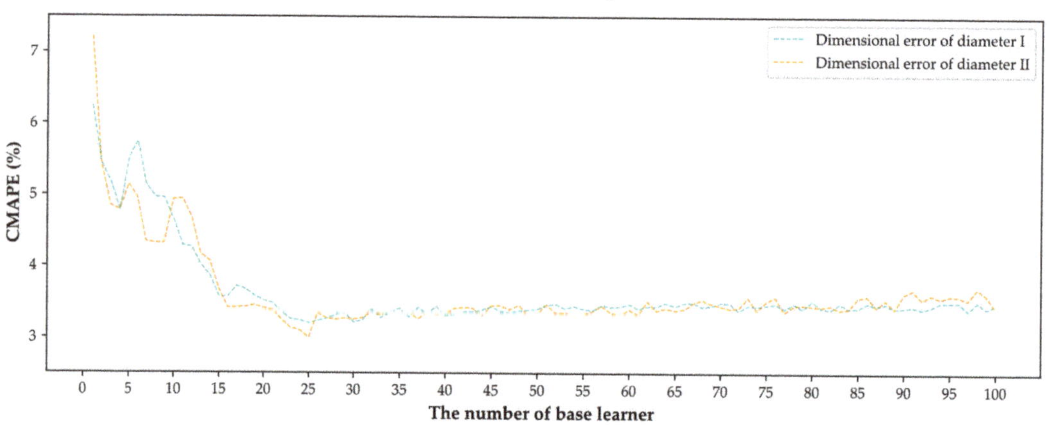

Figure 11. The prediction performance on the different number of base learners.

It can be seen from Figure 11, that when the number of base learners increases from 1 to 25, the value of CMAPE declines significantly. However, when the number of base learners exceeds 25, the curve tendency of CMAPE starts to slow down obviously, which means that further increase of the base learner number makes it difficult to obtain effective accuracy improvement. Finally, the optimal number of base learners is set as 25.

Furthermore, the robustness of bagging–GA–ELM needs to be investigated because of the high complexity and strong interference in the industrial environments. In order to simulate the noise of real industrial environments, white Gaussian noise is added to the raw data. The intensity of white Gaussian noise is estimated by the signal-to-noise ratio (SNR), as formed in Equations (18) and (19).

To evaluate the robustness of the proposed algorithm, the prediction accuracy of the bagging–GA–ELM model and GA–ELM model are analyzed under various white Gaussian noise (SNR ranging from 1 dB to 10 dB); results are shown in Figure 12.

$$SNR = 10log_{10}(P_r/P_n) \tag{18}$$

$$P_r = \sum_{i=1}^{n} A_{r,i}/n, \;\; P_n = \sum_{i=1}^{n} A_{noise,i}/n \tag{19}$$

where n denotes sample size, P_r and P_n denote the signal power of raw data and noise, respectively, $A_{r,i}$ and $A_{noise,i}$ represent the i-th amplitude value of raw data and noise, respectively.

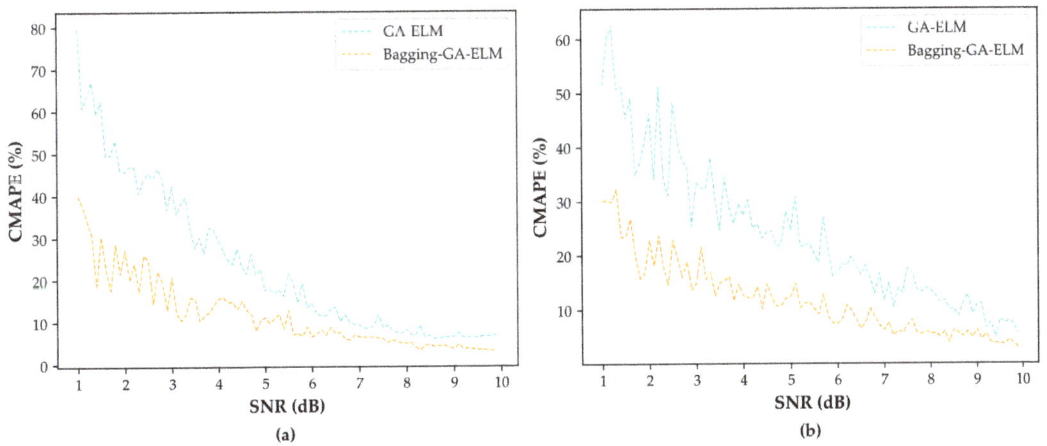

Figure 12. The performance of prediction models under different background noise. (**a**) Dimensional error of diameter I, (**b**) Dimensional error of diameter II.

It is clear from Figure 12, with the reduction of SNR, the interference of noise becomes obvious gradually and makes the CMAPE of both models start to increase. However, in comparison with the GA–ELM model, the CMAPE of bagging–GA–ELM is much lower than that of the GA–ELM model, which obviously shows the contribution of the bagging algorithm in the improvement of anti-interference ability.

3.2.4. The Structure of Proposed Algorithm

Based on the above discussions, the structure of the proposed algorithm has been determined as follows:

The raw data acquired from the grinding process is a matrix of N × 22, which means that there are N samples and each sample has 22 feature parameters. Then, the matrix will be normalized at first and the PCA is used to transform 22 feature parameters into five main relevant components to remove the redundant information as well as noise. Next step, the five main relevant components will be regarded as the inputs (N × 5) of the ELM model.

The ELM model has three layers (input layer, hidden layer, and output layer), in which the input layer has 5 neurons, the hidden layer has 150 neurons and the output layer has 2 neurons. Thus, the output of the proposed algorithm is a matrix of N × 2, which means that there are N samples and each sample has two prediction results (diameter I and diameter II).

3.3. *The Dimensional Error Prediction in Grinding Process*

In this section, the performance of bagging–GA–ELM is further discussed. Firstly, the bagging–GA–ELM will be compared with other prediction models (KSVM and ANN) to analyze its accuracy and calculation speed. Next, the proposed algorithm is applied to analyze the raw data acquired from the grinding process and shows the results of dimensional error prediction.

3.3.1. Comparisons of Prediction Models

Since KSVM and ANN are widely used in the field of industrial analysis, which are regarded to have characteristics of high accuracy and strong universality by practice. So as to evaluate the dimensional error prediction performance of the proposed algorithm in the grinding process, the above models are used to compare to the bagging–GA–ELM model, and the parameters of these models are recorded in Table 5. In addition, the comparisons among different prediction models are shown in Table 6.

Table 5. The parameters of different prediction models.

Model	Parameters	Value
KSVM	Kernel Penalty coefficient	Gaussian 1
ANN	Hidden layer Hidden neurons	5 25, 50, 150, 50, 25
Bagging–GA–ELM	Hidden neurons Max generation Number of base learners	150 100 25

Table 6. The comparison among different prediction models.

Model	CMAPE of Dimensional Error on Testing Set (%)	
	Diameter I	**Diameter II**
KSVM	12.03	11.54
ANN	5.78	6.33
Bagging–GA–ELM	3.20	3.01

According to Table 6, it is clear that the CMAPE of KSVM is higher than other models, which is related to the complexity of dimensional error prediction as well as the ability to deal with the non-linear problem. Moreover, it can be further found that the bagging–GA–ELM model achieves the optimal performance in prediction (the CMAPE of dimensional errors on the testing set are 3.20% for diameter I and 3.01% for diameter II, respectively).

3.3.2. Prediction Results Based on Proposed Algorithm

Based on the proposed algorithm, the dimensional error of products will be predicted during the grinding process. Once the predicted dimensional error exceeds the dimension tolerance, the process will be stopped and the alarm will be generated and relative information will be sent to operators to prevent a further quality problem. Thus, although the accuracy and robustness of the proposed algorithm have been verified in the above discussions, the ability of the proposed algorithm to identify the quality problem is further to be evaluated.

Considering that the difference between the number of products with a quality problem (namely positive samples) and the number of products without a quality problem (namely negative samples) is usually significant, it is necessary to investigate the identification rate of positive samples and negative samples, respectively. Therefore, the true positive ratio (TPR) and false positive ratio (FPR) are used as evaluation indicators of identification performance, in which TPR is used to evaluate the rate of positive samples predicted as positive, and the larger value of TPR denotes the higher identification performance on the positive samples. In addition, FPR is used to evaluate the rate of negative samples predicted as positive, so that the smaller value of FPR denotes the higher identification performance on the negative samples.

Therefore, based on the TPR and FPR, the identification performance of the proposed algorithm is evaluated and compared to other models, as shown in Table 7.

Table 7. The identification performance of quality problems under different background noise.

Model	Diameter I		Diameter II	
	TPR (%)	FPR (%)	TPR (%)	FPR (%)
KSVM	89.61	13.56	86.83	15.14
ANN	94.78	4.89	91.16	6.21
Bagging-GA-ELM	99.48	0.36	99.32	0.98

According to Table 7, it can be observed that bagging–GA–ELM achieves the best identification performance in both diameter I and diameter II. The values of TPR are higher than 99% and the values of FPR are lower than 1%, which means that the success rate of the proposed algorithm achieves more than 99% in the identification of the real-time collected data. Therefore, the proposed algorithm can be used to identify the quality problem accurately, which meets the requirement of the production line.

Finally, based on the proposed algorithm, the prediction results of data acquired from the real-time grinding process are shown in Table 8 and Figure 13.

Table 8. The comparison between the actual dimension error (ADE) and the predicted dimension error (PDE).

Data	Mean	Std
	10^{-3} mm	10^{-3} mm
ADE of diameter I	−5.32	3.03
PDE of diameter I	−5.35	2.37
The absolute difference between ADE and PDE in diameter I	0.79	1.27
ADE of diameter II	−5.46	3.48
PDE of diameter II	−5.47	2.89
The absolute difference between ADE and PDE in diameter II	0.75	1.49

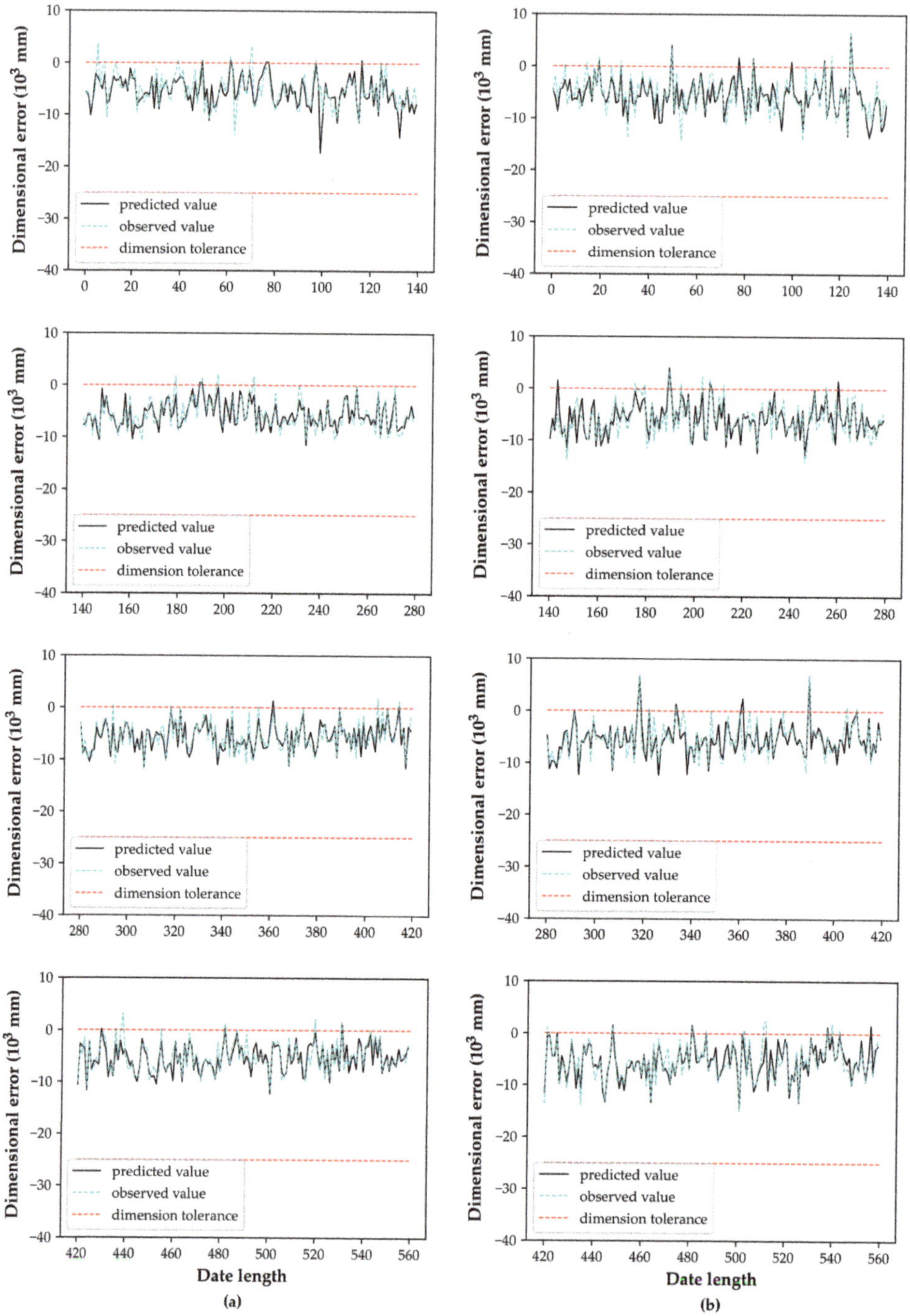

Figure 13. The prediction results of dimensional error. (**a**) The prediction of diameter I, (**b**) The prediction of diameter II.

4. Conclusions and Future Work

In this paper, an approach based on bagging–GA–ELM is proposed to predict the final part dimensional error during the grinding process with high accuracy and robustness. In particular, firstly, the raw data are pre-processed to eliminate the dimensionality of the data while avoiding negative noise. Then the bagging–GA–ELM, which combines the advantage of ELM with short training time, GA with global optimization ability, and bagging with good anti-interference ability, is used to predict the dimensional error of parts. The experimental performance shows that the proposed algorithm achieves the best dimensional error prediction performance compared with the KSVM model and ANN model. In addition, the proposed algorithm has achieved good quality problem identification performance (success rate of more than 99%), which can satisfy customers' requirements. Additionally, the application of the algorithm in the bearing company helps to save additional investment in equipment.

In the future, we intend to continue our research in the parameters of the grinding process to improve the prediction success rate. By doing this, the grinding process can be better controlled. In addition, we will also study the proposed algorithm in other products.

Author Contributions: Conceptualization, L.Y. and Y.J.; methodology, L.Y. and Y.J.; software, L.Y. and Y.J.; validation, L.Y., Y.J., H.L. and X.Y.; analysis, L.Y. and Y.J.; investigation, L.Y. and Y.J., resources, L.Y., Y.J., H.L. and X.Y.; writing—review and editing, L.Y., Y.J., H.L. and X.Y.; visualization, L.Y., Y.J. and H.L.; supervision, L.Y., Y.J., H.L. and X.Y.; project administration, L.Y. All authors have read and agreed to the published version of the manuscript.

Funding: This research received no external funding.

Data Availability Statement: Not applicable.

Conflicts of Interest: The authors declare no conflict of interest.

References

1. Wegener, K.; Bleicher, F.; Krajnik, P.; Hoffmeister, H.-W.; Brecher, C. Recent developments in grinding machines. *CIRP Ann.* **2017**, *66*, 779–802. [CrossRef]
2. Ahmer, M.; Marklund, P.; Gustafsson, M.; Berglund, K. Integration of process monitoring and machine condition diagnostics to improve quality prediction in grinding. *Procedia CIRP* **2021**, *101*, 170–173. [CrossRef]
3. Yang, L.; Deuse, J.; Jiang, P. Multi-objective optimization of facility planning for energy intensive companies. *J. Intell. Manuf.* **2012**, *24*, 1095–1109. [CrossRef]
4. Yang, L.; Deuse, J.; Jiang, P. Multiple-attribute decision-making approach for an energy-efficient facility layout design. *Int. J. Adv. Manuf. Technol.* **2012**, *66*, 795–807. [CrossRef]
5. Yadav, R.N.; Yadava, V.; Singh, G.K. Application of non-dominated sorting genetic algorithm for multi-objective optimization of electrical discharge diamond face grinding process. *J. Mech. Sci. Technol.* **2014**, *28*, 2299–2306. [CrossRef]
6. Yang, G.M.; Zhang, Y.P.; Li, K.Y.; Chen, G.D. A Processing Predictive Model of Ultrasonic Vibration Grinding Assisted Electric Discharge Machining Based on Support Vector Machines. *Adv. Mater. Res.* **2014**, *941*, 1928–1931. [CrossRef]
7. Shen, N.; He, Y.; Li, J.; Fang, M. An Improved Differential Evolution (IDE) Based on Double Populations for Cylindrical Grinding Optimization. In Proceedings of the 2009 International Conference on Measuring Technology and Mechatronics Automation, Zhangjiajie, China, 11–12 April 2009; pp. 724–727.
8. Mondal, S.C.; Mandal, P. An Application of Particle Swarm Optimization Technique for Optimization of Surface Roughness in Centerless Grinding Operation. In *ICoRD'15–Research into Design Across Boundaries Volume 2; Smart Innovation, Systems and Technologies*; Springer: Berlin, Germany, 2015; pp. 687–697.
9. Chen, Z.; Li, X.; Wang, L.; Zhang, S.; Cao, Y.; Jiang, S.; Rong, Y. Development of a hybrid particle swarm optimization algorithm for multi-pass roller grinding process optimization. *Int. J. Adv. Manuf. Technol.* **2018**, *99*, 97–112. [CrossRef]
10. Rudrapati, R.; Bandyopadhyay, A.; Pal, P.K.; Rathod, L. Analysis, modeling and optimization of surface roughness in cylindrical traverse cut grinding using factorial design, RSM and simulated annealing algorithm. *IOP Conf. Ser. Mater. Sci. Eng.* **2020**, *814*, 012016. [CrossRef]
11. Yuan, Z.G.; Zhang, Q.; Zhang, X.L.; Su, Z. Modeling for cement combined grinding system with closed grinding process based on LS-SVM in typical working condition. In Proceedings of the 2015 IEEE International Conference on Cyber Technology in Automation, Control, and Intelligent Systems (CYBER), Shenyang, China, 8–12 June 2015; pp. 1050–1055.
12. Pandiyan, V.; Caesarendra, W.; Tjahjowidodo, T.; Tan, H.H. In-process tool condition monitoring in compliant abrasive belt grinding process using support vector machine and genetic algorithm. *J. Manuf. Process.* **2018**, *31*, 199–213. [CrossRef]

13. Hung, L.X.; Pi, V.N.; Hong, T.T.; Ky, L.H.; Lien, V.T.; Tung, L.A.; Long, B.T. Multi-objective Optimization of Dressing Parameters of Internal Cylindrical Grinding for 9CrSi Aloy Steel Using Taguchi Method and Grey Relational Analysis. *Mater. Today Proc.* **2019**, *18*, 2257–2264. [CrossRef]
14. Xu, L.M.; Fan, F.; Hu, Y.X.; Zhang, Z.; Hu, D.J. A vision-based processing methodology for profile grinding of contour surfaces. *Proc. Inst. Mech. Eng. Part B J. Eng. Manuf.* **2019**, *234*, 27–39. [CrossRef]
15. Hao, X.; Li, Y.; Li, M.; Liu, C. A part deformation control method via active pre-deformation based on online monitoring data. *Int. J. Adv. Manuf. Technol.* **2019**, *104*, 2681–2692. [CrossRef]
16. Wang, X.; Bi, Q.; Zhu, L.; Ding, H. Improved forecasting compensatory control to guarantee the remaining wall thickness for pocket milling of a large thin-walled part. *Int. J. Adv. Manuf. Technol.* **2016**, *94*, 1677–1688. [CrossRef]
17. Chen, J.; Chen, H.; Xu, J.; Wang, J.; Zhang, X.; Chen, X. Acoustic signal-based tool condition monitoring in belt grinding of nickel-based superalloys using RF classifier and MLR algorithm. *Int. J. Adv. Manuf. Technol.* **2018**, *98*, 859–872. [CrossRef]
18. Li, Z.H.; Tang, Q.; Liu, Y.; Li, X. A Material Removal State Prediction Method Based on Multi-Scale Attention Mechanism. In Proceedings of the Volume 2: 42nd Computers and Information in Engineering Conference (CIE), St. Louis, MI, USA, 14–17 August 2022.
19. Gopan, V.; Wins, K.L.D.; Surendran, A. Integrated ANN-GA Approach For Predictive Modeling And Optimization Of Grinding Parameters With Surface Roughness As The Response. *Mater. Today Proc.* **2018**, *5*, 12133–12141. [CrossRef]
20. Caesarendra, W.; Triwiyanto, T.; Pandiyan, V.; Glowacz, A.; Permana, S.D.H.; Tjahjowidodo, T. A CNN Prediction Method for Belt Grinding Tool Wear in a Polishing Process Utilizing 3-Axes Force and Vibration Data. *Electronics* **2021**, *10*, 1429. [CrossRef]
21. Guo, W.; Wu, C.; Ding, Z.; Zhou, Q. Prediction of surface roughness based on a hybrid feature selection method and long short-term memory network in grinding. *Int. J. Adv. Manuf. Technol.* **2021**, *112*, 2853–2871. [CrossRef]
22. Yin, S.; Xiao, H. Model of grinding-induced line/area roughness and subsurface damage in brittle material based on genetic algorithm and deep neural network. In Proceedings of the Seventh Asia Pacific Conference on Optics Manufacture and 2021 International Forum of Young Scientists on Advanced Optical Manufacturing, Hong Kong, China, 28–31 October 2021.
23. Huang, G.-B.; Zhu, Q.-Y.; Siew, C.-K. Extreme learning machine: A new learning scheme of feedforward neural networks. In Proceedings of the 2004 IEEE International Joint Conference on Neural Networks (IEEE Cat. No.04CH37541), Budapest, Hungary, 25–29 July 2004; Volume 2, pp. 985–990.
24. Wang, J.; Lu, S.; Wang, S.-H.; Zhang, Y.-D. A review on extreme learning machine. *Multimed. Tools Appl.* **2021**, *81*, 41611–41660. [CrossRef]
25. Xie, W.; Wang, J.S.; Xing, C.; Guo, S.S.; Guo, M.W.; Zhu, L.F. Extreme Learning Machine Soft-Sensor Model With Different Activation Functions on Grinding Process Optimized by Improved Black Hole Algorithm. *IEEE Access* **2020**, *8*, 25084–25110. [CrossRef]
26. Meng, L.Y.; Liu, M.L.; Wei, P.Y.; Qin, H.B. Rolling Bearing Fault Diagnosis Based on Improved VMD And GA-ELM. In Proceedings of the 2021 40th Chinese Control Conference (CCC), Shanghai, China, 26–28 July 2021; pp. 4414–4419.
27. Sun, H.; Li, L.; Yan, C.P.; Song, L.; Huang, Y.G.; Zhou, C. Engineering; Technology. Optimization of High-speed Dry Milling Process Parameters Based on Improved ELM and Genetic Algorithm. *Highlights Sci. Eng. Technol.* **2022**, *7*, 272–283. [CrossRef]
28. Liu, J.Z.; Qu, J.; Luo, G.J.; Zhang, Y. Fault Diagnosis of Rolling Bearing Based on Fast Spectral Kurtosis and PSO Optimized Multi Kernel Extreme Learning Machine. In Proceedings of the 2021 3rd International Conference on Applied Machine Learning (ICAML), Changsha, China, 23–25 July 2021; pp. 86–90.
29. Sun, C.; Yang, S.; Yu, T. Prediction Model of Needle Valve Body Extrusion Grinding Process Based on GA-ELM. In Proceedings of the 2nd International Conference on Artificial Intelligence and Advanced Manufacture, Manchester, UK, 15–17 October 2020; pp. 382–386.
30. Chan, C.-L.; Chen, C.-L. A cautious PSO with conditional random. *Expert Syst. Appl.* **2015**, *42*, 4120–4125. [CrossRef]
31. Agrawal, S.; Silakari, S. FRPSO: Fletcher–Reeves based particle swarm optimization for multimodal function optimization. *Soft Comput.* **2013**, *18*, 2227–2243. [CrossRef]
32. Guo, J.; Sato, Y.J. A hierarchical bare bones particle swarm optimization algorithm. In Proceedings of the 2017 IEEE International Conference on Systems, Man, and Cybernetics (SMC), Banff, AB, Canada, 5–8 October 2017; pp. 1936–1941.
33. Qin, Q.; Cheng, S.; Zhang, Q.; Li, L.; Shi, Y. Particle Swarm Optimization With Interswarm Interactive Learning Strategy. *IEEE Trans. Cybern.* **2016**, *46*, 2238–2251. [CrossRef] [PubMed]
34. Dhibi, K.; Mansouri, M.; Bouzrara, K.; Nounou, H.; Nounou, M. An Enhanced Ensemble Learning-Based Fault Detection and Diagnosis for Grid-Connected PV Systems. *IEEE Access* **2021**, *9*, 155622–155633. [CrossRef]
35. Gao, K.; Chen, H.; Zhang, X.; Ren, X.; Chen, J.; Chen, X. A novel material removal prediction method based on acoustic sensing and ensemble XGBoost learning algorithm for robotic belt grinding of Inconel 718. *Int. J. Adv. Manuf. Technol.* **2019**, *105*, 217–232. [CrossRef]
36. Bustillo, A.; Reis, R.; Machado, A.R.; Pimenov, D.Y. Improving the accuracy of machine-learning models with data from machine test repetitions. *J. Intell. Manuf.* **2020**, *33*, 203–221. [CrossRef]
37. Liao, J.; Lv, J.; Shao, Y.; Li, P.; Hu, X. Application of BP Neural Network Ensemble Model Based on Bagging Algorithm. *Int. J. Mach. Learn. Comput.* **2019**, *9*, 121–128. [CrossRef]
38. Mo, W.; Luo, X.; Zhong, Y.; Jiang, W. Image recognition using convolutional neural network combined with an ensemble learning algorithm. *J. Phys. Conf. Ser.* **2019**, *1237*, 022026. [CrossRef]
39. Pearson, K. On lines and planes of closest fit to systems of points in space. *Philos. Mag. Ser.* **1901**, *1*, 559–572. [CrossRef]

40. Holland, J.H. *Adaptation in Natural and Artificial Systems: An Introductory Analysis with Applications to Biology, Control, and Artificial Intelligence*; MIT Press: Cambridge, MA, USA, 1992.
41. Ali, J.; Sheng, S.Y.; Rehman Imran, A.; Hassan, A.S.U. Predictive Prognostic Model for Lithium Battery Based on A Genetic Algorithm (GA-ELM) Extreme Learning Machine. *Int. J. Sci. Res. Publ. (IJSRP)* **2020**, *10*, 213–219. [CrossRef]
42. Breiman, L. Bagging Predictors. *Mach. Learn.* **1996**, *24*, 123–140. [CrossRef]
43. Marinescu, L.D.; Dimitrov, B.; Gheorghiu, Z.; Grigorescu, A. Some Aspects Concerning Wear and Tool Life of Diamond Wheels. *CIRP Ann.* **1983**, *32*, 251–254. [CrossRef]
44. Du, Z.; Ge, G.; Xiao, Y.; Feng, X.; Yang, J. Modeling and compensation of comprehensive errors for thin-walled parts machining based on on-machine measurement. *Int. J. Adv. Manuf. Technol.* **2021**, *115*, 3645–3656. [CrossRef]
45. Ding, S.; Zhao, H.; Zhang, Y.; Xu, X.; Nie, R. Extreme learning machine: Algorithm, theory and applications. *Artif. Intell. Rev.* **2013**, *44*, 103–115. [CrossRef]
46. Huang, G.; Chen, Y.Q.; Babri, H.A. Classification ability of single hidden layer feedforward neural networks. *IEEE Trans. Neural Netw.* **2000**, *11*, 799–801. [CrossRef]
47. Patil, V.P.; Pawar, D.D. The Optimal crossover or Mutation Rates in Genetic Algorithm: A Review. *Int. J. Appl. Eng. Technol.* **2015**, *5*, 38–41.

machines

Article

A Case Study in Social Manufacturing: From Social Manufacturing to Social Value Chain

Guang-Yu Xiong [1,2], Petri Helo [1,*], Steve Ekstrom [3] and Tariku Sinshaw Tamir [4,5,6]

1 Industrial Management, Logistics Systems and the Networked Value Systems Research Group, School of Technology and Innovations, University of Vaasa, 65280 Vaasa, Finland
2 Ningbo Intelligent Manufacturing Technology Institute, Ningbo 315000, China
3 Noswing Ab Oy, Jakobsgatan 41D, 68600 Jakobstad, Finland
4 The State Key Laboratory for Management and Control of Complex Systems, Beijing Engineering Research Center of Intelligent Systems and Technology, Institute of Automation, Chinese Academy of Sciences, Beijing 100190, China
5 School of Artificial Intelligence, University of Chinese Academy of Sciences, Beijing 100049, China
6 School of Electrical and Computer Engineering, Debre Markos Institute of Technology, Debre Markos 269, Ethiopia
* Correspondence: phelo@uwasa.fi

Abstract: A new manufacturing mode, called social manufacturing, has been developing widely, and employed in many enterprises across the business value chain in recent years. Faced with this increasing dynamic, both enterprises and customers have to be more aware of the potential opportunity and benefit to be derived from this new manufacturing mode. One benefit is more value-adding potential for both enterprises upstream and customers downstream across the business value chain, compared with the normal mode. This research extends the application of social manufacturing to the entire business value chain system to bring new opportunities and value-adding potential for enterprises. This paper proposes a social value chain system that applies the social manufacturing mode to the entire value chain and contributes to three areas: (1) a new way of thinking for enterprises to create new opportunities to add value throughout the value chain by employing the social manufacturing mode; (2) establishing the social value chain system for all participants/enterprises across the chain in order to gain a win–win situation for all participants; and (3) suggesting some idea of a suitable performance measurement to monitor and evaluate the proposed social value chain system.

Keywords: social manufacturing; social value chain system; value-adding; key supporting technologies; digital-driven technologies

Citation: Xiong, G.-Y.; Helo, P.; Ekstrom, E.; Tamir, T.S. A Case Study in Social Manufacturing: From Social Manufacturing to Social Value Chain. *Machines* **2022**, *10*, 978. https://doi.org/10.3390/machines10110978

Academic Editor: Zhuming Bi

Received: 5 October 2022
Accepted: 24 October 2022
Published: 26 October 2022

1. Introduction

Social manufacturing (SocialM) was first introduced as a "third industry revolution" in the Economist magazine in 2012 [1]. From then, through the efforts of some pioneers who have advocated the SocialM mode such as Wang et al., this new manufacturing mode has been developing along with the development of relevant technology [2–5]. The discussion on SocialM has been becoming a topic of great interest with the growth of the sharing economy and Internet-based and digital-driven technology [3–8]. In practice, the business model of enterprises has been changed compared with the normal mode due to the rapid development of Internet-based and digital-driven technology. For instance, one topical sourcing mechanism in manufacturing—crowd-sourcing—has been applied to some enterprises in order to provide the potential to share capacity and ability by means of Internet-based technology [9]. With this SocialM mode, it is possible to connect different enterprises and even customers across the business value chain to enable an effective collaboration among enterprises and customers to the meet the customers' requirements [6–10]. With the revolution in technology and business mode in the context SocialM, value-adding activities by operational processes are not only contributed to by upstream and intermediate enterprises (suppliers/manufacturers), but contributions from all participants, including downstream customers [10–14].

Regarding the mode of business operation, a business value chain (VC) is a supply demand network that may include multiple layers of participants who are linked to the chain by business, and which can add value for their downstream customers by means of business process or activity. The operational activity of participants must aim at adding value to the product/service by each of their focused business processes or activities. In the context of new manufacturing mode-SocialM, the value-adding mode can also be changed with the new manufacturing mode, or extend to a more value-adding opportunity. Figure 1 shows a normal business value chain of an enterprise, with the major business process across the order delivery chain. The enterprise needs to make value-adding to customer orders through the process of the value chain. Regarding the functional process with customer orders, the Association for Supply Chain Management (ASCM) [15] points out that the entire supply chain/value chain includes multiple enterprises across the chain, and the associated process is "the integrated process of plan, source, make, deliver, return, and enable spanning from the suppliers' supplier to the customers' customer". According to the above illustration, an entire value chain contains multiple participants: different levels' suppliers from upstream and different levels' customers downstream, and we call all of them participants. The participants can be linked by three major flows: materials, information, and cash (see Figure 2), among which the information flow includes internal and external flows. For the single product order delivery process, in the context of the normal VC shown in Figure 2, the material flow is a forward flow that is from the upstream suppliers (higher layer participants) moving to the downstream participant (except product returns). The cash flow is a reverse flow from downstream to upstream. The information flow should be both ways in the chain as the involved participants should communicate or share necessary information. The management of the value chain is mainly associated with the three flows.

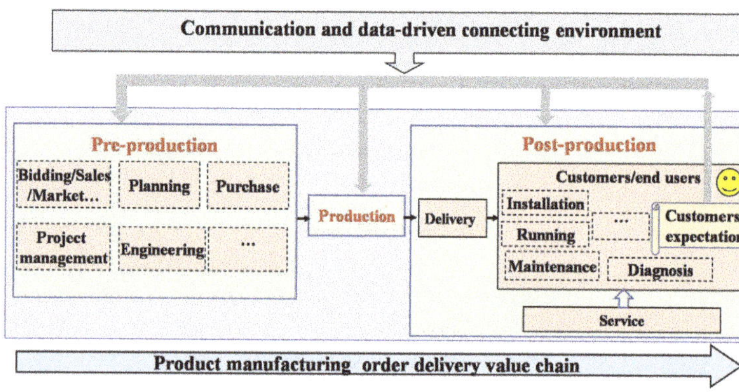

Figure 1. Manufacturing focused process for customer order delivery through the enterprise business value chain.

Figure 2 reflects the ASCM description, and the order delivery business VC consists of multiple participants across the chain, who are from upstream and intermediate (suppliers), to downstream participants and their end customers. In the normal value chain system, upstream participants need to create value (value-adding) to meet downstream participants' (enterprises and end customers/users) expectations; however, the customers make only a very limited contribution to the value-adding, except for providing the requirements or expectations associated with the order. In comparison, the SocialM mode offers more potential for downstream participants to be involved in value-adding activities across the value chain.

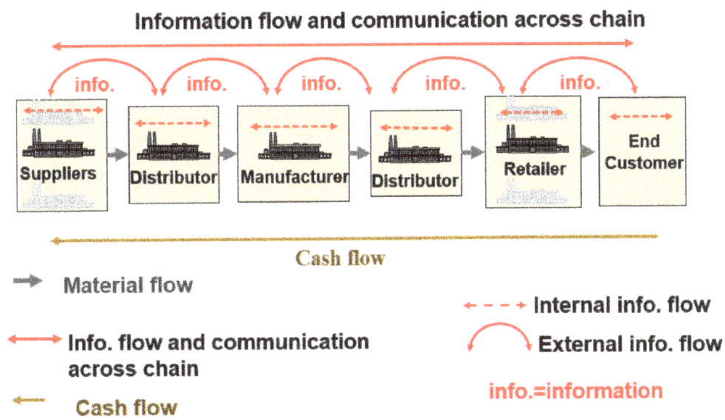

Figure 2. Three major flows through the value chain.

With the SocialM mode, new value-adding opportunities can be created by various socialized manufacturing resources (SMRs)/social manufacturing groups (SMGs)/prosumers (producer + consumer), including downstream participants and end customers who can connect and communicate through cross-enterprise-centered and Internet-based behaviors, supported by supporting technologies. Meanwhile, customers may obtain better quality, and better serviced product/service from upstream socialized suppliers/sub-suppliers. Upstream participants (socialized suppliers, socialized sub-suppliers) can gain stronger value-adding capacity to deliver product/service for their downstream participants (downstream suppliers and customers/users). In particular, cross enterprise-centered suppliers can extend more value-adding processes, such as after-sales service to end customers/users or maintenance for the whole product lifecycle, such as remote service through a cloud-platform, or even establishing service-oriented manufacturing in the context of SocialM.

Before proceeding, we need to clarify two main types of value chains in business. The first follows the order delivery process from CO (customers' order) to delivery of the product/service to the customer. We call this a CO-focused value chain, as shown in Figure 2. The second type follows the process of different phases of the product–lifecycle (PL), which is developed originally from Stan Shih's smiling curve [16], and then extended to illustrate value-adding and sharing benefits in different phases through the entire lifecycle of the product (see Figure 3). This type is called a PL-based value chain. From a continuous improvement perspective, in a PL-focused VC, all processes/activities of phases from starting a product development to the end of the product lifecycle can be taken into consideration in terms of value-adding, and in the CO-focused value chain, all processes from customer demand to delivery of the product/service to the customer can be taken into consideration in value-adding. This research emphasizes the first type of VC: the CO-focused value chain.

In recent years, numbers of scholars have been showing great interest in the SocialM mode, its application, and supporting technologies. However, there has been relatively limited research on the potential of value-adding creation of SocialM spreading to the entire value chain, when compared with the amount of research attention to the manufacturing mode itself and supporting technologies. This paper takes the perspective of the entire value chain, and attempts to fill the research gap by presenting a case study in SocialM: a new perspective for adding value through the social value chain system.

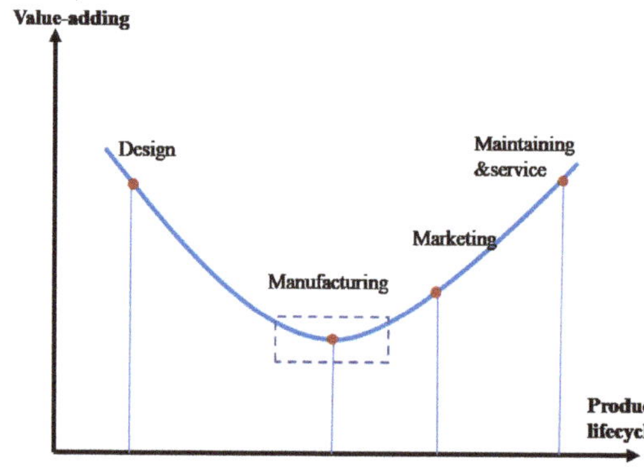

Figure 3. Smiling curve illustrating the relationship between the product life cycle and the benefits on different phases [16].

The purpose of the current research is to propose a new perspective on adding value for enterprises and customers through applying SocialM. A SocialVC (social value chain) system is presented that extends from SocialM mode to the entire business value chain. The rest of the paper is organized as follows: first, a definition of terms, and literature review and prior work are presented. Next, the research method is presented and the results introduced, which present a new perspective on value-adding, establishing the proposed social value chain system with some relevant key supporting technologies (KSTs) and a preliminary idea for a high-level measurement model for the performance of SocialVC. Finally, a summary of the study and some areas for further research are presented.

2. Literature and Prior Work

2.1. Definition of Terms

To better understand the new perspective of adding value through the SocialVC system, it is necessary to give definitions of key terms in the SocialVC system.

Definition 1. ***Value*** *is defined from the customers' perspective (externally focused) according to Lean Thinking [17]. Regarding the SocialVC system in this paper, the true value extends to the prosumers' perspective in the product or service. Enterprises in the SocialVC system should consider a strategy to identify participants, including prosumers who will provide the required work in achieving value for the product or service in the future through the SocialVC system.*

Definition 2. ***Value-adding/value-added (VA)*** *is defined as an enterprise adding value to its products or services before delivering them to customers so that value in a product or service is a true value only if the customer is willing to pay for it. Enterprises in the SocialVC system can use the SocialM mode with KSTs for participants in achieving VA in growing businesses through the sharing of opinions, capabilities and capacity for "win–win" scenarios.*

Definition 3. ***SocialVC system*** *is the demand–supply chain between the end customer and the suppliers in the context of the SocialM mode. This system not only considers both high level information and material flow which cover all nodes (including end customer and key suppliers) as normal VC, but also applies SocialM to involved participants. The value-adding not only aims to deliver maximum value to the end customer, but also achieve win–win scenarios for all involved participants.*

Defining the system aims to fix the scope, products, and nodes that a demand–supply chain will cover. For this target, the effective method is discussion with the management to take overall business units and their activities into account.

Definition 4. ***SocialVC network*** *is a network of chains that produce multiple products or services for prosumers according to the SocialVC system.*

Definition 5. ***Node*** *is a point among key participants and the main physical process where goods/data/knowledge comes to stay or pass in the context of the SocialVC system. Key roles include prosumers and other participants such as SMRs and SMGs across SocialVC, and the main physical process includes warehouse, factory, and any connecting points between the key roles and other participants.*

Definition 6. ***Prosumer*** *takes the roles of both producer and consumer at the same time. The role involves a combination of consumption and production according to Ritzer et al. [18].*

Definition 7. ***SMRs*** *comprise all kinds of property-type and consumable-supply-type manufacturing resources that can be geographically distributed across the prosumers on SocialVC in order to provide the activities of SocialM for a product or service [13,14].*

Definition 8. ***SMGs*** *are manufacturing resources grouped together in the context of SocialM to input SMRs which provide participants across SocialVC. The SMGs can be categorized according to the participants' wishes, manufacturing interests, resources, and social activities in the context of SocialVC. In the SocialVC system, SMG provides SMRs for SocialM in order to create value for products or services.*

Definition 9. ***Crowdsourcing*** *is a mechanism to provide for a distributed problem-solving work on a product or service through an Internet-based tool, such as Apps, Internet-based social media, or other internet platform [19]. Crowdsourcing allows everyone to have a chance to be involved in the work or activities of relevant problem-solving if capable of doing so. The involvement can include offering work, information, or opinions which people can submit online.*

Definition 10. ***KSTs*** *is the technology that may be adopted to support the realization of the SocialVC system. Some examples of technologies include digital-driven technologies such as Internet-based technology, RFID, social sensor, CPSS, digital-twins, blockchain, big data, AI; supporting mechanisms such as crowdsourcing and outsourcing mechanisms; or methods such as modeling and simulation, supporting management to achieve various goals, continuous improvement (CI), performance evaluation, etc.*

2.2. Evolution of Value Chain Concept

The normal value chain was not a very new concept, when first introduced by Professor M.E. Porter, who first used this concept as a decision support tool to the competitive strategies of enterprises [20,21], who has completed the important pioneering works on the value chain to both business and scholars. According to Porter's value chain [18], the process could be divided into two types of activities in enterprises: one is primary activities that cover production, marketing, transportation, after-sale services, etc., and these activities are directly relevant to how value is created to a product or service. The other type is supporting activities that cover raw material supply, technology, human resources, financials, etc., and these activities can implement and coordinate the primary activities. In the middle of the 1980s, Hopkins and Wallerstein proposed a CC (commodity chain) concept [22]. Then, Gereffi further developed the theory of global commodity chains during the mid-1990s, and he introduced an analytical and normative usage to the value chain [23]. Griffin pointed out that multinational enterprises act as "drivers" which can play a role in governance in the global value chain (GVC) by organizing, coordinating and controlling international production and supply activities.

After the 1990s, GVC have experienced a transition from rapid expansion, upgrading to the transition process with occasional small contractions. The focus of some scholars has extended from a VC or GVC concept to improvement in efficiency and profitability by optimizing the value-adding mode through management or the business process [24].

Regarding value-adding from business to customers, the lean concept has been applied widely since the 1970s by enterprises to remove non-valued-added (waste) activities from the business process, especially the seven wastes that Ohno defined [25]. As described by Womack and Jones [26], five key lean principles are important for business: value, value stream, flow, pull, and perfection. Among these five principles, value is first defined for enterprises as the customer's need for a specific product or service. Since then, the five principles provide clear systematic steps for improvement in value adding that can be widely applied for continuous improvement in the business process of enterprises moving towards an optimized process.

Extending from the business process perspective to the entire business value chain, involved participants in the value chain should also consider effective ways for adding value across the VC. In past years, in the context of normal manufacturing mode, enterprises/participants in the VC have been putting effort into considering how value is added through their business process of manufacturing, e.g., value might be added from removing non-value-adding activities according to the optimization of overall value streams. Comprehensive thinking on the combination of value-adding from the lean concept and value-adding across enterprises from the VC concept provides enterprises with more opportunities to reduce non-value-added activities in the VC process [26–30].

Regarding improvement in value-adding, in the past years, there have been significant achievements in value-adding through CI (continuous improvement) in industry and service businesses [24,31]. For a long time, identifying and eliminating waste has been a major CI (continuous improvement) task for value-adding activities in the context of any value chain. Even so, many enterprises are still facing challenges that limit their value-adding capability in normal manufacturing or normal value chain modes. Especially since the COVID-19 pandemic, even more challenges have prompted enterprises to re-think what new perspectives on value-adding for both enterprises and customers would produce more advantages than the normal way. The new perspectives should mitigate the difficulty of adding value, moving towards satisfying customers' expectations and requirements for both enterprises and customers in the new situation [31].

The main issues in value-adding across the normal VC based on normal manufacturing mode are listed as below: - Role of customers: due to customers having their ideas along with increasingly customized and personalized demand for products and service, customer participation in the value-adding process upstream of the value chain is a trend in the product or service chain. For instance, in the garment and fashion industry, customer value can be better reflected if the customers' creativity is added in the design stage. However, the customer participation is very limited by means of a normal value chain.

- Sourcing issues: due to the limited capacity of sourcing in enterprises and the supply-side market, it is not easy to select qualified suppliers at reasonable cost. Therefore, waste is caused by sourcing or supplier issues.
- Service issues: due to service having been considered a crucial factor in competitiveness in recent years, many enterprises must try to meet the requirements from customers. In particular, for the process of after-sales, such as installing, commissioning and maintenance on-site, service problems have existed for a long time, and in the normal mode are difficult to solve. For instance, the crane industry has suffered from timeliness, effectiveness, efficiency in after-sales service, such as delays, and declining productivity [30]. In particular, with an increased amount of customization and personalization, the demand for after-sales service or product maintenance has been increasing dramatically. Good service can provide more value-adding potential for enterprises; however, most enterprises lack effective service systems or appropriate service providers and support technology to meet the requirements through the normal mode.
- Networking issues: from the perspective of connecting with customers, customer feedback is an integral part of the business. There is no scope for improvement if enterprises do not get to know what the customer likes and does not like. From the per-

spective of interacting and communicating among manufacturing resources, such as people-to-people, people-to-machine or machine-to-people, and machine-to-machine, the communicating and interacting mode has great room for improvement.

- Mass customization (MC) issues: due to customized and personalized demand in markets becoming permanent trend, there is a lot of space to create value relevant to customized products and services. However, regardless of mass production, lean production, or other normal manufacturing mode, it has been increasingly difficult to satisfy increased individual requirements and specifications by means of normal VC.
- Sharing issues: limited sharing of information on sourcing, capacity, and other resources among all participants in VC brings about huge waste through normal VC. For instance, enterprise A has excess capacity, while enterprise B in the same industry a lack of capacity, but enterprise B cannot utilize the excess capacity from A. Similar issues have widely existed in the context of normal manufacturing mode and VC
- Technical issues: lack of support technology limits the value-added capacity throughout the VC process Technical issues are also obstacles to the creating of value from enterprises to customers.

In order to solve the above issues, there is a need to explore different ways of adding value to solve issues that are not easily solved in the normal value chain. In particular, the uncertainties caused by the COVID-19 pandemic in recent years have not only ruined the smoothness of flows in the supply–demand network, but also further exacerbated the anxiety of enterprises. Our research proposes a new perspective of value-adding for enterprises/participants in the context of the proposed SocialVC extended from SocialM. The following review is about the evolution of SocialM mode to SocialVC system and its use in the participants' manufacturing.

2.3. Evolution of SocialM

The theory of SocialM mode has been growing and developing along with the concept of the sharing economy over the last decade [32–34]. At an early stage, the emergence of the idea of the sharing economy has opened up people's horizons and brought enterprises more opportunities or potentials [34,35]. One important opportunity was that enterprises could have more possibilities to add value as the boundaries of manufacturing capabilities were expanding, bringing about a change in the customers' role from buyer to "prosumer" [18]. This new role can create more possibilities to create value to the ordered product/service in the context of SocialM. The prosumer concept in SocialM extends the responsibilities of customers and users, which can bring more value to consumers (prosumers in SocialM), such as better products and more professional services [13].

Regarding business operations, the sharing concept has shown power when it is applied in service operations, well-known examples of which are Uber and Airbnb [34]. Meanwhile, some public social media that need digital content operations have been growing rapidly, such as businesses pioneering Internet-based technology: Facebook, Twitter, and WeChat, amongst others. The operating mode of these public social media have not only influenced people's daily communicating, but also changed social manners that people have been used to using. Consequently, the change of communicating and social manners have significantly influenced the business model and promoted the development of the SocialM mode [6–13,36,37].

The emergence of the SocialM mode has changed the game rules among large and small enterprises in the business value chain. According to Hamalainen and Karjalainen, and Jiang et al., more individuals have participated in product manufacturing activities due to the evolution of Internet-based technology and communicating behaviors [6–12]. Clearly, not only enterprises in the traditional sense are affected by this change, but also MSMEs (micro, small and medium-sized enterprises), and even individuals. Moreover, KSTs also show the important power. For example, crowdsourcing and crowdfunding mechanisms and some service or product maintenance supported by Internet-based technology/systems bring new value-adding opportunities through customer order value chains and product value chains throughout the product lifecycle, through different forms of cooperation, interacting, and communication [9–14].

In practice, some enterprises have attempted to implement SocialM to solve some issues that the normal mode does not easily solve in order to create more value-adding opportunities. Some enterprises have adopted digital content or coding operations to achieve benefits from applying the SocialM mode. For example, RepRap has used the open-source 3D printers' manufacturer network as a new model for product development, and it has been beneficial from relevance open design platforms based on SocialM mode [38]; Vehicle Forge platform has provided a virtual collaborative environment for design work by the cloud infrastructure [39,40].

While SocialM mode has been applied and made progress in business, the academic discussion has also kept pace with the technological progress and gradual in-depth application. According to Wang et al., early in the development of SocialM, it was defined as a new manufacturing concept and emphasized social computing and considered social intelligence techniques in the configuration of outsourcing and crowdsourcing for the whole product life cycle [2,3]. Shang and Xiong et al. thought that SocialM used the smart-interactive connection of dynamic information to process manufacturing services, instead of the whole product life cycle focus. In their opinion, SocialM can be considered as a production process that consumers could be fully involved in through the internet, and relevant equipment could be connected directly on the network by the smart-interactive terminal [4,5]. Later, with the development of Internet-based technology, Hamalainen, Nyberg and Karjalainen et al. defined SocialM as a new collaborative manufacturing mode in which the relevant process can be facilitated by mobile technology, new digital manufacturing, and online social networks. They emphasized the application of new supporting technologies and advanced manufacturing concepts, such as 3D printing technology, mobile technology, customization concept, value chain concept and social networks. [6,7]. This paper uses the definition introduced by Hamalainen, Nyberg and Karjalainen as our study focuses on the CO-focused delivery value chain, which corresponds to this definition. Jiang et al. took into account the key factor in SocialM illustration, and pointed out that "SocialM is defined as a kind of Internet-based and service-oriented advanced manufacturing mode covering the whole stages of a product life cycle" [12–14]. In this definition, two important points were highlighted. The first was from a technical point of view, which was Internet-based technology, and the second was a distinct service characteristic, which met requirements across the product life cycle. Jiang's definition corresponds to PL-focused VC.

To have an overall understanding, Jiang et al. summarize the seven characteristics of SocialM as below [14]:

- Microlization and minimalization of manufacturing resources;
- Self-enterprise of socialized manufacturing resources;
- Virus-like propagation of enterprise structure;
- Sharing and competing capabilities and business benefits;
- Dynamically distributive infrastructure;
- Big-data driven decision-making and performance optimization;
- Industrial software model to be used.

The characteristics summarized above are not only associated with and impact the manufacturing, but also deeply affect how to accomplish value-adding activities when compared with the normal business value chain. The SocialM mode breaks the normal organizational mode, sourcing mode and ways of interacting and communicating among participants in the value chain. These changes have not only brought about innovation in the manufacturing mode, but have also deeply affected the value-added mode that participants in the business value chain are involved in every day. The resulting new value-added mode with new business value mode has not only been an interest for scholars, but also enterprises seeking out innovative ways to satisfy their customers through the new perspective of value-adding. Many customers and suppliers/enterprises have suffered from broken supply chains caused by the COVID-19 pandemic since 2019, so the new perspective of value-adding through the innovative value chain mode has a deeper practical significance for enterprises and customers.

To make the application of SociaM possible, support from technology is very crucial, and this has attracted the interest of academic scholars. We categorize three streams of supporting technologies for SocialM. The first stream is digital-technology-oriented, such as IT, the Internet, etc., and we call this stream hard technology in this paper. The second stream is called people-oriented and includes management, organization structure, and the like. We call this stream soft technology in this paper. The third stream is social computing relevant technology that is combined with both hard (the first stream) and soft technology (the second stream), and constitutes an extended version of the SocialM mode to realize the Societies 5.0 [40,41].

Regarding the digital-technology-oriented stream, a number of digital technologies have been developed since NC (numerical control) first occurred in manufacturing, some of which could support SocialM according to the manufacturing requirements. In the early stage of SocialM applications, Internet-based technology provided more possibilities for SocialM application [2–13]. Among Internet-based technology, 3D printing systems were considered to be a key enabler in implementing SocialM [10–14]. Recent developments have been the Internet of Things (IoT), cyber-physical-systems (CPS), RFID, social sensors, loud computing, blockchain, big data, digital-twins, machine learning, and deep learning [4,11–13,42,43]. Among these technologies, the social sensor is important in data transfer and in interacting and communicating among enterprises and various manufacturing resources in the context of SocialM, such as between "people–machine–material", "people–people", "people–machine", and "machine–machine" [44]. The adoption of these technologies has accelerated and enabled the application of SocialM, such as cloud service [4], open product design [6,36,37,39], healthcare systems [44], the crane industry [45], amongst others. In the reality of manufacturing, the application of digital-oriented technology still requires effective cooperation between people, and organization structure and corporate culture, which is called people-oriented technology in this paper.

Regarding the people-oriented stream, people's behavior, the organization's structure, and way of communication and cooperation among participants across the value chain must meet the new manufacturing mode from the social perspective [14,46]. As the environment of SocialM is different from the generic manufacturing environment, a corresponding change is required in the VC process. Consequently, a series of adjustments are needed in terms of people's behavior accordingly [6,47]. Meanwhile, the process change also brings about an organizational change [48], such as new roles, new mechanisms and new functions being established. The role of prosumers also brings about a new force to manufacturing resources for value–value capacity; a crowdsourcing mechanism can be applied to connect prosumers with upstream suppliers or manufacturers in the VC [4,8,47,49]. Self-organization corresponds to a dynamic resource community (DRC) in SocialM in order to provide suitable capacity for production and product-driven services to prosumers [13,49,50]. Moreover, all people/organizations have to have skills that align with SocialM requirements and take on the relevant responsibilities through VC. In reality, the behaviors involved with connecting and communicating between people in the VC could draw on some social-media-like technology. It requires people to cooperate skillfully through relevant Internet-based technology, which belongs to the digital-technology-oriented stream. Therefore, the effective application of SocialM is supported by either digital-technology-oriented or people-oriented supporting technologies, both of which are associated with the third stream of SocialM supporting technologies.

Regarding the third stream of supporting technologies—social computing relevant technology—advanced supporting technologies can be effective in combination with providing a cutting-edge manufacturing solution in moving towards the Society 5.0 era, as Wang et al. has pointed out [40,51]. A number of supporting technologies have emerged, including hard and soft technology that provide more power to promote the application of SocialM. To do so, SocialM in practice will need not only a Cyber-Physical-Social-System

(CPSS) that is extended from CPS (Cyber-Physical-Systems) and provides a socialized ecosystem for SocialM, but also the SMRs that can provide numbers of self-organized participants/prosumers in the VC for various specialized product-related, production-related or service-related capabilities to meet customers' requirements [2,49–51]. In social computing relevant technology, more functions can be realized through a combination of different digital-driven technologies. For instance, social sensor and RFID could be integrated with CPS to form a CPSS platform in the context of SocialM, in order to provide the socialized ecosystem. Big data technology combining the technology of digital-twins/modeling and simulation can support the digitalization in SocialM operation [52,53]. Regarding the Societies 5.0 era supported by SocialM, a number of emerging technologies have been effectively combined and configured for different manufacturing scenarios to meet the customers' requirements in products/service. Among the many emerged KSTs, a parallel system method has opened up in the research on SocialM, as presented by Wang, which provide effective methods in the control and management of a complex system. According to Wang, one common system can consist of a reality system and one or more virtual systems that can have some artificial system [54]. Alongside this idea, one KST—digital twin technology—can provide a tool for modelling and simulation and analyzing SocialM application, such as product design, production line design, and even design for social factory [55–57]. For the decision-making of enterprises, big-data, cloud computing, combined with deep learning, can support performance monitoring in the VC and optimization performance if needed, as well as support decision-making [58,59]. In short, AI-based, Internet-based KSTs and appropriate organizational arrangement in the value chain can provide more configurable technology in the application of SocialM. According to research in past years, it is clear that more new technology is bound to emerge to enable SocialM application for various enterprises such as AI and digital technology which will continue to advance in the near future. In the following discussion, we will collectively refer to AI, IT, internet-based technologies, as digital-driven technologies.

As reviewed above, many scholars have focused on SocialM manufacturing itself and technology. This paper applies the SocialM mode to the entire value chain, in order to create more value-adding potential, which will also fill one research gap in relation to SocialM.

2.4. Contributions of the Research

To explore the possibility of solving issues that present obstacles to value-adding in the context of the normal VC, this research proposes a new concept: the SocialVC system that extends the SocialM mode to the entire value chain, towards creating more potential for adding value across the value chain. The main contributions are listed as below:

- New thinking for enterprises in terms of having more opportunities to add value, as compared with the normal manufacturing mode;
- Establishing the social value chain system for all participants/enterprises across the chain by means of the value chain concept and SocialM mode with supporting technology;
- Considering a suitable performance measurement to monitor and evaluate whether the SocialVC system works efficiently.

3. Method

A new sight must be opened to for all participants on VC to adopt the new perspective on value-adding by SocialVC that is associated with SocialM mode. In past years, many scholars have been interested in SocialM mode and its application in enterprises; however, a little research focus on value-adding across entire business value chain when SocialM mode is applied across business process of enterprise, this research will fill this gap. A methodology of the research begins with a context of the theoretical reviews on previous research both in value chain and SocialM subject, follows with new idea of value-adding from the proposed SocailVC in this research, ends up with a conclusion on an innovative way for enterprises to add value for both customers and enterprises in the context of SocialVC system that is a different way from the normal VC.

The methodology mainly outlines four parts in this paper. In first part, a systematic review on normal business value chain concept associated with this research is made. After that, a review of the SocialM mode is followed to describe the relevant work on the area of the SocialM mode. Based on these, a literature gap is identified in the area of analysis on value-adding, especially relevant to value-adding in the context of value chain for customer order delivery. In the second part, a new perspective of value-adding across value chain that extend from SocialM to an entire value chain is illustrated, and a new potential of value-adding is followed. The third part introduces some important supporting technologies to support value-adding in the context of the proposed SocialVC system. The fourth part explores possible measurement metrics to measure performances of SocialVC. In the end, some conclusion and suggestions on the future work are presented based on the above parts.

One reminder is that this work is an academic research, and the technology adopted to the relevant SocialVC system is based on current relevant supporting technologies, and the paper just discusses some major supporting technologies; ltthere might be more existing technologies that are being adopted, but are not fully covered in paper or more innovative technology along with a continuing emergence of advanced technology may be applied for future SocialM or SocialVC, which is impossible to be covered in this research.

4. Results and Discussion

4.1. Establishing an Architecture of the SocialVC System

Regarding SocialVC system, it is associated with SocialM, the concept of value chain and the value-adding across chain, and some important supporting technologies mentioned in the previous sections. To establish the SocialVC system, this research proposes an architecture of the SocialVC system (See Figure 4). As Figure 4 shows, the architecture is layered and interconnected among each. The information and data through chain participants should be collected and shared by Internet-based social media platform, and the available manufacturing resources is identified. The required tasks for customer orders are distributed by means of supporting technologies. Meanwhile, SMRs are defined and matched to the SocialVC network. Finally, the ordered product/service and tasks are completed, and delivered by means of distributed SMRs/SMGs/prosumers. According to the above logic, the proposed architecture system is divided to five layers that are associated with the value-adding through the customer order delivery process in the context of the SocialVC system. The details of each layer are illustrated in the following subsections.

4.1.1. Layer1—Input Layer

Layer1 collects important data and information as input, which will be used for decision-making about the distributing of a resource with the required work. As Figure 4 shows, all input is from involved participants. Firstly, the involved multiple participants of each node on the VC is listed in layer1, and important resources of each participant are identified and collected, which should be sent to layer2—the support layer. In this research, we classify participants in three categories according to the manufacturing reality, namely enterprises, MSMEs (micro and-small-scale manufacturing enterprises)/Micro-enterprise [60] and individuals. Compared with the normal manufacturing mode and normal VC system, MSMEs and individuals have more opportunities to be involved more in the business to create value. Meanwhile, all participants can develop more flexible ways to be involved in the process of customer order fulfillment than in the normal value chain system, when these participants with relevant resources are included in the SMRs.

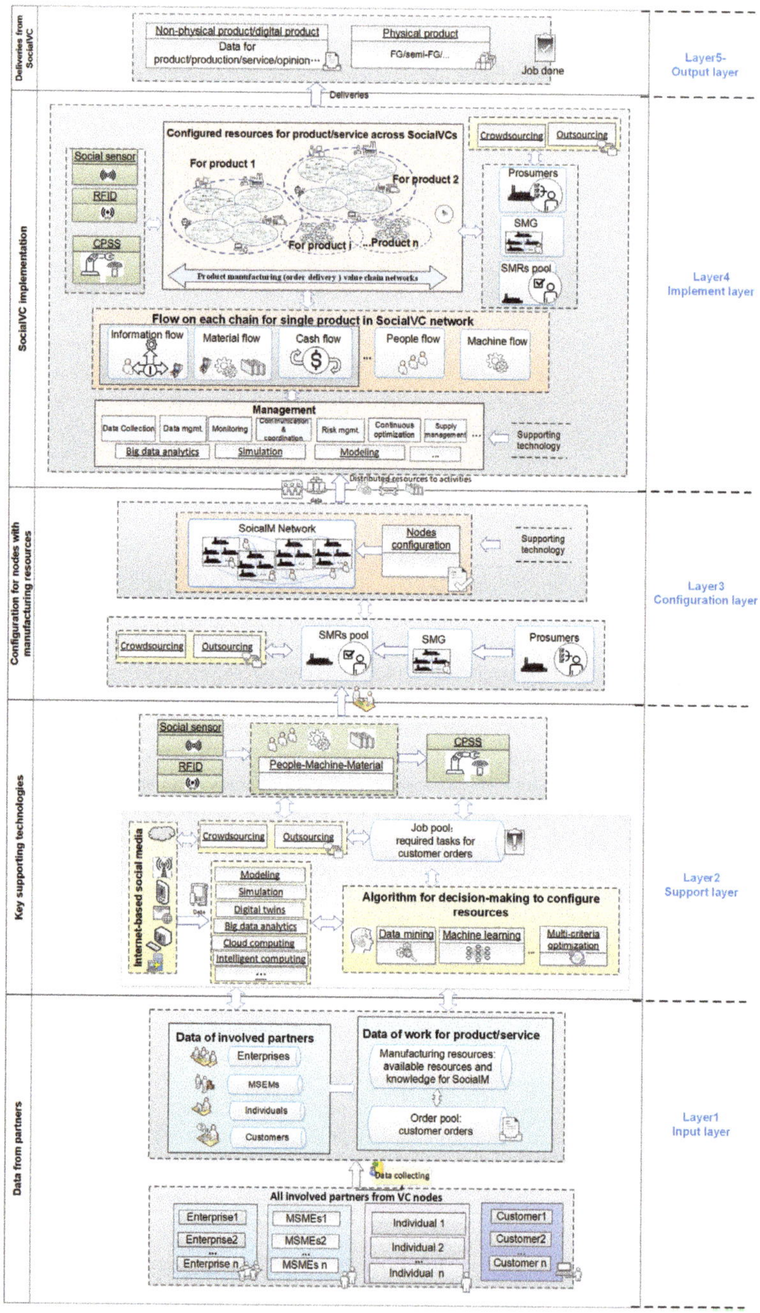

Figure 4. Architecture of the SocialVC system.

There are two main groups of data for input defined here. One group of data are manufacturing resources-related, including the available physical resources, nonphysical information, and knowledge. The collected data from each node/participant are a basis that present participants with the capacity to meet the required tasks that add value to product/service.

For the manufacturing business, the main resources can be classified as tangible and intangible resources. Tangible resources include resources such as the geographical sites, physical plants, available facilities, warehouse, local service or sales office, manpower, etc. These tangible resources can be adopted for value-adding activities through the manufacturing process in the value chain. Intangible resources include resources such as IP (intellectual property), brand value, trademark, self-developed software, knowledge and corporation culture. Intangible resources not only show competitiveness and advantages of the enterprise and power to create customer value, but also shareholder/stakeholder value. Besides data on resources, other important data such as key parameters also need to be collected, for example operations and supply chain KPI (key performance indictors) figures.

The second group of data are from customer orders for expected products/services. Due to the variability of the ordered product/service, the order fulfillment process varies widely, and relevant data and information need to be collected into a defined database system. All data/information are connected to layer2 through an Internet-based social media system and advance intelligent algorithm for configuration on SMRs. Participants become prosumers in the context of the SocialVC system, and they should be allocated to tasks of order fulfillment that are called task orders.

4.1.2. Layer2—Support Layer

Layer2 provides some KSTs which are mostly based on digital-driven technology to support the application of the SocialVC system. In the SocialVc system, each layer needs certain supporting technology to enable the defined function to satisfy the customers' requirements. The KSTs in layer2 not only support the manufacturing resource allocation for the manufacturing tasks in layer3, but also support the functions of all the other layers. The main KSTs include Internet-based social media, big data analytics, digital twins or modeling and simulation, and cloud computing. In addition, an algorithm to support intelligent decision-making is also important, containing big-data-driven data mining, machine learning, and multi-criteria optimization.

Through an Internet-based social media platform, the information and data can be transferred or even shared appropriately. Moreover, interactions and communication among nodes/participants can be performed through a social media platform, and behaviors of business process can be monitored and follow-up actions also communicated.

With support from the combination of social sensors and RFID, with CCPS, an intelligent configuration of nodes and SMRs for SocialVC is made and encapsulated. In particular, CCPS can also be widely adopted for effective interconnection and management among "people–machine–material" resources in multiple levels of manufacturing resources on the participants' site, including levels of machine, production line, shopfloor, and participants, involving for example the sharing and transferring of data and knowledge, dynamic scheduling, collaborating, and matching [12,50].

Through an advanced algorithm that integrates the data mining, machine/deep learning, and multi-criteria optimization, big-data-driven intelligent decision-making not only helps to identify and separate the customers' requirements into actionable job/activity, but also mines various relations and interactions among various resources, including SMRs and prosumers. The advanced algorithm also carries out configuration (configure layer) of the participants' resources by identifying the quantified SMRs with support from outsourcing and crowdsourcing mechanisms through SocialVC. All the capacity of configured SMRs needs to be distributed from the SMR pool to matched nodes (SMGs/prosumers) which will accomplish the required manufacturing tasks/task orders; hence, relevant SMRs/SMGs/prosumers will add value to the customer order through the business process in the context of SocialVC.

Through outsourcing and crowdsourcing mechanisms in Internet-based social media, more flexible capacity for production or problem-solving in value-adding activities can be gained in a competitive environment. The manufacturing capacity from outsourcing and crowdsourcing mechanisms is becoming one kind of socialized capacity, which not only creates value, but also may reduce manufacturing costs compared with corresponding activities/tasks in the context of normal manufacturing mode or normal value chain. Therefore, this type of outsourcing and crowdsourcing is also one type of value-adding mechanism.

Besides KSTs, there could be other manufacturing functions through the SocialM process, which are not included here. For instance, blockchain technology can be adopted as a credit and security mechanism, and 3D printing plays an important role in promoting decentralized social manufacturing [61,62]. 3D printing applied in SocialM not only achieves big cost savings, but also offers more potential for outsourcing, crowdsourcing and crowdfund mechanisms and the participation of MSEMs and individuals. In this research, we only focus on the more common KSTs relevant to establishing layers of SocialVC.

4.1.3. Layer3—Configuration Layer

Layer3 focuses on distributing manufacturing resources for the nodes/participants of each value chain in the SocialVC network. Various products are derived from various chains, which form the SocialVC network. Nodes on participants need to be distributed SMRs that match the required tasks in the job pool from the customer orders.

By supporting technically, the relations and interactions between participants in the value chain are mined and identified, in order to match SMRs from the SMR pool and required capacities of the relevant algorithm in support layer2.

From the operations prospective, the dynamic operation of outsourcing and crowdsourcing mechanism continuously collects updated information of manufacturing resources from prosumers, then matches them with the required tasks in the job pool. Finally, the task order can be made according to the successful match. During operation, the prosumers provide SMRs to SMGs, whose role is acting as an "agent" of various prosumers and works like a provider to the SMR pool. SMRs are made up of various types of enterprises, including big/normal sized enterprises, MSEMs, or even individuals. Meanwhile, as SMRs are characterized by socialization, decentralization, self-organization and specialization, they have the ability to provide the appropriate manufacturing resource accordingly [46], including tangible and intangible resources.

As various SMGs are grouped by similar business interests and common social activities, and provide various dynamic production capability to SMR, so numerous SMRs provide the configured capacity to nodes of participants along the chain. The task orders are performed by the configured capacity in SMGs, and relevant value-adding activities are reflected in products across the corresponding value chain. For instance, some enterprises with CNC (computer numerical control) as the main equipment can form an SMG, in order to take the kinds of task orders that need CNC.

All configured manufacturing resources are distributed to relevant nodes within the SocialVC network so that the prosumer on the node will complete the task orders or customer orders by means of the allocated capacity.

4.1.4. Layer4—Implement Layer

Layer4 is for implementing the customers' orders including all task orders or from the job pool in layer2, through configured SMRs in Social VC. The configured SMRs provided from the SMRs pool come from various different SMGs and are formed by prosumers. In order to fulfill an order, the configured resources are connected by process and supported by a defined management system for various goals.

Flows in the SocialVC become more complicated, compared with the normal VC. As the definition previously, the prosumers' role takes a dual role of customer and manufacturer, that is, prosumers are not only the product/service receiver, but also the supplier or manufacturer for product. Consequently, the material flow is changed to multiple paths for a single product in the context of SocialVC. For the same reason, the cash flow is also changed, and is not a single direction of flow, but a multiple path flow. The information flow is changed to multiple paths as well. Besides the major flows, there are also some other flows, such as people flow and machine flow. Supported by KSTs, the resource flows interconnect among "people–machine–material", "people–people", "people–machine", "machine–machine", as described in support layer2. For multiple products, the flows are even more multi-directional according to how the SMRs/SMGs/prosumers form. All flows are intertwined and more complicated, and can be called flow-networks (see Figure 5), compared with flows in the normal VC. With support from KSTs and formed from the social environment in the SocialVC system, all flows through flow-networks across each node can run in adaptive, autonomy and decentralization modes, as shown in Figure 5.

Meanwhile, management support is also important and should be defined to ensure that all processes operate smoothly and avoid uneven flows, with continuous optimization to make all resources produce the best capacity for value-adding. As shown in layer4 in Figure 5, the management system can contain data collection, data management, monitoring, communication, coordination, optimization, risk management, continuous optimization and supply management (not limited here), defined for various goals. Comparing with the normal VC, the connecting and interaction among nodes are not only dependent on process, but also interaction supported by the social sensor, RFID and CPSS in the context of SocialVC. Digital-driven technology enables various nodes/participants/prosumers to effectively connect and interact during order fulfilment, and avoid more non-value-adding activities due to poor communication, or untimely reaction, which happen often in the normal VC. With digital technology and management support, all SMRs work in an orderly and efficient manner to add value to the end product by means of the configured tasks.

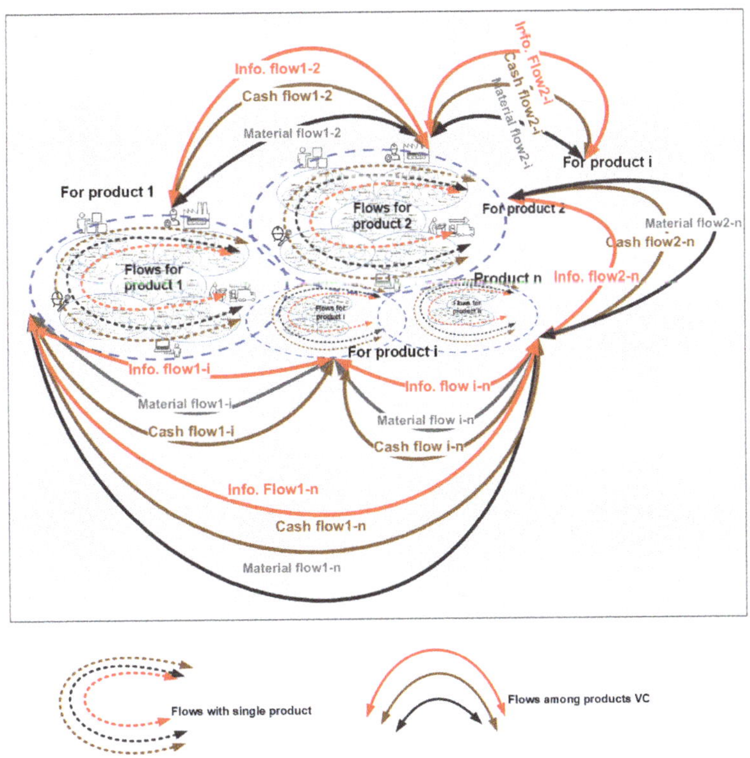

Figure 5. Flow-networks of the SocialVC system.

4.1.5. Layer5—Output Layer

Layer5 is for delivering the product/service to end customers appropriately. There are two kinds of deliveries from order fulfillment to customers, namely non-physical products/digital products and physical products. Digital products include all products which use data for the product, service, production and customer feedback, such as a software, data-based application or tool, a set of information, data for service, online training, and drawing and BOM (bill of materials) from engineering work. Physical products include all tangible items, such as books, various physical equipment, instruments, cranes, etc.

4.2. SocialVC Measurement Framework

To ensure that the SocialVC system operates in a stable and reliable way, and at the same time continues to optimize value-adding, measurement of the SocialVC system should be well defined and take into account monitoring of the performance of SocialVC. However, the performance measurement of SocialVC is still a new subject, even though some scholars have shown an interest in social performance (SP) [63]. For the normal value chain system, ASCM has defined the performance system in the SCOR (Supply Chain Operations Reference) model [64], which is a comprehensive system to measure the normal value chain. Most measurement of normal VC mainly focuses on delivery performance and cost performance, for instance SCOR performance. However, SocialVC needs the appropriate performance to measure how SocialVC meets the requirements of customers, how KSTs support the new system—SocialVC—and how manufacturing resources across the chain are efficiently used for adding value to the customers' order. The normal measurement system for the performance of normal VC is not able to cover the new requirements for the performance of the SocialVC.

4.3. Critical Success Factors for SocialVC Operation

To develop the performance of the SocialVC by measuring how the overall process behaves to create value and how the participants in the chain work together to meet the customers' requirements, this research uses a CSF tool as the basis for a performance measurement framework to measure the performance of the SocialVC.

CSF is an effective management tool that is strongly relevant to the strategic goals of enterprises [65]. This research extends the application of CSF to achieve success in the operation of SocialVC. CSF of SocialVC should take into account both the characteristics of SocialVC and value-adding tasks/activities for the required product/service through the value chain process.

Based on the characteristics and management requirements of SocialVC, this research suggests some main CSF to ensure operation of the process in the context of SocialVC. A logical relation between outputs from the SocialVC system and defined as three grouped CSF is illustrated in Figure 6. As Figure 6 shows, the value of output is relevant to the customers' satisfaction and products/services that meet the customers' expectations and achieve a win–win scenario. Regarding the requirement of success in the SIVC, we classify CSF as three groups that provide the core factors for success in the context of SocialVC, as below:

Group 1: information/data-driven CSF mainly includes two primary CSF: information/data transparency/sharing and communicating efficiently, which provide the major key basis for other CSF;

Group 2: technology and management-driven CSF, which is a supporting group and includes technology and management solutions to provide the relevant process and management skill/methods for value-adding capability;

Group 3: deliverability-driven CSF, which are directly linked to ensuring value-adding to product/service to meet the customers' requirements, supported by group 1 and group 2. The CSF of group 3 shows the key value-adding capacity to provide product/service to customers.

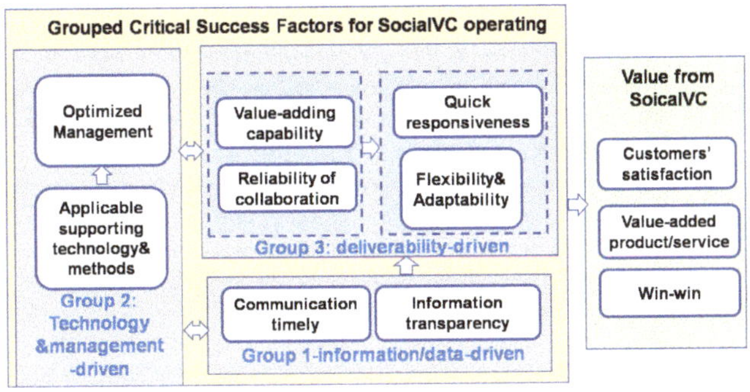

Figure 6. Grouped major CSF and relations with output from SocialVC.

These CSF lead to success in deliveries that create added value to the customers' requirements from SocialVC. Some CSF in SocialVC are similar to the normal VC; for example, information transparency/sharing is a key for the cooperation of participants across the supply chain. However, the even stricter requirements for SocialVC should be supported in the management process and supporting technologies to enable responsiveness and adaptability due to the characteristics of the social mode: being decentralized, self-adaptive, and needing self-organization and more involvement from MSMEs and individuals. Therefore, the process for information transparency/sharing of the relevant management should have a much higher requirement for SMRs to respond in a timely manner to the demands of the tasks or of customers. In addition, there are more highly prioritized factors in SocialVC to support flexibility and adaptability, which also relate to quicker responsiveness and more management commitment, and effective communication from all SMRs. As stated previously, CSFs are reflected in performance measurement, and some CSFs have a strong relation to KPIs, for instance quick responsiveness and flexibility strongly reflecting operational behavior of SocialVC. According to customer demand and characteristics in SocialVC, the above grouped CSF are defined as the most important factors according to SocialVC implementing. From a practical perspective, involved participants should develop and extend more customized levels of CSF based on the specific situation.

Based on the identified CSF, the relevant management system (see layer 4 in Figure 4) can be developed, including the relevant management process and approaches. Therefore, systematic management should be a research topic. It is noticed that management details are not a focus of this research, even though certain management elements are listed in the proposed architecture in Figure 4. Meanwhile, in order to quantify the CSF and measure the SocialVC system, the framework of performance is considered and discussed in the following subsection.

4.4. Performance Measurement Framework of VC

Comparing SCOR metrics for normal VC and taking into account CSF and characteristics of SocialVC, we propose a performance measurement framework in which high-level KPI (key performance indicators) are developed to evaluate and measure how the SocialVC system behaves in terms of value-adding activities/tasks for the relevant production and service according to customer requirements.

In the performance measurement framework, a primary/ high-level KPI is defined based on CSF and the strategic intentions of SocialVC (see Table 1). In Table 1, the primary SocialVC performance is listed and comprises a level of data/information supportability, cooperating-ability, technology supportability, management, delivery, value-adding capability and maturity of SC. Definitions of each primary performance and relevant examples are also made and shown in Table 1. Regarding the normal value chain, the KPI mainly focuses on delivery-driven performance to the customer and cost-driven performance for enterprises, such as the SCOR metrics developed by ASCM [64]. Regarding SocialVC in this research, more factors should be considered due to certain changes; for instance, involved participants become prosumers that link to SMRs/SMGs, communication is under the social environment, and various KSTs play an important role in the application of the SocialVC system. Consequently, CSF extends to more factors, and performance also includes more elements accordingly. As shown in Table 1, besides delivery performance and cost-driven performance, which are similar to with those in the normal value chain, more factors are taken into consideration in the SocialVC performance, such as more focus on data/information, communication/coordination, support technology, management, and maturity; and the definition of performance of value-adding capability is also expanded from the normal value chain.

Table 1. Primary measurement metrics of the social value chain.

No.	Metrics	Definitions	Examples
1	Level of data/ information supportability	Measurement that focuses on how effectively the relevance data system/process/program/activity behaves in data management towards the business goal, such as data collection, data analysis, sharing, etc. across the value chain	Accuracy, Completeness, Timeliness of data collection, Consistency of data, Revise timely, Quality of data analysis, Traceability, data/information richness for carrying out required tasks/delivery, Level of privacy & security of data
2	Level of cooperating-ability	Measurement that focuses on how effectively communication/coordination is carried out among participants (prosumers/SMRs/SMGs) across the value chain	Supporting data/information, Timeliness of responses and feedback, Efficiency of interaction process/techniques/social media/channels for communication/coordination/sharing
3	Level of technology supportability	Measurement that focuses on how the defined support technology supports relevant goals of process/management/prosumers	Convenience, Stability, Maintainability, Connectivity of multiple techniques, Connectivity for requirement, Enableability, Compatibility
4	Level of Management	Measurement that focuses on how defined management system with relevant process/role effectively to achieve relevant management goal	Achievement of goal, Maturity of management process, Development of people's skills, method of CI (continuous improvement) for optimization
5	Service level of delivery	Measurement that focuses on operational factors across the value chain to achieve both satisfaction of customers/prosumers, and win–win for all participants/prosumers of the value chain	Responsiveness to customer, Flexibility and adaptability to change, Rate of perfect order fulfillment
6	Value-adding capability	Measurement that focuses on the ability from all participants/prosumers across the value chain to carry out value-adding activity/task on product/service to meet the expectations of customers/prosumers	Total cash-to-cash cycle time, Total order delivery cost, Total of asset utilization of prosumers, Value of customer/prosumer; perceived value of product/service, Ability of innovation to create value for both customers and prosumers
7	Maturity of SocialVC	Measurement that focuses on how SocialVC operates in a resilient, stable, and healthy manner to deliver a product/service. Sl VC may define different maturity assessment models according to different management systems/processes with specific goals towards value-adding and win–win across the value chain	Maturity of data/information, Maturity of management/process across the chain, Maturity of overall cooperation/commitment of all prosumers/SMRs/SMGs

Well-defined KPI with multiple-levels not only can provide a comprehensive evaluation for understanding on the effectiveness and efficiency of the entire value chain system and value-adding capability in the context of SocialVC, but also provide a gap between the baseline (current KPI) and target (future KPI) by continuous improvement of systems/structures/processes. Therefore, a more detailed multiple-level KPI needs to be developed, guided by the proposed high-level KPI. Examples in Table 1 give more selections to define the detailed-level KPI even though this topic is excluded in this research. For instance, to achieve good performance, responsiveness to customers, flexibility and adaptability to change, and rate of perfect order fulfillment are associated with primary performance-delivery. All seven primer KPI can be broken down into more detailed KPI levels to develop the appropriate calculation accordingly, for better in-depth understanding of the entire SocialVC.

5. Conclusions and Future Research

As described above, this research presents a novel way of adding value via SocialVC, which provides more potential for enterprises to create value and strengthen their competition. The major contributions from this research work are summarized in this section, and directions for future research are also suggested.

5.1. Conclusions

Compared with the value-adding capability of the normal value chain, the proposed new perspective of value-adding in the context of SocialVC system can leverage the new value-adding mode provided by the SocialVC system in order to promote opportunities and capabilities for adding value for all involved participants. The proposed architecture of SocialVC can play an important role in forming the SocialVC system, which helps to understand how enterprises and customers can use the SocialVC system in dealing with coordinating various manufacturing resources, and how the value-adding activities/tasks made by SMRs/SMGs/prosumers are supported from various KETs and management systems. In reality, the SocialVC can mitigate the non-value-adding factors caused by certain issues in the context of the normal value chain. It can also promote a win–win scenario for all involved participants (prosumers/SMRs/SMGs) in the Social VC.

It can be concluded that there is plenty of value-adding potential from the enterprise perspective, especially for those MSMEs and individuals not easily involved in many manufacturing tasks in the context of the normal value chain. From the perspective of former customers in the normal value chain system, the role of customers changes to that of prosumers who can also be involved in the upstream manufacturing process to mitigate waste due to the ideas and manufacturing capacity of customers being neglected or not incorporated into the manufacturing system. As the manufacturers become SMRs/SMGs/prosumers, there is no clear line among enterprises/MSEMs/individuals and customers in the SocialVC system, and some issues that have been difficult to solve for a long time in the normal VC system must be greatly reduced, such as sourcing, service, and the roles of customers as discussed in Section 2.

Regarding the networking issue in the normal value chain, the new perspective in SocialVC is adopting social networking, such as Internet-based social media or platforms. Social networking not only allows feedback from all SocialVC participants, but also provides an effective way of interacting and communicating for collaborative value-adding activity with its downstream consumers (prosumers in SocialVC) across the chain. Social networking with other supporting technologies can minimize waste in networking in the context of the normal VC.

Regarding the high demand for mass customization, the customized product/service can be dealing with the SocialVC system, which provides an effective social network for stronger interacting and collaborating among various resources. Meanwhile, prosumers are one resource in the SocialVC; they have a better understanding of customization and individualization. Consequently, waste caused by understanding customers can be minimized in the context of SocialVC. With supporting technologies, such as social networking, big data, intelligent decisions, etc., the networked resources collaborate to contribute a lot of capacity/idea/knowledge for more value-adding activities/tasks in the manufacturing process, driven by much customization and individualization.

Regarding measurement of the SocialVC system's performance, the authors present a performance measurement framework considering the identified main CSF and characteristics of SocialVC. The primary/high-level KPI are defined in the performance measurement framework, and some examples of lower-level KPI are provided. With the high-level performance metrics, detailed low-level KPI with the relevant calculation should be further developed for evaluating how the SocialVC operates.

In brief, the SocialVC brings more potential for value-adding for all involved participants while reducing waste that is difficult to remove in the context of the normal VC. For this purpose, the SocialVC architecture provides a social networked manufacturing environment in which participants (enterprises/MSMEs/individuals) across the chain can collaborate to carry out order fulfillment, supported by KSTs and defined management systems. The theory of and knowledge about the SocialVC system still needs to be developed towards a more systematic and mutual SocialVC which fosters the capability of enterprises to add value and address issues relevant to the concerns of value-adding in the context of the normal VC.

5.2. Future Research

Deepening and exploring the various potentials of value-adding in the context of SocialVC remains a major area for future research. It is also important to draw on certain existing lean tools to analyze or identify the further potential of value-adding, such as value stream analysis/value stream mapping. In the context of digital technology, it would be crucial to search out more value-adding potential supported by advanced digital technology applications.

Regarding SocialVC networking, exploring more support technology and relevant computing methods for decision-making on the optimization of the shape of SMRs is needed. As described in Section 4, SMRs are characterized by socialization, decentralization, self-organized and specialization, which are different from the normal organization with centralized management. Another important research area in the future will focus on intelligent decision-making in more optimized modes for interacting and collaborating with SMRs' shape, in order to carry out customers' order fulfillment, in particular, orders with mass customization and mass individualization. For example, a topology model for enterprise relationship network in the context of the social manufacturing is introduced by Jiang et al. [54].

Regarding the high demand for mass customization, this is associated with the above research area: more optimized SMR shape with more effective interaction and collaboration to help in dealing with mass customization or mass individualization. In addition, from the business perspective, there is a need to improve the overall production planning mode in the context of the SocialVC system, including a socialized integrated planning mechanism in an ecosystem of the socialized business model.

Regarding the performance measurement of the SocialVC system, the original intention of this research was to propose a new perspective on value-adding for enterprises/customers, so performance-focused attention is limited in this paper, although the authors propose a high-level performance measurement framework which is helpful in developing detailed multiple low level KPIs with calculations to measure and evaluate the SocialVC system in order to optimize the system and create more value-adding potential. Liu et al. paid attention to social factors in evaluation of the supply chain management system. From the perspective of preferred suppliers, they developed an SSSE indicator system that considered economic, environmental, and social factors, and the cloud probability dominance relations (PDR) method was used for the selection of the optimal sustainable supplier [66]. From the perspective of comprehensively evaluating the performance of the SocialVC, a bigger effort is still needed to develop detailed performance measures to evaluate the major business process across the SocialVC. Further research on the SocialVC system is needed on the behavior of involved participants, such as behaviors among enterprises, MSEM and individuals who are involved in value-adding activity in the SocialVC system as prosumers, as their behavior impacts both the efficiency of the delivery of the SocialVC system and the value-adding capacity for all involved participants.

Regarding the product-lifecycle-based value chain that is actually not the focus in this paper, it is nevertheless still an important topic in SocialVC, for instance, how to build a sustainable socialized manufacturing value chain that covers the product lifecycle. More effort should be put into the product-lifecycle-based value chain: for example, Liu et al. proposed a graph-matching model that can calculate the similarity score to solve the matching problem in the crowdsourcing mechanism between the collaborative design crowdsourcing task network graph and the designer network graph [67]. There are more issues in the product-lifecycle-based value chain which could be an area focused on in the future. Moreover, future research may also focus on different industries to establish various industrial social systems according to the needs or specific purposes of the particular industry. In this area, some attempts have been made. For instance, a digital monitoring and management platform for urban construction crane machinery has been established in Ningbo, China [45]. More than 10,000 various construction crane equipment items have been connected to the digital platform, with 869 linked enterprises and 23,139 operators, and more than 3000 construction sites in 2022. This platform creates a city-wide socialized entire value chain management for the construction industry. This case from the crane industry provides a sample of the SocialVC of various industries which have a special purpose, and this can also be one future research direction.

Author Contributions: Conceptualization, G.-Y.X. and P.H.; methodology, G.-Y.X.; software, G.-Y.X.; validation, G.-Y.X., P.H., S.E. and T.S.T.; formal analysis, G.-Y.X. and P.H.; investigation, G.-Y.X., P.H., S.E. and T.S.T.; resources, G.-Y.X.; data curation, G.-Y.X., P.H., S.E. and T.S.T.; writing—original draft preparation, G.-Y.X.; writing—review and editing, G.-Y.X., P.H., S.E. and T.S.T.; visualization, G.-Y.X., P.H., S.E. and T.S.T.; supervision, G.-Y.X. and P.H.; project administration, G.-Y.X. and P.H.; funding acquisition, G.-Y.X. and P.H. All authors have read and agreed to the published version of the manuscript.

Funding: This work was supported in part by the project "New Service-oriented Business Models and a Sustainable Manufacturing Ecology for the Life Cycle of Smart Cranes", the Foundation of Business Finland under Grants 43898/31/2020.

Data Availability Statement: Not applicable.

Conflicts of Interest: The authors declare no conflict of interest.

References

1. Markillie, P. A Third Industrial Revolution. *The Economist*, 21 April 2012.
2. Wang, F.Y. From social computing to social manufacturing: The coming industrial revolution and new frontier in cyber-physical-social space. *Bull. Chin. Acad. Sci.* **2012**, *6*, 658–669.
3. Park, J.M.; Kim, B.S. Review of social manufacturing technology on product life cycle management (PLM) base. *J. Korean Inst. Ind. Eng.* **2013**, *39*, 156–162.
4. Shang, X.; Liu, X.; Xiong, G.; Cheng, C.; Ma, Y.; Nyberg, T.R. Social manufacturing cloud service platform for the mass customization in apparel industry. In Proceedings of the 2013 IEEE International Conference on Service Operations and Logistics, and Informatics, Dongguan, China, 28–30 July 2013; pp. 220–224.
5. Xiong, G.; Chen, Y.; Shang, X.; Liu, X.; Nyberg, T.R. AHP fuzzy comprehensive method of supplier evaluation in social manufacturing mode. In Proceedings of the 11th World Congress on Intelligent Control and Automation, Shenyang, China, 29 June–4 July 2014; pp. 3594–3599.
6. Mohajeri, B.; Nyberg, T.; Karjalainen, J.; Tukiainen, T.; Nelson, M.; Shang, X.; Xiong, G. The impact of social manufacturing on the value chain model in the apparel industry. In Proceedings of the 2014 IEEE International Conference on Service Operations and Logistics, and Informatics, Beijing, China, 12–15 October 2014; pp. 378–381.
7. Mohajeri, B. Paradigm Shift from Current Manufacturing to Social Manufacturing. Master's Thesis, Aalto University, Espoo, Finland, 2015.
8. Shang, X.; Shen, Z.; Xiong, G.; Wang, F.Y.; Liu, S.; Nyberg, T.R.; Wu, H.; Guo, C. Moving from mass customization to social manufacturing: A footwear industry case study. *Int. J. Comput. Integr. Manuf.* **2019**, *32*, 194–205. [CrossRef]

9. Hamalainen, M.; Karjalainen, J. Social manufacturing: When the maker movement meets interfirm production networks. *Bus. Horizons* **2017**, *60*, 795–805. [CrossRef]
10. Cao, W.; Jiang, P.Y. Cloud machining community for social manufacturing. In *Applied Mechanics and Materials*; Trans Tech Publications Ltd.: Baech, Switzerland, 2012; Volume 220, pp. 61–64.
11. Cao, W.; Jiang, P.; Jiang, K. Demand-based manufacturing service capability estimation of a manufacturing system in a social manufacturing environment. *Proc. Inst. Mech. Eng. Part B J. Eng. Manuf.* **2017**, *231*, 1275–1297. [CrossRef]
12. Jiang, P.; Ding, K.; Leng, J. Towards a cyber-physical-social-connected and service-oriented manufacturing paradigm: Social Manufacturing. *Manuf. Lett.* **2016**, *7*, 15–21. [CrossRef]
13. Jiang, P.; Leng, J. The configuration of social manufacturing: A social intelligence way toward service-oriented manufacturing. *Int. J. Manuf. Res.* **2017**, *12*, 4–19. [CrossRef]
14. Jiang, P. *Social Manufacturing: Fundamentals and Applications*; Springer: Berlin/Heidelberg, Germany, 2019.
15. ASCM. Association for Supply Chain Management. Available online: https://www.ascm.org/ (accessed on 20 March 2022).
16. The Smiling Curve, "Wikipedia". 2017. Available online: https://en.wikipedia.org/wiki/Smiling_curve#cite_note-1 (accessed on 15 January 2022).
17. Pettersen, J. Defining lean production: Some conceptual and practical issues. *TQM J.* **2009**, *21*, 127–142. [CrossRef]
18. Ritzer, G.; Dean, P.; Jurgenson, N. The coming of age of the prosumer. *Am. Behav. Sci.* **2012**, *56*, 379–398. [CrossRef]
19. Brabham, D.C. Crowdsourcing as a model for problem solving: An introduction and cases. *Convergence* **2008**, *14*, 75–90. [CrossRef]
20. Porter, M.E. How competitive forces shape strategy. *Harv. Bus. Rev.* **1979**, *57*, 137–145.
21. Porter, M.E. Techniques for Analysing Industries and Competitors. In *Competitive Strategy*; Free: New York, NY, USA, 1980.
22. Hopkins, T.K.; Wallerstein, I. *Commodity Chains in the World-Economy Prior to 1800; Research Foundation of State University of New York: Albany, NY, USA,* 1986; Volume 10, pp. 157–170.
23. Gereffi, G. The organization of buyer-driven global commodity chains: How US retailers shape overseas production networks. In *Commodity Chains and Global Capitalism*; ABC-CLIO: Santa Barbara, CA, USA, 1994; pp. 95–122.
24. Kaplinsky, R.; Morris, M. *A Handbook for Value Chain Research Brighton*; Institute for Development Studies: Brighton, UK, 2001.
25. Ohno, T. *Toyota Production System on Audio Tape: Beyond Large Scale Production*; Productivity Press: New York, NY, USA, 2001.
26. Womack, J.P.; Jones, D.T. From lean production to lean enterprise. *Harv. Bus. Rev.* **1994**, *72*, 93–103.
27. Krafcik, J.F. Triumph of the lean production system. *Sloan Manag. Rev.* **1988**, *30*, 41–52.
28. Womack, J.P.; Jones, D.T. *Lean Thinking*; Simon & Schuster: London, UK, 2005.
29. Xiong, G.; Wu, H.; Helo, P.; Shang, X.; Xiong, G.; Qin, R.; Wang, F.Y. A Kind of Change Management Method for Global Value Chain Optimization and Its Case Study. *IEEE Trans. Comput. Soc. Syst.* **2021**, *9*, 1060–1074. [CrossRef]
30. Rajnoha, R.; Sujová, A.; Dobrovič, J. Management and economics of business processes added value. *Procedia-Soc. Behav. Sci.* **2012**, *62*, 1292–1296. [CrossRef]
31. WTO. World Trade Report 2021: Economic Resilience and Trade. 2021. Available online: https://www.wto.org/english/res_e/booksp_e/wtr21_e/00_wtr21_e.pdf (accessed on 15 July 2022).
32. Olearczyk, J.; Lei, Z.; Ofrim, B.; Han, S.; Al-Hussein, M. Intelligent Crane Management Algorithm for Construction Operation. In Proceedings of the 2015 32nd ISARC, Oulu, Finland, 15–18 June 2015.
33. Acquier, A.; Daudigeos, T.; Pinkse, J. Promises and paradoxes of the sharing economy: An organizing framework. *Technol. Forecast. Soc. Chang.* **2017**, *125*, 1–10. [CrossRef]
34. Mont, O.; Palgan, Y.V.; Bradley, K.; Zvolska, L. A decade of the sharing economy: Concepts, users, business and governance perspectives. *J. Clean. Prod.* **2020**, *269*, 122215. [CrossRef]
35. Zervas, G.; Proserpio, D.; Byers, J.W. The rise of the sharing economy: Estimating the impact of Airbnb on the hotel industry. *J. Mark. Res.* **2017**, *54*, 687–705. [CrossRef]
36. RepRap Community. Available online: http://reprap.org (accessed on 15 May 2022).
37. Juracz, L.; Lattmann, Z.; Levendovszky, T.; Hemingway, G.; Gaggioli, W.; Netterville, T.; Pap, G.; Smyth, K.; Howard, L. VehicleFORGE: A Cloud-Based Infrastructure for Collaborative Model-Based Design. In *MDHPCL@ MoDELS*; Citeseer: Princeton, NJ, USA, 2013; pp. 25–36.
38. Zhou, J.; Yao, X. Hybrid teaching–learning-based optimization of correlation-aware service composition in cloud manufacturing. *Int. J. Adv. Manuf. Technol.* **2017**, *91*, 3515–3533. [CrossRef]
39. VehicleFORGE. Available online: http://www.vehicleforge.org (accessed on 8 April 2022).
40. Wang, F.Y.; Yuan, Y.; Wang, X.; Qin, R. Societies 5.0: A new paradigm for computational social systems research. *IEEE Trans. Comput. Soc. Syst.* **2018**, *5*, 2–8. [CrossRef]
41. Gladden, M.E. Who will be the members of Society 5.0? Towards an anthropology of technologically posthumanized future societies. *Soc. Sci.* **2019**, *8*, 148. [CrossRef]
42. Ding, K.; Jiang, P.; Su, S. RFID-enabled social manufacturing system for inter-enterprise monitoring and dispatching of integrated production and transportation tasks. *Robot. Comput.-Integr. Manuf.* **2018**, *49*, 120–133. [CrossRef]
43. Wang, S.; Wang, J.; Wang, X.; Qiu, T.; Yuan, Y.; Ouyang, L.; Guo, Y.; Wang, F.Y. Blockchain-powered parallel healthcare systems based on the ACP approach. *IEEE Trans. Comput. Soc. Syst.* **2018**, *5*, 942–950. [CrossRef]
44. Ding, K.; Jiang, P.Y. Social sensors (S2ensors): A kind of hardware-software-integrated mediators for social manufacturing systems under mass individualization. *Chin. J. Mech. Eng.* **2017**, *30*, 1150–1161. [CrossRef]

45. Insight Ningbo. Available online: http://news.cnnb.com.cn/system/2022/01/19/030322900.shtml (accessed on 25 June 2022).
46. Jiang, P.; Leng, J.; Ding, K. Social manufacturing: A survey of the state-of-the-art and future challenges. In Proceedings of the 2016 IEEE International Conference on Service Operations and Logistics, and Informatics (SOLI), Sydney, Australia, 24–27 July 2016; pp. 12–17.
47. Xiong, G.; Wang, F.Y.; Nyberg, T.R.; Shang, X.; Zhou, M.; Shen, Z.; Li, S.; Guo, C. From mind to products: Towards social manufacturing and service. *IEEE/CAA J. Autom. Sin.* **2017**, *5*, 47–57. [CrossRef]
48. Xiong, G.; Shang, X.; Xiong, G.; Nyberg, T.R. A kind of lean approach for removing wastes from non-manufacturing process with various facilities. *IEEE/CAA J. Autom. Sin.* **2019**, *6*, 307–315. [CrossRef]
49. Xiong, G.; Tamir, T.S.; Shen, Z.; Shang, X.; Wu, H.; Wang, F.Y. A Survey on Social Manufacturing: A Paradigm Shift for Smart Prosumers. *IEEE Trans. Comput. Soc. Syst.* **2022**, *Early Access*. [CrossRef]
50. Guo, W.; Li, P.; Yang, M.; Liu, J.; Jiang, P. Social Manufacturing: What are its key fundamentals? *IFAC-PapersOnLine* **2020**, *53*, 65–70. [CrossRef]
51. Wang, F.Y.; Shang, X.; Qin, R.; Xiong, G.; Nyberg, T.R. Social manufacturing: A paradigm shift for smart prosumers in the era of societies 5.0. *IEEE Trans. Comput. Soc. Syst.* **2019**, *6*, 822–829. [CrossRef]
52. Rosen, R.; Von Wichert, G.; Lo, G.; Bettenhausen, K.D. About the importance of autonomy and digital twins for the future of manufacturing. *IFAC-Papersonline* **2015**, *48*, 567–572. [CrossRef]
53. Ding, K.; Jiang, P.; Leng, J.; Cao, W. Modeling and analyzing of an enterprise relationship network in the context of social manufacturing. *Proc. Inst. Mech. Eng. Part B J. Eng. Manuf.* **2016**, *230*, 752–769. [CrossRef]
54. Wang, F.Y. Parallel system methods for management and control of complex systems. *Control Decis.* **2004**, *19*, 485–489.
55. Schleich, B.; Anwer, N.; Mathieu, L.; Wartzack, S. Shaping the digital twin for design and production engineering. *CIRP Ann.* **2017**, *66*, 141–144. [CrossRef]
56. Tao, F.; Sui, F.; Liu, A.; Qi, Q.; Zhang, M.; Song, B.; Guo, Z.; Lu, S.C.Y.; Nee, A.Y. Digital twin-driven product design framework. *Int. J. Prod. Res.* **2019**, *57*, 3935–3953. [CrossRef]
57. Tao, F.; Zhang, H.; Liu, A.; Nee, A.Y. Digital twin in industry: State-of-the-art. *IEEE Trans. Ind. Inf.* **2018**, *15*, 2405–2415. [CrossRef]
58. Leng, J.; Jiang, P. A deep learning approach for relationship extraction from interaction context in social manufacturing paradigm. *Knowl.-Based Syst.* **2016**, *100*, 188–199. [CrossRef]
59. Xue, X.; Wang, S.; Zhang, L.; Feng, Z.; Guo, Y. Social learning evolution (SLE): Computational experiment-based modeling framework of social manufacturing. *IEEE Trans. Ind. Inf.* **2018**, *15*, 3343–3355. [CrossRef]
60. European Union Commission. Commission recommendation of 6 May 2003 concerning the definition of micro, small and medium-sized enterprises. *Off. J. Eur. Union* **2003**, *46*, 36–41.
61. Tamir, T.S.; Xiong, G.; Dong, X.; Fang, Q.; Liu, S.; Lodhi, E.; Shen, Z.; Wang, F.Y. Design and Optimization of a Control Framework for Robot Assisted Additive Manufacturing Based on the Stewart Platform. *Int. J. Control. Autom. Syst.* **2022**, *20*, 968–982. [CrossRef]
62. Tamir, T.S.; Xiong, G.; Fang, Q.; Dong, X.; Shen, Z.; Wang, F.Y. A feedback-based print quality improving strategy for FDM 3D printing: An optimal design approach. *Int. J. Adv. Manuf. Technol.* **2022**, *120*, 2777–2791. [CrossRef]
63. Henao, R.; Sarache, W.; Gomez, I. Social Performance Metrics for Manufacturing Industry. In Proceedings of the 24th EurOMA Conference, Edinburgh, Scotland, 1–5 July 2017; Volume 10.
64. ASCM. Structure of SCOR. Available online: https://www.ascm.org/learning-development/certifications-credentials/scor-p/ (accessed on 15 July 2022).
65. Bullen, C.V.; Rockart, J.F. A primer on critical success factors. In *The Rise of Managerial Computing: The Best of the Center for Information Systems Research*; Massachusetts Institute of Technology, Sloan School of Management: Cambridge, MA, USA, 1981.
66. Zhu, Q.; Liu, A.; Li, Z.; Yang, Y.; Miao, J. Sustainable Supplier Selection and Evaluation for the Effective Supply Chain Management System. *Systems* **2022**, *10*, 166. [CrossRef]
67. Liu, D.; Wu, D.; Wu, S. A Graph Matching Model for Designer Team Selection for Collaborative Design Crowdsourcing Tasks in Social Manufacturing. *Machines* **2022**, *10*, 776. [CrossRef]

MDPI
St. Alban-Anlage 66
4052 Basel
Switzerland
Tel. +41 61 683 77 34
Fax +41 61 302 89 18
www.mdpi.com

Machines Editorial Office
E-mail: machines@mdpi.com
www.mdpi.com/journal/machines

www.ingramcontent.com/pod-product-compliance
Lightning Source LLC
Chambersburg PA
CBHW061616220926
51850CB00005B/525
9783036577562